Advances in

INORGANIC CHEMISTRY
AND
RADIOCHEMISTRY

―――――

Volume 17

CONTRIBUTORS TO THIS VOLUME

C. D. Garner

R. J. Gillespie

V. Gutmann

S. M. L. Hamblyn

B. Hughes

F. H. Jardine

C. A. McAuliffe

U. Mayer

A. H. Norbury

J. Passmore

B. G. Reuben

Advances in
INORGANIC CHEMISTRY
AND
RADIOCHEMISTRY

EDITORS

H. J. EMELÉUS

A. G. SHARPE

University Chemical Laboratory
Cambridge, England

VOLUME 17

1975

ACADEMIC PRESS New York San Francisco London

A Subsidiary of Harcourt Brace Jovanovich, Publishers

ACADEMIC PRESS, INC.
111 Fifth Avenue, New York, New York 10003

United Kingdom Edition published by
ACADEMIC PRESS, INC. (LONDON) LTD.
24/28 Oval Road, London NW1

LIBRARY OF CONGRESS CATALOG CARD NUMBER: 59-7692

ISBN 0−12−023617−6

PRINTED IN THE UNITED STATES OF AMERICA

CONTENTS

Complexes of Open-Chain Tetradentate Ligands Containing Heavy Donor Atoms

C. A. McAuliffe

The Functional Approach to Ionization Phenomena in Solutions

U. Mayer and V. Gutmann

Coordination Chemistry of the Cyanate, Thiocyanate, and Selenocyanate Ions

A. H. Norbury

LIST OF CONTRIBUTORS

Numbers in parentheses indicate the pages on which the authors' contributions begin.

C. D. GARNER (1), *Department of Chemistry, Manchester University, Manchester, England*

R. J. GILLESPIE (49), *Department of Chemistry, McMaster University, Hamilton, Ontario, Canada*

V. GUTMANN (189), *Institut für Anorganische Chemie, Technische Hochschule Wien, Vienna, Austria*

S. M. L. HAMBLYN* (89), *Borax Research Centre, Chessington, Surrey, England*

B. HUGHES† (1), *Department of Chemistry, Manchester University, Manchester, England*

F. H. JARDINE (115), *Department of Chemistry, North East London Polytechnic, London, England*

C. A. MCAULIFFE‡ (165), *Department of Chemistry, Auburn University, Auburn, Alabama*

U. MAYER (189), *Institut für Anorganische Chemie, Technische Hochschule Wien, Vienna, Austria*

A. H. NORBURY (231), *Department of Chemistry, Loughborough University of Technology, Loughborough, Leicestershire, England*

J. PASSMORE (49), *Chemistry Department, University of New Brunswick, Fredericton, New Brunswick, Canada*

B. G. REUBEN (89), *Chemistry Department, University of Surrey, Guildford, England*

* Present address: Lonza A.G., 3930 Visp, Switzerland.

† Present address: The Radiochemical Centre Ltd., Amersham, Buckinghamshire, England.

‡ Permanent address: Department of Chemistry, University of Manchester, Institute of Science and Technology, Manchester M60 1QD, England.

Advances in
INORGANIC CHEMISTRY
AND
RADIOCHEMISTRY

———

Volume 17

INORGANIC COMPOUNDS CONTAINING THE TRIFLUOROACETATE GROUP

C. D. GARNER and B. HUGHES*

Department of Chemistry, Manchester University, Manchester, England

* Present address: The Radiochemical Centre Ltd., Amersham, Buckinghamshire, England.

I. Introduction

Inorganic trifluoroacetates have been known for some 50 years, the first studies being completed by Swarts (247) in 1922. Trifluoroacetate derivatives have now been prepared for nearly all of the elements, the majority of them involving normal oxidation states; for example, there are a large number of trifluoroacetato complexes of chromium(III), nickel(II), copper(II), and tin(IV). Although simple electronegativity arguments would lead one to expect that this group could maintain its identity when bonded to atoms in high oxidation states, this does not appear to be the case. Thus no trifluoroacetato complexes of platinum(IV) or gold(III) have been reported, and lead(IV) trifluoroacetate readily decomposes to the lead(II) compound (192). Complexes involving the trifluoroacetate group coordinated to metal atoms in low oxidation states are, however, well known. For instance, $(\pi\text{-Cp})W(CO)_3$-(O_2CCF_3) (57), $Re(CO)_5(O_2CCF_3)$ (57, 137), and $Ir(CO)(PPh_3)_2(O_2CCF_3)$ (237) have been characterized, but clearly the carbonyl and the other ligands play the major role in stabilizing these low oxidation states.

Although a considerable amount of data has now been reported concerning the physical and chemical properties of inorganic compounds containing the trifluoroacetate group, little attempt has been made previously to report and examine these properties in a systematic manner. This review will attempt to do this and to show that, although trifluoroacetate is a member of the carboxylate family, it has individual characteristics which lead to some unique and interesting chemistry.

II. Trifluoroacetic Acid

Before describing the preparation and properties of trifluoroacetate compounds in general, particular consideration will be given to the parent acid. The normal commercial preparation of trifluoroacetic acid is the electrolytic fluorination of acetyl fluoride or chloride (210). Other preparative methods include the oxidation of trifluorotoluidine or other trifluoromethyl aryl derivatives (247) and the hydrolysis of trifluoroacetylchloride (6).

The physical properties of liquid trifluoroacetic acid are given in Table I. At normal temperatures it is a colorless liquid which fumes readily in moist air and has a powerful odor.

Trifluoroacetic acid is known to form cyclic dimers in the vapor phase which appear to be analogous to those formed by acetic acid. An electron diffraction study of gaseous trifluoroacetic acid identified monomeric and dimeric species, the O—H---O bond in the latter being estimated as 2.76(6) Å, which is essentially the same value as found for the acetic acid dimer (133). The heats of dissociation of $(CH_3CO_2H)_2$ and $(CF_3CO_2H)_2$ in the gas phase have been determined as 15.3 (205) and 14.1 kcal mole^{-1} (250) respectively. In a mixture of the two acid vapors, a heterodimer is formed which is more stable than either of the homodimers, with a heat of dissociation of 17 kcal mole^{-1} (159). An elegant vibrational study of isolated trifluoroacetic acid monomers and

TABLE I

PHYSICAL CONSTANTS OF LIQUID TRIFLUOROACETIC ACID

Property	Value	Reference
Melting point	$-15.2°C$	(115)
Boiling point	72.4°C (760 Torr)	(247)
Liquid range	87.6°C	
Density	1.535 gm/ml (25°C)	(247)
Viscosity	8.76×10^{-3} poise (25°C)	(201)
Dielectric constant	8.2 (30°C)	(56)
Specific conductance	3.5×10^{-7} mho (25°C)	(251)
K_{as}	4×10^{-14} mole^2 liter^{-2}	(109)
pK_a	± 0.25 (25°C)	(115)
Enthalpy of formation	253 kcal mole^{-1}	(146)
Heat of vaporization	8.3 kcal mole^{-1} (72.4°C)	(199)
Trouton constant	24 cal deg^{-1} mole^{-1}	
Dipole moment	2.3 D	(199)

dimers trapped in argon and neon matrices has been accomplished by Redington and Lin (202). This study not only allowed the vibrational spectra of both species to be unequivocally identified but also resolved some difficulties concerning the vibrational spectra of the trifluoroacetate group. Furthermore, the results indicate that a very low barrier hinders internal rotation of the CF_3 group in the acid.

The nature of the association of trifluoroacetic acid in inert and slightly basic solvents is still disputed. Murty and Pitzer (178) have suggested that infrared spectra indicate that a linear association of trifluoroacetic acid molecules occurs in such media; however, further such measurements obtained by Kirszenbaum et al. (141) appear to conflict with this interpretation.

The ^1H and ^{19}F NMR line parameters obtained for a powdered sample of trifluoroacetic acid have been interpreted (72) on the basis of dimeric molecules in this phase. However, the arrangement of these dimers in the crystal lattice appears to differ from that in trichloroacetic acid.

Trifluoroacetic acid is a nonoxidizing acid whose aqueous solutions are comparable in strength with those of the mineral acids (*115*), but as a bulk solvent it is weakly acidic and does not even protonate water (*236*). Trifluoroacetic acid is also weakly basic and is a nonelectrolyte in 100% sulfuric acid (*21*). One of the major difficulties of using pure trifluoroacetic acid is its remarkable affinity for water and, to maintain an anhydrous medium, a small quantity of trifluoroacetic anhydride is usually added. This affinity for water is so pronounced that trifluoroacetic acid will dehydrate oxyacids and, for example, converts sulfuric acid into polysulfuric acid.

Trifluoroacetic acid is quite a good solvent for organic and inorganic materials, and its ability to function as a nonaqueous solvent has been reviewed (*199*). A large range of organic solvents are completely miscible with the acid, which is also capable of dissolving proteins (*135*) and certain polymers (*248*). The ^{1}H NMR spectra of many organic substances may be conveniently recorded in trifluoroacetic acid since, not only do they have a reasonable solubility, but also the solvent has only one resonance absorption at low fields (-1.71 ppm) (*128*). Trifluoroacetic acid is a useful reaction medium for many organic reactions (*37*, *79*); it allows selective substitution to proceed very readily (*35*), and it is perhaps the best solvent for the platinum-catalyzed hydrogenation of ketones (*195*). Trifluoroacetic acid and, indeed, many of its salts have been used to initiate stereospecific polymerization of unsaturated hydrocarbons such as butadiene (*98*, *269*).

Trifluoroacetic acid is also a good solvent for most organometallic compounds, and the preparation of hydrido complexes by oxidative addition of the solvent to these solutes is a useful route to such compounds. These reactions will be considered in more detail in Section III since they clearly lead to the formation of metal trifluoroacetates. Trifluoroacetic acid is also a good solvent for the halogens, although it does react with them to a limited extent (*38*). The ionic behavior of several simple electrolytes in 100% trifluoroacetic acid has been studied by Simons and Lorentzen (*236*) and Harriss and Milne (*107*). The rather low conductivities of these solutions indicate that it is not a very good ionizing solvent, as would be expected in view of its relatively low dielectric constant. Furthermore, the dependence of ionic mobility on cation radius is different from that for aqueous media, suggesting that ion solvation is much less important in trifluoroacetic acid than in water. The large spread in the values of the limiting equivalent conductivities for the alkali metal trifluoroacetates in trifluoroacetic acid (*107*) indicates that the cation makes a large contribution to the total conductivity and, thus, the anion probably conducts by diffusion rather than by the self-dissociation, proton-transfer mechanism operating (207) for the solvent anions in water or sulfuric acid.

III. Synthesis of Metal Trifluoroacetates

This section will describe the range of preparative methods that have been used to obtain metal trifluoroacetates, each being illustrated by selected examples. However, no attempt will be made to give a comprehensive list of the compounds synthesized by any particular route.

In the first detailed investigations of metal trifluoroacetates, Swarts (247) described the preparations of the simple trifluoroacetate salts of Na(I), ammonium, Ba(II), Al(III), Tl(I), Pb(II), Fe(II), Cu(II), Ag(I), Hg(I), and Hg(II) by reacting the corresponding metal oxide, hydroxide, carbonate, or sulfate with an excess of an aqueous solution of trifluoroacetic acid. All of these preparations produced an aqueous solution of the trifluoroacetate salt and, by carefully heating these solutions under reduced pressure, the anhydrous metal trifluoroacetate was obtained in each case. Purification of the compounds by sublimation proved to be unsuccessful, the compounds having very low volatilities; even at 180° to 200°C and 1 Torr, very little sublimate was obtained and this was usually contaminated with decomposition products. Swarts observed that the anhydrous salts were very hygroscopic and frequently had to be handled in a dry atmosphere. This involatility and moisture sensitivity are general features of anhydrous inorganic trifluoroacetate compounds.

Hara and Cady (105) obtained many simple trifluoroacetate salts, using methods similar to those of Swarts, in a study that extended the range of known trifluoroacetates to include those of the lanthanide [La(III), Ce(III), Pr(III), and Nd(III)] and actinide [Th(IV) and U(VI)] metals. These workers prepared anhydrous aluminum(III) trifluoroacetate by treating freshly amalgamated aluminum in aqueous trifluoroacetic acid and decanting the solution off the amalgam before evaporating to dryness. Magnesium(II) and zinc(II) trifluoroacetates were similarly prepared from the corresponding metal. Compound $UO_2(O_2CCF_3)_2$ was obtained by the addition of U_3O_8 to aqueous trifluoroacetic acid and by the direct conversion from the corresponding acetate by treatment with trifluoroacetic acid. This was the first use of carboxylate exchange for the synthesis of a metal trifluoroacetate, a route which has also been employed for the synthesis of $Mo_2(O_2CCF_3)_4$ (52) and $Rh_2(O_2CCF_3)_4$ (127, 267). The mechanism of this latter carboxylate exchange reaction has been followed by ^1H NMR and mass spectrometry. The reaction involves a stepwise exchange of trifluoro-acetate for the acetate cage: the first substitution has a labilizing effect and the second substitution occurs twice as fast as the first. After the formation of $Rh_2(O_2CCH_3)_2(O_2CCF_3)_2$ the reaction proceeds at a much slower rate until the completely exchanged $Rh_2(O_2CCF_3)_4$ is formed (23).

Sartori and Weidenbruch developed several synthetic methods in preparing Group IV metal trifluoroacetates (213–215) which have been widely employed in other studies. Halide exchange with trifluoroacetic acid at ambient temperatures was used to prepare $Si(O_2CCF_3)_4$, $3TiO(O_2CCF_3)_2 \cdot 2CF_3CO_2H$, $Zr(O_2CCF_3)_4$, $Hf(O_2CCF_3)_4$, and $Th(O_2CCF_3)_4$ from the corresponding tetrachlorides. Metathesis with $Hg(O_2CCF_3)_2$ in trifluoroacetic acid afforded $Ge(O_2CCF_3)_4$ and $Sn(O_2CCF_3)_4$ from $GeCl_4$ and $SnCl_4$, respectively. These latter two trifluoroacetates were also obtained by the action of trifluoroacetic acid on the appropriate tetraphenyl derivative dissolved in benzene. Some extensions of the synthetic procedures developed by Sartori and Weidenbruch include the following. Halide exchange may be effected by the action of trifluoroacetic acid or anhydride on covalent fluorides [Eqs. (1) (184) and (2) (74, 179).] Compound $Hg(OSeF_5)_2$ has been

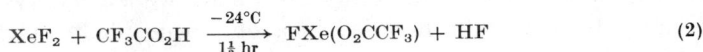

$$Cs[IF_4] + 4(CF_3CO)_2O \longrightarrow Cs[I(O_2CCF_3)_4] + 4CF_3COF \tag{1}$$

$$XeF_2 + CF_3CO_2H \xrightarrow[1\frac{1}{4}\ hr]{-24°C} FXe(O_2CCF_3) + HF \tag{2}$$

shown to react metathetically with CF_3COCl to afford trifluoroacetyl-pentafluoroselenate, $SeF_5(O_2CCF_3)$ (221). Sharp et al. (224–228) have prepared a large number of anhydrous transition metal trifluoroacetates by salt elimination reactions using stoichiometric quantities of the anhydrous metal chloride and a solution of silver trifluoroacetate in dried nitromethane or ether. Similarly, King and Kapoor (137) have obtained several organometallic trifluoroacetates by reacting the corresponding organometallic metal halide with a 5% excess of silver trifluoroacetate in methylene chloride solution at room temperature. Among the compounds prepared by these latter authors were $Mn(CO)_5(O_2CCF_3)$, $(\pi\text{-Cp})_2Ti(O_2CCF_3)_2$, and $(\pi\text{-allyl})Fe(CO)_3(O_2CCF_3)$, and several similar compounds involving other perfluorocarboxylato groups were also characterized in this study. Cleavage of metal–carbon σ-bonds by trifluoroacetic acid has allowed the preparation of $Me_2In(O_2CCF_3)$ (42), $(CH_2{=}CH)_2Sn(O_2CCF_3)_2$ (89), and $Re(CO)_5$-(O_2CCF_3) (57) from Me_3In, $(CH_2{=}CH)_4Sn$, and $MeRe(CO)_5$, respectively.

Trifluoroacetato complexes have also been obtained by the displacement of ligands, other than those described in the foregoing, using trifluoroacetic acid or its salts. Thus hydrido displacement has been achieved to prepare $(\pi\text{-}Cp)_2Zr(O_2CCF_3)_2$ from $(\pi\text{-}Cp)_2ZrH_2$ (262), and $(Ph_3P)_3Os(H)(O_2CCF_3)$ from $(Ph_3P)_4OsH_4$ (208), by the action of trifluoroacetic acid. Carbonyl groups may also be substituted by trifluoroacetate groups; thus, for example, the reaction of $W(CO)_6$ with $Et_4N(O_2CCF_3)$ in diglyme at 120°C affords $[W(CO)_5O_2CCF_3]^-$ (218). The dimeric complex $[(\pi\text{-}Cp)V(O_2CCF_3)_2]_2$, which has interesting magnetic properties (see Section V, B), has also been obtained by carbonyl displacement, in this instance by refluxing $(\pi\text{-}Cp)V(CO)_4$ in trifluoroacetic acid (150).

Compound $[Cr(CO)_5(O_2CCF_3)]^-$ may be prepared by oxidative substitution of $Ag(O_2CCF_3)$ or $Hg(O_2CCF_3)_2$ on $[Cr_2(CO)_{10}]^{2-}$ (218). Transition metal hydrido complexes have been obtained by oxidative addition reactions of trifluoroacetic acid which often lead to the formation of trifluoroacetato complexes. Equation (3) is an example of such a

$$(Ph_3P)_2\,Pt\{C(CF_3)_2NMe\} + CF_3CO_2H \longrightarrow (Ph_3P)_2Pt(H)\{C(CF_3)_2NMe\}(O_2CCF_3) \quad (3)$$

complex formed by the simple addition of the acid to the metal center (13). Oxidative addition may also occur for certain complexes containing acetylenic or olefinic ligands, the hydrogen adding to the unsaturated ligand which then undergoes a π- to σ-bonded rearrangement, and the trifluoroacetate groups coordinating to the metal, for example, Eqs. (4) and (5) (19).

$$(F_3C\cdot C\equiv CF_3)Pt(PPh_3)_2 + CF_3CO_2H \longrightarrow (F_3C\cdot CH\equiv C\cdot CF_3)Pt(PPh_3)_2(O_2CCF_3) \quad (4)$$

$$(F_2C\equiv CF_2)Pt(PPh_3)_2 + CF_3CO_2H \longrightarrow (HF_2C\cdot CF_2)Pt(PPh_3)_2(O_2CCF_3) \quad (5)$$

Trifluoroacetato complexes of transition metals have also been obtained from oxidative elimination reactions in which neutral π-acceptor ligands have been displaced in reactions with trifluoroacetic acid. These reactions may lead to hydrido complexes [Eqs. (6) (253) and (7) (26)] or they may not [Eqs. (8) (39) and (9) (19)].

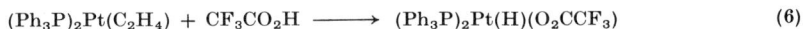

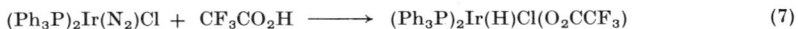

$$(Ph_3P)_2Pt(C_2H_4) + CF_3CO_2H \longrightarrow (Ph_3P)_2Pt(H)(O_2CCF_3) \quad (6)$$

$$(Ph_3P)_2Ir(N_2)Cl + CF_3CO_2H \longrightarrow (Ph_3P)_2Ir(H)Cl(O_2CCF_3) \quad (7)$$

$$\{(MeO)_3P\}_2Ru(CO)_3 + 2CF_3CO_2H \longrightarrow \{(MeO)_3P\}_2Ru(CO)_2(O_2CCF_3)_2 \quad (8)$$

$$(Ph_3P)_2Pt(PhC\equiv CPh) + 2CF_3CO_2H \longrightarrow (Ph_3P)_2Pt(O_2CCF_3)_2 + trans\text{-stilbene} \quad (9)$$

Reaction (9) presumably proceeds via an intermediate analogous to that produced in Eq. (4). Allyltrifluoroacetate effects oxidative displacement with biscyclooctatetraenenickel(O) to form bis(π-allylnickeltrifluoro-acetate) whose ESR spectrum is consistent with its formulation as a nickel(II) complex (59, 60). This latter complex is a versatile poly-merization catalyst for conjugated dienes (see Section VIII, C) (166).

IV. Trifluoroacetate Compounds of the s- and p- Block Elements

A. GROUP I

The simple salts $M(O_2CCF_3)$ (where M = Li, Na, K, Rb, Cs, NH_4, or NEt_4) have been isolated and characterized in a variety of studies (4, 28, 54, 84, 105, 109, 243, 247), and their infrared spectra have been described in some detail. The crystal structure of $NH_4(O_2CCF_3)$ has been determined (54) and each oxygen atom shown to be involved in two N—H⋯O hydrogen bonds.

Hydrogen bonding between two trifluoroacetate ions has constituted an important aspect of the study of this bonding since very short bonds may be obtained (143, 144, 173, 242). The ^{1}H NMR spectra of solutions of the alkali metal and some quaternary ammonium trifluoroacetates in trifluoroacetic acid over a range of concentrations have been interpreted in terms of the formation of the hydrogen ditrifluoroacetate anion $[H(O_2CCF_3)_2]^-$ (128). The salts $M[H(O_2CCF_3)_2]$ (where M = Na, K, Rb, or Cs) and some of their deuterium analogs have been isolated and characterized (99, 143, 144, 163). A recent neutron diffraction study (163) on these potassium salts has provided the most accurate structural data presently available for the trifluoroacetate group. The structure of the centrosymmetric hydrogen ditrifluoroacetate anion identified in $K[H(O_2CCF_3)_2]$ is illustrated in Fig. 1. The dimensions of these tri-fluoroacetate groups, most of which agree within experimental error with those reported for $NH_4(O_2CCF_3)$ (54), are presented in Table XI (see Section VII, B).

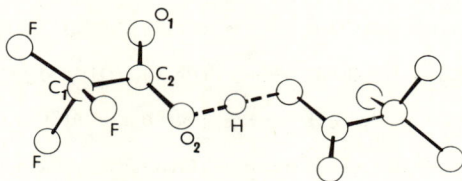

FIG. 1. The structure of the hydrogen ditrifluoroacetate anion in $K[H(O_2CCF_3)_2]$. From Macdonald et al. (163) with permission.

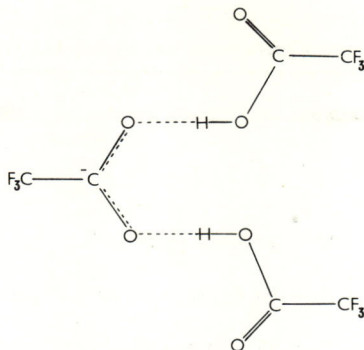

FIG. 2. Suggested structure for the anions in $Na(O_2CCF_3) \cdot 2CF_3CO_2H$ (143).

The salt $Na(O_2CCF_3) \cdot 2CF_3CO_2H$ and its deutero analog have been characterized, and the structure shown in Fig. 2 has been suggested for their anions (143).

B. GROUP II

The simple trifluoroacetates of all the alkaline earth elements have been reported (36, 105, 247, 268). Compound $Be(O_2CCF_3)_2$ was obtained in a study of several beryllium haloacetates and, although a deliberate attempt was made to prepare basic beryllium trifluoroacetate, $Be_4O(O_2CCF_3)_6$, by the thermal decomposition of the simple salt, none could be obtained (268). Also $Be_4O(O_2CCF_3)_6$ does not appear to be produced when beryllium carbonate is treated with trifluoroacetic acid and the solution extracted with chloroform (122, 174).

C. GROUP III

The known trifluoroacetates of boron, aluminum, indium, and thallium are summarized in Table II.

The preparation of $H[B(O_2CCF_3)_4]$ [Eq. (10)] (25), is directly analogous to the preparation of the superacid $H[B(HSO_4)_4]$ in oleum (83). Attempts to isolate $B(O_2CCF_3)_3$ from these trifluoroacetic acid

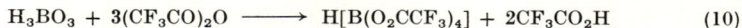

$$H_3BO_3 + 3(CF_3CO)_2O \longrightarrow H[B(O_2CCF_3)_4] + 2CF_3CO_2H \qquad (10)$$

solutions afforded only $B_2O(O_2CCF_3)_4$. However, this simple tris-trifluoroacetate may be obtained by the reaction between BCl_3, and CF_3CO_2H in n-pentane; the compound is unstable at its melting point (88°C) and decomposes to $B_2O(O_2CCF_3)_4$ (92).

TABLE II

TRIFLUOROACETATO COMPLEXES OF GROUP III ELEMENTS

Compound	Reference
$B(O_2CCF_3)_3$	(92)
$B_2(O_2CCF_3)_4$	(92)
$B_2O(O_2CCF_3)_4$	(109)
$\{(CF_3CO_2)_2B\}_2O$	(92)
$M[B(O_2CCF_3)_4]$	M = H or Cs (109)
$PhB(O_2CCF_3)_2$	(92)
$R_2B(O_2CCF_3)$	R = Et (256), Bu^n, or Ph (92)
$Al(O_2CCF_3)_3$	(12, 105, 247)
$Al(X)(O_2CCF_3)_2$	X = Cl (12) or OH (3, 12)
$AlCl_2(O_2CCF_3)$	(247)
$In(OH)(O_2CCF_3)_2$	(125)
$Me_2In(O_2CCF_3)$	(42)
$Tl(O_2CCF_3)$	(203, 247)
$Tl(O_2CCF_3)_3$	(168, 249)
$PhTl(O_2CCF_3)_2{}^a$	(154)
$PhTl(O_2CCF_3)_2 \cdot L$	L = bipy or o-phen (154)
$(C_6F_5)_2Tl(O_2CCF_3)$	(63)
$(C_6F_5)_2Tl(O_2CCF_3) \cdot L$	L = Ph_3PO, Ph_3AsO (64), bipy, or o-phen (63)

^{a} A very large number of other arylthallium bistrifluoroacetates have been obtained by the direct thallation of aromatic compounds (40, 168, 169, 170, 203) (see Section VIII, B).

Compound $Me_2In(O_2CCF_3)$ sublimes rapidly at 100°C under reduced pressure and is notable as one of the few metal trifluoroacetato complexes that has a reasonable volatility. Consistent with this volatility, $Me_2In(O_2CCF_3)$ dissolves in nitromethane as the covalent monomer (42). Also $(C_6F_5)_2Tl(O_2CCF_3)$ dissolves as discrete molecules in this solvent; however, in methanol it behaves as a 1:1 electrolyte (63).

The colorless, photosensitive, hygroscopic solid, $Tl(O_2CCF_3)_3$ is conveniently prepared by heating under reflux a suspension of Tl_2O_3 in trifluoroacetic acid. A solution of this salt in the acid constitutes a powerful reagent for the direct thallation of aromatic compounds [Eq. (11)] (168). Such products are usually soluble in most organic solvents

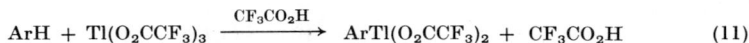

$$ArH + Tl(O_2CCF_3)_3 \xrightarrow{CF_3CO_2H} ArTl(O_2CCF_3)_2 + CF_3CO_2H \qquad (11)$$

and are powerful and versatile synthetic intermediates in organic chemistry (see Section VIII, B) (40, 169, 170, 203). These $ArTl(O_2CCF_3)_2$ derivatives show a tendency to decompose to the more stable $Ar_2Tl(O_2CCF_3)$ derivatives if their solutions are stored for long periods (169).

D. Group IV

Table III lists the trifluoroacetato complexes of silicon, germanium, tin, and lead that have been characterized. Although a large number of organotin(IV) trifluoroacetates have been reported, the only complete series $R_nSn(O_2CCF_3)_{4-n}$ (where $n = 0$–4) is the R = vinyl series. Peruzzo *et al.* (*194*) obtained $(CH_2{=}CH)_3Sn(O_2CCF_3)$ by the exchange reaction of $(CH_2{=}CH)_4Sn$ and $Na(O_2CCF_3)$. The dropwise addition of $(CH_2{=}CH)_4Sn$ to anhydrous CF_3CO_2H at room temperature results

TABLE III

TRIFLUOROACETATO COMPLEXES OF THE GROUP IV ELEMENTS

Compound	Reference
$Si(O_2CCF_3)_4$	(*213, 214, 215*)
$SiO(O_2CCF_3)_2$	(*213, 214*)
$MeSi(O_2CCF_3)_3$	(*9*)
$Et_2Si(O_2CCF_3)_2$	(*8*)
$X_3Si(O_2CCF_3)$	X = Cl (*214*) or Me (*9*)
$Ge(O_2CCF_3)_4$	(*120, 215*)
$RGe(O_2CCF_3)_3$	R = Me or CF_3 (*120*)
$Me_2Ge(O_2CCF_3)_2$	(*120*)
$R_3Ge(O_2CCF_3)$	R = Et (*10*) or Ph (*215*)
$Ag_2[Ge(O_2CCF_3)_6]$	(*120*)
$Ag_2[RGe(O_2CCF_3)_5]$	R = Me or CF_3 (*120*)
$Sn(O_2CCF_3)_2$	(*68*)
$Sn(O_2CCF_3)_4$	(*215*)
$(CH_2{=}CH)Sn(O_2CCF_3)_3$	(*114, 130*)
$R_2Sn(O_2CCF_3)_2$	R = Me (*211*), Et (*8, 11*) or $CH_2{=}CH$ (*123*)
$Me(n\text{-}C_5H_{11})Sn(O_2CCF_3)_2$	(*11*)
$R_3Sn(O_2CCF_3)$	R = Me (*197, 259*), Et (*171, 217*), Pr^n (*216*), Bu^n (*211, 216*), $CH_2{=}CH$ (*94*), Ph (*85, 164*), or trineophenyl, Me_2PhCCH_2 (*164*)
$Me_2RSn(O_2CCF_3)$	R = Bu^n or $n\text{-}C_5H_{11}$ (*11*)
$R_2ClSn(O_2CCF_3)$	R = Me, Bu^n, or Ph (*232*)
$(CH_2{=}CH)_2Sn(O_2CCF_3)_2 \cdot 2L$	L = Ph_3PO or hexamethylphosphoramide, or 2L = bipy, *o*-phen, en, or N,N'-ethylenebis(salicylideneimine) (*123*)
$PhSnO(O_2CCF_3)$	(*198*)
$Ph_4Sn_2(O_2CCF_3)_2$	(*196*)
$(CH_2{=}CH)_2Sn\{Mn(CO)_5\}(O_2CCF_3)$	(*123*)
$[SnF_n(O_2CCF_3)_{6-n}]^{2-}$	n = 2, 3, or 4 (*66*)
$Pb(O_2CCF_3)_2$	(*247*)
$Pb(O_2CCF_3)_4$	(*49, 118, 162, 192*)

in an immediate and exothermic reaction that yields $(CH_2{=}CH)_2$-$Sn(O_2CCF_3)_2$ (*123*). This latter compound has remarkably stable vinyl–tin bonds. It can be recovered unchanged from CF_3CO_2H solution after refluxing at atmospheric pressure for several days, and, only under forcing conditions, does this reaction lead to loss of a further vinyl group (*114, 130*). This contrasts with the ready replacement of all the vinyl groups of $(CH_2{=}CH)_4Sn$ by alkanoic acids (*114*).

Infrared, Mössbauer, and ^1H NMR spectra suggest that solid $Me_3Sn(O_2CCF_3)$ (*197, 259*) and $Ph_3Sn(O_2CCF_3)$ (*85*) are polymeric, with pentacoordinated tin(IV) atoms; the structure is shown in Fig. 3. These

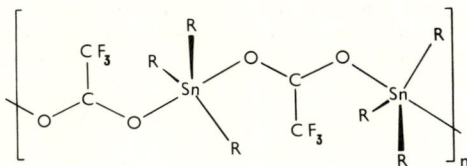

FIG. 3. Suggested structure for solid $R_3Sn(O_2CCF_3)$ where R = Me or Ph (*197*).

compounds are soluble in carbon tetrachloride and the shifts in the carboxylato stretching frequencies, for the former compound, have been interpreted in terms of a breakdown of the polymeric chain to give monomeric complexes containing unidentate trifluoroacetato groups (*197*).

Compound $Pb(O_2CCF_3)_4$ is only stable in the solid state or in fluorinated solvents such as CF_3CO_2H or C_6F_6, and it appears to have unique powers as an oxidant toward hydrocarbons (see Section VIII, A) (*192*).

E. Group V

Table IV summarizes the trifluoroacetato complexes of the Group V elements that have been characterized. The simple trifluoroacetates of arsenic, antimony, and bismuth(III) have been isolated only recently, following the reaction of the corresponding trichloride with $Ag(O_2CCF_3)$ in CH_2Cl_2 (*123*) and, for the bismuth derivative, by the reaction of Bi_2O_3 with $(CF_3CO)_2O$ (*200*). These are reasonably volatile compounds which are extremely moisture-sensitive.

Compound $Me_2As(O_2CCF_3)$ is unusual in that it may be thermally decarboxylated to the corresponding perfluoromethyl derivative (*55*); normally such reactions are unsuccessful (*61, 123*).

The compounds $R_3Sb(O_2CCF_3)_2$ appear to be covalent monomers and the positions of their infrared carboxylato stretching frequencies have been used as evidence that the molecules involve five-coordinate antimony(V) with two unidentate trifluoroacetato groups (97).

The ^{19}F NMR spectra have shown that SbF_5 acts as an acceptor acid in CF_3CO_2H to form $H[SbF_5(O_2CCF_3)]$ (108); the reaction is analogous to the behavior of SbF_5 in liquid HF (94).

TABLE IV

TRIFLUOROACETATO COMPLEXES OF THE
GROUP V ELEMENTS

Compound	Reference
$PF_2(O_2CCF_3)$	(82)
$As(O_2CCF_3)_3$	(123)
$As(O_2CCF_3)_3 \cdot bipy$	(123)
$Me_2As(O_2CCF_3)$	(55)
$Na[AsO(O_2CCF_3)_2]$	(200)
$Sb(O_2CCF_3)_3$	(123)
$R_3Sb(O_2CCF_3)_2$	R = Me or Ph (97)
$Ph_4Sb(O_2CCF_3)$	(96)
$H[SbF_5(O_2CCF_3)]$	(108, 109)
$Bi(O_2CCF_3)_3$	(123, 200)
$Na[Bi(O_2CCF_3)_4]$	(200)

F. GROUP VI

Trifluoroacetate derivatives of the Group VI elements are confined to the monosubstituted selenium(VI) fluoride, $SeF_5(O_2CCF_3)$ (221), and the hexatrifluoroacetatotellurate(IV) anion, $[Te(O_2CCF_3)_6]^{2-}$, which has been isolated as the sodium salt (200).

G. GROUP VII

Iodine is the only halogen that has been shown to form trifluoroacetato complexes. These have been identified for iodine(I), $I(O_2CCF_3)$ (32, 112, 119); iodine(III), $I(O_2CCF_3)_3$ (212), $Cs[I(O_2CCF_3)_4]$, and Cs_3-$[IF_n(O_2CCF_3)_{6-n}]$ (where $n = 2$, 4, or 6) (183, 184); and iodine(V), $IO_2(O_2CCF_3)$ (200) and $K[IO_2(O_2CCF_3)_2]$ (183). These complexes may be prepared by the action of trifluoroacetic anhydride on an appropriate iodine oxide or fluoride derivative. For example, KIO_3 reacts with $(CF_3CO)_2O$ in MeCN at ca. -40^aC to afford $K[IO_2(O_2CCF_3)_2]$; in the analogous reaction with $KBrO_3$ an explosion occurs with evolution of Br_2, CO_2, and C_2F_6 (183).

H. Group VIII

As would be anticipated, the noble gas trifluoroacetates are confined to xenon.

Compounds $Xe(O_2CCF_3)_2$ and $XeF(O_2CCF_3)$ have been obtained in several studies (74, 179, 239) by the reactions of XeF_2 with CF_3CO_2H. These compounds are thermally unstable with respect to Xe, XeF_2, CO_2, and C_2F_6 and are, thus, potentially explosive (76). Iskraut et al. (124) have claimed the preparation of $Xe(O_2CCF_3)_4$ and $Xe(O_2CCH_3)_4$; however, these results await confirmation (179).

V. Trifluoroacetato Complexes of the d-Transition Metals

A. Titanium, Zirconium, Hafnium (and Thorium)

The titanium(II) derivatives $Ti(O_2CCF_3)X$ (where X = Cl or Br) have been prepared by the reaction between $Ti(O_2CCF_3)_2$ and the corresponding thionyl halide in ether. All of these titanium(II) trifluoro-acetates are effective as catalysts for the stereospecific polymerization of butadienes (44).

The other trifluoroacetates of these elements that have been characterized are $Ti(O_2CCF_3)_3$ (225), $TiO(O_2CCF_3)_2$, $Zr(O_2CCF_3)_4$, $Hf(O_2CCF_3)_4$, $Th(O_2CCF_3)_4$ (122, 213, 215), and the bis-π-cyclopenta-dienyl complexes $(\pi\text{-Cp})_2Ti(O_2CCF_3)_2$ and $(\pi\text{-Cp})_2Zr(O_2CCF_3)_2$ which have been prepared in several studies (22, 31, 71, 73, 137, 262). The absence of $Ti(O_2CCF_3)_4$ is interesting since, not only are the corresponding compounds known for the other Group IV elements, Zr, Hf, Th (122, 213, 215), Si, Ge, Sn (214, 215), and Pb (49, 118, 162, 192) but also titanium(IV) tetraalkanoates, $Ti(O_2CR)_4$, have been reported (126). The tetrakistrifluoroacetato complexes of Zr(IV), Hf(IV), and Th(IV) are hygroscopic solids that are nonvolatile at temperatures up to 180°C (122) and, if maintained at higher temperatures, they decompose to yield $(CF_3CO)_2O$, CF_3COF, COF_2, CO_2, and CO as the volatile products (215). The mass spectra of these $M(O_2CCF_3)_4$ (where M = Zr, Hf, or Th) compounds contain peaks up to 800 mass units, and in each spectrum no peak corresponding to the $M(O_2CCF_3)_4{}^+$ parent ion is observed. These data show that the compounds are polymeric in the solid state and are, thus, presumed to involve bridging trifluoroacetato groups. However, $Zr(O_2CCF_3)_4$, $Hf(O_2CCF_3)_4$, and $Th(O_2CCF_3)_4$ all dissolve in polar organic solvents as discrete molecules. Since, in each case, there are only very slight shifts (± 2 cm^{-1}) in the carboxylato stretching frequencies between the solid and the solution phases, it seems reason-able to propose (see Section VII, D) that the monomeric solution species contain only bidentate trifluoroacetato groups (122). In this context it is interesting to note that several other compounds of formula ML_4 (where M = Zr, Hf, or Th, and L is a bidentate ligand) are eight coordinate (160, 191) with the arrangement of the ligand donor atoms about the metal being either square antiprismatic, for example, $Zr(acac)_4$ (233) and $Th(acac)_4$ (100), or trigonal dodecahedral, for example $Na_4[Zr(C_2O_4)_4] \cdot 3H_2O$ (95) and $Th(S_2CNEt_2)_4$ (34). Considerations of ligand–ligand repulsions for these two stereochemistries have led to the suggestion that for these ML_4 complexes, which involve bidentate ligands of a short "bite" such as nitrate, oxalate, or trifluoroacetate, the trigonal dodecahedral structure is favored over the square anti-prismatic one (3, 27). Therefore, it would seem reasonable to suggest that the tetrakistrifluoroacetates of zirconium, hafnium, and thorium-(IV) contain eight-coordinated metal atoms, with the monomeric solution species containing four bidentate ligands arranged to give a dodecahedral MO_8 unit.

B. Vanadium

Four trifluoroacetato complexes of vanadium have been charac-terized: $[(\pi\text{-Cp})V(O_2CCF_3)_2]_2$ (150), $[Me_2(stearyl)_2N][VCl_4(O_2CCF_3)]$ (20), $NH_4[VO(O_2CCF_3)_3]$, and $VO_2(O_2CCF_3)$ (200). The first of these is of the most interest. This compound has a dimeric structure (Fig. 4) with four bridging trifluoroacetato groups but no metal–metal bond

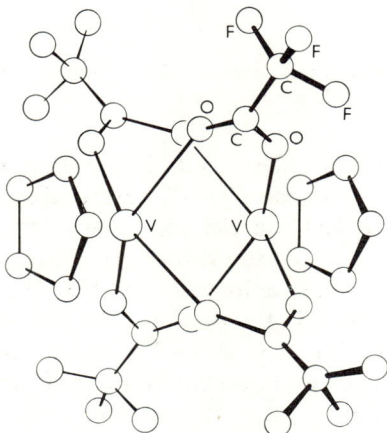

Fig. 4. The molecular structure of the (π-cyclopentadienyl)ditrifluoroaceta-tovanadium(III) dimer, $[(\pi\text{-Cp})V(O_2CCF_3)_2]_2$. From Larin et al. (150) with permission.

(V---V = 3.7 Å). Compound $[(\pi\text{-Cp})V(O_2CCF_3)_2]_2$ and its acetate analog have abnormally low room-temperature magnetic moments (μ_{eff} = 1.55 and 1.71 μ_B per vanadium, respectively) which decrease with decreasing temperature. These magnetic properties thus arise because of the interaction of the d^2 configurations of each vanadium(III) center via the π orbitals of the carboxylato groups.

C. CHROMIUM, MOLYBDENUM, AND TUNGSTEN

The known trifluoroacetato complexes of chromium, molybdenum, and tungsten are given in Table V. An X-ray structure determination (52) has shown that $Mo_2(O_2CCF_3)_4$ possesses the centrosymmetric structure illustrated in Fig. 5. The Mo—Mo distance of 2.090(4) Å is not

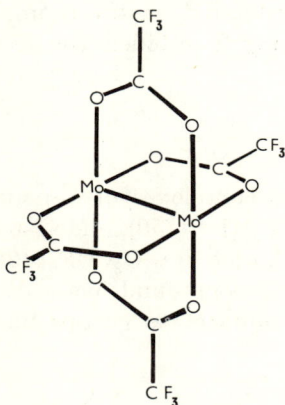

FIG. 5. Structure molybdenum(II) trifluoroacetate, $Mo_2(O_2CCF_3)_4$. From Cotton and Norman (52) with permission.

significantly different from that of 2.11(1) Å found in the analogous dimeric acetate (152), and the Mo—Mo stretching frequencies in the Raman spectra occur at 393 and 406 cm^{-1}, respectively. A quantitative assessment of the effect of coordination in the vacant axial positions of such a structure has been achieved for $Mo_2(O_2CCF_3)_4$, where the pyridine adduct, $Mo_2(O_2CCF_3)_4 \cdot 2py$, has been prepared and its crystal structure determined (51). The Mo—Mo separation of 2.129(2) Å is slightly longer than in the parent compound, with weak coordination of pyridine [Mo—N = 2.548(8) Å] compared to that of trifluoroacetate [Mo—O = 2.116(4) Å], and the Mo—Mo—N interbond angles = 171.0(2)°. Consistent with the slightly greater metal–metal separation, the Mo—Mo stretching frequency in the Raman spectrum of

TABLE V

TRIFLUOROACETATO COMPLEXES OF CHROMIUM, MOLYBDENUM, AND TUNGSTEN

Compound	Reference
$Et_4N[Cr(CO)_5(O_2CCF_3)]$	(218)
$Cr(O_2CCF_3)_2 \cdot nEt_2O$	$n = 0$ or 1 (116, 117)
$Cr(O_2CCF_3)_3$	(28, 91, 123, 156, 228, 255)
$[Cr(NH_3)_5(O_2CCF_3)](ClO_4)_2$	(58, 272)
$CrO_2(O_2CCF_3)_2$	(91)
$Et_4N[Mo(CO)_5(O_2CCF_3)]$	(218)
$Mo(CO)_2H(O_2CCF_3) \cdot 2THF$	(98)
$(\pi\text{-Cp})Mo(CO)_2(L)(O_2CCF_3)$	L = CO (57) or $(PhO)_3P$ (139)
$[(\pi\text{-Cp})Mo(NO)(O_2CCF_3)_2]_2$	(137)
$Mo_2(O_2CCF_3)_4$	(52)
$Mo_2(O_2CCF_3)_4 \cdot 2L$	L = py (51), Ph_3P, MeOH (190), bipy, or phen (123)
$(\pi\text{-Cp})_2Mo(O_2CCF_3)_2$	(102, 106)
$Et_4N[W(CO)_5(O_2CCF_3)]$	(218)
$(\pi\text{-Cp})W(CO)_3(O_2CCF_3)$	(57)
$(\pi\text{-Cp})_2W(O_2CCF_3)_2$	(102, 106)

$Mo_2(O_2CCF_3)_4 \cdot 2py$ is lowered to 367 cm^{-1}. A weakening of the Mo—Mo bond by coordination of the axial donor ligands might be expected since these ligands share the same metal orbitals as those used to form the metal–metal σ bond. Further evidence for this effect is provided by the Mo—Mo stretching frequencies for $Mo_2(O_2CCF_3)_4$ in several solvents, which decrease from the solid state value in a manner that roughly correlates with increasing donor strength of the solvent (51, 190).

Compound $Cr_2(O_2CCF_3)_4$ and its 1:1 ether adduct (116, 117) appear to be structurally similar to $Mo_2(O_2CCF_3)_4$ and $Mo_2(O_2CCF_3)_4 \cdot 2py$, respectively, but with a significantly weaker metal–metal interaction. This interaction is probably also less than that in $Cr_2(O_2CCH_3)_4 \cdot 2H_2O$ [where the Cr—Cr separation is 2.362(1) Å (50)] since the chromium(II) trifluoroacetates are paramagnetic, whereas the acetate is diamagnetic. The extent of the paramagnetism (μ_{eff} of $Cr_2(O_2CCF_3)_4 \cdot 2Et_2O$ = 0.85 μ_B at room temperature) is, however, consistent with significant spin exchange between the two d^4 centers, presumably via a direct Cr—Cr interaction and, as in $[(\pi\text{-Cp})V(O_2CCF_3)_2]_2$ (150), the bridging trifluoroacetato groups.

D. MANGANESE AND RHENIUM

The only trifluoroacetato complex of rhenium that has been prepared is $Re(CO)_5(O_2CCF_3)$ (57, 137). The corresponding manganese compound has been reported by King and Kapoor (137) who prepared it by the metathetical reaction between $Mn(CO)_5Br$ and $Ag(O_2CCF_3)$ in CH_2Cl_2. Green et al. (102) also claimed to have prepared this compound by reacting (σ-allyl)$Mn(CO)_5$ with CF_3CO_2H; however, the properties reported differ significantly from those of King and Kapoor whose results have been confirmed by a further study (89). The ESCA spectrum of $Mn(CO)_5(O_2CCF_3)$ has been recorded and compared with those of other $Mn(CO)_5X$ (where X = Br, I, Me, or CF_3) molecules (16). The values of the carbonyl carbon and oxygen 1s-electron-binding energies increase as X = Me < I < Br < CF_3 < O_2CCF_3, consistent with the trifluoroacetate group being the strongest electrophilic substituent in this series (89).

Substitution of the carbonyl ligands of $Mn(CO)_5(O_2CCF_3)$ has been achieved with several nitrogen-, oxygen-, phosphorus-, and sulfur-donor ligands ($89, 138$). Compounds $Mn(CO)_3L_2(O_2CCF_3)$ [where L = $P(OR)_3$ or $L_2 = R_2PCH_2CH_2PR_2$ and R = Me or Ph, or $L_2 = cis$-$Ph_2PCH=$ $CHPPh_2$] have been obtained by photochemical substitution using ultraviolet irradiation of a mixture of $Mn(CO)_5(O_2CCF_3)$ and the ligand in benzene–hexane solution at room temperature. Under these conditions, PPh_3 and $Mn(CO)_5(O_2CCF_3)$ afford a mixture of $Mn(CO)_3(PPh_3)_2$-(O_2CCF_3) and $Mn(CO)_2(PPh_3)_3(O_2CCF_3)$, and attempts to separate these complexes by chromatography on alumina resulted in decomposition with liberation of PPh_3 (138). However, when the corresponding reaction is carried out in refluxing chloroform the complex $Mn(CO)_3(PPh_3)_2$-(O_2CCF_3) is obtained exclusively. These latter conditions have also led to the isolation of other $Mn(CO)_3L_2(O_2CCF_3)$ complexes [where L = py, Ph_3PO, or $(Me_2N)_3PO$ and L_2 = bipy, phen, dithiohexane, or $C_6H_5SCH_2CH_2SC_6H_5$] (89). The stable isomer for the majority of these $Mn(CO)_3L_2(O_2CCF_3)$ complexes is the expected facial one in which the three carbonyl groups are mutually cis and, thus, no carbonyl group competes with any other such group trans to it, for the electron density of any d_π orbital. However, for the disubstituted triphenylphosphine complex, the steric requirements of these ligands favors that they adopt a trans arrangement; thus $Mn(CO)_3(PPh_3)_2(O_2CCF_3)$ is isolated from chloroform as the meridional isomer.

Manganese(II) trifluoroacetate ($225, 228$) and its adducts $Mn(O_2CCF_3)_2 \cdot nL$ (where L = py and n = 2 or 4; L = pyridine-N-oxide and n = 1; or L = γ-picoline and n = 1, 2, or 4) (7) have been characterized. Compound $Mn(acac)_2(O_2CCF_3)$ aids the polymerization of methylmethacrylate and vinyl acetate and is reduced to $Mn(acac)$-(O_2CCF_3) in these processes ($187, 188$).

E. Iron, Ruthenium, and Osmium

Table VI lists the known trifluoroacetato complexes of iron, ruthenium, and osmium. Photochemical substitution reactions of various tricovalent phosphorus-donor ligands with $(\pi\text{-allyl})Fe(CO)_2(O_2CCF_3)$ and $(\pi\text{-Cp})Fe(CO)_3(O_2CCF_3)$, analogous to those described in the preceding for $Mn(CO)_5(O_2CCF_3)$, have afforded $(\pi\text{-allyl})Fe(CO)(cis\text{-}Ph_2PCH{=}CHPPh_2)(O_2CCF_3)$, $(\pi\text{-Cp})Fe(CO)(PR_3)(O_2CCF_3)$ (where

TABLE VI
Trifluoroacetato Complexes of Iron, Ruthenium, and Osmium

Compound	Reference
$(\pi\text{-allyl})Fe(CO)L_2(O_2CCF_3)$	$L_2 = (CO)_2$ (137) or $cis\text{-}Ph_2PCH{=}CHPPh_2$ (138)
$(\pi\text{-Cp})Fe(CO)L(O_2CCF_3)$	$L = CO$ (137) or PR_3 (where $R = Ph$, OMe, or OPh) (138)
$(\pi\text{-Cp})FeL_2(O_2CCF_3)$	$L_2 = R_2PCH_2CH_2PR_2$ (where $R = Me$ or Ph) or $cis\text{-}Ph_2PCH{=}CHPPh_2$ (138)
$Fe(O_2CCF_3)_2 \cdot nH_2O$	$n = 0$ (105, 225, 228, 247, 255) or 2 (247)
$Fe(O_2CCF_3)_3$	(227)
$FeL(O_2CCF_3)$	$L = N,N'\text{-ethylenebis(salicylideneiminato)}$, $N,N'\text{-ethylenebis-(5,6-benzosalicylidene-iminato)}$, or $N,N'\text{-phenylenebis(salicylidene-iminato)}$ (158)
$Fe_3O_4(O_2CCF_3)$	(247)
$[Fe_3O(O_2CCF_3)_6]O_2CCF_3$	(247)
$Ru(H)(PPh_3)_3(O_2CCF_3)$	(209)
$Ru(H)(PPh_3)_2L(O_2CCF_3)_n$	$n = 1$, $L = PPh_3$, CO, or $n = 3$, $L = NO$ (208)
$Ru(Cl)(PPh_3)_2(CO)(O_2CCF_3)$	(208)
$Ru(PPh_3)_2(CO)(O_2CCF_3)_2$	(208)
$Ru(CO)_2(PR_3)_2(O_2CCF_3)_2$	$R = Ph$ (45) or OMe (39)
$[Ru(NH_3)_5(O_2CCF_3)](ClO_4)_2$	(246)
$Os(H)(PPh_3)_3(O_2CCF_3)$	(208)
$Os(Cl)(PPh_3)_3(CO)(O_2CCF_3)$	(208)
$Os(PPh_3)_2(CO)_n(O_2CCF_3)_2$	$n = 1$ or 2 (208)

$R = Ph$, OMe, or OPh), and $(\pi\text{-Cp})FeL_2(O_2CCF_3)$ (where $L_2 = Me_2PCH_2CH_2PMe_2$, $Ph_2PCH_2CH_2PPh_2$, or $cis\text{-}Ph_2PCH{=}CHPPh_2$), the ^1H and ^{19}F NMR spectra of which were reported (138).

Compound $[Fe_3O(O_2CCF_3)_6](CF_3CO_2)$ (247), in common with the other basic carboxylates of iron(III) (189), probably has a structure closely related to that of the basic chromium(III) carboxylato complexes (80).

F. COBALT, RHODIUM, AND IRIDIUM

The trifluoroacetato complexes of cobalt, rhodium, and iridium that have been isolated and characterized are listed in Table VII. An X-ray diffraction study of crystalline $(Ph_4As)_2[Co(O_2CCF_3)_4]$ has shown that the trifluoroacetato groups are essentially unidentate and give approximately tetrahedral coordination about the cobalt(II), as shown in Fig. 6 (25). Although the bulk magnetic and electronic spectral properties are similar to those of tetrahedral cobalt(II) complexes, the interpretation of the polarized single crystal d-d spectra and the temperature variation of the principal magnetic moments require that the elongation of the

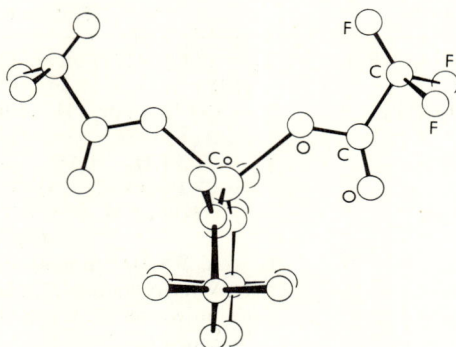

FIG. 6. Structure of the tetratrifluoroacetatocobalt(II) anion in $(Ph_4As)_2$-$[Co(O_2CCF_3)_4]$. Reprinted with permission from *Inorganic Chemistry* **5**, 1422 (1966). Copyright by the American Chemical Society.

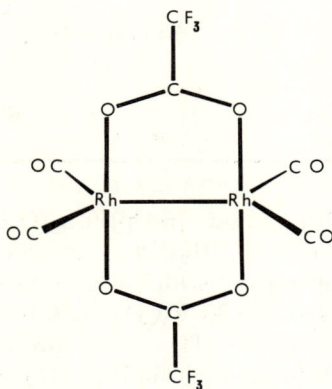

FIG. 7. Suggested structure for dicarbonyl trifluoroacetatorhodium(I) [Rh-$(CO)_2(O_2CCF_3)]_2$. From Lawson and Wilkinson (151) with permission.

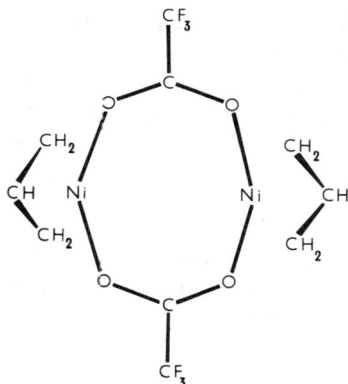

FIG. 8. Suggested structure for the dimer of π-allyl trifluoroacetatonickel(II), $[(\pi\text{-allyl})\text{Ni}(O_2CCF_3)]_2$ (60).

H. COPPER, SILVER, AND GOLD

A large number of trifluoroacetato complexes of copper have been prepared (Table IX), the majority of these being N-donor ligand adducts of $\text{Cu}(O_2CCF_3)_2$. The dimeric configuration adopted by copper(II) acetate and its adducts (104, 260) does not appear to be adopted as a general rule by the analogous trifluoroacetato complexes. Thus, anhydrous copper(II) trifluoroacetate has a room-temperature magnetic moment of 1.81 μ_B and its magnetic susceptibility conforms to the Curie–Weiss law between 94° and 297°K; the small negative θ value thus obtained suggests a weak antiferromagnetic exchange interaction between the metal centers, perhaps indicative of a polymeric structure (255). The majority of copper(II) trifluoroacetate adducts possess two or four other ligands per copper atom and are thus presumed not to have a dimeric structure (1, 2, 4). Also, those adducts that involve only one addended ligand per copper atom may not be dimeric. Thus, of all the 1:1 α-picoline adducts of the copper(II) salts of halogen-substituted acetates, only the trifluoroacetate is monomeric, all of the other haloacetate derivatives being dimeric (136). Evidence for antiferromagnetic exchange between copper(II) centers has been obtained for $\text{Cu}(O_2CCF_3)_2 \cdot \text{quinoline}$ (1) and $\text{Cu}(O_2CCF_3)_2 \cdot \text{dioxan}$ (258); however, in the latter instance the ESR spectrum also shows the presence of a mononuclear copper(II) species. The compound $\text{Cu}_2(O_2CCF_3)_4 \cdot 1,4\text{-di-}$ (2'-pyridyl)aminophthalizine may have a dimeric structure with two bridging and two terminal trifluoroacetato groups (254).

Attempts have been made to correlate the occurrence or non-occurrence of a binuclear structure for copper(II) carboxylates with the

pK_a of the parent acid (134, 157, 167). For parent acids with a pK_a of less than ca. 4, the binuclear structure would appear to be unstable because the residual positive charge on the copper atoms is too great to allow their close proximity favored in this structure. The apparent non-appearance of a dimeric structure for the majority of copper(II) derivatives of trifluoroacetic acid ($pK_a = \pm 0.25$) is thus consistent with this view.

TABLE IX

TRIFLUOROACETATO COMPLEXES OF COPPER

Compound	Reference
$Cu(O_2CCF_3)$	(29, 30, 67, 219, 225)
$Cu(CO)(O_2CCF_3) \cdot nCF_3CO_2H$	$n = 0$ or 1 (219)
$Cu(PPh_3)_2(O_2CCF_3)$	(67)
$Cu(Ph_2PR)_3(O_2CCF_3)$	R = Me (67), Ph (67, 103), or p-MeC_6H_4 (177)
$Cu(Ph_2PCH_2CH_2PPh_2)(O_2CCF_3)$	(67)
$Cu(PPh_3)(2,9\text{-}Me_2phen)(O_2CCF_3)$	(103)
$Cu(O_2CCF_3)_2$	(134, 225, 227, 228, 247, 255)
$Cu(Cl)(O_2CCF_3)$	(226)
$Cu(O_2CCF_3)_2 \cdot L$	L = H_2O (258), isoquinoline, pyridine-N-oxide, 3-Brpy, 2,6- or 3,5-lutidine (4), 2-Clpy (2, 4), α-picoline (136), quinoline (1), 1,2- or 1,3-propanediamine (172), or dioxan(258)
$Cu(O_2CCF_3)_2 \cdot 2L$	L = py, α-, β-, or γ-picoline (2, 4, 136, 155, 224), 2-Etpy, 2-Clpy (4), quinoline (1), 2,6-lutidine (4), pyridine-N-oxide, or γ-picoline-N-oxide (224), amines (172),
$Cu(O_2CCF_3)_2 \cdot 4L$	L = py (14, 155, 224), 3-Brpy, β-picoline, quinoline (4), or γ-picoline (224)
$Cu_2(O_2CCF_3)_4 \cdot L$	L = 1,4-di-(2'-pyridyl)aminophthalizine (254)
$[Cu_2OH(quinoline)_2(O_2CCF_3)_3]_2$	(161)

The only copper(II) trifluoroacetate shown by diffraction studies to involve trifluoroacetato groups bridging two metal atoms in relatively close proximity is $[Cu_2OH(quinoline)_2(O_2CCF_3)_3]_2$ (161). X-Ray analysis has shown that the compound has a novel tetranuclear structure, each copper atom having a distorted square pyramidal environment. Four of the trifluoroacetato groups in the unit form bridges from the apical position of one copper atom to a basal site of another such atom; the other two are each coordinated as a unidentate ligand in the basal plane of a copper atom. The magnetic susceptibility data show that a substantial copper–copper interaction is present in this compound. The relatively large metal–metal separations of 2.996(4) and 3.347(5) Å, together with the relative orientations of the copper coordination polyhedra, do not favor a direct metal–metal interaction. Thus, as in the case of $[(\pi\text{-Cp})V(O_2CCF_3)_2]_2$ (150), magnetic exchange via the π-orbitals of the bridging trifluoroacetato groups seems to be indicated.

Compound $Cu(CO)(O_2CCF_3)\cdot CF_3CO_2H$ is obtained by reacting copper(I) oxide with a trifluoroacetic acid–anhydride mixture under an atmosphere of carbon monoxide; the acid solvate molecule may be removed by pumping at room temperature and further such treatment then affords $Cu(O_2CCF_3)$ (219).

In a large number of studies (e.g., 110, 119, 228, 247), $Ag(O_2CCF_3)$ has been prepared, and its reactions with metal and nonmetal halides have been used widely for the preparation of the corresponding trifluoroacetato complexes (see Section III). Swarts (247) has also reported the preparation of $AgL(O_2CCF_3)$ (where $L = I$ or C_6H_6). Compound $Ag(O_2CCF_3)$ has been shown to react with halogens in fluorinated solvents according to Eq. (12) and $AgI(O_2CCF_3)$ has been isolated and

$$2Ag(O_2CCF_3) + X_2 \longrightarrow AgX(O_2CCF_3) + AgX \qquad (12)$$

characterized (53). A mixture of I_2 and $Ag(O_2CCF_3)$ is a powerful iodinating reagent (79) and iodinates aromatic compounds in positions expected for attack by an electrophilic reagent. Similarly, Br_2 and $Ag(O_2CCF_3)$ form a powerful brominating mixture, which, at low temperatures, reacts by an electrophilic mechanism (112). Furthermore, the reactions of Cl_2, Br_2, and I_2 with $Ag(O_2CCF_3)$ at elevated temperatures afford a convenient synthesis of the corresponding trifluoromethyl-halides (53, 110).

Compound $Au(PPh_3)(O_2CCF_3)$ has been prepared by the reaction of $Ag(O_2CCF_3)$ with Ph_3PAuCl, the complex being monomeric in chloroform solution (185). Also $Ag(PPh_3)(O_2CCF_3)$ has been reported (93). The complex $\{(p\text{-MeC}_6H_4)_3P\}_3Ag(O_2CCF_3)$ has been shown by ^{31}P NMR studies to disproportionate into $(p\text{-MeC}_6H_4)_2Ag(O_2CCF_3)$ and $[(p\text{-MeC}_6H_4)_4Ag]CF_3CO_2$ in dichloromethane solution (177).

I. Zinc, Cadmium, and Mercury

Table X lists the trifluoroacetato complexes of zinc, cadmium, and mercury that have been prepared. Mercury(II) trifluoroacetates in trifluoroacetic acid have been used widely in organic syntheses since they effect fast mercuration reactions of aromatic rings (35, 241),

TABLE X

TRIFLUOROACETATO COMPLEXES OF ZINC, CADMIUM, AND MERCURY

Compound	Reference
$Zn(O_2CCF_3)_2$	(225, 227)
$Zn(O_2CCF_3)_2 \cdot L$	L = pyridine-N-oxide or γ-picoline-N-oxide (224)
$Zn(O_2CCF_3)_2 \cdot 2L$	L = H_2O (180), py, or γ-picoline (224)
$Zn(py)_4(O_2CCF_3)_2$	(224)
$Cd(O_2CCF_3)_2$	(105)
$Hg_2(O_2CCF_3)_2$	(238, 247)
$Hg(O_2CCF_3)_2$	(65, 229, 230)
$Hg(O_2CCF_3)_2 \cdot nL$	n = 1 and L = PPh_3 (172), bipy, or phen (61, 186); or n = 2 and L = H_2O (247) or PPh_3 (62)
$HgX(O_2CCF_3)$	X = OH (247), Me (77), CF_3 (178, 229), Ph (223), C_6F_5 (65), or CF_3S (69)

analogous to the reactions described for thallium(III) trifluoroacetate; these complexes also undergo specific addition across non-conjugated double bonds (see Section VIII, B) (223, 229, 230). The thermal decarboxylation of $Hg(O_2CCF_3)_2$ (5) and $Hg(O_2CCF_3)_2 \cdot L$ (where L = bipy or phen) (48) are described in Section VIII, A.

VI. Trifluoroacetato Complexes of the f-Block Elements

The tris-trifluoroacetates, $M(O_2CCF_3)_3$, have been reported for the lanthanides La, Ce, Pr, Nd, and Sm (105, 206, 263). However, the most extended range of lanthanide trifluoroacetate complexes are the hexafluoroacetylacetonato derivatives $Ln(hfac)_2(O_2CCF_3) \cdot 2H_2O$ (where Ln = La, Ce, Pr, Nd, Sm, Gd, Dy, Ho, Er, Yb, or Y) (204). Compound $Sm(NH_3)_2(O_2CCF_3)_3$ has been isolated (263) and $Nd(phen)(O_2CCF_3)_3$, just as analogous carboxylato complexes of neodymium, is used as a liquid laser, the medium being a deuterated solvent (113). Other laser systems of this type employ $Nd(O_2CCF_3)_3$–$AlCl_3$ or $Nd(O_2CCF_3)_3$–$ZrCl_4$ in $POCl_3$ or $PSCl_3$ (81, 264).

The properties of $Th(O_2CCF_3)_4$ (122, 213, 215) have already been described (Section V, A). Also $U(O_2CCF_3)_4$ has been reported (220), but since few data were given concerning its properties, further characterization would be desirable. Compound $Pu(O_2CCF_3)_4 \cdot xH_2O$ has been prepared by allowing aerial oxidation of a solution of Pu(III) in CF_3CO_2H (43).

Hara and Cady (105) prepared $UO_2(O_2CCF_3)_2$ by carboxylate exchange from the corresponding acetate and showed that the compound is very moisture-sensitive and virtually insoluble in trifluoroacetic acid.

VII. Physical Properties of Trifluoroacetates

A. STRUCTURES

The possible modes of bonding of carboxylate groups to metal atoms have been described by Oldham (*189*) and thus far, for the trifluoroacetate group, three of these modes have been identified by diffraction studies. The $NH_4(O_2CCF_3)$ is composed of ions linked by hydrogen bonds (*54*); $(Ph_4As)_2[Co(O_2CCF_3)_4]$ (Fig. 6) (*25*) and $[CuOH(quinoline)_2-(O_2CCF_3)_3]_2$ (*161*) contain unidentate trifluoroacetato groups. The last-mentioned compound also possesses syn-syn-bridging trifluoroacetato groups, which have also been identified in $[(\pi\text{-}Cp)V(O_2CCF_3)_2]_2$ (see Fig. 4) (*150*), $Mo_2(O_2CCF_3)_4$ (see Fig. 5) (*52*), $Mo_2(O_2CCF_3)_4\cdot2py$ (*51*), and $Co_3(Cl)SO_4(dimethoxyethane)_3(O_2CCF_3)_3$ (*76*). The anions of $K[H(O_2CCF_3)_2]$ and $K[D(O_2CCF_3)_2]$ (see Fig. 1) (*163*) are perhaps best considered as examples of unidentate trifluoroacetato groups since the hydrogen bonding is strong and specific to one oxygen atom of each carboxylate group.

Although the definitive characterization of trifluoroacetate structures is limited to the foregoing data, the modes of coordination of this group in other compounds may be inferred with some confidence. Thus, bonding considerations would appear to restrict the trifluoroacetato group to unidentate coordination in $PF_2(O_2CCF_3)$ (*82*), $AuPPh_3-(O_2CCF_3)$ (*185*), $[M(NH_3)_5(O_2CCF_3)]^{2+}$ [where M = Cr (*58, 272*), Co (*147, 148*), Rh (*78, 175*), or Ir (*78*)], and $[M(CO)_5(O_2CCF_3)]^{n-}$ [where $n = 0$ and M = Mn (*89, 102, 137*) or Re (*57, 137*); or $n = 1$ and M = Cr, Mo, or W (*218*)]. No compounds have yet been shown by diffraction studies to contain symmetrically bidentate trifluoroacetato groups. However, since they are known to occur for the related ligands acetate (e.g., *265, 271*), carbonate (e.g., *18, 231*), and nitrate (*3*), it is likely that this mode of coordination will be characterized for the trifluoroacetate group. Such coordination was favored earlier (Section V, A) for monomeric $M(O_2CCF_3)_4$ [where M = Zr, Hf, or Th (*122, 213, 215*)] and, since $Cr(O_2CCF_3)_3$ also dissolves as a molecular unit in acetone (*122*) where its d-d spectrum is typical of an octahedral chromium(III) complex (*122, 228*), it too would seem to involve bidentate trifluoroacetato groups.

The non-volatility of most metal trifluoroacetates clearly implies the presence of bridging trifluoroacetato groups in the solid state, presumably with the anti–anti or syn–anti configurations. Such arrangements have already been favored for the structures of solid $M(O_2CCF_3)_4$ (where M = Zr, Hf, or Th) (Section V, A) and $Me_3Sn(O_2CCF_3)$ (Section IV, D) (see Fig. 3). The tendency for acetato groups to bridge metal centers has also been noted (32). A particular structural feature of the transition metal trifluoroacetates is the formation of dimeric units with two or four trifluoroacetato ligands bridging the two metal centers which may be as close as 2.1 Å, e.g., in $Mo_2(O_2CCF_3)_4$ (52), or as far apart as 3.7 Å, e.g., in $[(\pi\text{-Cp})V(O_2CCF_3)_2]_2$ (150). However, it seems probable that trifluoroacetate is less able than acetate to support such a structure, presumably because of the higher residual charge on the two metal atoms in close proximity. Thus the Mo—Mo stretching frequencies are 393 and 406 cm^{-1} in $Mo_2(O_2CCF_3)_4$ and $Mo_2(O_2CCH_3)_4$, respectively; $Cr_2(O_2CCF_3)_4 \cdot 2Et_2O$ is paramagnetic to the extent of 0.87 μ_B at room temperature, whereas $Cr_2(O_2CCH_3)_4 \cdot 2H_2O$ is diamagnetic, and $Cu(O_2CCF_3)_2$ and its adducts generally appear not to possess the dimeric structures characteristic of the corresponding acetato complexes.

B. DIMENSIONS

The most accurate structural data for trifluoroacetate groups are those obtained from neutron diffraction studies of $K[H(O_2CCF_3)_2]$ and $K[D(O_2CCF_3)_2]$ (see Fig. 1) (163), and their average dimensions are given in Table XI. The dimensions obtained in the other diffraction studies of trifluoroacetato compounds agree within experimental error with those quoted in Table XI in all respects except the length of the C—O bonds and the magnitude of the C—C—O interbond angles. In $NH_4(O_2CCF_3)$ (54), no distinction between the two C—O bonds of length 1.269(5) Å, or between the two C—C—O interbond angles of 115.8(3)°, was observed. The different geometries of the carboxylate

TABLE XI

DIMENSIONS AND THEIR ESTIMATED STANDARD DEVIA-
TIONS OF THE TRIFLUOROACETATE GROUPS IN
$K[H(O_2CCF_3)_2]$ AND $K[D(O_2CCF_3)_2]^{a,b}$

Bond	Length (Å)	Interbond Angle	Value (degrees)
C—F	1.325(6)	F—C—F	107.7(6)
C—C	1.541(2)	F—C—C	111.3(7)
C—O(1)	1.215(2)	C—C—O(1)	119.6(1)
C—O(2)	1.268(3)	C—C—O(2)	111.3(1)
O(2)····O(2')	2.437(4)	O—C—O	129.2(2)

[a] Data from Ref. 163.
[b] The dimensions quoted are averaged over the two salts.

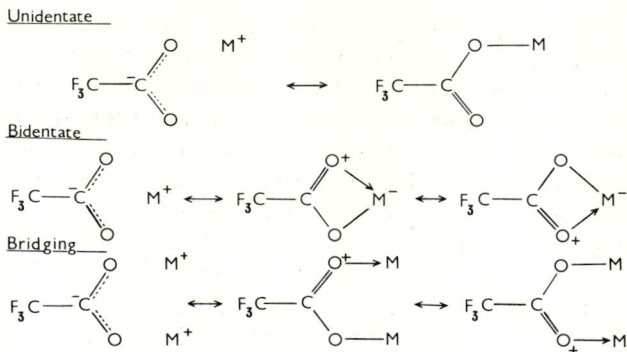

FIG. 9. Principal valence bond structures of unidentate, bidentate, and bridging trifluoroacetato groups.

groups in these potassium and ammonium salts are expected since, in the $[H(O_2CCF_3)_2]^-$ and $[D(O_2CCF_3)_2]^-$ ions, there is strong hydrogen bonding which only involves the O(2) atom of each trifluoroacetate ion, whereas in $NH_4(O_2CCF_3)$ both oxygen atoms of the anions participate in two N—H···O bonds of length ca. 2.9 Å. By applying these results to trifluoroacetato complexes, it is anticipated that, for unidentate coordination, the C—O bond involving the noncoordinated oxygen atom will be significantly shorter than that involving the coordinated one and that the C—C—O interbond angles may be different, whereas, for bidentate and bridging trifluoroacetato groups, no real difference in C—O bond lengths or C—C—O interbond angles is expected. These views are also consistent with simple valence bond arguments (Fig. 9). The relatively large estimated standard deviations of dimensions reported for $(Ph_4As)_2 [Co(O_2CCF_3)_4]$ (25) do not allow for any distinction between the C—O bond lengths and C—C—O interbond angles within the trifluoroacetato groups. The oxygen atoms of each of the bridging trifluoroacetato groups of $[(\pi\text{-}Cp)V(O_2CCF_3)_2]_2$ (150), $Mo_2(O_2CCF_3)_4$ (52), $Mo_2(O_2CCF_3)_4 \cdot 2py$ (51), and $Co_3(Cl)SO_4(\text{dimethoxyethane})_3(O_2CCF_3)_3$ (76) appear to be equivalent within the accuracy of these structure determinations. However, considerably more structural data are required before the dimensions of the carboxylate portion of coordinated trifluoroacetate groups can be regarded as established.

C. ^{19}F NUCLEAR MAGNETIC RESONANCE SPECTRA

The ^{19}F NMR spectra have been recorded for only a small percentage of the compounds described in Sections IV, V, and VI. However, the compounds studied in this respect have ranged from high oxidation states, e.g., $CrO_2(O_2CCF_3)_2$ (91) and $Pb(O_2CCF_3)_4$ (122), through intermediate, e.g., $Zn(O_2CCF_3)_2$ and $Hg(O_2CCF_3)_2$ (122), to low oxidation states, e.g., $Au(PPh_3)(O_2CCF_3)$ (185) and $Re(CO)_5$-(O_2CCF_3) (137), and have involved some twenty different metal atoms. In each case the room-temperature ^{19}F NMR spectrum consists of a single resonance clearly indicating essentially free rotation of the CF_3 moiety about the C—C bond at normal temperatures, as expected in view of the spectral results of Redington and Lin (202). The position of this resonance is in the range 74–79 ppm upfield of $CFCl_3$ for all the compounds studied with the exception of $Au(PPh_3)(O_2CCF_3)$ where it was observed (185) at 69.1 ppm upfield of $CFCl_3$. The spread of ^{19}F chemical shifts for trifluoroacetates is, therefore, very small when compared to the known range (266) for such nuclei, which extends from CH_3F at 278 to UF_6 at -746 ppm relative to $CFCl_3$. This, together with the difficult interpretation of ^{19}F chemical shifts, clearly limits the utility of ^{19}F NMR data for the interpretation of effects such as the nature and/or strength of the coordination of the trifluoroacetate group. However, it should be noted that recent studies have claimed that the nature of hydroxyl groups in natural products may be ascertained by converting them into trifluoroacetyl ones with trifluoroacetic anhydride; then the ^{19}F chemical shifts of such groups are capable of differentiating between primary, secondary, tertiary, and aromatic environments (261).

D. Vibrational Spectra

The highest symmetry that a trifluoroacetate group can possess, if free rotation about its C—C bond is not possible, is C_s. The ^{19}F NMR studies described earlier indicated that essentially free rotation about the C—C bond of trifluoroacetate groups occurs in solution at room temperature; in this situation the symmetry may be approximated to C_{2v} if both oxygen atoms are equivalent or to C_s if they are not. Although C_{2v} symmetry might be appropriate for an isolated trifluoroacetate group, in a crystal lattice and/or an individual molecule, vibrational coupling will occur between a trifluoroacetate group and other similar and/or different groups. This coupling will result in a splitting of the degenerate modes and a relaxation of selection rules. Therefore, the infrared and Raman spectra of compounds containing a trifluoroacetate group would be expected to contain peaks corresponding to virtually all of the fifteen normal vibrational modes of this group. Thus the number of infrared and Raman bands observed for a trifluoroacetate group is expected to be independent of its environment, and distinctions

TABLE XII

VIBRATIONAL FREQUENCIES OF TRIFLUOROACETATE
GROUPS IN $K[H(O_2CCF_3)_2]$[a]

Mode	Frequency (cm^{-1})	
	Infrared	Raman[b]
CO_2 asym. stretch	1792	1725
CO_2 sym. stretch	1420	1441
CF_3 stretch	1226	1190
CF_3 stretch	1160	1164
CF_3 stretch	1128	1150
C—C stretch	828	846
OCO def.	791	734
CF_3 bend	706	710 sh
CF_3 bend	590	630 vb
CF_3 bend	520	518
CCO γ bend	440	428
CCO δ bend	385	324
ω CF_3 wag	270	272
ρ CF_3 wag	265	269

[a] Data from Ref. *173*.

[b] sh = shoulder; vb = very broad.

between the various modes of bonding must be made on the basis of band positions and/or their relative intensities.

The vibrational spectra exhibited by trifluoroacetate groups have been the subject of much discussion (*17, 41, 87, 88, 132, 178*) and their assignment has only recently been clarified by the elegant matrix isolation studies of trifluoroacetic acid monomers and dimers by Redington and Lin (*202*) and the detailed study of the infrared and Raman spectra of $K[H(O_2CCF_3)_2]$ and $Cs[H(O_2CCF_3)_2]$ (and their deuterated analogs) by Miller *et al.* (*173*). Table XII lists the trifluoro-acetate vibrational frequencies of the latter salts. The vibrational spectra of the trifluoroacetate groups of other compounds are usually very similar to those presented in this table except that, as would be anticipated from the earlier discussion of the dimensions of this group, some variation in the carboxylate stretching frequencies is observed. Several attempts have been made to correlate the stretching frequencies of oxyanions with the extent and nature of their coordination (*3, 75, 86, 131, 155, 181, 252*). In the case of carboxylates, attention has focused on the positions and separations of the antisymmetric and symmetric carboxylate stretching frequencies. The data collected in Table XIII (although lacking the necessary diffraction studies for confirmation of

TABLE XIII

INFRARED CARBOXYLATE STRETCHING FREQUENCIES OF VARIOUS
TYPES OF TRIFLUOROACETATE GROUPS

Compound	Nature of tri-fluoroacetate group	$\nu(CO_2)$ (cm^{-1})		Δ	References
		Anti-sym-metric	Sym-metric		
$NH_4(O_2CCF_3)$	Ionic	1667	1465	202	(122)
$K(O_2CCF_3)$	Ionic[a]	1678	1437	241	(122)
$Cs(O_2CCF_3)$	Ionic[a]	1678	1427	251	(122)
$Cr(O_2CCF_3)_3$	Bidentate[a]	1710	1490	220	(122)
$Zr(O_2CCF_3)_4$(solution)	Bidentate[a]	1662	1483	179	(122)
$Mo_2(O_2CCF_3)_4$	Bridging	1592⎫ 1572⎬	1459	∼133	(52)
$Rh_2(O_2CCF_3)_4 \cdot 2EtOH$	Bridging[a]	1664	1467	197	(267)
$Zr(O_2CCF_3)_4$(solid)	Bridging[a]	1660	1481	179	(122)
$Me_3Sn(O_2CCF_3)$(solid)	Bridging[a]	1652	1340	312	(197)
$(Ph_4As)_2[Co(O_2CCF_3)_4]$	Unidentate	1692	1421	271	(122)
$Mn(CO)_5(O_2CCF_3)$	Unidentate[a]	1694	1416	278	(89)
$Au(PPh_3)(O_2CCF_3)$	Unidentate[a]	1695	1406	289	(185)
$Me_3Sn(O_2CCF_3)$(solution)	Unidentate[a]	1720	1290	430	(197)
CF_3CO_2H	Unidentate[a]	1819	1415	404	(202)

[a] Unconfirmed by diffraction studies.

many of the structures) indicate that the positions of the infrared carboxylate stretching frequencies do not afford a general distinction between the various types of trifluoroacetate groups. This is not surprising since, not only do simple valence bond arguments (see Fig. 9) anticipate that little distinction between ionic, bridging, and bidentate trifluoroacetate groups should be achieved in this manner, but also that the vibrational frequencies of coordinated trifluoroacetate groups will depend on the nature (182) and oxidation state of the central atom and the other groups bonded to this atom. However, in situations where the latter parameters are unchanged, the C—O stretching frequencies of a trifluoroacetato group may differentiate between unidentate coordination, on the one hand, and ionic, bidentate, or bridging, on the other. This view has been used to support the inference that $Me_3Sn(O_2CCF_3)$ is polymeric in the solid state with bridging trifluoroacetato groups and monomeric in CCl_4 solution with unidentate trifluoroacetato groups (197). The shift in the asymmetric (CO_2) stretching frequency from 1652 to 1720 cm^{-1} would appear to be genuine evidence of this effect, however, the values of the symmetric (CO_2) stretching frequency of 1340 and 1290 cm^{-1} appear to be unusually low in comparison with the other values in Table XIII.

It remains to be seen whether further work will establish other means of identifying the nature of trifluoroacetate groups from their vibrational spectra. A possibility that does not appear to have been explored fully in this respect is the relative intensity of the anti-symmetric and symmetric CO_2 stretching frequencies, for both infrared and Raman spectra.

VIII. Reactions of Metal Trifluoroacetates

A. Decarboxylation

The reactions of Na(I), K(I), Ba(II), Ag(I), Hg(II), and Pb(II) trifluoroacetates with iodine at elevated temperatures has been shown to be a convenient synthesis of trifluoroiodomethane, CF_3I, by Haszeldine (110). Similar reactions of $Ag(O_2CCF_3)$ with chlorine or bromine give the corresponding trifluoromethylhalide.

Compound $Me_2As(O_2CCF_3)$ may be thermally decarboxylated to give Me_2AsCF_3 (55). Thermal decarboxylation of $Hg(O_2CCF_3)_2$ proceeds readily at 300°C to give $CF_3Hg(O_2CCF_3)$ (5) which may be converted into CF_3HgI by treatment with $NaI \cdot 2H_2O$. This latter mercury(II) derivative is a convenient source of difluorocarbene (222). Also $Hg(O_2CCF_3)_2 \cdot L$ (where L = bipy or phen) may be decarboxylated at their melting points (ca. 200°C) to afford the corresponding trifluoro-methyl derivatives (48).

The preceding reactions represent rare conversions of trifluoro-acetates into trifluoromethyl derivatives; such thermal decarboxylation reactions are usually unsuccessful (61, 123, 193). Thus, attempts to prepare $M(CF_3)_n$ (where $n = 2$ and M = Hg; or $n = 3$ and M = P or As) by heating the appropriate element with $Ag(O_2CCF_3)$ failed in their objective (111). Studies by Sharp et al. (228) and by Sartori and Weidenbruch (215) on trifluoroacetato complexes have shown that, although decarboxylation frequently does occur, the products obtained are M, MO_x, CO_2, $(CF_3CO)_2O$, C_2F_6, COF_2, and CF_3COF. Swarts (247) and Simons et al. (235) also produced $(CF_3CO)_2O$ and CF_3COF by thermal decomposition of $Na(O_2CCF_3)$ and $Ba(O_2CCF_3)_2$, respectively.

The failure of trifluoroacetate decarboxylation as a general synthetic procedure is perhaps not surprising since such reactions are usually of limited utility. Furthermore, trifluoroacetate groups are quite resistant to decarboxylation which is more difficult to achieve for $Na(O_2CCF_3)$ than for $Na(O_2CCCl_3)$ or $Na(O_2CCBr_3)$ (33). Auerbach et al. (15) have shown that the activation energies for the decarboxylation of $Na(O_2CCF_3)$ and $Na(O_2CCCl_3)$ are 42 and 32 kcal mole^{-1}, respectively. The former reaction proceeds by a first-order mechanism in ethylene glycol and a convenient rate is only achieved at ca. 180°C.

Lead(IV) trifluoroacetate trapped in a benzene matrix at liquid nitrogen temperatures has been shown to afford CF_3· radicals when irradiated by ultraviolet light (162). Thus photolytic decarboxylation of trifluoroacetates would appear to merit further investigation.

B. METAL–TRIFLUOROACETATE BOND CLEAVAGE

Since trifluoroacetic acid is a relatively strong acid, its conjugate base $CF_3CO_2^-$ is relatively weak and $M—O_2CCF_3$ bonds are usually quite labile. This lability is indicated by the moisture sensitivity of most trifluoroacetato complexes. Therefore, these complexes have potential value as synthetic intermediates as illustrated by the following examples.

1. Elimination of Trifluoroacetic Acid

Although thallium(III) acetate fails to react with aromatic compounds under mild conditions, thallium(III) trifluoroacetate will effect rapid electrophilic thallation of a wide range of aromatic substances (40, 168, 169, 170, 203) according to the general reaction [Eq. (11)]. These reactions are very rapid and are usually completed in a few minutes at room temperature for activated aromatic nuclei, to give stable colorless solids that generally crystallize from solution. The orientation of thallation in such reactions may be influenced by temperature [Eq. (13)], or by the substituents on the aromatic nucleus.

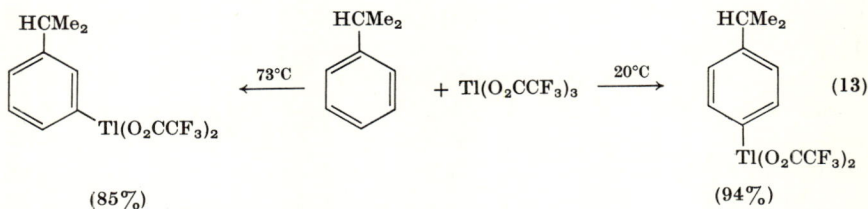

$$\underset{(85\%)}{\underset{HCMe_2}{\text{[ring]}}\text{Tl}(O_2CCF_3)_2} \xleftarrow{73°C} \underset{HCMe_2}{\text{[ring]}} + \text{Tl}(O_2CCF_3)_3 \xrightarrow{20°C} \underset{(94\%)}{\underset{HCMe_2}{\text{[ring]}}\text{Tl}(O_2CCF_3)_2} \qquad (13)$$

Thus bulky groups usually prevent ortho substitution [e.g., Eq. (13)]. However, substituents such as benzoic acid and its esters, which are able to complex the incoming thallium atom, result in almost exclusive ortho substitution [Eq. (14)],

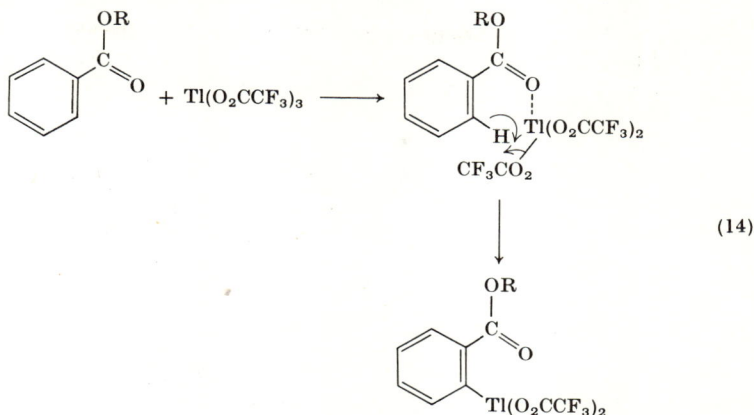

$$(14)$$

At present such metallation reactions have only been developed for the trifluoroacetates of thallium(III) and mercury(II) (*35, 241*), and they clearly offer considerable scope for the extension of the organometallic chemistry of these elements.

Displacement of the metal from the aromatic nucleus may be effected by a variety of reactants, as illustrated for thallium(III) by Eqs. (15,) (16), and (17).

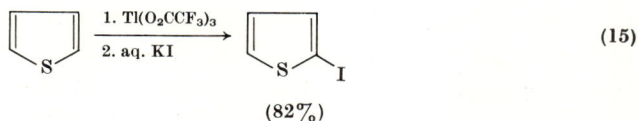

$$(15)$$

(82%)

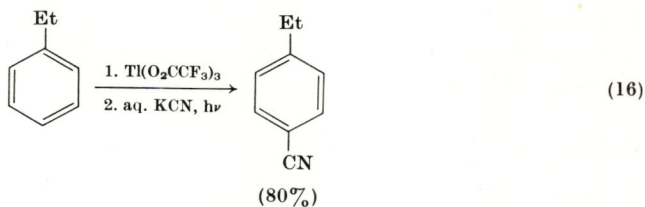

$$(16)$$

(80%)

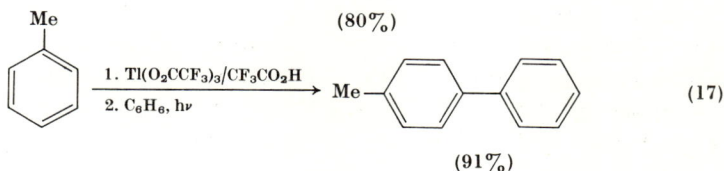

$$(17)$$

(91%)

This sequential process of thallation followed by displacement represents a new and versatile method for aromatic substitution.

Thallium(III) trifluoroacetate will also eliminate trifluoroacetic acid in reactions with compounds containing N—H bonds (40, 240). Thus, brief treatment of octaethylporphyrin in CH_2Cl_2 and tetrahydrofuran with an excess of $Tl(O_2CCF_3)_3$ followed by chromatography on deactivated alumina affords the complex shown in Fig. 10. This was the first unambiguous characterization of a thallium(III) porphyrin (240).

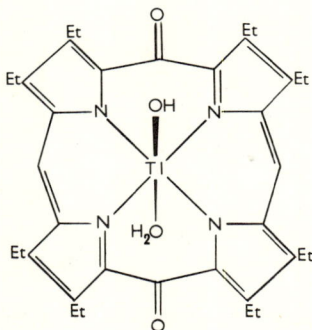

Fig. 10. Thallium(III) octaethylporphyrin complex. From Smith (240) with permission.

Lead(IV) trifluoroacetate undergoes reductive elimination of trifluoroacetic acid in the presence of both alkyl and aryl C—H bonds [Eq. (18)].

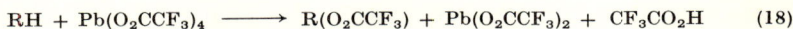

$$RH + Pb(O_2CCF_3)_4 \longrightarrow R(O_2CCF_3) + Pb(O_2CCF_3)_2 + CF_3CO_2H \qquad (18)$$

In this respect $Pb(O_2CCF_3)_4$ appears to have unique powers as an oxidant toward hydrocarbons, and, thus, nonactivated molecules, such as heptane or benzene, are converted to their trifluoroacetoxy-substitution products, $C_7H_{15}O_2CCF_3$ and $C_6H_5O_2CCF_3$, respectively. Subsequent mild hydrolysis completes a convenient route to the corresponding alcohol or phenol (192).

2. Other Trifluoroacetate Elimination Reactions

Salt elimination reactions have been used to obtain the divinyl tin(IV) derivatives, $(CH_2{=}CH)_2Sn(O_2CCF_3)(Mn(CO)_5)$, $(CH_2{=}CH)_2Sn(Mn(CO)_5)_2$, $[(CH_2{=}CH)_2SnFe(CO)_4]_2$, and $(CH_2{=}CH)_2Sn(Co(CO)_4)_2$, by reacting $(CH_2{=}CH)_2Sn(O_2CCF_3)_2$ with the sodium salt of the appropriate carbonyl anion, as exemplified in Eq. (19) (123). Such

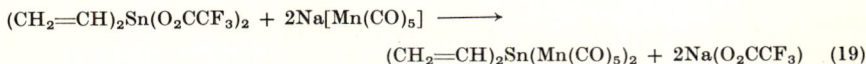

$$(CH_2{=}CH)_2Sn(O_2CCF_3)_2 + 2Na[Mn(CO)_5] \longrightarrow$$

$$(CH_2{=}CH)_2Sn(Mn(CO)_5)_2 + 2Na(O_2CCF_3) \qquad (19)$$

divinyltin derivatives are usually difficult to obtain since in reactions of $(CH_2\!=\!CH)_2SnR_2$ (where R = alkyl or aryl) compounds, it is usually the tin–vinyl bonds that are cleaved. These salt elimination reactions are, of course, directly analogous to those commonly used to form metal–metal bonds from halogeno complexes (e.g., Refs. *121, 145*). However in the foregoing system, the trifluoroacetato complex is easier to prepare and handle than the corresponding halogeno derivatives.

The elimination of the trifluoroacetate ion has been used synthetically in ligand substitution reactions of the $[M(CO)_5(O_2CCF_3)]^-$ (where M = Cr, Mo, or W) anions; Eq. (20) is typical of these reactions.

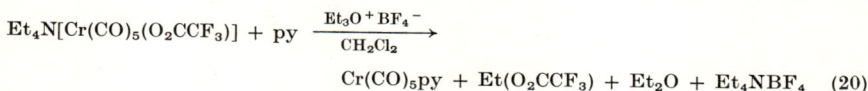

$$Et_4N[Cr(CO)_5(O_2CCF_3)] + py \xrightarrow[\substack{CH_2Cl_2}]{Et_3O^+BF_4^-}$$

$$Cr(CO)_5py + Et(O_2CCF_3) + Et_2O + Et_4NBF_4 \quad (20)$$

Such substitutions are completed within a few seconds at room temperature and afford the neutral product $M(CO)_5L$ in high yield for a wide variety of neutral Lewis bases (L) (*89*).

The base hydrolysis of $[Co(NH_3)_5(O_2CCF_3)]^{2+}$ and the analogous complexes of rhodium, iridium, and chromium(III) appears to involve the concerted attack of two hydroxide ions—one bonding to the acyl carbon atom of the trifluoroacetato group, and the other deprotonating the first (*58, 129*).

3. Addition across Olefinic Bonds

Mercury(II) trifluoroacetate finds frequent employment in organic synthesis (*223, 229, 230*) since it adds rapidly, stereospecifically, and quantitatively across nonconjugated olefinic bonds, for example, Eq. (21). The mercurated products thus prepared may be readily converted

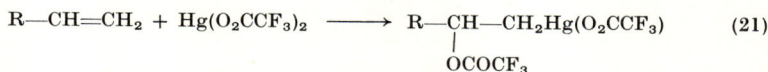

$$R\!-\!CH\!=\!CH_2 + Hg(O_2CCF_3)_2 \longrightarrow \underset{\underset{OCOCF_3}{|}}{R\!-\!CH\!-\!CH_2Hg(O_2CCF_3)} \quad (21)$$

to a variety of compounds that retain this stereospecificity, for example, Eq. (22)

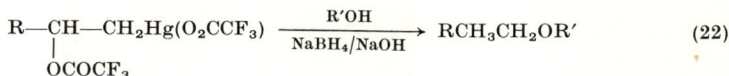

$$\underset{\underset{OCOCF_3}{|}}{R\!-\!CH\!-\!CH_2Hg(O_2CCF_3)} \xrightarrow[\substack{NaBH_4/NaOH}]{R'OH} RCH_3CH_2OR' \quad (22)$$

C. Catalysis

Metal trifluoroacetates are used as catalysts in several industrial processes. The use of $[(\pi\text{-allyl})Ni(O_2CCF_3)_2]_2$ in the stereospecific polymerization of butadiene has been studied in some detail (*59, 60, 166,*

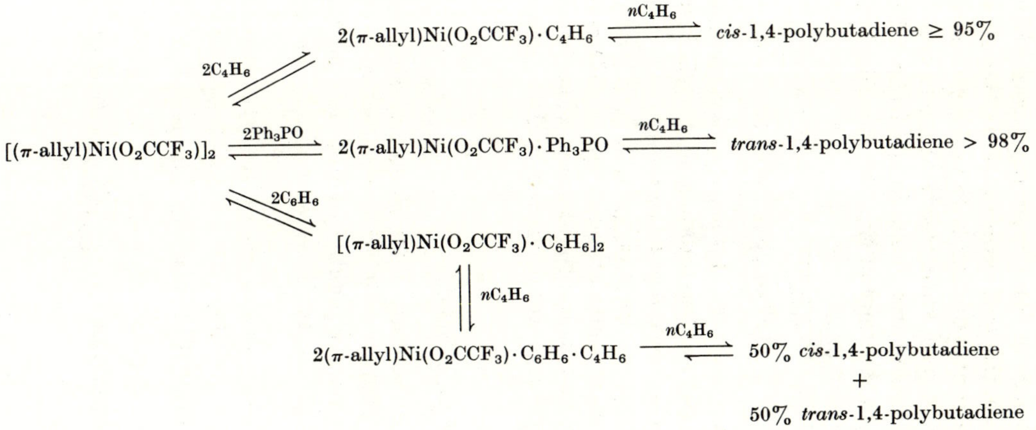

$[(\pi\text{-allyl})\text{Ni}(\text{O}_2\text{CCF}_3)]_2$

$\xrightarrow[2\text{C}_4\text{H}_6]{}$

$2(\pi\text{-allyl})\text{Ni}(\text{O}_2\text{CCF}_3)\cdot\text{C}_4\text{H}_6 \xrightarrow{n\text{C}_4\text{H}_6} cis\text{-}1,4\text{-polybutadiene} \geq 95\%$

$\xrightarrow[2\text{C}_6\text{H}_6]{2\text{Ph}_3\text{PO}}$

$2(\pi\text{-allyl})\text{Ni}(\text{O}_2\text{CCF}_3)\cdot\text{Ph}_3\text{PO} \xrightarrow{n\text{C}_4\text{H}_6} trans\text{-}1,4\text{-polybutadiene} > 98\%$

$[(\pi\text{-allyl})\text{Ni}(\text{O}_2\text{CCF}_3)\cdot\text{C}_6\text{H}_6]_2$

$\xrightleftharpoons{n\text{C}_4\text{H}_6}$

$2(\pi\text{-allyl})\text{Ni}(\text{O}_2\text{CCF}_3)\cdot\text{C}_6\text{H}_6\cdot\text{C}_4\text{H}_6 \xrightarrow{n\text{C}_4\text{H}_6}$ 50% $cis\text{-}1,4\text{-polybutadiene}$
+
50% $trans\text{-}1,4\text{-polybutadiene}$

SCHEME 1.

The effect of ligands on the type of polymer obtained from 1,3-butadiene for $[(\pi\text{-allyl})\text{Ni}(\text{O}_2\text{CCF}_3)]$ catalysis (166).

270). In saturated hydrocarbons, this compound induces the poly-merization of 1,3-butadiene to yield polymers with cis-1,4 contents as high as 95% of the total saturation. By adding appropriate amounts of other ligands, it is possible to affect the nature of the solution species and, hence, to change the type of polymer obtained (Scheme 1).

Also $Mo(CO)_2H(O_2CCF_3)\cdot 2THF$ (*98*), $M(O_2CCF_3)_2$, and $MX(O_2CCF_3)_2$ (where M = Ti, Co, or Ni and X = Cl or Br) (*44*) have been employed to catalyze stereospecific butadiene polymerization. Compounds $Mn(acac)_2(O_2CCF_3)$ (*187*) and $Mn(acac)(O_2CCF_3)$ (*187, 188*), have been shown to be efficient catalysts in the preparation of polymers and copolymers of vinyl compounds from the corresponding monomers; $Cr(O_2CCF_3)_3$ is used in vulcanization, such as the production of carboxynitroso rubbers (*156*).

In so far as metal trifluoroacetates have been used as catalysts, the features of trifluoroacetate groups that would appear to be valuable in this respect include their ability to support a structure in which two or more metal centers are in close proximity, the ease with which they may be displaced from the coordination of an atom to generate sites for catalysis, together with their thermal stability and chemical inertness that minimize unwanted side reactions.

REFERENCES

1. Ablov, A. V., Milkova, L. N., and Yablokov, Yu.V., *Russ. J. Inorg. Chem.* **14**, 358 (1969).
2. Ablov, A. V., Milkova, L. N., and Yablokov, Yu. V., *Russ. J. Inorg. Chem.* **15**, 1523 (1970).
3. Addison, C. C., Logan, N., Wallwork, S. C., and Garner, C. D., *Quart. Rev., Chem. Soc.* **25**, 289 (1971).
4. Agambar, C. A., and Orrell, K. G., *J. Chem. Soc., A* p. 897 (1969).
5. Aldrich, P. E., U.S. Patent 3,043,859 (1962).
6. Allied Chemical Corp., Netherlands Appl. Patent 6,408,620 (1965).
7. Amasa, S., Brown, D. H., and Sharp, D. W. A., *J. Chem. Soc., A* p. 2892 (1969).
8. Anderson, H. H., *J. Amer. Chem. Soc.* **74**, 2371 (1952).
9. Anderson, H. H., *J. Org. Chem.* **19**, 1767 (1954).
10. Anderson, H. H., *J. Amer. Chem. Soc.* **79**, 326 (1957).
11. Anderson, H. H., *Inorg. Chem.* **1**, 647 (1962).
12. Appell, H. R., U.S. Patent 2,882,289 (1959).
13. Ashley-Smith, J., Green, M., and Stone, F. G. A., *J. Chem. Soc., A* p. 3161 (1970).
14. Atkins, P. W., Green, J. C., and Green, M. L. H., *J. Chem. Soc., A* p. 2275 (1968).
15. Auerbach, I., Verhoek, F. H., and Henne, A. L., *J. Amer. Chem. Soc.* **72**, 299 (1950).

16. Barber, M., Connor, J. A., Hillier, I. H., Meredith, W. N. E., and Herd, Q., *J. Chem. Soc., Faraday Trans. 2*, p. 1677 (1973).
17. Barcelo, J. R., and Otero, C., *Spectrochim. Acta* **18**, 1231 (1962).
18. Barclay, G. A., and Hoskins, B. F., *J. Chem. Soc.*, p. 586 (1962).
19. Barlex, D. M., Kemmitt, R. D. W., and Littlecott, G. W., *Chem. Commun.* p. 613 (1969).
20. Barney, A. L., and Morgan, R. L., U.S. Patent 3,418,303 (1968).
21. Barr, J., Gillespie, R. J., and Robinson, E. A., *Can. J. Chem.* **39**, 1266 (1961).
22. Beachell, H. C., and Butter, S. A., *Inorg. Chem.* **4**, 1133 (1965).
23. Bear, J. L., Kitchens, J., and Willcott, M. R., *J. Inorg. Nucl. Chem.* **33**, 3479 (1971).
24. Bergman, J. G., and Cotton, F. A., *J. Amer. Chem. Soc.* **86**, 2941 (1964).
25. Bergman, J. G., and Cotton, F. A., *Inorg. Chem.* **5**, 1420 (1966).
26. Blake, D. M., and Kubota, M., *J. Amer. Chem. Soc.* **92**, 2578 (1970).
27. Blight, D. G., and Kepert, D. L., *Inorg. Chem.* **11**, 1556 (1972).
28. Bluhm, H. F., Donn, H. V., and Zook, H. D., *J. Amer. Chem. Soc.* **77**, 4406 (1955).
29. Blytas, G. C., German Patent 2,014,473 (1969).
30. Blytas, G. C., Slott, E. S., and Bell, E. R., U.S. Patent 3,524,754 (1970).
31. Brainina, E. M., and Freidlina, R. Kh., *Izv. Akad. Nauk SSSR, Otd. Khim. Nauk* p. 835 (1963).
32. Brandon, R. W., and Claridge, D. V., *Chem. Commun.* p. 677 (1968), and references therein.
33. Brower, K. R., Gay, B., and Konkol, T. L., *J. Amer. Chem. Soc.* **88**, 1681 (1966).
34. Brown, D., Holah, D. G., and Rickard, C. E. F., *J. Chem. Soc., A* p. 423 (1970).
35. Brown, H. C., and Wirkkala, R. A., *J. Amer. Chem. Soc.* **88**, 1447, 1453, and 1456 (1966), and references therein.
36. Bruce, J. M., U.S. Patent 2,927,941 (1960).
37. Bryce, H. G., *in* "Fluorine Chemistry" (J. H. Simons, ed.), Vol. V, p. 490 Academic Press, New York, 1964.
38. Buckles, R. E., and Mills, J. F., *J. Amer. Chem. Soc.* **75**, 552 (1953).
39. Burt, R., Cooke, M., and Green, M., *J. Chem. Soc., A* p. 2645 (1969).
40. Cavaleiro, J. A. S., and Smith, K. M., *Chem. Commun.* p. 1384 (1971), and references therein.
41. Clague, D., and Novak, A., *J. Chim. Phys.* **67**, 1126 (1970).
42. Clark, H. C., and Pickard, A. L., *J. Organometal. Chem.* **13**, 61 (1968).
43. Cleveland, J. M., *J. Inorg. Nucl. Chem.* **26**, 461 (1964).
44. Codet, G., De Charentenay, F., Dawans, F., and Teyssie, P., German Patent 1.949,242 (1970).
45. Collman, J. P., and Roper, W. R., *J. Amer. Chem. Soc.* **87**, 4008 (1965).
46. Collman, J. P., and Roper, W. R., *Advan. Organometal. Chem.* **7**, 72 (1968).
47. Commereuc, D., Douek, I., and Wilkinson, G., *J. Chem. Soc., A* p. 1771 (1970).
48. Connett, J. E., and Deacon, G. B., *J. Chem. Soc. C* p. 1058 (1966).
49. Convery, R. J., U.S. Patent 2,985,673 (1961).
50. Cotton, F. A., DeBoer, B. G., LaPrade, M. D., Pipal, J. R., and Ucko, D. A., *Acta Crystallogr., Sect. B* **27**, 1664 (1971).
51. Cotton, F. A., and Norman, J. G., Jr., *J. Amer. Chem. Soc.* **94**, 5697 (1972).
52. Cotton, F. A., and Norman, J. G., Jr., *J. Coord. Chem.* **1**, 161 (1972).

53. Crawford, G. H., and Simons, J. H., *J. Amer. Chem. Soc.* **77**, 2605 (1955), and references therein.
54. Cruikshank, D. W. S., Jones, D. W., and Walker, G., *J. Chem. Soc., A* p. 1302 (1964).
55. Cullen, W. R., and Walker, L. G., *Can. J. Chem.* **38**, 472 (1960).
56. Dannhauser, W., and Cole, R. H., *J. Amer. Chem. Soc.* **74**, 6105 (1952).
57. Davison, A., McFarlane, W., Pratt, L., and Wilkinson, G., *J. Chem. Soc., London* p. 3653 (1962).
58. Davies, R., Evans, G. B., and Jordan, R. B., *Inorg. Chem.* **8**, 2025 (1969).
59. Dawans, F., Marechal, J. C., and Teyssie, P., *J. Organometal. Chem.* **21**, 259 (1970).
60. Dawans, F., and Teyssie, P., *J. Polym. Sci., Part B* **7**, 111 (1969).
61. Deacon, G. B., and Connett, J. E., *J. Chem. Soc., C* p. 1058 (1966).
62. Deacon, G. B., and Green, J. H. S., *Spectrochim. Acta, Sect. A* **24**, 845 (1968).
63. Deacon, G. B., Green, J. H. S., and Nyholm, R. S., *J. Chem. Soc., London* p. 3411 (1965).
64. Deacon, G. B., and Nyholm, R. S., *J. Chem. Soc., London* p. 6107 (1965).
65. Deacon, G. B., and Taylor, F. B., *Aust. J. Chem.* **21**, 2675 (1968).
66. Dean, P. A. W., and Evans, D. F., *J. Chem. Soc., A* p. 1154 (1968).
67. Dines, M. B., *Inorg. Chem.* **11**, 2949 (1973).
68. Donaldson, J. D., and Jelen, A., *J. Chem. Soc., A* p. 1448 (1968).
69. Downs, A. J., Ebsworth, E. A. V., and Eméleus, H. J., *J. Chem. Soc., London* p. 1254 (1962).
70. Drinkard, W. C., Eaton, D. R., Jesson, J. P., and Lindsey, R. V., Jr., *Inorg. Chem.* **9**, 392 (1970); German Patent 1,808,434 (1970).
71. Drozdov, G. V., Klebanskii, A. L., and Bartashov, V. A., *Zh. Obshch. Khim.* **33**, 2422 (1963).
72. Dunnell, B. A., Reeves, L. W., and Strømme K. O., *Trans. Faraday Soc.* **57**, 381 (1961).
73. Dvoryantseva, G. G., Lazareva, N. A., Dubovitskii, V. A., Sheinker, Yu. N., Nogina, O. V., and Nesmayanov, A. N., *Dokl. Akad. Nauk SSSR* **161**, 603 (1965).
74. Eisenburg, M., and DesMarteneau, D. D., *Inorg. Nucl. Chem. Lett.* **6**, 29 (1970).
75. Ellis, B., and Pyszora, H., *Nature (London)* **181**, 181 (1958).
76. Estienne, J., and Weiss, R., *Chem. Commun.* p. 862 (1972).
77. Evans, D. F., Ridout, P. M., and Wharf, I., *J. Chem. Soc., A* p. 2127 (1968).
78. Fabrizio, M., *J. Inorg. Nucl. Chem.* **29**, 1079 (1967).
79. Fieser, L. F., and Fieser, M., "Reagents for Organic Synthesis." Wiley, New York, 1967.
80. Figgis, B. N., and Robertson, G. B., *Nature* (London) **205**, 694 (1965).
81. Fill, E. E., *J. Appl. Phys.* **41**, 4749 (1970).
82. Flaskerud, G. G., Pullen, K. E., and Shreeve, J. M., *Inorg. Chem.* **8**, 728 (1969).
83. Flowers, R., Gillespie, R. J., and Oubridge, J. V., *J. Chem. Soc., London* p. 1925 (1956).
84. Forcier, G. A., and Olver, J. W., *Electrochem. Acta* **15**, 1609 (1970).
85. Ford, B. F. E., and Sams, J. R., *J. Organometal. Chem.* **31**, 47 (1971).
86. Fujita, J., Martell, A. E., and Nakamoto, K., *J. Chem. Phys.* **36**, 339 (1962).
87. Fuson, N., and Josien, M. L., *J. Amer. Opt. Soc.* **43**, 1102 (1953).

88. Fuson, N., Josien, M. L., Jones, E. A., and Lawson, J. R., *J. Chem. Phys.* **20**, 1627 (1952).

89. Garner, C. D., and Hughes, B., *J. Chem. Soc. Dalton Trans.* p. 735 (1974).

90. Garner, C. D., and Mabbs, F. E., submitted for publication.

91. Gerlach, J. N., and Gard, G. L., *Inorg. Chem.* **9**, 1565 (1970).

92. Gerrard, W., Lappert, M. F., and Shafferman, R., *J. Chem. Soc., London* p. 3648 (1958).

93. Gibson, D., Ph.D. Thesis, University of Manchester (1967).

94. Gillespie, R. J., and Moss, K. C., *J. Chem. Soc., A* p. 1170 (1966).

95. Glen, G. L., Silverton, J. V., and Hoard, J. L., *Inorg. Chem.* **2**, 250 (1963),

96. Goel, R. G., *Can. J. Chem.* **47**, 4607 (1969)

97. Goel, R. G., and Ridley, D. R., *J. Organometal. Chem.* **38**, 83 (1972).

98. Goldenberg, E., Dawans, F., Dwand, J. P., and Martino, G., German Patent 2,232,767 (1973).

99. Golic, L., and Speakman, J. C., *J. Chem. Soc., London* p. 2530 (1965).

100. Grdenic, D., and Matkovic, B., *Acta Crystallogr.* **12**, 817 (1959).

101. Green, M., Osborn, R. B. L., Rest, A. J., and Stone, F. G. A., *J. Chem. Soc., A* p. 2525 (1968).

102. Green, M. L. H., Massey, A. G., Moelwyn-Hughes, J. T., and Nagy, P. L. I., *J. Organometal. Chem.* **8**, 511 (1967).

103. Hammond, B., Jardine, F. H., and Vohra, A. G., *J. Inorg. Nucl. Chem.* **33**, 1017 (1971).

104. Hanic, F., Stempelova, D., and Hanicova, K., *Acta Crystallogr.* **17**, 633 (1964), and references therein.

105. Hara, R., and Cady, G. H., *J. Amer. Chem. Soc.* **76**, 4285 (1954).

106. Harriss, M. G., Green, M. L. H., and Lindsell, W. E., *J. Chem. Soc., A* p. 1453 (1969).

107. Harriss, M. G., and Milne, J. B., *Can. J. Chem.* **49**, 1888 (1971).

108. Harriss, M. G., and Milne, J. B., *Can. J. Chem.* **49**, 2937 (1971).

109. Harriss, M. G., and Milne, J. B., *Can. J. Chem.* **49**, 3612 (1971).

110. Haszeldine, R. N., *J. Chem. Soc., London* p. 584 (1951).

111. Haszeldine, R. N., *J. Chem. Soc., London* p. 4259 (1952).

112. Haszeldine, R. N., and Sharpe, A. G., *J. Chem. Soc., London* p. 993 (1952).

113. Heller, A., U.S. Patent 3,466,568 (1969).

114. Henderson, A., and Holliday, A. K., *J. Organometal. Chem.* **4**, 377 (1965).

115. Henne, A. L., and Fox, C. J., *J. Amer. Chem. Soc.* **73**, 2323 (1951); Hood, G. C., Redlich, O., and Reilly, C. A., *J. Chem. Phys.* **23**, 2229 (1953).

116. Herzog, S., and Kalies W., *Z. Chem.* **6**, 344 (1966).

117. Herzog, S., and Kalies, W., *Z. Anorg. Allg. Chem.* **351**, 237 (1967).

118. Heusler, K., and Loeliger, H., *Helv. Chim. Acta* **52**, 1495 (1969).

119. Hey, D. G., Meakins, G. D., and Pemberton, M. W., *J. Chem. Soc., C* p. 1331 (1966).

120. Hota, N. K., and Willis, C. J., *Can. J. Chem.* **46**, 3921 (1968).

121. Hsieh, A. T. T., and Mays, M. J., *J. Organometal. Chem.* **22**, 29 (1970).

122. Hughes, B., M.Sc. Thesis, University of Manchester (1971).

123. Hughes, B., Ph.D. Thesis, University of Manchester (1974).

124. Iskraut, A., Taubenest, R., and Schumacher, E., *Chimia* **18**, 188 (1964).

125. Jaffe, M. S., U.S. Patent 2,849,339 (1958).

126. Jaura, K., Banga, H., and Kaushik, R. L., *J. Indian Chem. Soc.* **39**, 531 (1962), and references therein.

127. Johnson, S. A., Hunt, H. R., and Neumann, H. M., *Inorg. Chem.* **2**, 960 (1963).

128. Jones, R. G., and Dyer, J. R., *J. Amer. Chem. Soc.* **95**, 2465 (1973).

129. Jordan, R. B., and Taube, H., *J. Amer. Chem. Soc.* **88**, 4406 (1966).

130. Kaesz, H. D., and Stone, F. G. A., in "Organometallic Chemistry" (H. Zeiss, ed.), p. 123, Van Nostrand-Reinhold, Princeton, New Jersey, 1960.

131. Kagarise, R. E., *J. Phys. Chem.* **59**, 271 (1955).

132. Kagarise, R. E., *J. Chem. Phys.* **27**, 519 (1957).

133. Karle, J. and Brockway, L. O., *J. Amer. Chem. Soc.* **66**, 574 (1944).

134. Kato, M., Jonassen, H. B., and Fanning, J. C., *Chem. Rev.* **64**, 99 (1964).

135. Katz, J. J., *Nature (London)* **174**, 509 (1954).

136. Kettle, S. F. A., and Piolo, A. J. P., *J. Chem. Soc.*, A p. 1243 (1968).

137. King, R. B., and Kapoor, R. N., *J. Organometal. Chem.* **15**, 457 (1968).

138. King. R. B., and Kapoor, R. N., *J. Inorg. Nucl. Chem.* **31**, 2169 (1969).

139. King, R. B., and Pannell, K. H., *Inorg. Chem.* **7**, 2356 (1968).

140. Kircheiss, A., *Z. Anorg. Allg. Chem.* **378**, 80 (1970).

141. Kirszenbaum, M., Corset, J., and Josien, M. L., *J. Phys. Chem.* **75**, 1327 (1971).

142. Kitchens, J., and Bear, J. L., *Thermochim. Acta* **1**, 537 (1970).

143. Klemperer, W., and Pimentel, G. C., *J. Chem. Phys.* **22**, 1399 (1954).

144. Klotz, I. M., Russo, S. F., Hanlon, S., and Stake, M. A., *J. Amer. Chem. Soc.* **86**, 4774 (1964).

145. Knox, S. A. R., and Stone, F. G. A., *J. Chem. Soc.*, A p. 2559 (1969).

146. Kolesov, V. P., Slavutskaya, G. M., and Papina, T. S., *Zh. Fiz. Khim.* **46**, 815 (1972).

147. Kuo, K.-W., and Madan, S. K., *Inorg. Chem.* **8**, 1580 (1969).

148. Kuroda, K., and Gentile, P. S., *Bull. Chem. Soc. Jap.* **38**, 2159 (1965).

149. Kuroda, K., and Gentile, P. S., *J. Inorg. Nucl. Chem.* **29**, 1963 (1967).

150. Larin, G. M., Kalinnikov, V. T., Aleksandrov, G. G., Struchkov, Yu. T., Pasnskii, A. A., and Kolobova, N. E., *J. Organometal. Chem.* **27**, 53 (1971).

151. Lawson, D. N., and Wilkinson, G., *J. Chem. Soc.*, London p. 1900 (1965).

152. Lawton, D., and Mason, R., *J. Amer. Chem. Soc.* **87**, 921 (1965).

153. Lebedev, S. V., French Patent 1,550,097 (1968).

154. Lee, A. G., *J. Organometal. Chem.* **22**, 537 (1970).

155. Lever, A. B. P., and Ogden, D., *J. Chem. Soc.*, A p. 2041 (1967).

156. Levine, N. B., *Rubber Age (New York)* **41**, 299R (1969).

157. Lewis, J., Lin, Y. C., Royston, L. K., and Thompson, R. C., *J. Chem. Soc.*, London p. 6464 (1965), and references therein.

158. Lewis, J., Mabbs, F. E., Richards, A., and Thornley, A. S., *J. Chem. Soc.*, A p. 1993 (1969).

159. Ling, C., Christian, S. D., Affsprung, H. E., and Gray, R. W., *J. Chem. Soc.*, A p. 293 (1966), references therein.

160. Lippard, S. J., *Progr. Inorg. Chem.* **8**, 109 (1967).

161. Little, R. G., Yawney, D. B. W., and Doedens, R. J., *Chem. Commun.* p. 228 (1972).

162. Loeliger, H., *Helv. Chim. Acta.* **52**, 1516 (1969).

163. Macdonald, A. L., Speakman, J. C., and Hadži, D., *J. Chem. Soc.*, Perkin Trans. 2, 825 (1972).

164. MacFarlane, W., and Wood, R. J., *J. Organometal. Chem.* **40**, C17 (1972).

165. Mann, B. E., Shaw, B. L., and Tucker, N. I., *Chem. Commun.* p. 1333 (1970).

166. Marechal, J. C., Dawans, F., and Teyssie, P., *J. Polym. Sci.*, *Part A* **8**, 1993 (1970).
167. Martin, R. L., and Waterman, H., *J. Chem. Soc.*, *London* p. 2545 (1957); pp. 1359 and 2960 (1959).
168. McKillop, A., Fowler, J. S., Zelesko, M. J., Hunt, J. D., Taylor, E. C., and McGillivray, G., *Tetrahedron Lett.* **9**, 2423 (1969).
169. McKillop, A., Hunt, J. D., Zelesko, M. J., Fowler, J. S., Taylor, E. C., McGillivray, G., and Kienzle, F., *J. Amer. Chem. Soc.* **93**, 4841 (1971), and references therein.
170. McKillop, A., and Taylor, E. C., *Chem. Brit.* **9**, 4 (1973).
171. Mehotra, R. C., *J. Indian Chem. Soc.* **38**, 509 (1961).
172. Melnik, M., *Suom. Kemistilehti B* **43**, 256 (1970).
173. Miller, P. J., Butler, R. A., and Lippincott, E. R., *J. Chem. Phys.* **57**, 5451 (1972), and references therein.
174. Moeller, T., *Inorg. Syn.* **3**, 4 (1950).
175. Monacelli, F., Basolo, F., and Pearson, R. G., *J. Inorg. Nucl. Chem.* **24**, 1241 (1962).
176. Morehouse, S. M., Powell, A. R., Heffer, J. P., Stephenson, T. A., and Wilkinson, G., *Chem. Ind.* (*London*) **13**, 544 (1964).
177. Muetterties, E. L., and Alegranti, C. W., *J. Amer. Chem. Soc.* **92**, 4114 (1970).
178. Murty, T. S. S. R., and Pitzer, K. S., *J. Phys. Chem.* **73**, 1426 (1969).
179. Musher, J. I., *J. Amer. Chem. Soc.* **90**, 7371 (1968).
180. Mustafa, M. R., *Diss. Abstr. Int. B* **31**, 6543 (1971).
181. Nakamoto, K., Fujita, J., Tanaka, S, and Kobayshi, M., *J. Amer. Chem. Soc.* **79**, 4904 (1957).
182. Nakamoto, K., Morimoto, Y., and Martell, A. E., *J. Amer. Chem. Soc.* **83**, 4528 (1961).
183. Naumann, D., Dolhaine, H., and Stopschinski, W., *Z. Anorg. Allg. Chem.* **394**, 133 (1972).
184. Naumann, D., Schmeisser, M., and Scheele, R., *J. Fluorine Chem.* **1**, 321 (1972).
185. Nichols, D. I., and Charleston, A. S., *J. Chem. Soc.*, *A* p. 2581 (1969).
186. Nielson, A. T., *J. Org. Chem.* **31**, 624 (1966).
187. Nikolaev, A. F., Belogorodskaya, K. V., and Shibalovich, V. G., U.S.S.R. Patent 276,413 (1970).
188. Nikolaev, A. F., Belogoradskaya, K. V., and Shibalobich, V. G., *Vyoskomol. Soedin.*, *Ser. B* **13**, 837 (1971).
189. Oldham, C., *Progr Inorg. Chem.* **10**, 223 (1968)
190. Oldham, C., and Ketteringham, A. P., *J. Chem. Soc.*, *Dalton Trans.* p. 1067 (1973).
191. Parish, R. V., *Coord. Chem. Rev.* **1**, 439 (1966), and references therein.
192. Partch, R., *J. Amer. Chem. Soc.* **89**, 3662 (1967).
193. Pearlson, W. H., *in* "Fluorine Chemistry" (J. H. Simons, ed.), Vol. 1, p. 480. Academic Press, New York, 1950.
194. Peruzzo, V., Plazzogna, G., and Tagliavini, G., *J. Organometal. Chem.* **24**, 347 (1970).
195. Peterson, P. E., and Casey, C., *J. Org. Chem.* **29**, 2325 (1964).
196. Plazzogna, G., Peruzzo, V., and Tagliavini, G., *J. Organometal. Chem.* **24**, 667 (1970).
197. Poder, C., and Sams, J. R., *J. Organometal. Chem.* **19**, 67 (1969).

198. Poller, R. C., Ruddock, J. N. R., Taylor, B., and Toley, D. L. B., *J. Organometal. Chem.* **24**, 341 (1970).
199. Popov, A. I., *in* "The Chemistry of Nonaqueous Solvents" (J. J. Lagowski, ed.), Vol. 3, pp. 366–375. Academic Press, New York, 1970, and references therein.
200. Radeshwa, P. V., Dev, R., and Cady, G. H., *U.S. Nat. Tech. Inform. Serv., AD Rep.* **AD-736682** (1972).
201. Randles, J. E. B., Tatlow, J. C., and Tedder, J. M., *J. Chem. Soc., London* p. 436 (1954).
202. Redington, R. L., and Lin, K. C., *Spectrochim. Acta, Sect. A* **27**, 2445 (1971).
203. Reid, D. H., and Webster, R. G., *Chem. Commun.* p. 1283 (1972), and references therein.
204. Richardson, M. F., Sands, D. E., and Wagner, W. F., *J. Inorg. Nucl. Chem.* **30**, 1275 (1968).
205. Ritter, H. L., and Simons, J. H., *J. Amer. Chem. Soc.* **67**, 757 (1945).
206. Roberts, J. E., *J. Amer. Chem. Soc.* **83**, 1087 (1961).
207. Robinson, R. A., and Stokes, R. H., "Electrolyte Solutions," p. 171. Butterworth, London, 1965.
208. Robinson, S. D., and Uttley, M. F., *Chem. Commun.* p. 1047 (1972).
209. Rose, D., Gilbert, J. D., Richardson, R. P., and Wilkinson, G., *J. Chem. Soc., A* p. 2610 (1969).
210. Rudge, A. J., "The Manufacture and Uses of Fluorine and its Compounds," p. 76. Oxford Univ. Press, London and New York, 1962.
211. Saitow, A., Rochow, E. G., and Seyferth, D., *J. Org. Chem.* **23**, 116 (1958).
212. Sartori, P., Schmeisser, M., and Naumann, D., *Chem. Ber.* **103**, 312 (1970).
213. Sartori, P., and Weidenbruch, M., *Angew. Chem., Int. Ed. Engl.* **3**, 376 (1964).
214. Sartori, P., and Weidenbruch, M., *Angew. Chem., Int. Ed. Engl.* **4**, 1079 (1965).
215. Sartori, P., and Weidenbruch, M., *Chem. Ber.* **100**, 2049 (1967).
216. Sasin, R., and Sasin, G. S., *J. Org. Chem.* **20**, 388 (1955).
217. Sasin, R., and Sasin, G. S., *J. Org. Chem.* **20**, 770 (1955).
218. Schlientz, W. J., Lavender, Y., Welcman, W., King, R. B., and Ruff, J. K., *J. Organometal. Chem.* **33**, 357 (1971).
219. Scott, A. F., Wilkening, L. L., and Rubin, B., *Inorg. Chem.* **8**, 2533 (1969).
220. Selbin, J., Schober, M., and Ortego, J. D., *J. Inorg. Nucl. Chem.* **28**, 1385 (1966).
221. Seppelt, K., *Chem. Ber.* **105**, 3131 (1972).
222. Seyferth, D., and Hopper, S.P., *J. Organometal. Chem.* **26**, C62 (1971).
223. Seyferth, D., Prokai, B., and Cross, R. J., *J. Organometal. Chem.* **13**, 169 (1968).
224. Sharp, D. W. A., Brown, D. H., and Amasa, S., *J. Chem. Soc., A* p. 2892 (1969).
225. Sharp, D. W. A., Brown, D. H., Moss, K. C., and Baillie, M. J., *Proc. Int. Conf. Coord. Chem., 8th, 1964* p. 322 (1964).
226. Sharp, D. W. A., Brown, D. H., Moss, K. C., and Baillie, M. J., *Chem. Commun.* p. 91 (1965).
227. Sharp, D. W. A., Brown, D. H., Moss, K. C., and Baillie, M. J., *J. Chem. Soc., A* p. 104 (1968).
228. Sharp, D. W. A., Brown, D. H., Moss, K. C., and Baillie, M. J., *J. Chem. Soc., A* p. 3110 (1968).

229. Shearer, D. A., and Wright, G. F., *Can. J. Chem.* **33**, 1002 (1955).
230. Shearer, D. A., Wright, G. F., and Rodgman, A., *Can. J. Chem.* **35**, 1377 (1957).
231. Shinn, D. B., and Eick, H. A., *Inorg. Chem.* **7**, 1340 (1968).
232. Shreeve, J. M., and Wang, C. S. C., *Chem. Commun.* p. 151 (1970).
233. Silverton, J. V., and Hoard, J. L., *Inorg. Chem.* **2**, 243 (1963).
234. Simanova, N. P., Lobach, M. I., Babitskii, B. D., and Kormer, V. A., *Vysokomol. Soedin., Ser. B* **10**, 588 (1968).
235. Simons, J. H., Bond, R. L., and McArthur, R. E., *J. Amer. Chem. Soc.* **62**, 3477 (1940).
236. Simons, J. H., and Lorentzen, K. E., *J. Amer. Chem. Soc.* **74**, 4746 (1952).
237. Singer, H., and Wilkinson, G., *J. Chem. Soc., A* p. 2516 (1968).
238. Sipkes, J., Netherlands Appl. Patent 6,508,523 (1966).
239. Sladky, F. O., *Monatsh. Chem.* **101**, 1571 (1970).
240. Smith, K. M., *Chem. Commun.* p. 540 (1971).
241. Sokolov, V. I., Bashilov, V. V., and Rentov, O. A., *Dokl. Akad. Nauk SSSR* **197**, 101 (1971).
242. Speakman, J. C., "Structure and Bonding," Vol. XII, p. 141, and references therein. Springer-Verlag, Berlin and New York, 1972.
243. Spinner, E., *J. Chem. Soc., London* p. 4217 (1964).
244. Stephenson, T. A., Morehouse, S. M., Powell, A. R., Heffer, J. P., and Wilkinson, G., *J. Chem. Soc., London* p. 3632, 1965.
245. Stephenson, T. A., and Wilkinson, G., *J. Inorg. Nucl. Chem.* **29**, 2122 (1967).
246. Stritar, J. A., and Taube, H., *Inorg. Chem.* **8**, 2281 (1969).
247. Swarts, F., *Bull. Cl. Sci., Acad. Roy. Belg.* [5] **8**, 343 (1922), *Bull. Acad. Roy. Belg., Extr. Bull. Cl. Sci.* p. 341 (1922); *Bull. Soc. Chim. Belg.* **48**, 176 (1939).
248. Sweet, S. S., Van Horn, M. H., and Newsome, P. T., U.S. Patent 2,710,848 (1955).
249. Taylor, E. C., and McKillop, A., German Patent 2,000,880, (1970).
250. Taylor, M. D., and Templeman, M. B., *J. Amer. Chem. Soc.* **78**, 2950 (1956).
251. Tedder, J. M., *J. Chem. Soc., London* p. 2646 (1954).
252. Theimer, R., and Theimer, O., *Monatsh. Chem.* **81**, 313 (1950).
253. Thomas, K., Dumler, J. T., Renoe, B. W., Nyman, C. J., and Roundhill, D. M., *Inorg. Chem.* **11**, 1795 (1972) and references therein.
254. Thompson, L. K., Chacko, V. T., Elvidge, J. A., Lever, A. B. P., and Parish, R. V., *Can. J. Chem.* **47**, 4141 (1969).
255. Thompson, R. C., and Yawney, D. B. W., *Can. J. Chem.* **43**, 1240 (1965).
256. Toporcer, L. H., Dessy, R. E., and Green, S. I. E., *J. Amer. Chem. Soc.* **87**, 1236 (1965).
257. Tripathy, P. B., and Roundhill, D. M., *J. Organometal. Chem.* **24**, 247 (1970).
258. Uggla, R., and Melnik, M., *Suom. Kemistilehi B* **45**, 16 (1972).
259. Van den Berghe, E. V., Van der Kelen, G. P., and Albrecht, J., *Inorg. Chim. Acta* **2**, 89 (1968).
260. Van Niekerk, J. A., and Schoening, F. R. L., *Acta Crystallogr.* **6**, 227 (1953).
261. Völter, W., Jung, G., and Breitmaier, E., *Chim. Ther.* **7**, 29 (1972).
262. Wailes, P. C., and Weigold, H., *J. Organometal. Chem.* **24**, 413 (1970).
263. Watt, W. G., and Nuga, M. L., *J. Inorg. Nucl. Chem.* **9**, 166 (1959).
264. Weichselgartner, H., and Perchemeier, J., *Z. Naturforsch. A* **25**, 1244 (1970).
265. Whimp, P. O., Bailey, M. F., and Curtis, N. F., *J. Chem. Soc., A* p. 1956 (1970).

266. Williams, D. H., and Fleming, I., "Spectroscopic Methods in Organic Chemistry," McGraw-Hill, New York, 1966.
267. Winkhaus, G., and Ziegler, P., *Z. Anorg. Allg. Chem.* **350**, 51 (1969).
268. Wynne, K. J., and Bauder, W., *Inorg. Chem.* **9**, 1985 (1970).
269. Yagi, Y. Narisawa, F., Oshima, T., and Hata, K., British Patent 2,147,851 (1972).
270. Yakovlev, V. A., Makovetskii, K. L., Dolgoplosk, B. A., and Tinyakova, E. J., *Dokl. Akad. Nauk SSSR* **187**, 354 (1969).
271. Zachariasen, W. H., and Plettinger, H. A., *Acta Crystallogr.* **12**, 526 (1959).
272. Zinato, E., Lindolm, R., and Adamson, A. W., *J. Inorg. Nucl. Chem.* **31**, 449 (1969).

HOMOPOLYATOMIC CATIONS OF THE ELEMENTS

R. J. GILLESPIE

Department of Chemistry, McMaster University, Hamilton, Ontario, Canada

and

J. PASSMORE

Chemistry Department, University of New Brunswick, Fredericton, New Brunswick, Canada

I. Introduction

During the last 10 years a number of homopolyatomic cations (M_y^{x+}, where $x \leqslant y$) have been prepared and characterized. For a long time the only known example of this type of species was the mercurous

49

ion Hg_2^{2+} but this can no longer be regarded as a chemical oddity, as it has now been joined not only by the analogous species Zn_2^{2+} and Cd_2^{2+} but also by Hg_3^{2+} and many cations of the nonmetals such as I_2^+, O_2^+, S_8^{2+}, and Te_4^{2+}. It is not surprising that some of the earliest examples of this type of cation, e.g., O_2^+ and Bi_9^{5+}, were discovered quite accidentally, but it is perhaps surprising that some of the species have been known for at least 150 years but were not recognized as such. For example, during the early nineteenth century it was reported that sulfur, selenium, and tellurium dissolve in concentrated sulfuric acid or in oleum (H_2SO_4–SO_3) to give various highly colored solutions. The origin of these colors was never clearly established, but it has now been shown that they are due to various polyatomic cations of these elements such as S_8^{2+}, Se_4^{2+}, and Te_4^{2+}. Chemists have long been fascinated by the possibility that elements such as iodine might be obtained in the cationic form I^+ as well as in the well-known anionic form I^-. However, although there is no evidence for the existence of I^+ or of Br^+ or Cl^+ as stable species in solution or in the solid state, the search for such species has led to the discovery of polyatomic cations of the halogens such as I_2^+, Br_3^+, and Cl_3^+ which under appropriate conditions are quite stable.

The structures of the homopolyatomic cations are of obvious interest particularly because of their simplicity in that they contain only one kind of atom. Thus, although homonuclear clusters of atoms are well-known among the transition metals in "cluster" compounds, such as $Mo_6Cl_8^{4+}$, and in the boron hydrides, e.g., $B_{12}H_{12}^{2-}$, the description of the bonding in these compounds is somewhat complicated by the presence of the ligands that are at least partially responsible for holding the metal atoms together.

These new cations, particularly those of the nonmetals, are "electron-deficient" with respect to the element itself and, thus, they are highly electrophilic. They are, accordingly, only stable in the absence of bases with which they readily react, generally disproportionating to more stable valency states. Water is, for example, a sufficiently strong base to react with these ions, which, in general, disproportionate to the element and one of the familiar oxidation states of the element that is stable in aqueous solution. It is not surprising, therefore, that the discovery of these cations owes much to recent developments in the chemistry of nonaqueous solvent systems, particularly highly acidic systems, including acidic fused salt media. Some of the cations, e.g., Br_2^+, are stable only in the most highly acidic and most weakly basic solvent media known, e.g., HSO_3F–SbF_5. In the solid state, stable crystalline salts can only be obtained with the anions of very strong acids, e.g., SO_3F^-, $Sb_2F_{11}^-$, and $AlCl_4^-$, typically large singly charged anions containing the electronegative elements F, O, and Cl. In addition to highly acidic media, the very weakly basic and rather unreactive solvent SO_2 has proved to be very useful in the preparation and study of these cations.

II. Polyatomic Cations of Group VII

A. IODINE CATIONS

The existence of I_3^+ and I_5^+ was deduced over 30 years ago by Masson (1) from his studies of aromatic iodination reactions, but it is only recently that his conclusions have been confirmed by physical measurements. The controversy over the nature of the blue solutions of iodine in various highly acidic media has now been resolved, and it has been shown conclusively that these solutions contain I_2^+ (2–4) and not I^+ as suggested earlier (5). There is, moreover, no convincing evidence for the existence of Cl^+ or Br^+ as stable species in solution or in the solid state. There is, however, evidence for polyatomic cations of chlorine and bromine analogous to the iodine cations, i.e., Cl_3^+, Br_3^+, and Br_2^+.

1. I_3^+ and I_5^+

The first evidence for the existence of a stable iodine cation was obtained by Masson (1) in 1938. He postulated the presence of I_3^+ and I_5^+ in solutions of iodine and iodic acid in sulfuric acid in order to explain the stoichiometry of the reaction of such solutions with chlorobenzene to form both iodo and iodoso derivatives. Later, Symons and co-workers (6) gave conductometric evidence for I_3^+ formed from iodic acid and iodine in 100% sulfuric acid and suggested that I_5^+ may be formed on the basis of changes in the UV and visible spectra when iodine is added to I_3^+ solutions. Gillespie and co-workers (7) on the basis of detailed conductometric and cryoscopic measurements confirmed that I_3^+ is formed from HIO_3 and I_2 in 100% sulfuric acid according to Eq. (1). The I_3^+ cation may also be prepared in fluoro-

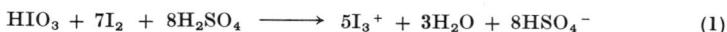

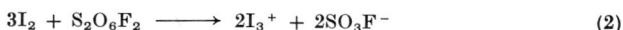

$$HIO_3 + 7I_2 + 8H_2SO_4 \longrightarrow 5I_3^+ + 3H_2O + 8HSO_4^- \tag{1}$$

$$3I_2 + S_2O_6F_2 \longrightarrow 2I_3^+ + 2SO_3F^- \tag{2}$$

sulfuric acid (2) by the reaction in Eq. (2). Solutions of red–brown I_3^+ in H_2SO_4 or HSO_3F have characteristic absorption maxima at 305 and 470 nm, with a molar extinction coefficient of 5200 at 305 nm.

Solutions of I_3^+ in 100% H_2SO_4 (7), or in fluorosulfuric acid (2), dissolve at least 1 mole of iodine per mole of I_3^+, and a new absorption spectrum is obtained which has bands at 270, 340, and 470 nm. At the same time, there is no change in either the conductivity or the freezing point of the solutions; therefore, it has to be concluded that I_5^+ is formed according to Eq. (3). Some further iodine will dissolve in solutions of I_5^+, indicating possible formation of I_7^+.

$$I_3^+ + I_2 \longrightarrow I_5^+ \tag{3}$$

Recently, Corbett et al. (8) have prepared the compounds $I_3^+AlCl_4^-$ and $I_5^+AlCl_4^-$, which they characterized by phase equilibria studies and nuclear quadrupole resonance spectroscopy. The shiny black phase $I_{3.0 \pm 0.15}AlCl_4$ melts congruently at $45° \pm 1°C$, and the green "metallic" I_5AlCl_4 ($4.8 < I/AlCl_4 < 5.3$) melts slightly incongruently at $50°$ to $50.5°$. Chung and Cady (8a) have determined the melting points for the system I_2–$S_2O_6F_2$ and confirmed the previously known solids $I(SO_3F)_3$, ISO_3F, and I_3SO_3F. A new compound, I_7SO_3F, was also established. No evidence for I_5SO_3F or I_2SO_3F was obtained, and the nature of the compound ISO_3F remains uncertain.

In 1906, Ruff (9) reported that excess iodine and SbF_5 react to form a brown solid which he formulated as SbF_5I. Kemmitt et al. (4) have since shown from the absorption spectrum of the solid in liquid AsF_3 that it contains some I_3^+ cation. However, it must be concluded from the method of preparation that this material is not a single compound, and almost certainly contains Sb(III).

2. I_2^+

Gillespie and Milne (2) have shown, by conductometric, spectrophotometric, and magnetic susceptibility measurements in fluorosulfuric acid, that the blue iodine species observed in strong acids is I_2^+. When iodine was oxidized by peroxodisulfuryl difluoride in fluorosulfuric acid, the concentration of the blue iodine species reached a maximum at the 2:1 $I_2/S_2O_6F_2$ mole ratio [Eq. (4)] and not at the 1:1 mole ratio as would

$$2I_2 + S_2O_6F_2 \longrightarrow 2I_2^+ + 2SO_3F^- \tag{4}$$

$$I_2 + S_2O_6F_2 \longrightarrow 2I^+ + 2SO_3F^- \tag{5}$$

be anticipated for the formation of I^+ [Eq. (5)]. The conductivities of 2:1 solutions of iodine–$S_2O_6F_2$ at low concentrations were found to be very similar to solutions of KSO_3F at the same concentration, showing that 1 mole of SO_3F^- had been formed per mole of iodine. The magnetic moment of the blue species in fluorosulfuric acid was found to be $2.0 \pm 0.1 \mu_B$ which agrees with the value expected for the $^3\Pi_{3/2}$ ground state of the I_2^+ cation. The I_2^+ has characteristic peaks in its absorption spectrum at 640, 490, and 410 nm and has a molar extinction coefficient at 640 nm of 2560.

The I_2^+ cation is not completely stable in fluorosulfuric acid and undergoes some disproportionation to the more stable I_3^+ species and $I(SO_3F)_3$ according to Eq. (6). This disproportionation is largely pre-

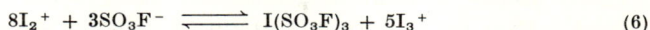

$$8I_2^+ + 3SO_3F^- \rightleftharpoons I(SO_3F)_3 + 5I_3^+ \tag{6}$$

vented in a 1:1 I_2–$S_2O_6F_2$ solution in which $I(SO_3F)_3$ is also formed [Eq.

$$5I_2 + 5S_2O_6F_2 \longrightarrow 4I_2^+ + 4SO_3F^- + 2I(SO_3F)_3 \tag{7}$$

(7)]. The disproportionation can also be prevented if the fluorosulfate ion concentration in fluorosulfuric acid is lowered by addition of antimony pentafluoride [Eq. (8)] or by using the less basic solvent 65% oleum.

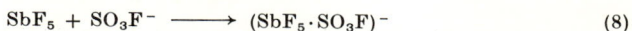

$$SbF_5 + SO_3F^- \longrightarrow (SbF_5 \cdot SO_3F)^- \tag{8}$$

In 100% H_2SO_4 the disproportionation of I_2^+ to I_3^+ and an iodine(III) species, probably $I(SO_4H)_3$, is essentially complete, and only traces of I_2^+ can be detected by means of its resonance Raman spectrum.

Solutions of the blue iodine cation in oleum have been reinvestigated (3) by conductometric, spectrophotometric, and cryoscopic methods confirming the formation of I_2^+. In 65% oleum, iodine is oxidized to I_2^+ according to Eq. (9).

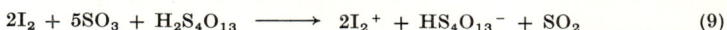

$$2I_2 + 5SO_3 + H_2S_4O_{13} \longrightarrow 2I_2^+ + HS_4O_{13}^- + SO_2 \tag{9}$$

Adhami and Herlem (10) have carried out a coulometric titration at controlled potential of iodine in fluorosulfuric acid and have shown that iodine is quantitatively oxidized to I_2^+ by removal of one electron per mole of iodine.

The blue solid prepared by Ruff et al. (9) in 1906 and thought to be $(SbF_5)_2I$ was probably a mixture of an I_2^+ fluoroantimonate salt, and some Sb(III)-containing material. Pure crystalline $I_2^+Sb_2F_{11}^-$ has recently been prepared by the reaction of iodine with antimony pentafluoride in liquid sulfur dioxide as solvent (11). After removal of insoluble SbF_3, deep blue crystals of $I_2^+Sb_2F_{11}^-$ were obtained from the solution. An X-ray crystallographic structure determination showed the presence of the discrete ions I_2^+ and $Sb_2F_{11}^-$. Crystalline solids that can be formulated as $I_2^+Sb_2F_{11}^-$ and $I_2^+Ta_2F_{11}$ have also been prepared by Kemmitt et al. (4) by the reaction of iodine with antimony or tantalum pentafluorides in iodine pentafluoride solutions.

B. Bromine Cations

1. Br_3^+

A compound formulated as SbF_5Br was prepared by Ruff (9) in 1906 by the reaction of Br_2 and SbF_5, but the nature of this compound remained a mystery. Later McRae (12) reported evidence that Br_3^+ was formed in this system. Gillespie and Morton (13, 14) showed more recently that Br_3^+ is formed quantitatively in the superacid medium $HSO_3F-SbF_5-SO_3$ according to Eq. (10). These solutions are brown

$$3Br_2 + S_2O_6F_2 \longrightarrow 2Br_3^+ + 2SO_3F^- \qquad (10)$$

and have a strong absorption at 300 nm with a shoulder at 375 nm. Solutions of Br_3^+ can also be obtained in a similar way in fluorosulfuric acid; however, they are not completely stable in this more basic solvent and undergo some disproportionation according to Eq. (11).

$$Br_3^+ + SO_3F^- \rightleftharpoons Br_2 + BrOSO_2F \qquad (11)$$

Glemser and Smalc (15) have prepared the compound $Br_3^+AsF_6^-$ by the displacement of oxygen in dioxygenyl hexafluoroarsenate by bromine [Eq. (12)] and by the reaction of bromine pentafluoride,

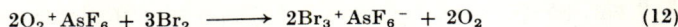

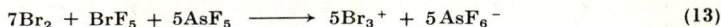

$$2O_2^+AsF_6 + 3Br_2 \longrightarrow 2Br_3^+AsF_6^- + 2O_2 \qquad (12)$$

$$7Br_2 + BrF_5 + 5AsF_5 \longrightarrow 5Br_3^+ + 5AsF_6^- \qquad (13)$$

bromine, and arsenic pentafluoride [Eq. (13)]. The compound is chocolate-brown and in solution has absorption bands at 300 and 375 nm; it has fair thermal stability and can be sublimed at 30° to 50° under an atmosphere of nitrogen.

2. Br_2^+

The Br_2^+ cation can be prepared (14) by oxidation of bromine by $S_2O_6F_2$ in the superacid $HSO_3F-SbF_5-3SO_3$; however, even in this very weakly basic medium, the Br_2^+ ion is not completely stable as it undergoes appreciable disproportionation according to Eq. (14).

$$2Br_2^+ + 2HSO_3F \rightleftharpoons Br_3^+ + BrOSO_2F + H_2SO_3F^+ \qquad (14)$$

Moreover, the $BrOSO_2F$ that is formed itself undergoes some disproportionation to Br_2^+, Br_3^+, and $Br(OSO_2F_3)_3$, so that the equilibria in these solutions are quite complex involving not only Eq. (14) but Eqs. (15) and (16) as well.

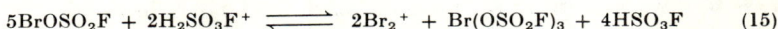

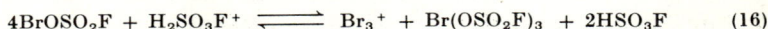

$$5BrOSO_2F + 2H_2SO_3F^+ \rightleftharpoons 2Br_2^+ + Br(OSO_2F)_3 + 4HSO_3F \qquad (15)$$

$$4BrOSO_2F + H_2SO_3F^+ \rightleftharpoons Br_3^+ + Br(OSO_2F)_3 + 2HSO_3F \qquad (16)$$

Solutions of Br_2^+ in superacid have a characteristic cherry red color with maximum absorption at 510 nm and a single band in the Raman spectrum at 360 cm^{-1}.

The paramagnetic scarlet crystalline compound $Br_2^+Sb_3F_{16}^-$ (16, 17) has been prepared by the reaction [Eq. (17)].

$$9Br_2 + 2BrF_5 + 30SbF_5 \longrightarrow 10Br_2^+Sb_3F_{16}^- \qquad (17)$$

It is a stable salt and can be sublimed at 200°.

C. CHLORINE CATIONS

1. Cl_3^+

There is no evidence for either Cl_2^+ or Cl_3^+ in superacid media (18); however, Cl_2, ClF, and AsF_5 react at $-70°$ to form Cl_3AsF_6 according to Eq. (18) (19). The Cl_3^+ cation has also been identified by its Raman

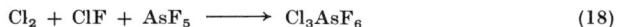

$$Cl_2 + ClF + AsF_5 \longrightarrow Cl_3AsF_6 \qquad (18)$$

spectrum in the yellow solid which precipitates from a solution of Cl_2 and ClF in HF–SbF$_5$ at $-76°$. At room temperature the Cl_3^+ cation completely disproportionates in this solvent to chlorine and ClF_2^+ salts. There is no evidence that Cl_3BF_4 is formed from mixtures of chlorine, chlorine monofluoride, and boron trifluoride at temperatures ranging from ambient to $-130°$.

2. Cl_2^+

The Cl_2^+ ion has been observed in the gas phase at very low pressures, and a value of ω_e of 645.3 cm^{-1} was obtained from the electronic absorption spectrum (20). More recently, Olah and Comisarow (21, 22) have claimed to have identified Cl_2^+ and ClF^+ in solutions on the basis of ESR spectra of chlorine fluorides in SbF$_5$, HSO$_3$F–SbF$_5$, or HF—SbF$_5$, but this claim has been disputed by various workers. Symons et al. (23) have argued that the ESR spectrum assigned to ClF^+ arises from $ClOF^+$, and that assigned to Cl_2^+ from $ClOCl^+$. Christe and Muirhead (24) have reported that they have not detected radicals in the reaction of highly purified SbF$_5$ and ClF_3, or ClF_5, and they suggest that the radicals observed by Olah and Comisarow must have been due to impurities. Gillespie and Morton (18) reported a very large increase in intensity of the ESR signal previously assigned to ClF^+ on adding a trace of water to a sample of $ClF_2^+SbF_6^-$ in SbF$_5$, supporting the assignment to an oxyradical, which, they argue, is probably $FClO^+$ which is isoelectronic with ClO_2 or ClO_2F^+ which is isoelectronic with ClO_3.

A simple calculation of the heats of formation of salts of Cl_2^+ and O_2^+, based on the ionization potentials and the lattice energies given by Kapustinskii's (25) second equation, gives values for the Cl_2^+ salts with hexafluoride anions only 3 kcal less favorable than the corresponding (26) O_2^+ salts. Although this indicates that the salts $Cl_2^+PtF_6^-$ and $Cl_2^+Sb_2F_{11}^-$ are thermodynamically feasible, we expect no kinetic barrier to fluorination via fluorine bridging to give salts of the Cl_2F^+ cation. Thus attempts to prepare salts of Cl_2^+ cations are analogous to attempts to prepare those of Xe^+ in which the product seems always to be the XeF^+ cation (27).

D. Relative Stabilities of Halogen Polyatomic Cations

The problem of stabilizing halogen cations appears to be essentially one of providing a sufficiently weakly basic medium to prevent negative ion transfer, the first step in the decomposition of the polyatomic cation. The more polarizing the cation the more difficult it is to effect stabilization. For example, whereas Cl_2^+ has not been prepared in solution or in the solid state, Br_2^+ exists in equilibrium with other species [see Eq. (14)] in the superacid HSO_3F–SbF_5–$3SO_3$ and as the crystalline salt $Br_2^+Sb_3F_{16}^-$. The larger I_2^+, on the other hand, is stable in the superacid and is only slightly disproportionated in the more basic solvent, HSO_3F. A similar trend is observed for the triatomic cations: I_3^+ is stable in 100% H_2SO_4; Br_3^+ is only stable in the more acidic HSO_3F–SbF_5–SO_3; whereas Cl_3^+ has only been detected in the solid state at $-78°$ as the AsF_6^- salt. In all cases the triatomic cation is more readily stabilized than the smaller, more polarizing, diatomic cation. In general the most stable environment for halogen polyatomic cations appears to be as a crystalline salt with the AsF_6^-, SbF_6^-, $Sb_2F_{11}^-$, or $Sb_3F_{16}^-$ anions.

E. Structures of Halogen Cations

1. Diatomic Cations

The structures of $Br_2^+Sb_3F_{16}^-$ and $I_2^+Sb_2F_{11}^-$ have been determined by X-ray crystallography (11, 16, 17). They both contain a discrete diatomic cation and a fluoroantimonate anion. The bond lengths in the Br_2^+ and I_2^+ cations were found to be 2.13 and 2.56 Å, respectively. The cations have a shorter bond length than the corresponding neutral diatomic molecules and this is consistent with an increase in bond order resulting from the loss of an antibonding electron from the neutral molecule.

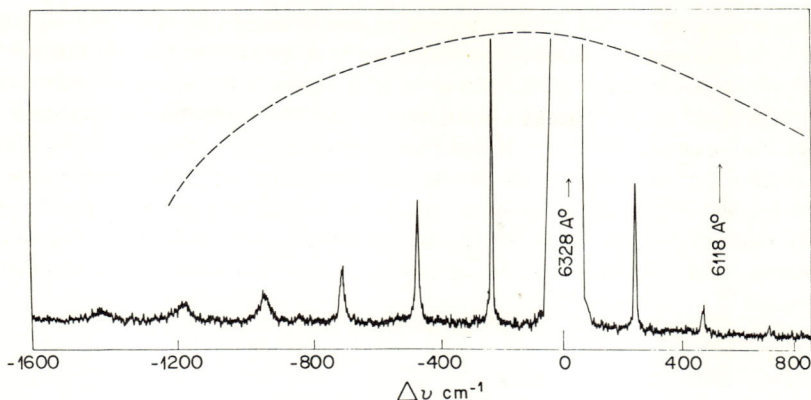

FIG. 1. Resonance Raman spectrum of I_2^+. Dashed line—contour of visible absorption band.

Initial attempts (4) to observe the vibrational frequency of the I_2^+ cation by Raman spectroscopy were unsuccessful owing to the absorption of the existing radiation by the highly colored solutions. Later it was shown (28, 29) that the resonance Raman spectrum of the I_2^+ cation can be observed using 6328 Å He–Ne excitation and very dilute solutions. The resonance Raman spectrum of a 10^{-2} M solution of the I_2^+ cation in fluorosulfuric acid (Fig. 1) shows in addition to the fundamental at 238 cm^{-1}, a number of intense overtones which gradually become progressively broader and weaker. In this particular case the relatively weak Raman scattering from the fluorosulfuric acid solvent is completely absorbed by the solution and only the very strong resonance

TABLE I

STRETCHING FREQUENCIES, ABSORPTION MAXIMA, AND BOND LENGTHS OF THE HALOGENS AND DIATOMIC HALOGEN CATIONS

Cation	Stretching frequency (cm^{-1})	Principal absorption (nm)	Bond length Å	Ionization energy[b] (eV)
Cl_2	564.9[a]	330	1.98	11.50
Cl_2^+	645.3[a]	—	1.89	—
Br_2	320	410	2.28	10.51
Br_2^+	360	510	2.13[a]	—
I_2	215	510	2.66	9.31
I_2^+	238	646	2.56	—

[a] Herzberg (20).
[b] Frost et al. (30).

Raman spectrum of I_2^+ is observed. A solution of the Br_2^+ cation also gives a resonance Raman spectrum with a fundamental of 360 cm^{-1} and strong overtones (14). Edwards and Jones (17) reported that solid $Br_2^+Sb_3F_{16}^-$ has a Raman band at 368 cm^{-1} which they attributed to the Br_2^+ cation. Table I shows the stretching frequencies, absorption maxima, and bond lengths of the halogens and the diatomic halogen cations (20, 30). The increase in stretching frequency of Cl_2^+, Br_2^+, and I_2^+ relative to Cl_2, Br_2, and I_2 is consistent with a decrease in bond distance and an increase in bond strength on removal of an antibonding electron.

2. Triatomic Halogen Cations

The Raman spectrum (19) of $Cl_3^+AsF_6^-$ shows bands due to the AsF_6^- ion, together with three relatively intense bands at 490 (split to 485 and 492), 225, and 508 cm^{-1} which have been assigned to $v_1, v_2,$ and v_3, respectively, of the bent Cl_3^+ cation. The assigned frequencies are very close to the vibrational frequencies of the isoelectronic SCl_2 molecule (31) (514, 208, and 535 cm^{-1}) which has a bond angle of 93°, and it is concluded that the Cl_3^+ cation has a similar structure. Using a simple valence force field, good agreement was obtained for the observed frequencies of the Cl_3^+ cation with a bond angle of $\sim 100°$ and a stretching force constant $f = 2.5$ mdyn Å^{-1} (Table II). For a solution of Br_3^+ in HSO_3F—SbF_5 the only band that can be definitely assigned to Br_3^+ is a relatively strong band at 290 cm^{-1} which is assigned as the symmetrical and asymmetrical stretching vibrations v_1 and v_3. However, ther seems no reason to doubt that Br_3^+ is a bent molecule the same as Cl_3^+ and I_3^+.

Solutions of I_3^+ in H_2SO_4 give Raman spectra (28) that have three bands, in addition to the solvent peaks, at 114, 207, and 233 cm^{-1} which may be assigned as the v_2, v_1, and v_3 vibrations of an angular

TABLE II

VIBRATIONAL FREQUENCIES AND FORCE CONSTANTS FOR THE TRIATOMIC HALOGEN CATIONS

Cation	v_1 (cm^{-1})	v_2 (cm^{-1})	v_3 (cm^{-1})	f (mdyn/Å^{-1})	d (mdyn/Å^{-1})
Cl_3^+	485, 493	225	508	2.5	0.36
Br_3^+	290	(140)^a	290	—	—
I_3^+	207	114	233	1.7	0.32

^{a} Calculated.

molecule. The force constants calculated from the frequencies are given in Table II. It may be noted that the average stretching frequency of 220 cm^{-1} in the I_3^+ molecule is appreciably lower than the stretching frequency of 238 cm^{-1} for the I_2^+ molecule and, in fact, closer to the frequency of 213 cm^{-1} for the neutral molecule. This is consistent with I_3^+ having a formal I–I bond order of 1.0 as in the simple valence bond formulation,

$$\underset{I}{\diagup}\overset{I^+}{}\underset{I}{\diagdown}$$

whereas that in I_2^+ is 1.5.

Recently, Corbett et al. (8) on the basis of ^{127}I nuclear quadrupole resonance (NQR) studies of $I_3^+ AlCl_4^-$ have predicted a bond angle of 97° between the two bonding orbitals on the central atom.

III. Polyatomic Cations of Group VI

A. The O_2^+ Cation

The existence of O_2^+ in the gas phase at low pressures has been well established (32). However, it was not until 1962 that a compound containing O_2^+ was identified (33). It was discovered as a reaction product of the fluorination of platinum in a silica apparatus. The product was first thought to be $PtOF_4$ (34), but later it was shown to be $O_2^+ PtF_6^-$ (33). It was then prepared by direct oxidation of molecular oxygen by platinum hexafluoride at room temperature. Bartlett speculated that, if oxygen [ionization potential (IP) = 12.2 eV] could be oxidized by platinum hexafluoride, then so could xenon (IP = 12.13 eV). Consequently, he reacted xenon and platinum hexafluoride and thus prepared $XePtF_6$ (35)—the first compound of the so-called inert gases.

It now appears that the dioxygenyl salt $O_2^+ BF_4^-$ may have been prepared prior to 1962 (36), although the nature of the material was not elucidated. This and other interesting related work was reviewed in 1966 (36) with extensive reference to sources that are not readily available in the literature. Several O_2^+ salts have now been prepared (see Table III) (37–43). In addition, there is a preliminary report of the preparation of $O_2^+ VF_6^-$ by the reaction of O_2F_2 and VF_5 (44) and a patent referring to $O_2^+ BiF_6^-$ prepared by the same method (45). Also, Bantov and co-workers have reported the reaction of O_2F_2 with various fluorides including SnF_4 which gives $(O_2)_2SnF_6$ (46). The antimony salt prepared by Young (38) has been reported by Nikitina and Rosolovskii (39) to be $O_2^+ Sb_2F_{11}^-$ rather than $O_2^+ SbF_6^-$.

TABLE III

Preparative Routes to O_2^+-Containing Compounds

Product	Reaction	Conditions	References
O_2PtF_6	$F_2 + O_2 + PtF_6$ (sponge)	425°–450°. Flow system	(37)
O_2PtF_6	$F_2O + Pt$(sponge)	Above 400°. Flow system	(37)
O_2PtF_6	$F_2 + PtCl_2, PtCl_4, PtBr_4, PtI_4$	Above 400° in glass. Flow system	(37)
O_2PtF_6	$O_2 + PtF_6$	Tensimetric titration at room temperature	(37)
O_2PF_6, O_2AsF_6	$O_2F_2 + PF_5, AsF_5$	Excess O_2F_2. Reaction at about $-163.5°$	(38)
O_2SbF_6	$O_2Sb_2F_{11}$	(Heat $O_2Sb_2F_{11}$ at 130° *in vacuo*)	(39)
$O_2Sb_2F_{11}$	$O_2F_2 + SbF_5$	Low temperatures	(39)
O_2BF_4, O_2PF_6	$O_2F_2 + BF_3, PF_5$	Excess BF_3, PF_5; $-126°$	(40, 41)
O_2BF_4	$O_4F_2 + BF_3$	Excess BF_3; $-138°$	(40)
O_2AsF_6, O_2SbF_6	$O_2 + F_2 + AsF_5, SbF_5$	$F_2/O_2/AsF_5$, SbF_5 ratio 0.5:1:1; 150 atm; 200°, 5 days	(42)
O_2AsF_5, O_2SbF_6	$O_2 + F_2 + AsF_5, SbF_5$	Excess F_2 and O_2. Pyrex or Kel-F vessel. Expose to sunlight	(43)
O_2AsF_6	$N_2FAsF_6 + O_2$	2 atm (O_2)	(36)

The most convenient route to O_2^+ salts appears to be the photochemical synthesis of $O_2^+AsF_6^-(SbF_6)^-$ from oxygen, fluorine, and arsenic (antimony) pentafluoride *(43)*. Most O_2^+ preparations involve the reaction of fluoride ion acceptors with O_2F_2 or O_4F_2 at low temperatures or with O_2 and F_2 mixtures under conditions favoring synthesis of the long-lived O_2F radical, e.g.,

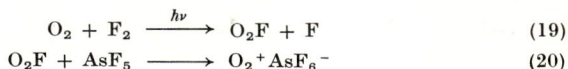

$$O_2 + F_2 \xrightarrow{h\nu} O_2F + F \tag{19}$$

$$O_2F + AsF_5 \longrightarrow O_2^+AsF_6^- \tag{20}$$

Compounds containing O_2^+ are colorless with the exception of $O_2^+PtF_6^-$ which is red due to the PtF_6^- ion. The compound $O_2^+PF_6^-$ decomposes slowly *(38)* at $-80°$, and rapidly at room temperature, giving oxygen, fluorine, and phosphorous pentafluoride; $O_2^+BF_4^-$ decomposes at a moderate rate at $0°$ into similar products. Kinetic data and ^{18}F tracer studies have led to the conclusion that the mechanism of the decomposition involves the equilibrium

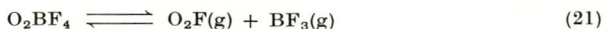

$$O_2BF_4 \rightleftharpoons O_2F(g) + BF_3(g) \tag{21}$$

followed by a bimolecular decomposition of O_2F *(40)*.

Dioxygenyl hexafluoroantimonate has been studied by differential thermal analysis *(39)*. Decomposition of $O_2^+SbF_6^-$ proceeds in two stages, according to the mechanism

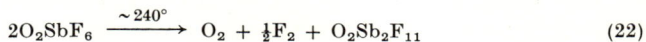

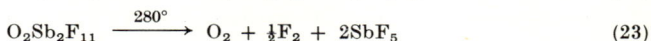

$$2O_2SbF_6 \xrightarrow{\sim 240°} O_2 + \tfrac{1}{2}F_2 + O_2Sb_2F_{11} \tag{22}$$

$$O_2Sb_2F_{11} \xrightarrow{280°} O_2 + \tfrac{1}{2}F_2 + 2SbF_5 \tag{23}$$

The $O_2^+Sb_2F_{11}^-$ was converted into $O_2^+SbF_6^-$ by heating at $130°$ *in vacuo*, and conversely, $O_2^+Sb_2F_{11}^-$ was prepared by reaction of $O_2^+SbF_6^-$ and SbF_5 at $180°$ to $200°$. Dioxygenyl hexafluoroarsenate is markedly less stable than the fluoroantimonate salts; it decomposes rapidly at $130°$ to $180°$ *(38)*. $O_2^+PtF_6^-$ can be sublimed above $90°$ *in vacuo* and melts with some decomposition at $219°$ in a sealed tube *(37)*.

X-Ray powder data obtained from the cubic form of O_2PtF_6 were consistent with the presence of O_2^+ and PtF_6^- ions *(37)*. The structure was refined using neutron diffraction powder data. The PtF_6^- ion was located unambiguously, but the length of the O—O bond could not be determined with certainty, probably because of disorder of the O_2^+ ion in the structure *(47)*. Table IV lists the crystal type and cell parameters of some O_2^+-containing salts *(37, 38, 43, 48)*. Confusing results on the powder diffraction of $O_2^+SbF_6^-$ have recently been cleared up by McKee and Bartlett *(43)*. In every case there is a structural relationship to the analogous nitrosyl salts.

TABLE IV

CRYSTAL TYPE AND CELL PARAMETERS OF SOME O_2^+-CONTAINING SALTS

Compound	Symmetry of unit cell	z	Cell parameters	Reference
O_2PtF_6	Cubic	8	$a = 10.032$	(37)
O_2PtF_6	Rhombohedral	1	$a \sim 4.96$; $97.5°$	(37)
O_2BF_4	Orthorhombic	4	$a = 8.777$, $b = 5.581$, $c = 7.036$	(48)
O_2AsF_6	Cubic	4	$a = 8.10$	(43)
O_2AsF_6	Cubic	4	$a = 8.00$	(38)
O_2SbF_6	Cubic	4	$a = 10.132$	(43)

The Raman spectra of various O_2^+ salts have been obtained and all show a strong absorption attributable to O_2^+ as well as those due to the corresponding anions; namely, $O_2^+PtF_6^-$ (49) 1837 cm^{-1}, $O_2^+AsF_6^-$ 1858 cm^{-1}, $O_2^+SbF_6^-$ 1862 cm^{-1}, and $O_2^+SbF_6^-$ in SbF_5 1860 cm^{-1} (50). The infrared spectrum of $O_2^+BF_4^-$ (51) at $-196°$ has a weak doublet at 1868 and 1866 cm^{-1}. The assignment of these bands to the O_2^+ vibration was confirmed by ^{18}O substitution, which led to a shift of the doublet to 1764 and 1762 cm^{-1}. These frequencies may be compared with the value of 1876 cm^{-1} determined (32) from the electronic band spectrum of gaseous O_2^+.

The magnetic behavior of O_2^+ in $O_2^+PtF_6^-$ over the temperature range 77°–298°K is similar to that of nitric oxide showing the presence of one unpaired electron ($^2\Pi$ ground state). The magnetic moment of O_2^+ was found to be $\mu_{eff} = 1.57 \, \mu_B$ at room temperature (52). A magnetic moment of 1.66 μ_B has been reported for $O_2^+SbF_6^-$ (53), and a value of 1.7 μ_B for $O_2^+BF_4^-$ (53a). An ESR spectrum has been observed for $O_2^+AsF_6^-$ with a single line with a g value at $-80°C$ of 1.9980 corresponding to one free electron.

The chemistry of the O_2^+ cation does not appear to have been extensively studied although various displacement reactions of the type

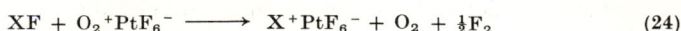

$$XF + O_2^+PtF_6^- \longrightarrow X^+PtF_6^- + O_2 + \tfrac{1}{2}F_2 \qquad (24)$$

$$XF = KF, ClF_3, IF_5$$

have been described (37). Another interesting oxygen displacement is the reaction of $O_2^+AsF_6^-$ and bromine leading to the preparation of $Br_3^+AsF_6^-$ (15):

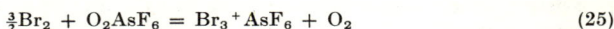

$$\tfrac{3}{2}Br_2 + O_2AsF_6 = Br_3^+AsF_6 + O_2 \qquad (25)$$

Various reactions of O_2^+ salts are listed in Ref. 36. Recently, the reaction of $O_2^+BF_4^-$ with xenon has been reported (54). At 173°K, oxygen and fluorine were liberated and a white solid formed, which, on the basis of analytical and vibrational spectroscopic data, is claimed to be $FXe-BF_2$.

B. OTHER OXYGEN POLYATOMIC CATIONS

Ozone [IP $= 12.3$ eV (55)] reacts with PtF_6 in the gas phase to give $O_2{}^+PtF_6{}^-$; no evidence for $O_3{}^+PtF_6{}^-$ was obtained (56). Goetschel and co-workers (57) reacted a mixture of oxygen fluorides, obtained by the radiolysis of F_2 and O_2 with boron trifluoride at low temperatures, and claim to have made O_4BF_4 and O_6BF_4 although reliable evidence for the existence of these interesting compounds has not been obtained.

C. SULFUR POLYATOMIC CATIONS

The nature of the colored solutions obtained on dissolving sulfur in oleum (58) has until recently remained a mystery since their discovery by Bucholz (59) in 1804. Red, yellow, and blue solutions have been prepared; however, particular attention has been given to the blue solutions. The species responsible for the blue color has been identified by various workers as S_2O_3 (60), S_2 (61), the radical ion $(X_2S-SX_2){}^+$ (62), and a species designated S_x (63). The confusing evidence concerning the blue compound "S_2O_3" has been reviewed (64). Recently, the various colors have been shown to be due to the cations $S_{16}{}^{2+}$, $S_8{}^{2+}$, and $S_4{}^{2+}$ $(65-67)$.

1. Preparation

Sulfur can be quantitatively oxidized by arsenic or antimony pentafluoride to red compounds of composition $S_{16}(AsF_6)_2$ and $S_{16}(SbF_6)_2$ or to the deep blue compounds, $S_8(AsF_6)_2$ and $S_8(Sb_2F_{11})_2$ according to Eq. (26)–(29). In addition the pale yellow compound $S_4(SbF_6)_2$ has been

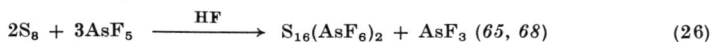

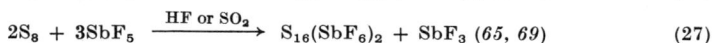

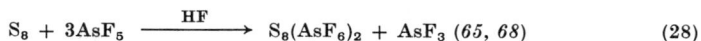

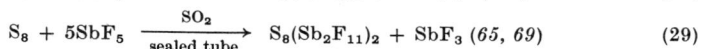

$$2S_8 + 3AsF_5 \xrightarrow{\text{HF}} S_{16}(AsF_6)_2 + AsF_3 \ (65,\ 68) \tag{26}$$

$$2S_8 + 3SbF_5 \xrightarrow{\text{HF or SO}_2} S_{16}(SbF_6)_2 + SbF_3 \ (65,\ 69) \tag{27}$$

$$S_8 + 3AsF_5 \xrightarrow{\text{HF}} S_8(AsF_6)_2 + AsF_3 \ (65,\ 68) \tag{28}$$

$$S_8 + 5SbF_5 \xrightarrow[\text{sealed tube}]{\text{SO}_2} S_8(Sb_2F_{11})_2 + SbF_3 \ (65,\ 69) \tag{29}$$

prepared by the reaction of sulfur $(65,\ 69)$ and SbF_5 at 140°. Solid materials were obtained by Ruff (9) and by Peacock (70), which were assigned the compositions SbF_5S and $(SbF_5)_2S$, respectively. It is probable, however, that the materials that they obtained were not pure compounds but contained SbF_3 or an $SbF_3 \cdot SbF_5$ complex in addition to cations of sulfur and an anion such as $Sb_2F_{11}^-$. A blue material obtained by the reaction of sulfur and SO_3 has been known for a long time (*64, 71*) and has been described as a lower oxide of sulfur with the composition S_2O_3. This material must contain S_8^{2+} and is probably $S_8(HS_3O_{10})_2$ (*66*) but may also contain $S_4(S_4O_{13})$.

Sulfur may also be oxidized by $S_2O_6F_2$ in fluorosulfuric acid at 0°C (*65, 69*). The results of conductometric and cryoscopic measurements carried out on this red solution were consistent with the formation of S_{16}^{2+} according to Eq. (30). Further oxidation by $S_2O_6F_2$ produces a

$$2S_8 + S_2O_6F_2 \longrightarrow S_{16}^{2+} + 2SO_3F^- \tag{30}$$

blue solution containing S_8^{2+}; however, these solutions are not stable and slowly deposit sulfur on standing. The pale yellow compound $S_4(SO_3F)_2$ has been prepared by carefully reacting $S_2O_6F_2$ with elemental sulfur in sulfur dioxide solvent at low temperatures. This compound is not stable in fluorosulfuric acid as the characteristic peak of the blue S_8^{2+} cation slowly appears and increases in intensity with time. However, a stable colorless solution is obtained in the stronger acid HSO_3F–SbF_5. The absorption spectra of S_{16}^{2+} and S_8^{2+} in HSO_3F, and of S_4^{2+} in HSO_3F–SbF_5 are shown in Fig. 2.

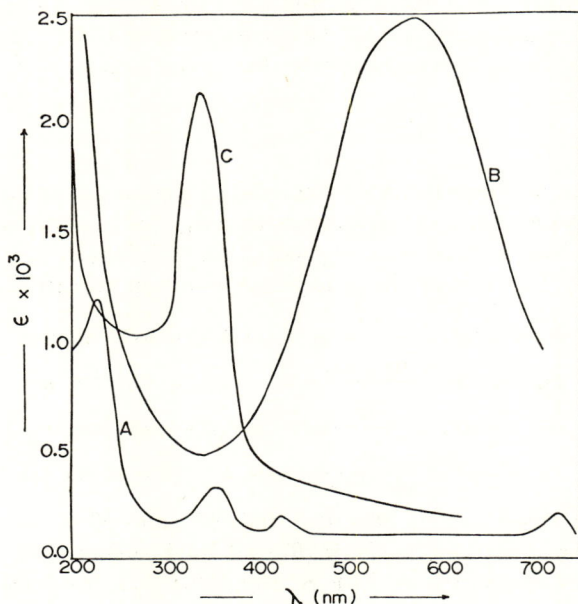

Fig. 2. Absorption spectra of S_{16}^{2+} (A) and S_8^{2+} (B) in HSO_3F and of S_4^{2+} in (C) HSO_3F–SbF_5.

Seel and co-workers (*72*) have reported that $S_2F^+AsF_6^-$ gives a mixture of sulfur polyatomic cations and $SF_3^+AsF_6^-$ on warming to 100°, or at room temperature in the presence of AsF_5.

The deeply colored solutions of sulfur in oleum have been known for a long time (*59*), but it was not until the identification of the sulfur cations S_4^{2+}, S_8^{2+}, and S_{16}^{2+} that the nature of these solutions became clear (*66*). In 95–100% H_2SO_4 sulfur forms a colloidal solution but after 12 hr at 75° the element dissolves as S_8 molecules. In 5% oleum, oxidation is observed and S_{16}^{2+} is formed. In 10% and 15% oleum, sulfur is oxidized first rather rapidly to a mixture of S_{16}^{2+} and S_8^{2+} and then very slowly to SO_2. In 30% oleum, S_{16}^{2+} and S_8^{2+} produced initially are further oxidized to S_4^{2+} and finally to SO_2. In more concentrated oleums (45 and 65%), S_8^{2+} and S_4^{2+} are the initial products, and as S_4^{2+} appears to be rather stable in these solvents further oxidation to SO_2 is very slow. Changes in concentration of the various species with time and with SO_3 concentration are complicated by disproportionation reactions. Thus, S_8^{2+} disproportionates to SO_2 and S_{16}^{2+} in oleum containing less than 15% SO_3, and S_4^{2+} disproportionates to S_8^{2+} and SO_2 in oleum containing less than 40% SO_3.

2. Structures of S_{16}^{2+}, S_8^{2+}, and S_4^{2+}

No structural data are available for S_{16}^{2+}. The crystal structure of $S_8(AsF_6)_2$ has been determined (*67*); it contains the S_8^{2+} ion which has the structure shown in Fig. 3. It consists of a folded ring with approximately C_s symmetry and has an *endo-exo* conformation. The average bond distance around the ring is 2.04 Å, which is identical with that in the S_8 molecule (*73, 74*). The three cross-ring distances, as determined in the two crystallographically different S_8^{2+} rings, are S_4–S_6 [2.942(10),

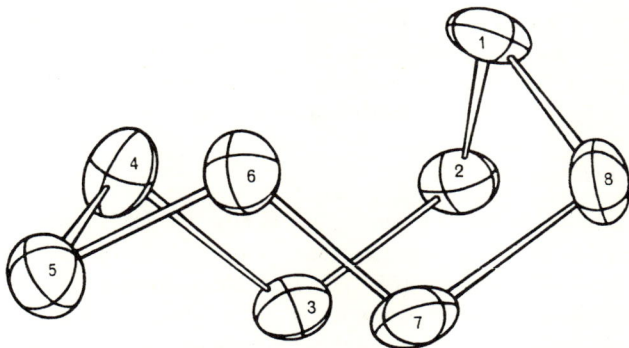

FIG. 3. Structure of S_8^{2+} in $S_8(AsF_6)_2$.

3.053(12) Å], S_3-S_7 [2.832(10), 2.889(12) Å], and S_2-S_8 [3.010(11), 2.866(11) Å], significantly shorter than in the S_8 ring (4.68 Å) or the van der Waals distance of 3.7 Å. These findings strongly suggest that there is weak transannular bonding. It is also noted that there are other sulfur–sulfur bond distances in the ring significantly shorter than the van der Waals distance, e.g., S_5-S_3 [3.082(9), 3.065(10) Å]. The bonding in the ion is, therefore, complex but may perhaps be described by the valence bond structures (I–IV) and in addition others, such as V and VI, where the dashed line indicates the plane of symmetry in the

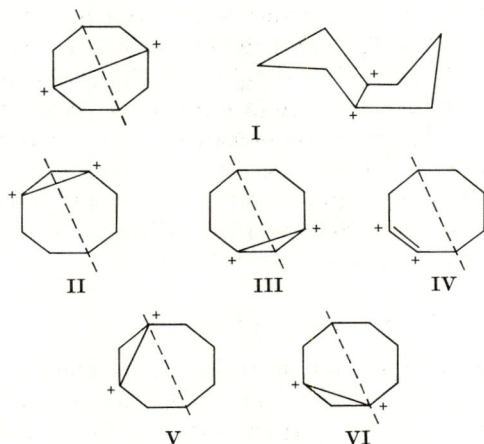

molecule. The distortion of the ring produced by this cross-ring bonding causes all the angles, which range from 91.5° to 104.3°, to be smaller than the angle of 107.9° found in the S_8 ring.

The ultraviolet and Raman spectra of S_4^{2+} (65) are very similar to those of Se_4^{2+} and Te_4^{2+} which have been shown to be planar, suggesting that S_4^{2+} has the same geometry (Tables VIII and IX). The results of a study of the magnetic circular dichroism of solutions of S_4^{2+}, Se_4^{2+}, and Te_4^{2+} also lead to the same conclusion (75).

3. Radical Cations

Solutions of sulfur in oleum give rise to ESR spectra, but the interpretation of these spectra has been the subject of some controversy in the literature (62, 76, 77). No progress was made in the interpretation of these spectra until it had been established that the main species present under various conditions are the sulfur cations S_4^{2+}, S_8^{2+}, and S_{16}^{2+}. It was then shown that solutions of S_8^{2+} in HSO_3F

are paramagnetic and give an ESR spectrum ($g = 2.014$) which is identical with that obtained from blue solutions of sulfur in 60% oleum. Since on cooling these solutions the intensity of the ESR signal decreases, it was proposed that there is an equilibrium between S_8^{2+} and the radical cation S_4^+, i.e.,

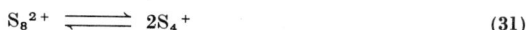

$$S_8^{2+} \rightleftharpoons 2S_4^+ \tag{31}$$

This has been confirmed by the observation of the ESR spectrum of a solution of ^{33}S in 60% oleum which was found (77) to consist of thirteen lines consistent with the presence of four equivalent sulfur atoms of spin $\frac{3}{2}$. Presumably S_4^+ has a square planar structure the same as S_4^{2+}. The observed g values were reported as 2.0163 for ^{33}S and 2.013 for ^{32}S which are to be compared with the value of $g = 2.014$ reported by Gillespie et al. (65). Symons and Wilkinson (78) have recently given a different interpretation of the spectrum of ^{33}S in oleum but this seems to be inconsistent with all the other information about these solutions. The conclusion that the radical species is S_4^+ also receives some support from the ESR spectra of frozen solutions reported by Giggenbach (79) which gave a typical glass spectrum of $g_\perp = 2.0004$ and $g_\parallel = 2.0192$, indicating that the species giving rise to this signal has axial symmetry. This is consistent with the proposed planar structure for S_4^+.

Solutions of sulfur in more dilute oleum, e.g., 15%, give ESR spectra with a second signal ($g = 2.027$). This signal is also obtained from a solution of S_{16}^{2+} in fluorosulfuric acid. It seems reasonable, therefore, to attribute this ESR signal to a radical associated with S_{16}^{2+}, presumably S_8^+, formed by dissociation [Eq. (32)].

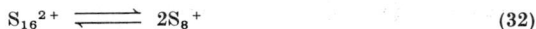

$$S_{16}^{2+} \rightleftharpoons 2S_8^+ \tag{32}$$

Consistent with this proposal, it was found that the intensity of the $g = 2.027$ signal decreased on cooling the oleum solutions and at the same time absorption bands at 430, 720, and 935 nm decreased in intensity. Presumably the foregoing equilibrium shifts to the left on cooling the solution, and the bands at 430, 720, and 935 nm as well as the ESR signal at $g = 2.027$ are to be attributed to S_8^+. The g values

TABLE V

THE g VALUES AND ABSORPTION MAXIMA FOR SULFUR CATIONS

Parameters	S_{16}^{2+}	S_8^+	S_8^{2+}	S_4^+	S_4^{2+}
g	—	2.027	—	2.014	—
Absorption max. (nm)	235	430	590	—	330
	335	720	—	—	—
	—	935	—	—	—

and absorption maxima for the various sulfur cations are summarized in Table V.

No ESR spectra have been observed for solutions of S_4^{2+}.

D. SELENIUM POLYATOMIC CATIONS

The colored solutions produced on dissolving elemental selenium in sulfuric acid were first observed by Magnus in 1827 (80). Since then a number of workers have investigated the nature of selenium solutions in sulfuric acid, oleum, and sulfur trioxide, providing (81) a substantial amount of data but little understanding of the system. Recently, it has been shown that these solutions contain the yellow Se_4^{2+} and green Se_8^{2+} polyatomic cations (82).

1. Preparation

Selenium polycations are less electrophilic than their sulfur analogs and give stable solutions in various strong acids (82). In fluorosulfuric acid, selenium can be oxidized quantitatively by $S_2O_6F_2$ to give yellow Se_4^{2+} [Eq. (33)]

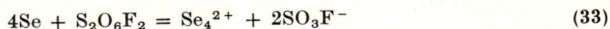

$$4Se + S_2O_6F_2 = Se_4^{2+} + 2SO_3F^- \tag{33}$$

A photometric titration of selenium and $S_2O_6F_2$ established the oxidation state of the yellow species as $+\frac{1}{2}$; conductometric measurements showed that two fluorosulfate ions are produced per four selenium atoms; and the molecular weight of Se_4^{2+} was established by cryoscopy. The absorption spectrum of the yellow Se_4^{2+} solution in HSO_3F is shown in Fig. 4.

The addition of selenium to the yellow solution up to a 8:1 ratio of $Se-S_2O_6F_2$ did not appreciably affect the conductivity. This indicated that the SO_3F^- ion concentration remained unchanged and that the Se_4^{2+} ion is reduced by selenium according to Eq. (34).

$$Se_4^{2+} + 4Se = Se_8^{2+} \tag{34}$$

Conductivity measurements of selenium in pure fluorosulfuric acid were also consistent with the formation of Se_8^{2+}. The absorption spectrum of the green solution is shown in Fig. 4.

Solutions of Se_8^{2+} in 100% H_2SO_4 may be prepared by heating selenium in the acid at 50° to 60°; the element is oxidized by sulfuric acid according to Eq. (35).

$$8Se + 5H_2SO_4 = Se_8^{2+} + 2H_3O^+ + 4HSO_4^- + SO_2 \tag{35}$$

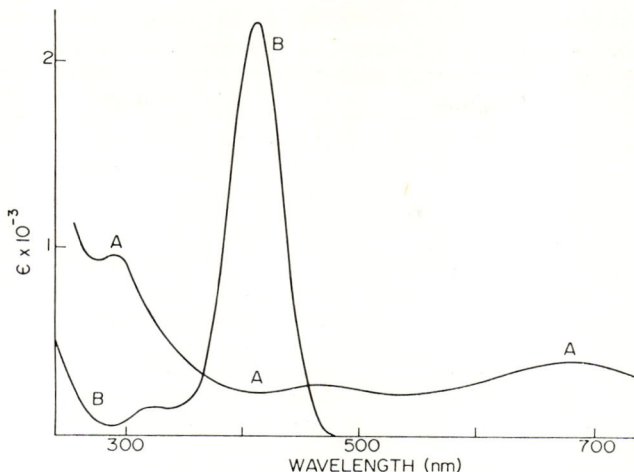

FIG. 4. Absorption spectra of Se_4^{2+} (A) and Se_8^{2+} (B) in HSO_3F.

The cation Se_4^{2+} was obtained on further oxidation of Se_8^{2+} with selenium dioxide:

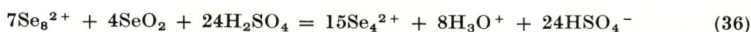

$$7Se_8^{2+} + 4SeO_2 + 24H_2SO_4 = 15Se_4^{2+} + 8H_3O^+ + 24HSO_4^- \tag{36}$$

The cations Se_4^{2+} and Se_8^{2+} can also be obtained in disulfuric acid by oxidation of elemental selenium by the solvent, first to Se_8^{2+} and with time to Se_4^{2+} according to Eq. (37) and (38).

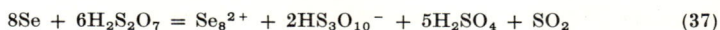

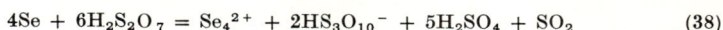

$$8Se + 6H_2S_2O_7 = Se_8^{2+} + 2HS_3O_{10}^- + 5H_2SO_4 + SO_2 \tag{37}$$

$$4Se + 6H_2S_2O_7 = Se_4^{2+} + 2HS_3O_{10}^- + 5H_2SO_4 + SO_2 \tag{38}$$

Various Se_4^{2+}- and Se_8^{2+}-containing compounds have been prepared by oxidizing selenium with $SeCl_4$ plus $AlCl_3$, sulfur trioxide, oleum, SbF_5, and AsF_5. These preparations are listed in Table VI (83–87). In addition to the compounds listed, Paul and co-workers (88) have reported the compounds $Se_4S_4O_{13}$, $Se_4S_3O_{10}$, and $Se_4S_2O_7$, prepared by the reaction of elemental selenium and sulfur trioxide for various periods of time. The compounds $Se_4(HS_2O_7)_2$ and $Se_4S_4O_{13}$ have very similar analyses and were both previously incorrectly described as $SeSO_3$ (89). The yellow material described by Aynsley, Peacock, and Robinson(70) as $Se(SbF_5)_2$, whatever its exact composition, very probably contains the Se_4^{2+} cation. All selenium polyatomic cations are diamagnetic and so far no evidence has been reported for radicals analogous to S_4^+ and S_8^+.

TABLE VI

PREPARATION OF COMPOUNDS CONTAINING POLYATOMIC CATIONS OF SELENIUM

Compound	Reaction	Conditions	References
$Se_4(HS_2O_7)_2$	Se + 65% oleum	50°–60°. Left until yellow-brown. Crystals given on standing	(83)
$Se_4S_4O_{13}$	Se + excess SO_3	0°. Left 24 hr	(83)
$Se_4(SO_3F)_2$	$4Se + S_2O_6F_2$	Solvent HSO_3F	(83)
$Se_4(Sb_2F_{11})_2$	Se + excess SbF_5	Heat at 100°–140° for 6 hr	(83)
$Se_4(AsF_6)_2$	$Se_8 + 6AsF_5$	Solvent SO_2; 80° for 8 days. Yellow solid deposited from green solution	(84)
$Se_8(Sb_2F_{11})_2$	$Se_8 + 5SbF_5$	Solvent SO_2; −23° for 3 days	(85)
$Se_8(AsF_6)_2$	$Se_8 + 3AsF_5$	Solvent HF. Warmed up slowly from −78° to 0° over 3 days	(85)
$Se_8(AlCl_4)_2$	Se + $SeCl_4$ + $2AlCl_3$	Fuse at 250° for 3 hr	(86, 87)
$Se_4(AlCl_4)_2$	Obtained from Se–($SeCl_4$–$4AlCl_3$) melts		(86)

2. Structures of Se_4^{2+} and Se_8^{2+}

The crystal structure of $Se_4(HS_2O_7)_2$ (90, 91) has shown Se_4^{2+} to be square planar with an Se–Se bond distance of 2.283(4) Å, significantly less than that of 2.34(2) Å found in the Se_8 molecule (92), indicating some degree of multiple bonding. Such a result is consistent with a valence bond description of the molecule involving four structures of type VII. Alternatively the structure can be understood in terms of molecule orbital theory. The circle in structure VIII denotes a closed-shell (aromatic?) six-π-electron system. Of the four π molecular orbitals,

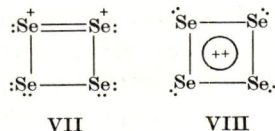

VII   VIII

the two almost nonbonding (e_g) orbitals and the lower-energy (b_{2u}) bonding orbital are occupied by the six π electrons, leaving the upper antibonding (a_{1g}) orbitals empty. The intense yellow-orange color of Se_4^{2+} has been attributed to the dipole allowed excitation of an electron from an e_g orbital to the lowest empty π orbital (b_{2u}). Stephens (75) has shown that the magnetic circular dichroism results are consistent with such a model.

The square planar structure was also found to be consistent with the infrared and Raman spectra of several compounds containing Se_4^{2+} (93). A normal coordinate analysis yielded a value of 2.2 mdynes $Å^{-1}$ for the Se–Se stretching constant, which is somewhat greater than the value of 1.67 mdynes $Å^{-1}$ obtained for the single Se–Se bond in $(CH_3)_2Se_2$.

The structure of Se_8^{2+} in $Se_8(AlCl_4)_2$ (86, 87) is similar to that of S_8^{2+} except that the cross-ring distance Se_3–Se_7 is relatively shorter than that found in the sulfur cation, and the other cross-ring distances,

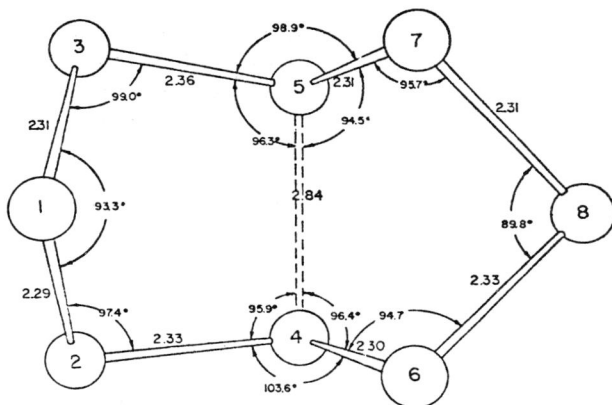

FIG. 5. Projection of Se_8^{2+} structure down the a axis.

Se_4–Se_6 and Se_2–Se_8, are relatively long (Fig. 5). The cation Se_8^{2+} is, therefore, reasonably well described by valence bond structure (IX),

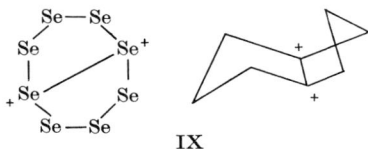

IX

although there may be small contributions from structures analogous to structures II to VI proposed for S_8^{2+}. The ring has an endo-exo conformation with approximately C_s symmetry. The bond lengths around the ring vary between 2.29 and 2.36 Å and do not differ significantly from those found in α- and β-selenium, but the bond angles are smaller than in the Se_8 ring. The bond distances and angles in Se_8^{2+} are given in Fig. 5.

E. TELLURIUM POLYATOMIC CATIONS

The red color produced when tellurium dissolves in concentrated sulfuric acid was first observed as long ago as 1798 (*94*), but the origin of this color has remained somewhat of a mystery until very recently. Much more recently Bjerrum and Smith (*95*) and Bjerrum (*96*) have studied the reaction of tellurium tetrachloride with tellurium in molten $AlCl_3$–$NaCl$. They obtained a purple melt which they concluded contained the species $Te_{2n}{}^{n+}$ (probably $Te_4{}^{2+}$) formed by reaction (39).

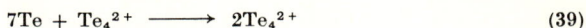

$$7Te + Te_4{}^{2+} \longrightarrow 2Te_4{}^{2+} \tag{39}$$

At about the same time solutions of tellurium in various acids were investigated in detail (*97, 98*). It was found that red solutions are produced when tellurium is dissolved in sulfuric acid, fluorosulfuric acid, or oleum with the simultaneous production of SO_2, indicating that

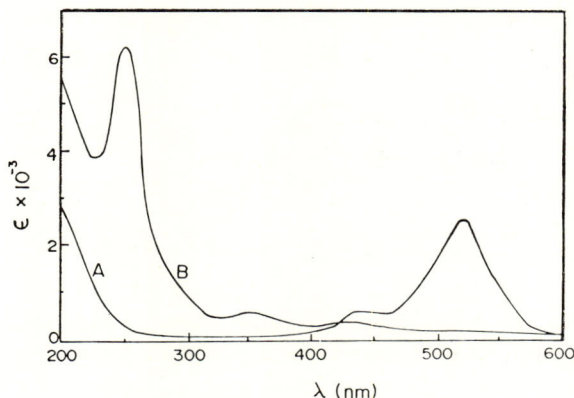

FIG. 6. Absorption spectra of HSO_3F solution of the red tellurium species A and the yellow tellurium species B.

the tellurium is oxidized. The spectra of the solutions (Fig. 6) were found to be identical with those obtained by Bjerrum and Smith from their melts. Conductometric and cryoscopic measurements of the acid solutions led to the conclusion that they contain a species Te_{2n}^{n+} which was certainly not $Te_2{}^+$ but probably $Te_4{}^{2+}$.

Reaction of tellurium with $S_2O_6F_2$ (*98*), SbF_5, and AsF_5 in SO_2 gave the compounds $Te_4(SO_3F)_2$, $Te_4(Sb_2F_{11})_2$, and $Te_4(AsF_6)_2$ and, from Te–($TeCl_4$–$AlCl_3$) melts, compounds $Te_4(AlCl_4)_2$ and $Te_4(Al_2Cl_7)_2$ (*99*) were obtained [Table VII (*98–101*)]. The formulation of the red species

TABLE VII

PREPARATION OF COMPOUNDS CONTAINING POLYATOMIC CATIONS OF TELLURIUM

Compound	Reaction	Conditions	References
$Te_4(Sb_2F_{11})_2$	$Te + SbF_5$	Solvent SO_2. Stirred for several days at $-23°C$. SO_2-soluble products extracted by the solvent	(98)
$TeSbF_6$	$Te + SbF_5$	$TeSbF_6$ is insoluble in SO_2, therefore readily separated from $Te_4(Sb_2F_{11})_2$	(98, 100)
$Te_4(AsF_6)_2$	$4Te + 3AsF_5$	Solvent SO_2. Stirred at $25°C$ for 1 day	(98)
Te_3AsF_6	$6Te + 3AsF_5$	Conditions as above	(98)
$Te(SO_3F)_2$	$4Te + S_2O_6F_2$	Solvent SO_2. Stirred at $-63°C$ and $-23°C$ for 1 day, respectively.	(98)
$TeSO_3F$	$4Te + S_2O_6F_2$ (excess)	Compound is unstable above $-20°C$.	(98)
$Te_4(AlCl_4)_2$ $Te_4(Al_2Cl_7)_2$ $Te_6(AlCl_4)_2$		Obtained from $Te–(TeCl_4–4AlCl_3)$ melts	(99)
$Te_4S_3O_{10}$	$Te + SO_3$	$0°C$; excess SO_3; 24 hr	(101)
$Te_2S_3O_{10}$	$Te + SO_3$	Room temp.; excess SO_3; several days	(100, 101)

as $Te_4{}^{2+}$ was finally confirmed by the determination of the crystal structures of these latter two compounds (102).

When the acid solutions are warmed above room temperature or in the case of solution in 45% oleum at room temperature the color of the solution changes slowly from red to orange and to yellow. The same change in color is produced by the addition of an oxidizing agent such as $S_2O_6F_2$ or peroxodisulfate. Absorption spectra and cryoscopic and conductometric measurements on the fluorosulfuric acid solutions established that the yellow species is $Te_n{}^{n+}$ and that it could not be $Te_2{}^{2+}$ and was probably $Te_4{}^{4+}$ although higher molecular weights, such as $Te_6{}^{6+}$ and $Te_8{}^{8+}$, were not excluded with certainty (98). Paul and co-workers (103) have, however, concluded from absorption spectra and from cryoscopic and conductometric measurements that the yellow species is $Te_2{}^{2+}$. A similar conclusion was made by Bjerrum (104) from spectrometric measurements of $TeCl_4$ and elementary tellurium in $KAlCl_4$ melts buffered with $KCl–ZnCl_2$. The equilibrium (40) was reported to occur under these conditions. The possibility that there are

$$3Te(II) \rightleftharpoons Te_2{}^{2+} + \quad 3Te(II) \rightleftharpoons Te_2{}^{2+} + Te(IV) \qquad (40)$$

various Te_n^{n+} species, depending on the nature of the solvent or accompanying ions, cannot be ruled out. Yellow solids of empirical formula $TeSO_3F$, $TeSbF_6$, and $Te_2S_3O_{10}$ have been obtained from the reactions of tellurium with $S_2O_6F_2$, SbF_5, and oleum, respectively (Table VII). A crystal structure of a Te_n^{n+} salt is badly needed to help resolve this problem.

The tellurium analog of Se_8^{2+} and S_8^{2+} has not been reported; however, a gray solid of empirical formula Te_3AsF_6 has been prepared (98) by reacting tellurium with a stoichiometric amount of arsenic pentafluoride in liquid SO_2. The compound is diamagnetic and is, therefore, probably $Te_6^{2+}(AsF_6)_2^-$. In phase diagram studies of the system $Te-(TeCl_4-4AlCl_3)$, Corbett et al. (99) found the phase $(Te_3AlCl_4)_n$ and were able to grow black crystals by vapor phase transport. The compound is diamagnetic, and the density and dimensions of the unit cell indicate that $n = 1$ or 2; hence, the compound is reasonably formulated as $Te_6(AlCl_4)_2$. Some evidence for a lower oxidation state of tellurium had been previously obtained by Bjerrum and Smith (95) from experiments in which they had added more than seven parts of tellurium to one part of $TeCl_4$ in molten $AlCl_3-NaCl$.

Structure of Te_4^{2+}

The structure of Te_4^{2+} has been determined (102) from the crystal structures of $Te_4(AlCl_4)_2$ and $Te_4(Al_2Cl_7)_2$. In both cases the Te_4^{2+} ion lies on a center of symmetry and is almost exactly square planar. The tellurium–tellurium distance of 2.66 Å is significantly shorter than the tellurium–tellurium distance of 2.864 Å within the spiral chain in elemental tellurium (105). This is consistent with a structure exactly analogous to that for Se_4^{2+} in which each bond has 25% double bond character. The Raman spectra of Te_4^{2+} in solution and the solid state are analogous to those of Se_4^{2+} and S_4^{2+} but shifted to lower frequency (Table VIII). The magnetic circular dichroism (75) and visible and

TABLE VIII

VIBRATIONAL FREQUENCIES OF THE S_4^{2+}, Se_4^{2+},

AND Te_4^{2+} IONS

Vibrational mode	S_4^{2+}	Se_4^{2+}	Te_4^{2+}
$\nu_1(A_{1g})$ (cm^{-1})	584	327	219
$\nu_2(B_{1g})$ (cm^{-1})	530	319	219
$\nu_3(E_u)$ (cm^{-1})	460	306	—
$\nu_4(B_{2g})$ (cm^{-1})	330	192	139

TABLE IX

COMPARISON OF ABSORPTION SPECTRA
OF Te_4^{2+}, Se_4^{2+}, AND S_4^{2+} CATIONS

Cation	max (nm)	
	Strong	Weak
Te_4^{2+}	510	420
Se_4^{2+}	410	320
S_4^{2+}	330	280

ultraviolet spectrum (Table IX) of solutions of Te_4^{2+} were also similar to those of Se_4^{2+} as expected on the basis of their structural similarity.

No structural information is available for Te_n^{n+} or Te_6^{2+}. It is interesting to note, however, that, if Te_n^{n+} is in fact Te_4^{4+}, it is iso-electronic with Sb_4 and would presumably have the same tetrahedral

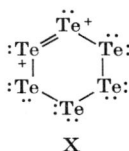

X

structure. It is also tempting to speculate that Te_6^{2+} might have the cyclic structure (X) or possibly six resonance structures of this type.

F. Reactions of Group VI Polyatomic Cations

The reactions of Group VI polyatomic cations are as yet almost completely uninvestigated, but this will no doubt be an area of activity in the future. The only reaction that has so far been studied is that of tetrafluoroethylene with various Group VI polyatomic cations in a solid-gas reaction and in SO_2. The results are given in Table X (106–108). It is possible that initially C_2F_4 acts as a diradical toward the centers of unsaturation and very weak bonds in the various polyatomic cations; e.g., the long S–S and Se–Se bonds in S_8^{2+} and Se_8^{2+}, respectively, and the double bond in Te_4^{2+}, to form active intermediates which may abstract fluoride ion from AsF_6^-. In sulfur dioxide solution the reaction products are more complicated and, in addition to the products in the neat reactions, OSF_2 and carbonyl fluorides are formed [e.g., C_2F_5Se-$SeCF_2COF$], suggesting that the solvent itself takes part in the reaction.

TABLE X

REACTION OF C_2F_4 WITH VARIOUS GROUP VI POLYATOMIC CATIONS

Compound	Conditions	Products[a]	Reference
$S_8(AsF_6)_2$	a. Room temp.; ambient pressures of C_2F_4 b. Solvent SO_2; room temp.; 3 atm pressure C_2F_4	$(C_2F_5)_2S_x$ $(x = 2\text{-}6)$ $(C_2F_5)_2S_x$ $(x = 2\text{-}3)$ $C_2F_5S_xCF_3$ $C_2F_5S_xCF_2COF$	(106)
$S_{16}(AsF_6)_2$	Room temp.; ambient pressures C_2F_4	$(C_2F_5)_2S_x$ $(x = 2\text{-}6)$	
$Se_8(AsF_6)_2$	a. Room temp.; excess C_2F_4 in pressure reactor b. In SO_2 solution; room temp.; about 4 atm pressure C_2F_4	$(C_2F_5)_2Se_x$ $(x = 2, 3)$ $(C_2F_5)_2Se_x$ $(x = 2, 3)$ $C_2F_5Se_2C_2F_5$ $C_2F_5Se_xCF_2COF$	(107)
$Te_4(AsF_6)_2$	a. 100°C; excess C_2F_4 in pressure reactor b. SO_2 solvent; 100°C; excess C_2F_4 in pressure reactor	$(C_2F_5)_2Te_x$ $(x = 1, 2)$ $C_2F_5Te_xC_4F_9$ $(C_2F_5)_2Te_x$ $(x = 1, 2)$ $C_2F_5Te_xC_4F_9$ $C_2F_5TeC_3F_6COF$	(108)

[a] In addition to these products, arsenic trifluoride and unidentified solids were obtained.

IV. Polyatomic Cations of Group V

A. BISMUTH POLYATOMIC CATIONS

1. Preparation and Structure of Bi_9^{5+}

The discovery of bismuth polycations arose out of an investigation into the nature of "BiCl", first prepared by reduction of bismuth trichloride by bismuth metal by Eggink (*110*) in 1908. More recently, Hershaft and Corbett (*109*), obtained black crystals of this material from the melt and by single crystal X-ray diffraction showed that the unit cell contained $4Bi_9^{5+}$, $8BiCl_5^{2-}$, and $2Bi_2Cl_8^{2-}$, i.e., it has the empirical composition Bi_6Cl_7. Recently the cation Bi_9^{5+} has also been

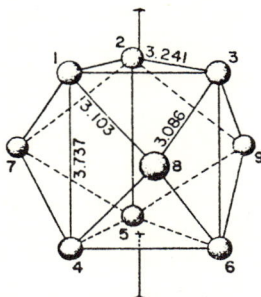

FIG. 7. Structure of Bi_9^{5+}.

identified in the compound $Bi_{10}HfCl_{18}$ (*111*) which was prepared by the reduction of a 3:2 mixture of hafnium tetrachloride and bismuth trichloride with elemental bismuth. By X-ray crystallography this compound was shown to be $Bi^+ Bi_9^{5+} (HfCl_6^{2-})_3$. The structure of the Bi_9^{5+} cation is shown in Fig. 7. It is a tricapped trigonal prism which ideally has D_{3h} symmetry but in this case is slightly distorted to give C_{3h} symmetry. The bonding (*112*) in this ion has been treated using D_{3h} symmetry orbitals obtained from the 6p atomic orbitals in a linear combination of atomic orbitals–molecular orbital (LCAO-MO) calculation. The stability and diamagnetism of the cation was explained by the closed-shell MO configuration of 22p electrons in eleven bonding MO's.

2. Preparation of Bi_5^{3+} and Bi_8^{2+}

Bjerrum and Smith have established the identity of Bi^+, Bi_5^{3+} (*113, 114*), and Bi_8^{2+} (*115*) in fused salts. The formulas Bi^+ and Bi_5^{3+} were determined by studying the equilibria (41)–(43). Equilibrium

(41) was studied in melts of $AlCl_3$–$NaCl$ eutectic, and equilibria (42) and (43) in molten $ZnCl_2$–KCl eutectic as solvent. The cation $Bi_8{}^{2+}$ was also prepared and identified in $AlCl_3$–$NaCl$ melts by reduction of Bi^{3+} according to Eq. (44) (115). Spectrophotometric measurements established

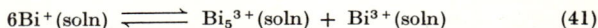

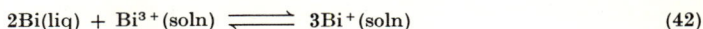

$$6Bi^{+}(soln) \rightleftharpoons Bi_5{}^{3+}(soln) + Bi^{3+}(soln) \tag{41}$$

$$2Bi(liq) + Bi^{3+}(soln) \rightleftharpoons 3Bi^{+}(soln) \tag{42}$$

$$4Bi(liq) + Bi^{3+} \rightleftharpoons Bi_5{}^{3+}(soln) \tag{43}$$

$$22Bi(metal) + 2Bi^{3+}(soln) = 3Bi_8{}^{2+}(soln) \tag{44}$$

that equilibrium (44) is displaced strongly to the right, so that the oxidation state was readily established as 0.25 by the uptake of bismuth by a known amount of $BiCl_3$. In molten $NaAlCl_4$ saturated with $NaCl$ as solvent, the $Bi_8{}^{2+}$ was shown to be in equilibrium with Bi^{+} and bismuth metal. Spectrophotometric measurements on various mixtures yielded the complete reaction stoichiometry and definitely fixed the formula as $Bi_8{}^{2+}$.

The compounds $Bi_5(AlCl_4)_3$ and $Bi_8(AlCl_4)_2$ were prepared (116, 117) by reaction of $BiCl_3$–$AlCl_3$ with a stoichiometric quantity of bismuth and with an excess quantity of bismuth, respectively, in liquid $NaAlCl_4$. The compounds are diamagnetic and have electronic spectra very similar to those of $Bi_8{}^{2+}$ and $Bi_5{}^{3+}$ in solution. Trigonal bipyramidal (D_{3h}) and square antiprismatic (D_{4h}) structures have been predicted for $Bi_5{}^{3+}$ and $Bi_8{}^{2+}$ on the basis of LCAO-MO calculations (116), although direct evidence is lacking.

Reports of $Bi_3{}^{3+}$ (118) and $Bi_4{}^{4+}$ (119, 120) have been shown to be incorrect (116, 121).

B. THE POLYATOMIC CATION $Sb_n{}^{n+}$

Antimony metal has been oxidized by arsenic pentafluoride (122) to the compound $SbAsF_6$ according to Eq. (45).

The compound $SbAsF_6$ may contain the $(Sb^{+})_n$ cation, but it would be difficult on the basis of analysis alone to rule out other stoichiometries such as $Sb_5{}^{4+}(AsF_6{}^{-})_4$ where the antimony is in an oxidation state close to but not equal to $+1$.

Metallic antimony dissolves slowly in fluorosulfuric acid (123) at room temperature according to Eq. (46) to give the compound $SbSO_3F$ which has been isolated as a pure solid.

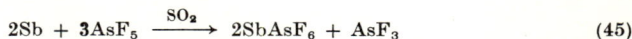

$$2Sb + 3AsF_5 \xrightarrow{SO_2} 2SbAsF_6 + AsF_3 \tag{45}$$

$$2Sb + 4HSO_3F \longrightarrow 2Sb(SO_3F) + H_3O^{+} + SO_3F^{-} + SO_2 + HF \tag{46}$$

C. Other Polyatomic Cations of Group V

The cations Sb_4^{2+}, Sb_8^{2+} (124), As_4^{2+}, As_2^{2+} (125), P_4^{2+}, and P_8^{2+} (126) have been reported as products of the reaction of the elements with $S_2O_6F_2$ in HSO_3F or with oleum. However, the ultraviolet spectra reported for these species are very similar to those found for S_{16}^{2+}, S_8^{2+}, or S_4^{2+}, and it seems very probable that antimony, arsenic, and phosphorus reduce HSO_3F and $H_2S_2O_7$ to elemental sulfur, which is then oxidized to S_{16}^{2+}, S_8^{2+}, or S_4^{2+}. Indeed, it has been demonstrated that elemental sulfur is one of the products of the reduction of oleum by antimony (123). Thus there is at present no reliable evidence for any polyatomic cations of P, As, or Sb, with the exception of $(Sb^+)_n$.

V. Polyatomic Cations of Group IIb

A. Hg_2^{2+}, Cd_2^{2+}, Zn_2^{2+}

The mercurous ion Hg_2^{2+} is by far the most stable of the known polyatomic cations, and its existence in acidic aqueous solution and in a variety of simple crystalline salts, e.g., Hg_2X_2 (X = F, Br, Cl, I) is well documented (127). The corresponding cadmium ion Cd_2^{2+} is less well established but evidence for the compound $Cd_2(AlCl_4)_2$ has been obtained from a study of the $Cd–CdCl_2–AlCl_3$ phase diagram (128). Evidence has been obtained for the cation Zn_2^{2+} (129) in solutions of zinc in zinc chloride and in zinc chloride–cerium chloride melts, although compounds containing this cation have not been isolated. The Raman spectra of solutions containing Hg_2^{2+}, Cd_2^{2+}, or Zn_2^{2+} show peaks at 169 (130) [more recently 182 (131)], 183 (132), and 175 (129) cm^{-1}, respectively, which have been attributed to the metal–metal stretching vibrations. Force constants of 2.52 (132), 1.68 (132), and 0.6 (129) mdyn $Å^{-1}$ have been estimated for Hg_2^{2+}, Cd_2^{2+}, and Zn_2^{2+}, respectively. The value for Hg_2^{2+} is probably somewhat higher than 2.52 mdyn $Å^{-1}$ as it was based on the earlier 169 cm^{-1} value of the Hg–Hg stretching frequency. It has been suggested (132) that the higher Hg metal–metal bond strength in Hg_2^{2+} is a consequence of the higher electron affinity of Hg^+ relative to Cd^+ (first IP Hg = 10.43 eV, Cd = 8.99 eV). However, although zinc has an ionization potential [9.39 eV] intermediate between mercury and cadmium, Zn_2^{2+} has a very low force constant (0.6 mdyn).

B. Hg_3^{2+}

Although the Hg_2^{2+} ion has been known for a very long time, it is only very recently that evidence for other ions of the general formula

Hg_n^{2+} has been obtained. The compound $Hg_3(AlCl_4)_2$ (*133, 134*) has been prepared by reacting a 1:2:2 molar mixture of $HgCl_2$, Hg, and $AlCl_3$ at 240° for 6 days. The absorption spectra of a mixture of Hg_2^{2+} and Hg in molten $AlCl_3$–$NaCl$ at 175° gave an absorption due to a mercury species of lower oxidation state than Hg_2^{2+} which was attributed to Hg_3^{2+}. Polarograms for the reduction of Hg^{2+} in molten $AlCl_3$–$NaCl$ show three waves consistent with the reaction scheme:

$$2Hg^{2+} + 2e^- = Hg_2^{2+} \qquad (47)$$

$$3Hg_2^{2+} + 2e^- = 2Hg_3^{2+} \qquad (48)$$

$$Hg_3^{2+} + 2e^- = 3Hg \qquad (49)$$

Equilibrium constants for the reactions $Hg^{2+} + Hg_3^{2+} = 2Hg_2^{2+}$ and $Hg_2^{2+} + Hg = Hg_3^{2+}$ have been obtained (*134*) by linear sweep voltammetry and chronopotentiometry for several $AlCl_3$–$NaCl$ composition ratios at various temperatures.

The yellow compound $Hg_3(AsF_6)_2$ has been prepared in sulfur dioxide solution (*131, 136*) either by oxidizing mercury with AsF_5,

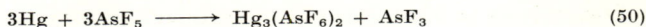

$$3Hg + 3AsF_5 \longrightarrow Hg_3(AsF_6)_2 + AsF_3 \qquad (50)$$

or by reacting mercurous hexafluoroarsenate with mercury,

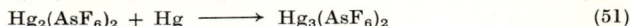

$$Hg_2(AsF_6)_2 + Hg \longrightarrow Hg_3(AsF_6)_2 \qquad (51)$$

The compound $Hg_3(Sb_2F_{11})_2$ can also be prepared by the similar reaction of mercury with SbF_5 in SO_2 solution:

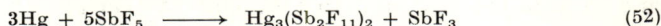

$$3Hg + 5SbF_5 \longrightarrow Hg_3(Sb_2F_{11})_2 + SbF_3 \qquad (52)$$

The Raman spectrum of $Hg_3^{2+}(AsF_6)_2$ in sulfur dioxide solution shows in addition to peaks attributable to AsF_6^- and the solvent, a single strong polarized band at 118 cm^{-1} which was assigned to a Hg–Hg stretch indicating that Hg_3^{2+} has the linear structure Hg^+–Hg–Hg^+.

The structure of $Hg_3(AsF_6)_2$ has been determined by X-ray crystallography and the Hg_3^{2+} ion has been found to be linear and symmetric. The mercury–mercury distance was found to be 2.552(4) Å (*135*). The crystal structure of $Hg_3(AlCl_4)_2$ has also been determined (*136*). In this case the two mercury–mercury bond distances were found to be almost equal [2.551(1) and 2.562(1) Å], but the ion is not quite linear having a bond angle of 174.4°. The mercury–mercury bond distance of 2.55 Å in both compounds is somewhat longer than the range of 2.49 to 2.54 Å reported for the Hg–Hg bond lengths in several halides (*137*) and salts of Hg_2^{2+} (*138–143*). The rather short $Hg\cdots Cl$ distance of 2.54 Å (cf. Hg–Cl = 2.43 Å in Hg_2Cl_2) indicates considerable covalent interaction between the Hg_3^{2+} "ion" and the $AlCl_4^-$.

Accordingly, Ellison *et al.* (*136*) have preferred to describe the compound as molecular rather than ionic.

C. Hg_4^{2+}

By using more mercury than is necessary to prepare Hg_3^{2+} in the reaction with AsF_5 in SO_2 solution, the dark red crystalline compound $Hg_4(AsF_6)_2$ can be obtained:

$$4Hg + 3AsF_5 \longrightarrow Hg_4(AsF_6)_2 + AsF_3$$

A determination of the structure of this compound by X-ray crystallography has shown that the Hg_4^{2+} ion has a centrosymmetric almost linear structure with the following dimensions (*144*):

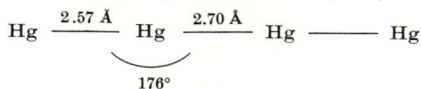

$$Hg \xrightarrow{\text{2.57 Å}} Hg \xrightarrow{\text{2.70 Å}} Hg \rule{2cm}{0.4pt} Hg$$
$$176°$$

D. $Hg_n^{0.35n+}$

When mercury is allowed to react with a solution of arsenic pentafluoride in SO_2 at room temperature, a remarkable reaction is observed in which the mercury crystallizes over a period of 10 to 15 min to a golden-yellow solid with a striking metallic appearance. If excess AsF_5 is present the solid eventually dissolves to give a yellow solution of Hg_3^{2+}. When a limited amount of AsF_5 is used (i.e., $AsF_5/Hg = 1:2$) the gold solid, which is quite insoluble in SO_2, can be obtained in a pure state. When this compound was first analyzed, it was believed (*145*) to have the composition Hg_3AsF_6 and to have been formed according to the following reaction:

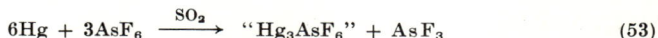

$$6Hg + 3AsF_6 \xrightarrow{SO_2} \text{``}Hg_3AsF_6\text{''} + AsF_3 \tag{53}$$

Supporting this assumption, it was found possible to prepare the compound by reacting $Hg_2(AsF_6)_2$ with the appropriate amount of mercury according to Eq. (54). However, the determination of the structure of

$$Hg_2(AsF_6)_2 + 4Hg \longrightarrow 2Hg_3AsF_6 \tag{54}$$

this compound by X-ray crystallography (*146*) has shown that it, in fact, has the composition $Hg_{2.85}AsF_6$ and that it has a remarkable structure in which the octahedral AsF_6^- ions are stacked in such a manner that the fluorines occupy three-quarters of the sites of a cubic close-packed lattice and so that there are channels running through the lattice in two mutually perpendicular directions. Within these channels are infinite chains of mercury atoms, each with an average formal charge of $+0.35$, and with an average mercury–mercury distance of 2.64(1) Å (see Fig. 8). The crystals have a conductivity of the order of magnitude of that expected for a metal. It is noteworthy that this interesting compound contains covalently bonded AsF_6^- ions, metallically bonded $Hg_n^{0.35n+}$ chains, and ionic bonding between the metallic chains and the AsF_6^- ions. Each mercury chain constitutes a one-dimensional metal.

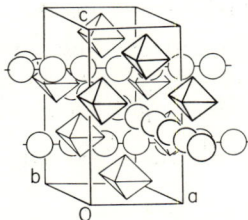

FIG. 8. Structure of $Hg_{2.85}AsF_6$.

VI. Polyatomic Cations of Other Elements

We have not discussed the evidence for polyatomic cations formed in the mass spectrometer or as transient reaction intermediates; instead we have concentrated on recent work on those polyatomic cations that exist as stable entities in solution or in the solid state. We may add that evidence has also been given for the formation of Pb_2^{2+} (*147*), Mg_2^{2+} (*148*), Ca_2^{2+} (*149*), Sr_2^{2+} (*150*), and Ba_2^{2+} (*150*) on addition of the respective element to the corresponding MCl_2 melt at high temperatures, and ESR evidence has been presented for Ag_4^+ or Ag_4^{3+} (*151*), Ag_2^+, Cd_2^{3+} (*152*), and Hg_2^{3+} or Hg_2^+ (*153*).

VII. Conclusion

Salts of homopolyatomic cationic clusters now constitute a well-established class of compound—there are at least twenty-six fairly well-characterized examples. It is probable that many other elements will also be shown to form polyatomic cations. As yet reactions of these species have been little studied, and there is obviously a wide open field here awaiting exploration. Many structures of known cations, as well as those that have not yet been prepared, remain to be investigated, and there is a need for theories that can predict the stability and geometry of these cations and provide a description of the bonding.

REFERENCES

1. Masson, I., *J. Chem. Soc., London* p. 1708 (1938).
2. Gillespie, R. J., and Milne, J. B., *Inorg. Chem.* **5**, 1577 (1966).
3. Gillespie, R. J., and Malhotra, K. C., *Inorg. Chem.* **8**, 1751 (1969).
4. Kemmitt, R. D. W., Murray, M., McRae V. M., Peacock, R. D., and Symons, M. C. R., *J. Chem. Soc., London* p. 862 (1968).
5. Arotsky, J., and Symons, M. C. R., *Quart. Rev., Chem. Soc.* **16**, 282 (1962), and references therein.
6. Arotsky, J., Mishra, H. C., and Symons, M. C. R., *J. Chem. Soc., London* p. 2582 (1962).
7. Garrett, R. A., Gillespie, R. J., and Senior, J. B., *Inorg. Chem.* **4**, 563 (1965).
8. Merryman, D. J., Edwards, P. A., Corbett, J. D., and McCarley, R. E., *Chem. Commun.* p. 779 (1972).
8a. Chung, C., and Cady, G. H., *Inorg. Chem.* **11**, 2528 (1972).
9. Ruff, O., Graf, H., Heller, W., and Knock, *Ber.* **39**, 4310 (1906).
10. Adhami, G., and Herlem, M., *J. Electroanal. Chem.* **26**, 363 (1970).
11. Davies, C., Gillespie, R. J., and Sowa, J. M., *Can. J. Chem.* **52**, 791 (1974).
12. McRae, V. M., Ph.D. Thesis, University of Melbourne (1966).
13. Gillespie, R. J., and Morton, M. J., *Chem. Commun.* p. 1565 (1968).
14. Gillespie, R. J., and Morton, M. J., *Inorg. Chem.* **11**, 586 (1972).
15. Glemser, O., and Smalc, A., *Angew. Chem., Int. Ed. Engl.* **8**, 517 (1969).
16. Edwards, A. J., Jones, G. R., and Sills, R. J. C., *Chem. Commun.* p. 1527 (1968).
17. Edwards, A. J., and Jones, G. R., *J. Chem. Soc., A* p. 2318 (1971).
18. Gillespie, R. J., and Morton, M. J., *Inorg. Chem.* **11**, 591 (1972).
19. Gillespie, R. J., and Morton, M. J., *Inorg. Chem.* **9**, 811 (1970).
20. Herzberg, G., "Molecular Spectra and Molecular Structure," Vol. I. Van Nostrand-Reinhold, Princeton, New Jersey, 1960.
21. Olah, G. A., and Comisarow, M. B., *J. Amer. Chem. Soc.* **90**, 5033 (1968).
22. Olah, G. A., and Comisarow, M. B., *J. Amer. Chem. Soc.* **91**, 2172 (1969).
23. Eachus, R. S., Sleight, T. P., and Symons, M. C. R., *Nature (London)* **222**, 769 (1969).
24. Christe, K. O., and Muirhead, J. S., *J. Amer. Chem. Soc.* **91**, 7777 (1969).
25. Kapustinskii, A. F., *Quart. Rev., Chem. Soc.* **10**, 284 (1956).
26. Bartlett, N., Beaton, S. P., and Jha, N. K., *Chem. Commun.* p. 168 (1966).
27. McRae, V. M., Peacock, R. D., and Russel, D. R., *Chem. Commun.* p. 62 (1969).
28. Gillespie, R. J., Morton, M., and Sowa, J. M., *Advan. Raman Spectrosc.* **1**, 530 (1972).
29. Gillespie, R. J., and Morton, M., *J. Mol. Spectrosc.* **30**, 178 (1969).
30. Frost, D. C., McDowell, C. A., and Vroom, D. A., *J. Chem. Phys.* **46**, 4255 (1967).
31. Siebert, H., "Anwendungen der Schwingungs spektroskopie in der Anorganischen Chemie." Springer-Verlag, Berlin and New York, 1966.
32. Herzberg, G., "The Spectra of Diatomic Molecules." Van Nostrand-Reinhold, Princeton, New Jersey, 1950.
33. Bartlett, N., and Lohmann, D. H., *Proc. Chem. Soc., London* p. 115 (1962).
34. Bartlett, N., and Lohmann, D. H., *Proc. Chem. Soc., London* p. 14 (1960).

35. Bartlett, N., *Proc. Chem. Soc., London* p. 218 (1962).
36. Lawless, E. W., and Smith, I. C., "Inorganic High-Energy Oxidizers," and references therein. Dekker, New York, 1968.
37. Bartlett, N., and Lohmann, D. H., *J. Chem. Soc. London* p. 5253 (1962).
38. Young, A. R., II, Hirata, T., and Morrow, S. I., *J. Amer. Chem. Soc.* **86**, 20 (1964).
39. Nikitina, Z. K., and Rosolovskii, V. Ya., *Izv. Akad. Nauk SSSR, Ser. Khim.* p. 2173 (1970).
40. Keith, J. N., Solomon, I. J., Sheft, I., and Hyman, H. H., *Inorg. Chem.* **7**, 230 (1968).
41. Solomon, I., Brabets, R. I., Uenishi, R. K., Keith, J. N., and McDonough, J. M., *Inorg. Chem.* **3**, 457 (1964).
42. Beal, J. B., Jr., Pupp, C., and White, W. E., *Inorg. Chem.* **8**, 828 (1969).
43. McKee, D. E., and Bartlett, N., *Inorg. Chem.* **12**, 2738 (1973).
44. Solomon, I. J., *U.S. Govt. Res. 8 Develop. Rep.* **69**, 62 (1969); *Chem. Abstr.* **71**, 18410j (1969).
45. Young, A. R., II, Hirata, T., and Morrow, S. I., U.S. Patent 3,385,666 (1968); *Chem. Abstr.* **69**, 20801q (1968).
46. Bantov, D. V., Sukhoverkhov, V. F., and Mikhailov, Yu. N., *Izv. Sib. Otd. Akad. Nauk SSSR, Ser. Khim. Nauk* **2**, 184 (1968).
47. Ibers, J. A., and Hamilton, W. C., *J. Chem. Phys.* **44**, 1748 (1966).
48. Wilson, J. W., Curtis, R. M., and Goetschel, C. T., *J. Appl. Crystallogr.* **4**, 260 (1971).
49. Bartlett, N., *Angew. Chem., Int. Ed. Engl.* **7**, 433 (1968).
50. Shamir, J., Binenboyn, J., Claasen, H. H., *J. Amer. Chem. Soc.* **90**, 6223 (1968).
51. Loos, K. R., Campanile, V. A., and Goetschel, C. T., *Spectrochim. Acta, Part A* **26**, 365 (1970).
52. Bartlett, N., and Beaton, S. P., *Chem. Commun.* p. 167 (1966).
53. Belova, V. I., Rosolovskii, V. Ya., and Nikitina, E. K., *Russ. J. Inorg. Chem.* **16**, 772 (1971).
53a. Belova, V. I., Syrkin, Ya. K., Bantov, D. V., and Sukhoverkhov, V. F., *Zh. Neorg. Khim.* **13**, 1457 (1968); *Russ. J. Inorg. Chem.* **13**, 765 (1968).
54. Goetschel, C. T., and Loos, K. R., *J. Amer. Chem. Soc.* **94**, 3018 (1972).
55. Radwan, T. N., and Turner, D. W., *J. Chem. Soc., A* p. 85 (1966).
56. Paige, H., and Passmore, J., private communication.
57. Goetschel, C. T., Campanile, V. A., Wagner, C. D., and Wilson, J. N., *J. Amer. Chem. Soc.* **91**, 4702 (1969).
58. Mellor, J. W., "Comprehensive Treatise on Inorganic and Theoretical Chemistry," Vol. 10, pp. 184–186 and 992. Longmans, Green, New York, 1930.
59. Bucholz, C. F., *Gehlen's Neues J. Chem.* **3**, 7 (1804).
60. Weber, R., *Ann. Phys. (Leipzig)* [2] **156**, 531 (1875).
61. Auerbach, R., *Z. Phys. Chem., Abt.* **121**, 337 (1926).
62. McNeil, D. A. C., Murray, M., and Symons, M. C. R., *J. Chem. Soc., A* p. 1019 (1967).
63. Lux, H., Bohm, E., *Chem. Ber.* **98**, 3210 (1965).
64. Nickless, G., ed., "Inorganic Sulphur Chemistry," p. 412. Elsevier, Amsterdam, 1968.
65. Gillespie, R. J., Passmore, J., Ummat, P. K., and Vaidya, O. C., *Inorg. Chem.* **10**, 1327 (1971).

66. Gillespie, R. J., and Ummat, P. K., *Inorg. Chem.* **11,** 1674 (1972).
67. Davies, C., Gillespie, R. J., Park, J. J., and Passmore, J., *Inorg. Chem.* **10,** 2781 (1971).
68. Gillespie, R. J., and Passmore, J., *Chem. Commun.* p. 1333 (1969).
69. Barr, J., Gillespie, R. J., and Ummat, P. K., *Chem. Commun.* p. 264 (1970).
70. Aynsley, E. E., Peacock, R. D., and Robinson, P. L., *Chem. Ind. (London)* p. 1117 (1951).
71. Vogel, I., and Partington, J. D., *J. Chem. Soc., London* **127,** 1514 (1925).
72. Seel, F., Hartmann, V., Molnar, I., Budenz, R., and Gombler, W., *Angew. Chem., Int. Ed. Engl.* **10,** 186 (1971).
73. Abrahams, S. C., *Acta Crystallogr. Engl.* **8,** 66 (1955).
74. Caron, A., and Donohue, J., *Acta Crystallogr.* **18,** 562 (1965).
75. Stephens, P. J., *Chem. Commun.* p. 1496 (1969).
76. Gardner, D. M., and Fraenkel, G. K., *J. Amer. Chem. Soc.* **28,** 6411 (1956).
77. Beaudet, R. A., and Stephens, P. J., *Chem. Commun.* p. 1083 (1971).
78. Symons, M. C. R., and Wilkinson, J. G., *Nature (London)* **236,** 126 (1972).
79. Giggenback, W. F., *Chem. Commun.* p. 852 (1970).
80. Magnus, G., *Ann. Phys. (Leipzig)* [2] **10,** 491 (1827); **14,** 328 (1828).
81. J. W. Mellor, "Comprehensive Treatise on Inorganic and Theoretical Chemistry," Vol. 10, pp. 922–923. Longmans, Green, New York, 1930.
82. Barr, J., Gillespie, R. J., Kapoor, R., and Malhotra, K. C., *Can. J. Chem.* **46,** 149 (1968).
83. Barr, J., Crump, D. B., Gillespie, R. J., Kapoor, R., and Ummat, P. K., *Can. J. Chem.* **46,** 3607 (1968).
84. Gillespie, R. J., and Ummat, P. K., unpublished results.
85. Gillespie, R. J., and Ummat, P. K., *Can. J. Chem.* **48,** 1239 (1970).
86. Mullen, R. K., Prince, D. J., and Corbett, J. D., *Inorg. Chem.* **10,** 1749 (1971).
87. Mullen, R. K., Prince, D. J., and Corbett, J. D., *Chem. Commun.* p. 1438 (1969).
88. Paul, R. C., Arora, C. L., Virmani, R. N., and Malhotra, K. C., *Indian J. Chem.* **9,** 368 (1971).
89. Divers, E., and Shimose, M., *J. Chem. Soc., London* **43,** 329 (1883).
90. Brown, I. D., Crump, D. B., Gillespie, R. J., and Santry, D. P., *Chem. Commun.* p. 853 (1968).
91. Brown, I. D., Crump, D. B., and Gillespie, R. J., *Inorg. Chem.* **10,** 2319 (1971).
92. March, R. E., Pauling, L., and McCullough, J. D., *Acta Crystallogr., Sect. B* **6,** 71 (1953).
93. Gillespie, R. J., and Pez, G. P., *Inorg. Chem.* **8,** 1229 (1969).
94. Klaproth, M. H., *Phil. Mag.* **1,** 78 (1798).
95. Bjerrum, N. J., and Smith, G. P., *J. Amer. Chem. Soc.* **90,** 4472 (1968).
96. Bjerrum, N. J., *Inorg. Chem.* **9,** 1965 (1970).
97. Barr, J., Gillespie, R. J., Kapoor, R., and Pez, G. P., *J. Amer. Chem. Soc.* **90,** 6855 (1968).
98. Barr, G., Gillespie, R. J., Pez, G. P., Ummat, P. K., and Vaidya, O. C., *Inorg. Chem.* **10,** 362 (1971).
99. Prince, D. J., Corbett, J. D., and Garbisch, B., *Inorg. Chem.* **9,** 2731 (1970).
100. Barr, J., Gillespie, R. J., Pez, G. P., Ummat, P. K., and Vaidya, O. C., *J. Amer. Chem. Soc.* **92,** 1081 (1970).
101. Paul, R. C., Arora, C. L., Puri, J. K., Virmani, R. N., and Malhotra, K. C., *J. Chem. Soc., Dalton Trans* p. 781 (1972).

102. Couch, T. W., Lokken, D. A., and Corbett, J. D., *Inorg. Chem.* **11**, 357 (1972).
103. Paul, R. C., Puri, J. K., and Malhotra, K. C., *Chem. Commun.* p. 776 (1970).
104. Bjerrum, N. J., *Inorg. Chem.* **11**, 2648 (1972).
105. Straumanis, M., *Z. Kristallogr., Kristallgeometrie, Kristallphys. Kristallchem.* **102**, 432 (1946).
106. Paige, H. L., and Passmore, J., *Inorg. Chem.* **12**, 593 (1973).
107. Desjardins, C. D., Paige, H. L., and Passmore, J., Abstr. *164th Meet., Amer. Chem. Soc., New York,* (1972).
107a. Desjardins, C. D., and Passmore, J., *J. Chem. Soc., Dalton Trans.* 2314 (1973)
108. Paige, H. L., and Passmore, J., *Inorg. Nucl. Chem. Lett.* **9**, 277 (1973).
109. Hershaft, A., and Corbett, J. D., *Inorg. Chem.* **2**, 979 (1963).
110. Eggink, B. G., *Z. Phys. Chem. Abt. A* **64**, 449 (1908).
111. Friedman, R. M., and Corbett, J. D., *Chem. Commun.* p. 422 (1971).
111a. Friedman, R. M, and Corbett, J. D., *Inorg. Chim. Acta* **7**, 525 (1973).
112. Corbett, J. D., and Rundle, R. E., *Inorg. Chem.* **3**, 1408 (1964).
113. Bjerrum, N. J., Boston, C. R., Smith, G. P., and Davies, H. L., *Inorg. Nucl. Chem. Lett.* **1**, 141 (1965).
114. Bjerrum, N. J., Boston, C. R., and Smith, G. P., *Inorg. Chem.* **6**, 1162 (1967).
115. Bjerrum, N. J., and Smith, G. P., *Inorg. Chem.* **6**, 1968 (1967).
116. Corbett, J. D., *Inorg. Chem.* **7**, 198 (1968).
117. Corbett, J. D., *Inorg. Nucl. Chem. Lett.* **3**, 173 (1967).
118. Levy, H. A., Bredig, M. A., Danford, M. D., and Agron, P. A., *J. Phys. Chem.* **64**, 1959 (1960).
119. Topol, L. E., Yosim, S. J., and Osteryoung, R. A., *J. Phys. Chem.* **65**, 1511 (1961).
120. Boston, C. R., Smith, G. P., and Howick L. C., *J. Phys. Chem.* **67**, 1849 (1963).
121. Boston, C. R., *Inorg. Chem.* **9**, 389 (1970).
122. Dean, P. A. W., and Gillespie, R. J., *Chem. Commun.* p. 853 (1970).
123. Gillespie, R. J., and Vaidya, O. C., *Chem. Commun.* p. 40 (1972).
124. Paul, R. C., Paul, K. K., and Malhotra, K. C., *Chem. Commun.* p. 453 (1970).
125. Paul, R. C., Puri, J. K., Paul, K. K., Sharma, R. D., and Malhotra, K. C., *Inorg. Nucl. Chem. Lett.* **7**, 725 (1971).
126. Paul, R. C., Puri, J. K., and Malhotra, K. C., *Chem. Commun.* p. 1031 (1971).
127. Roberts, H. L., *Advan. Inorg. Chem. Radiochem.* **11**, 309 (1968).
128. Corbett, J. D., Burkhard, W. J., and Druding, L. F., *J. Amer. Chem. Soc.* **83**, 76 (1961).
129. Kerridge, D. H., and Turig, S. A., *J. Chem. Soc., A* p. 1122 (1967).
130. Woodward, L. A., *Phil. Mag.* [7] **18**, 823 (1934).
131. Davies, C. G., Dean, P. A. W., Gillespie, R. J., and Ummat, P. K., *Chem. Commun.* p. 782 (1971).
132. Corbett, J. D., *Inorg. Chem.* **1**, 700 (1962).
133. Torsi, G., and Mamantov, G., *Inorg. Nucl. Chem. Lett.* **6**, 843 (1970).
134. Torsi, G., Fung, K. W., Begun, G. M., and Mamantov, G., *Inorg. Chem.* **10**, 2285 (1971).
135. Cutforth, B. D., Davies, C. G., Dean, P. A. W., Gillespie, R. J., Ireland, P., and Ummat, P. K., *Inorg. Chem.,* **12**, 1343 (1973).

136. Ellison, R. D., Levy, H. A., and Fung, K. W., *Inorg. Chem.* **11**, 833 (1972).
137. Dorm, E., *Chem. Commun.* p. 466 (1971).
138. Grdenic, D., *J. Chem. Soc., London* p. 1312 (1956).
139. Johansson, G., *Acta Chem. Scand.* **20**, 553 (1966).
140. Dorm, E., *Acta Chem. Scand.* **21**, 2834 (1967).
141. Lindh, B., *Acta Chem. Scand.* **21**, 2743 (1967).
142. Elder, R. C., Halpern, I., and Pond, J. S., *J. Amer. Chem. Soc.* **89**, 6877 (1967).
143. Dorm, E., *Acta Chem. Scand.* **23**, 1607 (1969).
144. Cutforth, B. D., Gillespie, R. J., and Ireland, P. R. *Chem. Commun.* p. 723 (1973).
145. Gillespie, R. J., and Ummat, P. K., *Chem. Commun.* p. 1168 (1971).
146. Brown, I. D., Cutforth, B. D., Davies, C. G., Gillespie, R. J., Ireland, P., and Vekris, J. E., *Can. J. Chem.* **52**, 791, (1974).
147. Van Norman, J. D., Bookless, J. S., and Egan, J. J., *J. Phys. Chem.* **70**, 1276 (1966).
148. Krumpelt, M., Fischer, J., and Johnson, I., *J. Phys. Chem.* **72**, 506 (1968).
149. Dworkin, A. S., Bronstein, H. R., and Bredig, M. A., *J. Phys. Chem.* **70**, 2384 (1966).
150. Dworkin, A. S., Bronstein, H. R., and Bredig, M. A., *J. Phys. Chem.* **72**, 1892 (1968).
151. Eachus, R. S., and Symons, M. C. R., *J. Chem. Soc., A* p. 1329 (1970).
152. Eachus, R. S., Marov, I., and Symons, M. C. R., *Chem Commun.* p. 633 (1970).
153. Booth, R. J., Starkie, H. C., and Symons, M. C. R., *J. Chem. Soc., A* p. 3198 (1971).

USE OF RADIO-FREQUENCY PLASMA IN CHEMICAL SYNTHESIS

S. M. L. HAMBLYN*

Borax Research Centre, Chessington, Surrey, England

and

B. G. REUBEN

Chemistry Department, University of Surrey, Guildford, England

* Present address: Lonza A.G., 3930 Visp, Switzerland.

I. Introduction

Investigation of chemical reactions in thermal plasma devices operating at atmospheric pressure involves the use of the plasma as an energy source for the activation of endothermic reactions, which can also be carried out by more conventional high-temperature techniques. It is hoped that the use of plasma will result in the formation of either a cheaper product than the conventional route, or one with superior physical or chemical properties.

The first application of a radio-frequency (RF) energy source to sustain a plasma in a flowing stream of argon at atmospheric pressure by Reed (43), triggered a resurgence of interest in plasma chemistry in general. This interest increased rapidly during the 1960s, as evidenced by the large number of articles and patents which have since been published. Several general reviews of the plasma chemistry field have appeared in the last few years. McTaggart (39) reviewed the use of low-pressure microwave discharges in chemical synthesis. The use of arc plasma devices in chemical synthesis was reviewed by Landt (35), and Sayce (45) has described the use of plasma devices in extractive metallurgy. Radio-frequency plasma torches used for chemical syntheses have been mentioned in some review articles (29, 58), but no comprehensive review of chemical syntheses in RF plasmas has as yet been published.

II. Thermal Decomposition of Gases and Liquids

A. METHANE

The pyrolysis of methane in an RF plasma discharge was first investigated by Grosse et al. (22) using a helium plasma of 0.5-kW power input. They obtained high yields of carbon and hydrogen at high methane concentrations (1:1 CH_4/He), but as the concentration of methane decreased, they were able to obtain 30% conversion to acetylene (1:7 CH_4/He). By extrapolation of these results, Grosse et al. predicted that 100% conversion of methane to acetylene would be possible at a He/CH_4 ratio of 28:1. They concluded, however, that although the RF plasma was useful for acetylene formation, they obtained better results with a direct-current arc plasma.

Teresawa (49) studied the influence of operating variables, such as argon and methane flow rates and the dimensions of the RF torch, on the conversion of methane to acetylene. He observed that, under certain conditions, methane decomposed completely to 85% acetylene, when passed through the argon plasma. Teresawa also observed that the conversion to acetylene dropped and the level of reaction decreased when the volume of the reactor was doubled. This was presumably caused either by a lower reaction rate due to decreased energy density in the larger torch or by a residence time effect. Besombes-Vailhé (5) established a relationship between RF torch volume, methane concentration, and input power level in methane pyrolysis. He found that, at power levels between 2 and 5 kW, maximum conversions of methane to acetylene were obtained when a methane concentration of 27% in argon was used and when the contact time of gases in the plasma was 1.7×10^{-2} sec. The minimum power requirement for acetylene production in Besombes-Vailhé's work was given as 23.5 kWh/m^3 C$_2$H$_2$.

Amouroux and Talbot (1) studied the decomposition of methane to acetylene in an RF plasma employing a fluidized bed of refractory material to quench the reaction products. By using a constant power and constant flow rate argon plasma, they studied the effects of methane flow, fluid bed material, and excess hydrogen on the formation of acetylene. They found that the conversion of methane to acetylene decreased from 50 to 30% when the initial methane concentration was increased from 5 to 25%, and with a fluidized bed of sand at 190°C. Slightly better conversions to acetylene were obtained with graphite as fluidizing material, and Amouroux and Talbot postulated that this was due to the inhibition of carbon formation in the reaction by the partial pressure of carbon already in the graphite bed. By using the graphite fluidized bed and C/H ratios of 0.15:1, they were able to achieve up to 90% conversion of methane to acetylene at power requirements of 28 kWh/m^3 C$_2$H$_2$.

B. HIGHER HYDROCARBONS

Nishimura et al. (41) studied the pyrolysis of propane in a 15-kW argon induction plasma. They fed in propane countercurrent to the tail of the argon plasma and found that the conversion of propane decreased with increasing distance between plasma and feed point. Nishimura et al. considered this phenomenon to be due simply to an effective decrease of the reaction temperature. Maximum yields of acetylene (28%) and ethylene (5%) were obtained at a feed distance of 2.5 cm. The yields of acetylene, ethylene, and carbon were found to decrease with increasing propane flow rates, which was considered to be caused by a decreased residence time of the propane in the reaction zone.

Nishimura et al. also found that they could successfully inhibit carbon formation by adding hydrogen to the plasma argon. They were thus able to obtain a maximum yield of 40% acetylene while converting

60% of the original propane. This yield, however, was achieved at an extremely high energy consumption of 780 kWh/m³ C_2H_2.

A high surface-area carbon black (90 m²/gm) was obtained by Jordan (32) by decomposition of butene-1 in an argon induction plasma. Butene-1 was injected into the plasma at a rate of 0.007 m³/hr and after 1 hr operation 15 gm of carbon black was collected. A process for high surface-area carbon black (12) has also been developed in which a carbon feed is vaporized in an RF plasma and the vapor subsequently quenched in a halogen-rich environment.

C. Halogen Compounds

Kana'an and Margrave (33) investigated the decomposition of carbon tetrachloride in an RF plasma and reported the formation of low yields of the benzene derivative, C_6Cl_6. Using spectroscopic techniques, Kana'an and Margrave (34) also identified free radicals such as CCl, CF, and CF_2 in a chlorofluorocarbon plasma. They postulated that the C_6Cl_6 present in the CCl_4 decomposition products was formed by polymerization of reactive species such as CCl_3, CCl_2, and CCl.

Bequin et al. (4) extended this work and investigated the decomposition of CCl_4, CCl_3F, CCl_2F_2, and CF_4 in a 10-kW argon RF plasma. They also obtained benzene derivatives such as C_6Cl_6, C_6Cl_5F, $C_6Cl_4F_2$, and $C_6Cl_2F_4$ in low yields, together with a variety of chlorofluoroethane and ethylene derivatives. They reported, however, that the major decomposition products were chlorine, fluorine, and carbon.

Bequin et al. also identified the molecule CCl_3—$CClF_2$ as one of the decomposition products of $CClF_3$, and postulated that as there was no CCl_3 present in the reactant, the CCl_3—$CClF_2$ must have been formed by recombination of CCl_3, already formed by recombination of simpler radicals. They suggested the following mechanism for CCl_3 formation:

$$C + Cl \longrightarrow CCl \tag{1}$$
$$CCl + Cl \longrightarrow CCl_2 \tag{2}$$
$$CCl_2 + Cl \longrightarrow CCl_3 \tag{3}$$

The radical $CClF_2$ could have been formed by thermal decomposition of $CClF_3$ and further reaction could then occur to the ethane derivative as follows:

$$CCl_3 + CClF_2 \longrightarrow CCl_3—CClF_2 \tag{4}$$

Bequin et al. stated that all identified products could be accounted for by such radical recombination reactions, but produced no other evidence for such mechanisms. They also postulated that, because no acetylene derivatives were identified as products, it was probable that the benzene derivatives were formed by trimerization of acetylene compounds such as C_2Cl_2, C_2F_2, and C_2FCl. It is worth noting that the occurrence of such compounds would have represented stronger evidence.

Work carried out on the behavior of boron trichloride in an RF plasma (23) showed that, although it dissociates almost completely when it is passed into the plasma, after it leaves the plasma, complete recombination and charge neutralization of the dissociation products occurs to give BCl_3 again. No dissociation products could be found.

Spectrographic examination of the emission from an argon–BCl_3 plasma showed that the main decomposition products of BCl_3 that emit radiation are BCl, BCl_2, B, B^+, Cl_2, and Cl_2^+. The predominant emitting species was found to be BCl, which is considered to have been formed by the reaction,

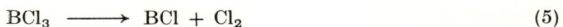

$$BCl_3 \longrightarrow BCl + Cl_2 \tag{5}$$

III. Reduction Reactions

A. ALUMINUM OXIDE

Rains and Kadlec (42) injected Al_2O_3 particles into an RF argon plasma and measured the conversions to aluminum under neutral conditions and when H_2, CH_4, and CO were added to the argon. They found that conversions to aluminum were dependent on parameters such as particle size of Al_2O_3, power input, type of reducing gas, and the mass flow rate through the plasma. For argon plasmas at 5 kW, a maximum conversion of 30% Al_2O_3 to aluminum was obtained at a feed rate of 6 gm Al_2O_3/min. As the feed rate was increased to 36 gm/min, the conversion to aluminum dropped rapidly to 2%. Similar conversions and behavior were observed when hydrogen was added to the plasma. However, when carbon monoxide or methane was used as reducing agent, conversions to aluminum of 2 and 4 times that obtained with pure argon or argon–hydrogen mixtures were reported. Rains and Kadlec postulated that this was caused by the presence of carbon vapor in the discharge, which would have a higher affinity for oxygen than for hydrogen.

B. TITANIUM TETRACHLORIDE

Miller and Ayen (40) investigated the hydrogen reduction of titanium tetrachloride in an RF plasma torch. They obtained conversions of $TiCl_4$ to $TiCl_3$ of between 60 and 80%, at feed rates of 1.5

gm/min of $TiCl_4$. They observed that the conversions were independent of $TiCl_4$ feed rate and of power input up to 5 kW, as well as of the $H_2/TiCl_4$ ratio when it was greater than stoichiometric. Miller and Ayen injected the $TiCl_4$ vapor into an argon plasma but found that the plasma could only tolerate up to 1 mole % $TiCl_4$ in the argon at 10 kW, due to low coupling efficiency. They, therefore, measured the coupling efficiency of their plasma induction circuit and found that, although an efficiency of between 60 and 70% was obtained for 2 mole % hydrogen in argon, changeover to 0.15 mole % $TiCl_4$ in argon reduced the efficiency to less than 40%. They did not report any attempt to increase the coupling efficiency to $TiCl_4$ by better impedance matching of the RF generator. Miller and Ayen also investigated various quenching devices for argon–$TiCl_4$ plasmas in an attempt to prepare lower titanium chlorides without hydrogen. No significant solid products were obtained under any of the quench conditions, presumably because of extremely rapid recombination and charge neutralization of dissociated species in the plasma tail flame.

C. Silicon Tetrachloride

Vurzel and Polak (55) carried out extensive kinetic studies of the reduction of silicon tetrachloride to silicon in plasma devices. They first decomposed $SiCl_4$ to $SiCl_3$ in an adiabatic compression–expansion device and then completed the reduction in an RF plasma. They claimed that the decomposition of $SiCl_4$ to silicon proceeded by a two-stage mechanism of chlorine atom removal:

$$\text{stage 1.} \quad SiCl_4 \longrightarrow SiCl_3 \longrightarrow SiCl_2 \quad \text{(fast)} \tag{6}$$
$$\text{stage 2.} \quad SiCl_2 \longrightarrow SiCl \longrightarrow Si \quad \text{(slow)} \tag{7}$$

They postulated that the second stage was rate-controlling and determined the rate constants for both stages:

$$k_1 = 5 \times 10^8 \exp\left(-88{,}000 \pm 5{,}000/RT\right) \text{ sec}^{-1}$$
$$k_2 = 5 \times 10^7 \exp\left(-126{,}000 \pm 10{,}000/RT\right) \text{ sec}^{-1}$$

From additional calculations of the thermodynamics of the $SiCl_4$ decomposition, Vurzel and Polak predicted that the decomposition would be favored by reducing conditions and confirmed this experimentally, obtaining high-purity silicon by use of an RF plasma torch.

D. Boron Trichloride

The hydrogen reduction of boron trichloride to boron in an RF plasma has been investigated by Hamblyn et al. (25). By use of a special design of RF reactor system (24), it was found possible to produce up to 0.25 kg/hr of 99% boron at 30 kW. The conversion of BCl_3 to boron was found to be dependent on parameters such as BCl_3 concentration in the reaction mixture, residence time of the mixture in the reactor, and the configuration of the reactant feeds to the reactor.

The boron so formed resembled other amorphous borons in appearance and was found to be similar in structure to β-rhombohedral boron. Individual particles were random in shape, consisting mainly of platelets between 200 and 7500 Å in diameter. The presence of regular dodecagonal platelets of boron was also observed, in the product, which had typical diameters of 7000 Å and a unit cell lattice constant of 30 Å. This unit cell dimension is considerably larger than any previously reported form of boron and was considered to be a further modification of known boron structures.

The kinetics of the reduction of BCl_3 with hydrogen were further studied by Hamblyn (23), and the reaction was found to be approximately first order with respect to BCl_3 and very low order with respect to hydrogen. The low order for hydrogen suggested that its role in the reaction is merely as a scavenger for chlorine, produced by the dissociation of BCl_3 in the plasma. Experimental results were interpreted partly by a thermal mechanism involving species such as BCl, Cl, and BH, and partly in terms of an ionic mechanism.

IV. Oxidation Reactions

A. Mixed Oxides

Barry et al. (3) injected mixtures of chromium and aluminum chloride vapors and titanium and chromium chloride vapors into RF argon plasmas, and they obtained high yields of the respective oxide mixtures.

McPherson (38), using the same technique, prepared a series of 50–500 Å size range, single and mixed oxides. Spherical, metastable Al_2O_3 particles were formed by condensation of molten Al_2O_3 droplets in the tail of an oxygen plasma. The Al_2O_3 droplets were considered to freeze and undergo nucleation to the γ-Al_2O_3 form, which on further cooling transformed to stable δ-Al_2O_3.

By cocondensation of Al_2O_3–Cr_2O_3 mixtures and Al_2O_3–TiO_2 mixtures, McPherson obtained metastable solid solutions of Cr_2O_3 in θ-Al_2O_3 and an Al_2O_3–TiO_3 solid solution in which δ-Al_2O_3 and rutile were present. A third phase was also observed to be present in the latter product, which was tentatively identified as a metastable Al_2O_3–TiO_2 compound.

B. Titanium Dioxide

The preparation of a pigmentary TiO_2 by oxidation of $TiCl_4$ in chlorine oxygen and argon RF plasmas is described in two British patents (7, 8). In the former, the use of chlorine as plasma gas enabled a $TiCl_4$–O_2 mixture to be fed into the tail, producing approximately 2.6 kg/hr of pigmentary TiO_2. Reduced throughputs of $TiCl_4$–O_2 mixture were necessary when an oxygen plasma was operated because recirculation of $TiCl_4$ into the oxygen plasma, where it was oxidized, caused extinction of the plasma. This did not occur in the chlorine plasma. The second patent (8) describes a similar process for the preparation of pigmentary TiO_2 in an RF plasma. In this process, higher $TiCl_4$ throughputs were achieved by use of chlorine as a plasma-stabilizing additive in the $TiCl_4$ itself. With an argon plasma and a 5% chlorine in $TiCl_4$ mixture, up to 7 kg/hr of pigmentary TiO_2 could be prepared when quenching was by oxygen or air.

C. Silicon Dioxide

Audsley and Bayliss (2) injected $SiCl_4$ vapor into the tail of an RF oxygen plasma and obtained 99% oxidation of the chloride to SiO_2. Injection of the $SiCl_4$ directly into the oxygen plasma caused plasma extinction in each of four different torches tried, presumably due to the same phenomenon as in the oxidation of $TiCl_4$. However, from experiments with tail feeding of $SiCl_4$, Audsley and Bayliss concluded that the oxidation rate was only limited by kinetics, and they successfully oxidized up to 4 kg/hr of $SiCl_4$ to SiO_2.

A process for the preparation of high surface-area SiO_2 by use of an RF plasma is described in a recent patent (13). By the feeding of 200-mesh sand particles into a mixed argon–oxygen plasma, up to 0.25 kg/hr of 260 m^2/gm silica could be prepared when the SiO_2 vapor was quenched with air. When hydrogen was used as quench gas instead of air, an activated silica with hydrophilic properties was prepared. The requirements for the formation of a hydrophilic active silica were that free hydrogen should be present in the quench region during the condensation of the SiO_2 particles. This condition could be achieved by the inclusion of hydrogen or hydrogen-containing compounds with the plasma gas, so that dissociation to free hydrogen could occur. When the SiO_2 condensation was carried out in the presence of a hydrogen compound with a hydrophobic group (such as a straight-chain alcohol or a chlorosilane) and the compound was injected into the quench region at a low enough temperature, decomposition of the hydrogen compound did not occur and the activated silica had hydrophobic properties.

Bush and Sterling (15) have extended the foregoing work on activated silica. They developed a rotating RF plasma furnace into which were fed mixtures of sand and carbon in the form of paste or rods. Operating with an argon plasma and speeds of furnace rotation of up to 1000 rpm, they were able to maintain a semiplastic ring of molten silica around the plasma, held in place by centrifugal force. As the reaction of SiO_2 with carbon occurred,

$$SiO_2 + C \longrightarrow SiO + CO \tag{8}$$

the melt could be replenished by new material. The SiO was oxidized at the exit of the furnace to give activated silica of high surface area. When operating with an air plasma, Bush and Sterling found it necessary to feed the sand and carbon to the melt in the form of rods, formed by mixing sand with an oil and using a cellulose thickener. This was necessary in order to avoid rapid oxidation of the carbon before reaction with the molten sand could occur. By appropriate control of the composition and feed position of the quench gas, either hydrophilic or hydrophobic activated silica could be prepared.

D. OTHER OXIDES

Chase and Potter (18) have carried out extensive investigations of the preparation of high surface area oxides, by oxidation of metal and metal oxide powders. Table I summarizes the experimental conditions and results obtained in this work. The powders were fed to an RF argon plasma in a stream of argon or oxygen, were vaporized in the plasma, and the condensing species oxidized in the plasma tail flame. The condensing vapors were also rapidly quenched in the tail flame to prevent extensive particle growth and to achieve high surface areas. The surface area of Sb_2O_3 powder so prepared was found to be greatly influenced by the dilution of the condensing vapors in the gas mixture. For example, a decrease in particle density from 1.1×10^{-4} to 0.2×10^{-4} gm Sb_2O_3/liter gas decreased the average particle diameter of the final product from 820 to 150 Å. The antimony oxide was found to have excellent fire-retarding properties in acrylic fibers and not to decrease the brightness of the fiber to the same extent as normal commercial Sb_2O_3. Zinc oxide so produced was claimed to have an ultraviolet absorption efficiency equivalent to normal commercial UV absorbers when tested in polyurethane film. Films using this zinc oxide also proved more fade-resistant than commercial absorbers when tested in accelerated radiation conditions over 40 hr.

TABLE I
PREPARATION OF SUBMICRON OXIDES[a]

Feed powder	Feed rate	Argon plasma (kW)	Quench gas	Quench rate (liter/min)	Conversion (%)	Product	Particle size (Å)	Surface area (m²/gm)
Zinc dust	0.078 gm/min	1.35	O_2	280	100	ZnO white	268	40
Antimony (<44 μm)	0.58 gm/min	1.35	O_2	280	90	Sb_2O_3 white	340	—
Antimony 0.15–0.2 μm	0.43 gm/min	1.35	Air	292	100	Sb_2O_3 white	142	—
WO_3 0.2–0.4 μm	4 gm/hr	4	Air	425	68	WO_3 light yellow	315	26.6
MoO_3 5–20 μm	2.7 gm/hr	3	Air	509	100	MoO_3 blue	—	26.5

[a] Data taken from Chase and Potter (18).

V. Preparation of Refractory Compounds

A. CARBIDES

A process for the formation of subpigmentary (< 0.5 μm) silicon carbide has been described in a recent British patent (9). Silica powder (44–150 μm) was injected into an argon–methane plasma maintained by a 10-kW RF induction torch. Feed rates of up to 1.5 gm/min silica were used, which yielded a high conversion of micron-sized silicon carbide. An interesting feature of the plasma torch used in this work was the porous plasma-containing wall through which hydrogen was passed, said to prevent carbon building up and to increase plasma enthalpy. Also, the RF coil used had a reverse turn at the lower end to effect magnetic confinement of the plasma tail.

Salinger (44) reported the successful conversion of methyltrichlorosilanes to silicon carbide in a 50-kW RF plasma torch. The liquid methyltrichlorosilanes were fed to the tail flame of various plasmas and the solid products were recovered in an acid-resistant bag filter. Up to 85% recovery of theoretical solid product was reported, which was subsequently heated at 500°C to remove elemental carbon. Under the best condition (20–25% vol. hydrogen in argon plasma at 36 kW), up to 70% conversion to β-SiC was obtained with ca. 10% conversion to amorphous SiC. Salinger suggested that the good crystallinity of the β-SiC so obtained meant that the reaction occurred in a gas temperature range in which β-SiC was the stable crystalline form (i.e., < 2300°C).

MacKinnon and Wickens (37) have prepared boron carbide in the form of submicron particles by injecting BCl_3, H_2, and CH_4 into the tail of a 20-kW argon RF plasma. A 3^3 factorial experiment to study the effect of operating variables was carried out; a preliminary analysis of which gave the following indications:

1. The maximum percentage boron conversion to B_4C (93%) was obtained at a low (20 liters/min) BCl_3 flow rate, at a relatively high (8 : 1) H_2/BCl_3 ratio, and at a stoichiometric (1 : 4) CH_4/BCl_3 ratio.

2. The complete set of results obtained indicated the presence of interactions among the effects of BCl_3 flow rate, H_2/BCl_3 ratio, and CH_4/BCl_3 ratio on the conversions obtained.

Hartl (26) treated vanadium, chromium, and titanium nitrides in argon–5% acetylene RF plasmas. The nitrides were dropped as powders into a vertical plasma torch in which the gas stream was flowing upwards. Product was collected both as wall deposits and as loose powder at the lower end of the torch (i.e., the plasma gas inlet). In the case of vanadium, the products collected were identified by X-ray analysis to be a mixture of vanadium nitride and carbide (VC–VN) with

TABLE II

PREPARATION OF Si_3N_4 AND TiN[a]

RF plasma gas	Flow rate to plasma	NH_3 feed rate	Power (kW)	Conversion to product	Surface area (particle size)
1. $SiCl_4$	0.5 mole/min	0.65 mole/min	36	65% Si_3N_4	30.5 m^2/gm
2. Ar + $SiCl_4$	25, 0.47 liters/min	0.69 mole/min	26	67% Si_3N_4	47.5 m^2/gm
3. Argon	80 liters/min	0.5 liter/min	26.5	70% TiN	0.05–0.4 μm
probe $TiCl_4$	0.25 mole/min	—	—	—	(spherical particles)

[a] Data from British Titan Products (*11*).

the presence of small amounts of the carbonitride $VC_{0.4}N_{0.4}$. Chromium nitride (CrN) was converted to the chromium carbide (Cr_3C_2), and titanium nitride to titanium carbide (TiC).

B. Nitrides

Nitridations of silicon, magnesium, and aluminum powder in a 2-kW, 37-MHz argon–nitrogen plasma have been attempted by Fletcher et al. (20). Virtually no magnesium or aluminum nitrides were detected in the products. The authors suggested that products had been formed but had been rapidly destroyed by atmospheric hydrolysis. However, 7% nitridation of silicon metal was obtained with 14% nitrogen in argon plasma. Another RF plasma process for the nitridation of metals has also been patented (11). Silicon nitride so produced was described as having an opacity and brightness similar to aluminum silicate when evaluated as a filler and extender in paper. Table II summarizes the reported results from this work. The titanium nitride powder was reported to be in the form of 0.05–0.4 μm spherical particles. This could be possibly due to the use of a relatively low quenching rate, allowing the formation of a liquid TiN phase which, on cooling, formed spheroids.

Hartl (26) reported attempted nitridations of various metal oxides by means of countercurrent feeding of the oxide powders into an RF nitrogen plasma. Although nitrides were identified by X-ray analysis, their separation was not possible. Table III summarizes the resulting oxynitrides obtained for vanadium, titanium, and chromium.

TABLE III

NITRIDATION OF METAL OXIDES[a]

Feed material	Powder product	Torch wall deposit
V_2O_5	$VO_{1.4}N_{0.2}$; $VO_{1.2}N_{0.1}$	$VN_{0.7}O$; $VN_{0.8}O_{0.1}$
TiO_2	$TiO_{1.8}N_{0.1}$; $TiO_{1.6}N_{0.1}$	$TiO_{1.5}N_{0.3}$; $TiN_{0.8}O_{0.1}$
Cr_2O_3	$CrO_{1.4}$; $CrO_{1.5}$	$CrO_{1.2}N_{0.2}$; $CrN_{0.7}O_{0.1}$

[a] Data from Hartl (26).

C. Borides

Triché et al. (54) reported the successful preparation of titanium diboride in an RF plasma reactor. They exposed compressed pellets of various combinations of reactants to the tail flame of a 12-kW, 6.3-MHz argon plasma. They carried out the following reactions, in each of which titanium boride was successfully prepared:

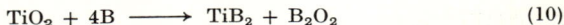

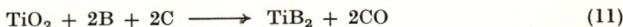

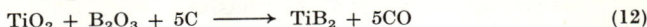

$$Ti + 2B \longrightarrow TiB_2 \tag{9}$$

$$TiO_2 + 4B \longrightarrow TiB_2 + B_2O_2 \tag{10}$$

$$TiO_2 + 2B + 2C \longrightarrow TiB_2 + 2CO \tag{11}$$

$$TiO_2 + B_2O_3 + 5C \longrightarrow TiB_2 + 5CO \tag{12}$$

The progress of each reaction was followed by spectrographic measurement of the emission of the Ti^+ ion in the plasma tail flame. Triché et al. established that in the reactions involving TiO_2 as reactant, a maximum in titanium ion emission was reached within 1 min, followed by a rapid decrease to a minimum emission between 2 and 3 min. Thereafter, the emission rose again to an equilibrium value. They postulated that the initial maximum in Ti^+ emission corresponded to TiO_2 evaporating from the pellet before reaction commenced, and the minimum in emission to the completion of the reaction. The subsequent increase in Ti^+ emission was found to correspond to Ti^+ emission from the tail flame when a TiB_2 pellet was heated. Times for complete reaction were found to be of the order of 2 min in each case. The progress of one of the reactions, Eq. (10), was followed in more detail by quantitative X-ray analysis of a series of pellets exposed to the plasma for increasing time intervals of up to 3 min. Conversion data obtained correspond approximately to first-order kinetics.

Foex (21) reports the successful formation of borides in a rotating batch melting furnace, fired by an RF induction plasma. An amount of 0.8 kg from a stoichiometric mixture of ZrO_2 and boron, previously compressed in the rotating vessel to a pressure of 40 kg/cm², is treated by the action of an argon plasma (1 m³/hr) for 0.5 hr. Zirconium diboride is formed by the reaction,

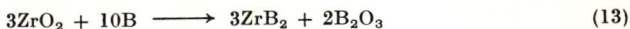

$$3ZrO_2 + 10B \longrightarrow 3ZrB_2 + 2B_2O_3 \tag{13}$$

which proceeds to about 50% conversion. During the reaction considerable evolution of B_2O_3 occurs.

VI. Nitrogen Fixation

Timmins and Ammann (53) reviewed the application of plasma discharge devices in general to the fixation of atmospheric nitrogen. The bulk of investigations on the preparation of compounds such as nitric oxide, hydrogen cyanide, cyanogen, and hydrazine in plasmas has been carried out in DC devices, either arc torches or glow discharges.

There have been several investigations of nitrogen fixation reactions in RF plasma devices, however, and some patents have been granted protecting processes developed from these investigations.

A. NITRIC OXIDE

Table IV summarizes work carried out on nitric oxide preparation in RF plasmas. Two important investigations using DC plasmas have also been included for comparison.

Bequin *et al.* (*4*) passed nitrogen and oxygen through a 5–7 MHz induced plasma torch and collected nitric oxide, together with trace oxygen and argon, in a series of liquid nitrogen traps. At power inputs to the plasma of 9–10 kW, conversions of nitrogen to nitric oxide of up to 2.12% were obtained. This can be seen from Table IV to be similar to the results obtained by Grosse *et al.* (*22*) when they quenched a stream of plasma nitrogen from a DC plasma jet with an oxygen stream.

Stokes *et al.* (*47*) reported similar conversions of oxygen in air to nitrogen oxides in a low-pressure RF plasma. After a run time of 2.5 hr, 0.4 ml of material was collected in liquid nitrogen traps, which on analysis showed a 2% conversion to nitric oxide. LaRoche (*36*) had previously demonstrated the beneficial effect of quenching the reaction products. In experiments with a low-pressure RF discharge, he obtained an increase in conversion to nitric oxide from 2 to 4% by a rapid quenching technique.

Timmins and Ammann (*53*) report much higher conversions of oxygen to nitric oxide by use of a DC plasma device, "the constricted arc," in which the anode and cathode are separated by a series of water-cooled segments. A considerably larger plasma zone can be stabilized with this device than with the conventional plasma jet. By operation of this device in air at gas enthalpies greater than 270 cal/gm and with careful control of quench conditions, Timmins and Ammanns were able to obtain up to 7% nitric oxide in the exit gas. This amounts to 30% conversion of oxygen to nitric oxide. Still higher conversions were claimed by Jackson and Bloom (*30*) using a capacatively coupled RF plasma, in which conversions of oxygen to nitric oxide of up to 90% were possible.

It would appear from these series of investigations of nitric oxide preparation in plasma devices that, although reasonable conversions of oxygen to nitric oxide can be obtained under conditions of excess nitrogen, the final concentrations and production rates are low. The best conversions quoted in the foregoing are equivalent to approximately 200 kWh/lb nitric oxide produced, which is extremely high. The use of

TABLE IV

SUMMARY OF NITRIC OXIDE WORK

Plasma device	Reactants	Feed rate (liter/m)		Power (kW)	Percent conversion of $O_2 \rightarrow NO$	NO in product (%)	Reference
		N_2	O_2				
5–7 MHz RF (atm. pressure)	$N_2 + O_2$	17.7	14.3	10	2.12		Bequin et al. (4)
DC plasma jet	$N_2 + O_2$	5.5	5.5	15	2.03		Grosse et al. (22)
27-MHz low-press. RF plasma	Air	0.1		0.3	2.0		Stokes et al. (47)
RF plasma: With quench	Air (low pressure)				20.0	4	LaRoche (36)
Without quench					10.0	2	
RF plasma (1–2 atm)	Air	0.5			90.0	17	Jackson and Bloom (30)
		10.0			10.0	2	
DC constricted arc	Air	0.43×10^{-4} (lb/sec)		5.2	30	7	Timmins and Ammann (53)

higher-power level plasma devices and high surface area plasmas combined with controlled quenching would appear to offer the most promise for commercial processes.

B. HYDROGEN CYANIDE AND CYANOGEN

There have been two recent investigations of hydrogen cyanide preparation in RF plasma torches. In one investigation carried out at atmospheric pressure in a 1.5-kW nitrogen plasma by Stokes et al. (47), conversions of injected methane to HCN of up to 35% were obtained with additional formation of acetylene. The second investigation by Bronfin (14) was carried out mainly at pressures below 460 torr. Injection of N_2 and CH_4 mixtures into a 3.5-kW argon plasma produced conversions of N_2 to HCN of up to 70% at CH_4/N_2 ratios greater than 2:1. These results were reported to be in good agreement with thermodynamic equilibrium conversions. The preparation of HCN from CH_4 and N_2 in a capacitively coupled RF plasma was also carried out by Jackson and Bloom (30). In the exit gases, 0.22 mole% HCN was obtained when CH_4 was added to the tail of a nitrogen plasma. The same workers also claim the successful preparation of hydrazine in this RF plasma torch at atmospheric pressure. Ammonia and nitrogen injected into the torch in various configurations and flow rates produced a maximum yield of 0.02 mole% N_2H_4 in the product gas stream. This yield is slightly better in terms of the mass of hydrazine produced than other work involving low-pressure glow discharges (50). A cheap route to hydrazine would be of great importance in the development of economically viable fuel cells.

VII. Heat Treatment of Solids

By injection of refractory powders into RF discharges, it is possible to heat treat the particles. Decomposition, vaporization, melting, or surface modification of particles can be achieved in varying degrees depending on the physical properties of the material, particle residence time, particle surface area, and the enthalpy available in the plasma.

A. CRYSTAL GROWING

The first reported heat treatment of a solid in an RF plasma was carried out by Reed (43). He injected powders at high velocity through an atmospheric pressure argon plasma and deposited molten powder on a crystal boule in the plasma tail flame. A sapphire polycrystal, some

30 mm long and 10 mm diameter, was grown from sapphire powder by this technique. Zirconium oxide and niobium crystals were also grown in 50% oxygen–argon and 15% helium–argon plasmas, respectively. Reed concluded that the correct particle size range of the powder was crucial for successful crystal growth. Powders that were too fine vaporized and those that were too coarse led to surface bubbling effects on the crystal. Since this early work, many other applications of this technique of crystal growing have been studied and are well reported in other review articles. The most recent application of the technique is probably work carried out by Sienko and Young (46) on the growth of scandium oxide crystals. Scandium sesquioxide (Sc_2O_3) powder was passed through the axis of a 10-kW RF argon plasma. Crystals of Sc_2O_3 were grown on the end of a 5-mm diameter magnesia rod, supported in the plasma tail flame by a boron nitride rod. Crystal boules of up to 12 mm diameter were grown, the lower portions of which were sometimes covered with dendritic and whisker growth extensions. Chemical analysis of the crystals by reoxidation indicated an oxygen deficiency of up to 4% in the crystal. The formation of a substoichiometric scandium oxide, Sc_2O_{3-x} (where $x = 0.008$ and 0.115), was explained by the loss of oxygen from the crystal surface according to the reaction,

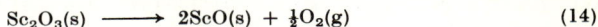

$$Sc_2O_3(s) \longrightarrow 2ScO(s) + \tfrac{1}{2}O_2(g) \qquad (14)$$

B. Spheroidization

An extension of the use of RF plasma for particle heating is the spheroidization of solids. By careful control of plasma enthalpy, particle size, feed rate, and feed position, it is possible to melt each particle as it passes through the plasma. The liquid droplet forms a sphere due to surface tension and, on cooling, retains its spherical shape. Spheroidized particles are commercially useful because they will flow easily.

Hedger and Hall (27) were first to study the spheroidization of metal and metal oxide powders in an RF plasma. They injected 100–150 μm size range powders into an argon plasma at rates up to 5 gm/min. Yields of between 50 and 70% spheroids were obtained for Cr, Mo, Ta, W, and Al_2O_3 and for several uranium compounds, the remaining material being random-shaped lumps. Magnesia powder was also successfully spheroidized in a 20% oxygen–argon plasma. No obvious correlation was observed between the yield of spheroids and the melting points of the parent materials. Hedger and Hall postulated that this lack of correlation was due to ejection of particles from the plasma which they had observed photographically. The ejection could have been caused by magnetic or viscous drag effects and could result in a percentage of particles failing to penetrate the plasma, the proportion involved varying unpredictably with the material. Plutonium dioxide, alone and mixed with other oxides (such as UO_2, ThO_2, and ZrO_2), has also been successfully spheroidized in an RF plasma by Jones et al. (31).

Waldie (56, 57), in an attempt to prepare ultrafine powders from coarser materials, obtained spheroids of oxide powders in low-power RF torches. When silica powder (50–72 μm) was injected into a 2.5-kW, 34-MHz argon plasma at 15 gm/hr, a 15% conversion to ultrafine particles (0.015–0.15 μm) and coarse spheroids were obtained. Ultrafine powders of barium oxide (50% < 0.1 μm) and alumina spheroids were also prepared by this technique. When alumina was injected cocurrently into a 3.5-kW, 10-MHz argon plasma (57), 48% spheroidization of a 180–250 μm powder was obtained at a feed rate of 36 gm/hr. Waldie obtained better results by use of countercurrent particle flow similar to the technique used by Hartl (26). Up to 26% spheroidization of a 300–500 μm powder was measured for an alumina feed rate of up to 140 gm/hr. It is evident from this work that countercurrent spheroidization can achieve not only higher yields of spheroids but also spheroidization of a larger size range of solid.

Boron powder has been spheroidized by Sullenger et al. (48) using an induced argon plasma. When 50–100 μm size range, β-rhombohedral, boron particles were injected into the plasma the main bulk of particles was spheroidized and had improved β-rhombohedral crystallinity. Negligible size reduction of the powder occurred, and microscopic examination of the powder revealed that no true spheres were in fact present, all the spheroids having small flattened faces. The presence of additional well-faceted crystals was also observed. These crystals were classified into four main types: (a) slightly elongated octohedra, (b) hexagonal and dodecagonal right prisms, (c) square right prisms, and (d) truncated tetrahedra. Crystal habits a, c, and d were attributed by Sullenger et al. to monocrystals of two known and one unknown form of boron. Type b was thought to be a polycrystalline growth of an unreported form of boron, consisting of hexagonal platelets stacked in an imprecise pattern. The unit cell lattice of 10 Å quoted by Sullenger et al. for this dodecagonal boron was 3 times as small as the lattice parameter of dodecagonal boron platelets formed in the hydrogen reduction of boron trichloride in an RF plasma (23).

C. DECOMPOSITION

Warren and Shimizu (59) injected single oxides and oxide concentrates into a 2-MHz, 7-kW argon plasma. Alumina, silica, and niobium

108   S. M. L. HAMBLYN AND B. G. REUBEN

oxide powders were nearly completely spheroidized. On a second pass
through the plasma, the spheroids scarcely changed size or lost weight
by vaporization. Mixed oxide concentrates, such as columbite (Fe,Mn)
$(Cb,Ta)_2O_5$, pyrochlore $CaNb_2O_8 \cdot Ca(Ti,Th)O_3$, scheelite $CaWO_4$, and
wolframite $(Fe,Mn)WO_4$, were injected into the argon plasma, resulting
invariably in plasma extinction even at very low feed rates. Warren
and Shimizu also reported that even low concentrations of diatomic
gases, such as N_2 and Cl_2, extinguished the plasma. These findings
indicate a very low coupling efficiency of the RF plasma coil and a high
degree of mismatching of the impedance in the RF circuit and would,
undoubtedly, explain the low tolerance of the plasma to solids.

Huska and Clump (28) investigated the decomposition of molybdenum disulfide in an RF plasma torch. Passage of 0.7 gm/hr of 74-μm
MoS_2 through a 5-kW argon plasma resulted in 70% conversion of
molybdenite to molybdenum. The conversion decreased rapidly, however, when feed rates were increased, and only 30% conversion to
molybdenum metal was obtained at 2.5 gm/hr. At feed rates greater
than 3 gm/hr MoS_2, the plasma became radially unstable, probably
caused by asymmetric powder injection. Charles et al. (17) have recently
reported the injection of up to 400 gm/hr of a 50-μm powder into an
argon plasma without it becoming unstable. Conversions of molybdenite
to molybdenum metal of between 60 and 70% were obtained but
details of particle size and feed rates were not given.

Several studies have been made of the behavior of oxide powders
injected into neutral or reducing RF plasmas. Borgianni et al. (6)
injected Al_2O_3, CuO, NiO, and TiO_2 powders along the axis of a 4-MHz
argon plasma run at atmospheric pressure. They measured the extent
of decomposition of the powders to lower oxide and metal as a function
of the axial distance traveled by different sizes of particles and varying
power inputs. For 60-μm Al_2O_3 particles, a maximum conversion to
aluminum of 12% was obtained at 20 cm from the injection point. The
plasma had a low tolerance for Al_2O_3 powder, however, and the
maximum usable mass flow was 72 gm/hr. The results for CuO, NiO,
and TiO_2 are summarized below in Table V.

Borgianni et al. assumed that the rate-determining step in the
decompositions was heat transfer across the particle boundary layer.
They developed an equation of motion for the particle in the plasma
with which they were able to explain their experimental results satisfactorily.

Capitelli et al. (16) investigated the decomposition of 60-μm Al_2O_3
in argon–N_2 plasmas further, using the same experimental and analytical techniques as did Borgianni et al. They observed a decrease in yield

TABLE V

DECOMPOSITION OF OXIDES IN RADIO-FREQUENCY ARGON PLASMAS[a]

Oxide	Feed particle size (μm)	Power (kW)	Maximum conversion to metal (%)	Optimum distance from injection point (cm)	Maximum mass flow (gm/hr)
CuO	60	5.5	60	13	28.8
NiO	60	5.5	30	20	—
TiO_2	60	5.5	20 (Based on O_2 deficiency)	17.5	—

[a] Data from Borgianni et al. (6).

of aluminum when nitrogen was added to the plasma. For pure argon, 12% of aluminum was formed 20 cm from the injection point. With 4 and 8% nitrogen–argon plasmas, the maximum conversion to Al metal decreased to 8%, which was attributed to shrinkage of the plasma and shortening of the plasma tail. The enthalpy available in the plasma would also be reduced due to the dissociation energy of nitrogen.

Charles and co-workers (17) obtained more encouraging results with decomposition work in an investigation of extractive metallurgical processing of various minerals in RF plasmas. Both rhodonite ($MnSiO_3$) and ilmenite ($FeTiO_3$) were successfully treated in a 37-mm diameter RF torch at feed rates up to 60 gm/hr. Maximum yields of 30 gm/hr of available MnO in the recovered solids were obtained from a 50-μm rhodonite feed at 60 gm/hr feed rate and 15 kW power level.

Manganese oxide-rich solids were found to be present in the cooler regions of the apparatus, whereas SiO-rich solids collected predominantly in the torch itself. A similar quench separation into component-rich products was obtained for ilmenite, where TiO_2-rich solids collected in the torch and FeO-rich solids in the quench tube. These quench separations were successfully correlated with the free energy of formation of the oxides. Charles et al. also processed zircon sand and measured negligible dissociation which they attributed to the high melting point of this mineral. Ferrous sulfide, however, was completely oxidized to Fe_2O_3 at feed rates of up to 400 gm/hr.

Two methods have recently been patented for the use of RF plasma torches for processing titaniferous and other minerals. Titanium dioxide can be recovered from rutile (9) by passage of the latter at low feed rates (0.1 gm/min) through an argon–oxygen plasma. After 3 hr of operation, titania particles of 0.18–0.3 μm size range could be recovered, the purest samples containing only 0.3% w/w iron impurity as Fe_2O_3. In a further development of the preceding technique (10), rutile and other minerals were processed at higher feed rates (10 gm/min for rutile) by injecting them into the tail of oxygen or chlorine plasmas.

VIII. Commercial Application of Radio-Frequency Plasmas

In contrast to the study of chemical syntheses in low-pressure, microwave, plasma discharges, the studies of chemical syntheses in atmospheric pressure, thermal plasmas have not given rise to any chemical compound that cannot be prepared by other techniques. The use of thermal plasma discharges does, however, offer a unique source of energized gas available at higher temperatures than normal chemical flames or other, indirect, electric heating techniques.

The philosophy behind the many investigations of chemical syntheses in thermal plasma devices must surely be that the availability of usable gas enthalpy at temperatures up to 10,000°C should permit the shortening of conventional production routes to the bulk chemicals that are prepared by endothermic reactions. This concept is pinpointed, for example, by the production of high surface-area activated silica. The present industrial process, which has been in use for the last 20 years, involves a multistage production route beginning with sand. The sand is first reduced to silicon or silicon carbide and is then chlorinated to silicon tetrachloride. The chloride is then oxidized in an oxyhydrogen flame and the oxide quenched to produce activated silica. Various workers (13–15) have shown that it is feasible to short-circuit this complicated route by evaporating or reducing sand to silicon monoxide in a plasma device and then quenching the monoxide to yield a surface-active silica. A similar chemical reaction in which the plasma could provide a shortened process is the direct formation of pigmentary titanium dioxide from rutile, bypassing the titanium tetrachloride step. This use of plasma devices in chemical synthesis is undoubtedly the most promising commercially, and it is to be expected that future development will be concentrated along these lines.

Radio-frequency plasma, in particular, offers additional advantages over DC arc plasma devices in that, because of the separation of the electrodes from the discharge chamber itself, gases normally corrosive to electrode materials (e.g., chlorine) can be readily processed. Possibly this could extend the application of RF plasma devices into the field of mineral processing, for example, by selective chlorination of rare metals. Contamination of products due to erosion of electrodes when noncorrosive gases are treated is also avoided by use of RF plasma devices and provides an additional advantage over DC devices in the field of ultrapure chemical production. The counteracting disadvantages that RF plasma efficiencies are usually some 20% lower than DC devices must be taken into account for each individual case, especially when scale-up is considered.

The use of thermal plasma in the chemical processing industries as a production tool has been relatively slow to develop. The last decade saw a rapid development of induction plasma hardware, undoubtedly accelerated by U.S. aerospace activity. For successful economic chemical processing in RF plasma to become a reality, however, scale-up of RF induction plasma to much higher-power levels must first be achieved. Various attempts have been made to attain this goal by operation of high-power torches at lower frequencies, as the capital investment required for high-power, high-frequency plasma generators is prohibitive.

Thorpe and Scammon (52) reported the first successful operation of a high-power, low-frequency induced plasma torch. They operated a 1-MW torch on argon at 450 kHz, producing a plasma of some 10 cm in diameter. Recently, Thorpe (51) reported the successful operation of an induced plasma at the still lower frequencies of 10 and 1 kHz.

The operation of induced plasma devices at line frequency (50 Hz) is the ultimate objective of development work on torch scale-up, and Dundas (19) recently published data predicting operating powers and torch diameters for low-frequency plasma. The expected power requirement for a line frequency-induced plasma torch is circa 15 MW, and the achievement of this will clearly require considerable time and expenditure.

IX. Conclusions

Although no economic plasma process has yet been developed in which a plasma reactor has either replaced a conventional high-temperature reactor or provided an alternative route to the production of a chemical by a well-established chemical process, this is probably only a matter of time. The commercial success of a plasma chemical process will depend partly on technological but mainly on economic factors. Taking electricity at 1¢ per kWh, the electricity cost of acetylene and boron production on the yields given earlier in this review comes, respectively, to approximately $250 and $1200 per ton. Such costs are unacceptable for any but high-cost products. They could be brought down by higher plasma efficiencies and could also drop if cheap electricity becomes available. Although nuclear power stations are unlikely to offer electricity at peak hours at a cost substantially lower than conventional stations, the technology of their operation might make off-peak electricity available very cheaply and thus make electrochemical processes in general that much more practicable.

On the technological side, a notable gap is the lack of any real understanding of chemical processes in plasmas. The studies reported in this review have been largely exploratory and qualitative and have not been concerned, in detail, with this topic. To a first approximation, the high temperature of a plasma must give rise to a "molecular soup" in which reactant molecules have been partly or wholly dissociated and from which products emerge as a result of sequential reactions in the plasma and the tail gases. What is less clear is whether the products result from a series of chemical equilibria set up as the gases cool, and then frozen at temperatures decreed by thermodynamic considerations, or whether kinetic factors are also involved. In either case, is the important region of the plasma that section of the tail flame where the temperature has dropped to 2000°–2500°K? If it is, then there is the possibility that products and yields can be profoundly influenced by reactants added to the tail flame. Hamblyn (23) has shown that if boron trichloride and hydrogen are added to the tail of an argon plasma, boron is obtained in yields not much lower than those when all the gases pass through the plasma. Furthermore, as the distance between the RF coil and the BCl_3–H_2 inlet is increased, the yield also increases, going through a maximum at a distance of 6 cm.

These experiments suggest that the factors affecting the yields from plasma reactors are not always the most obvious ones, and much remains to be learned about the molecular processes that occur. The difficulties involved in the study of kinetics and thermodynamics of chemical reactions in high-temperature plasma tails are formidable, but even a crude understanding might permit a degree of control over plasma reactors that does not at present exist.

REFERENCES

1. Amouroux, J., and Talbot, J., *Ann. Chim.* (*Paris*) [14] **3**, 219–233 (1968).
2. Audsley, A., and Bayliss, R. K., *J. App Chem.* **19**, 33–38 (1969).
3. Barry, T. I., Bayliss, R. K., and Lay, L. A., *J. Mater. Sci.* **3**, 229–239 (1968).
4. Bequin, C. P., Ezell, J. B., Salvemini, A., Thompson, J. C., Vickroy, D. G., and Margrave, J. L., *in* "The Application of Plasma to Chemical Synthesis" (R. F. Baddour and R. S. Timmins, eds.), pp. 49–50. MIT Press, Cambridge, Massachusetts, 1967.
5. Besombes-Vailhé, J., *Bull. Soc. Chim. Fr.* [6] No. 2, pp. 462–468 (1969).
6. Borgianni, C., Capitelli, M., Cramarossa, F., Triolo, L., and Molinari, E., *Combust. Flame* **13**, 181–194 (1968).
7. British Patent 1,085,450 to British Titan Products (1967).

8. British Patent 1,088,924 to Du Pont de Nemours (1967).
9. British Patents 1,093,441/3 to British Titan Products (1967).
10. British Patent 1,164,396 to British Titan Products (1968).
11. British Patent 1,199,811 to British Titan Products (1970).
12. British Patent 1,202,587 to Cabot Corporation (1970).
13. British Patent 1,211,702 to Unilever (1970).
14. Bronfin, B. R., Amer. Chem. Soc., *Div. Fuel Chem., Prepr.* **11**, 347–363 (1967).
15. Bush E. L., and Sterling, H. F., British Patent 1,313,467 (1973).
16. Capitelli, M., Cramarossa, F., Triolo, L., and Molinari, E., *Combust. Flame* **15** 23–32 (1970).
17. Charles, J. A., Davies, G. M., Jervis, R. M., and Thursfield, G., *Inst. Mining Met., Trans., Sect. C* **79**, C54–C59 (1970).
18. Chase, J. D., and Potter, R. L., German Patent 2,157,723 (1972).
19. Dundas, P. H., *Proc. Int. Conf. Phenomena Ionized Gases, 10th, 1971* p. 243 (1971).
20. Fletcher, F. J., Ibrahim, J., and Kerry, I. M., *in* "Symposium on Electrochemical Engineering," p. 145. Inst. Chem. Eng., London, 1971.
21. Foex, M., U.S. Patent 3,257,196 (1966).
22. Grosse, A. V., Stokes, C. S., Cahill, J. A., and Correa, J. J., "Final Annual Report, "ONR Contract No. 3082(02). Temple University, Philadelphia, Pennsylvania, 1961.
23. Hamblyn, S. M. L., Ph. D. Thesis, University of Surrey, Guildford, England (1972).
24. Hamblyn, S. M. L., MacKinnon, I. M., Money, A. P., and Trotter, J. E., British Patent 1,288,913 (1972).
25. Hamblyn, S. M. L., Reuben, B. G., and Thompson, R., *Spec. Ceram.* **5**, 147–155 (1972).
26. Hartl, K., *Chem.-Ing.-Tech.* **39**, 1253–1254 (1967).
27. Hedger, H. J., and Hall, A. R., *Powder Met.* **8**, 65–72 (1961).
28. Huska, P. A., and Clump, C. W., *Ind. Eng. Chem., Process Des. Develop.* **6**, 238–244 (1967).
29. Ibberson, V. J., *High Temp.—High Pressures* **1**, 243–269 (1969).
30. Jackson, K., and Bloom, M. S., British Patent 915,771 (1963).
31. Jones, L. V., Ofte, D., Phipps, K. D., and Tucker, P. A., *Ind. Eng. Chem., Prod. Res. Develop.* **3**, 78–82 (1964).
32. Jordan, J. E., French Patent 1,399,183 (1965)
33. Kana'an, A. S., and Margrave, J. L., *Int. Sci. Technol.* **8**, 75 (1962).
34. Kana'an, A. S., and Margrave, J. L., *Appl. Spectrosc.* **20**, p18 (1966).
35. Landt, U., *Angew. Chem., Int. Ed. Engl.* **9**, 780–792 (1970).
36. LaRoche, M. J., *in* "La chimie des hautes températures," p. 71–81. CNRS, Paris, 1955.
37. MacKinnon, I. M., and Wickens, A. J., Borax Consolidated Ltd., London (private communication), (1973).
38. McPherson, R., *Proc. Aust. Ceram. Conf., 4th, 1970* AB15, pp. 1–2 (1970).
39. McTaggart, F. K., *in* "Plasma Chemistry in Electrical Discharges." Elsevier, Amsterdam, 1967.
40. Miller, R. C., and Ayen, R. J., *Ind. Eng. Chem., Process Des. Develop.* **8**, 370–377 (1969).
41. Nishimura, Y., Takeshita, K., Adachi, Y., Nakashio, F., and Sakai, W., *Int. Chem. Eng.* **10**, 133–137 (1970).

42. Rains, R. K., and Kadlec, R. H., *Met. Trans.* **1**, 1501–1506 (1970).
43. Reed, T. B., *J. Appl. Phys.* **32**, 2534–2535 (1961).
44. Salinger, R. M., *Ind. Eng. Chem.*, *Prod. Res. Develop.* **11**, 230–231 (1972).
45. Sayce, I. G., *Conf. Advan. Extr. Met.*, *1971*, Paper No. 27.
46. Sienko, M. J., and Young, Y. E., *Nat. Bur. Stand.* (*U.S.*), *Spec. Publ.* **364**, 385–395 (1972).
47. Stokes, C. S., Correa, J. J., Streng, L. A., and Leutner, H. W., *J. Amer. Inst Chem.*, *Eng.* **11**, 370–384 (1965).
48. Sullenger, D. B., Phipps, K. D., Seabough, P. W., and Hudgens, C. L., *Science* **163**, 935–937 (1969).
49. Teresawa, S., *Sekiyu Gakkai Shi* **10**, 130–134 (1967).
50. Thornton, J. D., Charlton, W. D., and Spedding, P. L., *Advan. Chem. Ser.* **80**, 165 (1967).
51. Thorpe, M. L., *Conf. Advan. Extr. Met.*, *1971*, Paper No. 26.
52. Thorpe, M. L., and Scammon, L., *Contract. NASA Rep.* **NASA CR–1343** (1969).
53. Timmins, R. S., and Ammann, P. R., *in* "The Application of Plasma to Chemical Synthesis" (R. F. Baddour and R. S. Timmins, eds.), pp. 99–131. MIT Press, Cambridge, Massachusetts, 1967.
54. Triché, H., Talayrach, B., and Besombes-Vailhé, J., *Method. Phys. Anal.* **5**, 343–347 (1969).
55. Vurzel, F. B., and Polak, L. S., *Khim. Vys. Energ.* **1**, 268 (1967).
56. Waldie, B., *Trans. Inst. Chem. Eng.* **48**, T90–T93 (1970).
57. Waldie, B., *Trans. Inst. Chem. Eng.* **49**, T114–T116 (1971).
58. Waldie, B., *Chem. Eng.* (*London*) **259**, 92–97 (1972).
59. Warren, I. H., and Shimizu, H., *Trans. Can. Inst. Mining Met.* **58**, 169–178 (1965).

COPPER (I) COMPLEXES

F. H. JARDINE

Department of Chemistry, North East London Polytechnic, London, England

I. Introduction

A very large number of stable copper(I) complexes exist in a variety of stoichiometries. In few of these complexes does the formal coordination number of the metal atom exceed four. Indeed, along with silver(I) and gold(I), it is one of the few oxidation states to exhibit regularly the low coordination numbers two and three. The simple amine and halo complexes isolated from aqueous solution fortuitously contain linear copper(I) ions. As a result, coordination number two is erroneously considered to be a common coordination number for this oxidation state. In fact, two-coordinate complexes are probably outnumbered by the trigonally coordinated complexes, whereas against the vast host of tetrahedral complexes the two- and three-coordinate complexes are numerically insignificant.

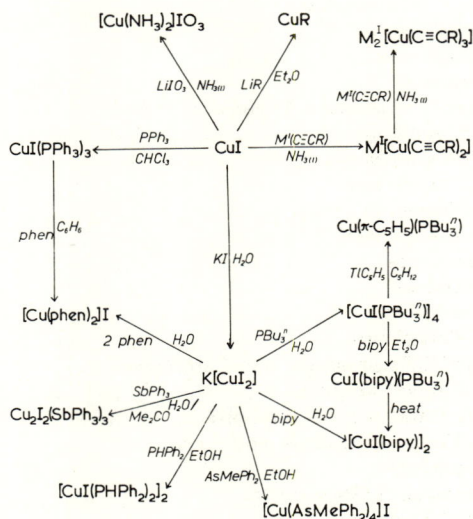

FIG. 1. Complexes derived from copper(I) iodide.

Most of the nonradioactive elements of Groups V, VI, and VII coordinate with the metal in this oxidation state, and there is also quite an extensive organometallic chemistry as befits an element of such importance in synthetic organic chemistry. An indication of the diversity of complexes [in this instance all derived from copper(I) iodide] is given in Fig. 1.

The thermodynamic factors involved in the relationship among

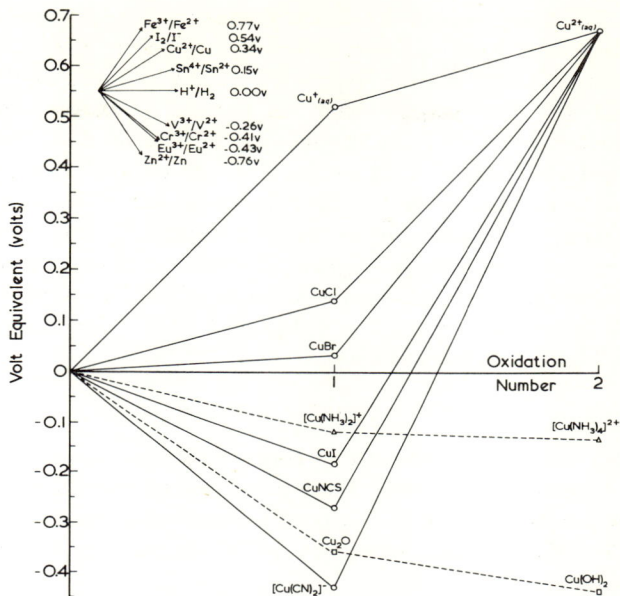

FIG. 2. Oxidation state diagram (285) for copper. (Data from Ref. 207.)

copper(I), copper(II), and copper(0) (207) are shown in Fig. 2. From this it can be observed that it is possible for copper(I) complexes to disproportionate:

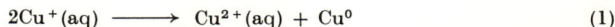

$$2Cu^+(aq) \longrightarrow Cu^{2+}(aq) + Cu^0 \tag{1}$$

The equilibrium constant of reaction (1), $K = [Cu^{2+}][Cu^0]/[Cu^+]^2$, is of the order of 10^6; thus, only vanishingly small concentrations of aquo-copper(I) species can exist at equilibrium. However, in the absence of catalysts for the disproportionation—such as glass surfaces, mercury, red copper(I) oxide (7), or alkali (311)—equilibrium is only slowly attained. Metastable solutions of aquocopper(I) complexes may be generated by reducing copper(II) salts with europium(II) (113), chromium(II), vanadium(II) (113, 274), or tin(II) chloride in acid solution (264). The employment of chromium(II) as reducing agent is best (113), since in most other cases further reduction to copper metal is competitive with the initial reduction (274).

Aquocopper(I) complexes are fairly powerful reducing agents and the kinetics of their reactions with iron(III) (275), vanadium(IV) (312), cobalt(III), and mercury(II) (113) have been studied. Further, the role of copper(I) species in the copper(II)-catalyzed reduction of cobalt(III) by vanadium(II) (112) has been confirmed with the reduction of

$[Co(NH_3)_5Br]^{2+}$ by copper(I) (277). Similarly Cu(I) species catalyze the reduction of cobalt(III) complexes by vanadium(III) (276).

Solutions of many simple copper(I) compounds are easily oxidized by air, and the reactions proceed via peroxo species which, in turn, form hydrogen peroxide from the solvent (157, 261, 353, 362).

Polarographic behavior of copper(II) and copper(I) species in the presence of various ligands has been investigated (154, 215, 258, 279, 343). In some instances, however, no copper(I) intermediate was observed in the reduction to copper amalgam at the electrode, e.g., at low concentrations of 1,2-diaminoethane (270) or similar diamines (205); this is in keeping with a value of 4×10^5 for the equilibrium constant (33, 291) of the reaction

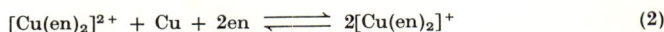

$$[Cu(en)_2]^{2+} + Cu + 2en \rightleftharpoons 2[Cu(en)_2]^+ \tag{2}$$

In other solvents, particularly those that are capable of solvating copper(I) ions more effectively than water, copper metal can be used to reduce copper(II) species to copper(I) complexes. Such is the case in acetonitrile (197), liquid ammonia (314), or in the molten eutectic $AlCl_3$–$NaCl$–KCl (9).

At high temperatures under oxygen-free conditions, copper(I) is the preferred oxidation state. Thus, above 725°K, copper(II) chloride decomposes to $(CuCl)_3$ vapor and chlorine (227), and CuO yields Cu_2O at ca. 1275° (318).

II. Anionic Complexes

A. DIHALOCUPRATE(I) COMPLEXES

The enhanced solubilities of copper(I) halides or pseudohalides in aqueous solutions of alkali or ammonium halides are due to the formation of anionic complexes (37, 121, 134, 263):

$$CuX + MX' \rightleftharpoons M^+[CuXX']^- \tag{3}$$

In ethanol, containing dissolved HCl, the formation of dichlorocuprate(I) salts is even more favored than in aqueous solution (187). This would seem to be generally true for most weakly polar, organic solvents since the cuprate(I) complexes are extracted preferentially into diethyl ether (91), tributylphosphate (321, 347), or cyclohexanone (321). In the first two solvents the ions have been shown to be linear and centrosymmetric by infrared and Raman spectroscopy (91, 347).

The anions form isolable salts with a variety of cations, e.g., ammonium (*102, 297, 304*), nitrogen bases (*70, 71, 96, 97, 172, 196, 297*), alkali metals (*137, 138*), copper(II) complex cations (*153, 236, 266, 286*), and the triethylsulfonium cation (*97*).

In the salt $[\{N_6P_6(NMe_2)_{12}\}Cu(II)Cl][Cu(I)Cl_2]$ (*236*) the anion is linear, but in

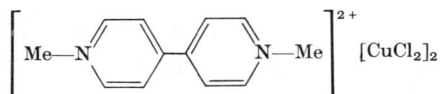

$$\left[\text{Me—N} \underset{}{\bigcirc}\underset{}{\bigcirc} \text{N—Me} \right]^{2+} [CuCl_2]_2$$

(*228*) the anion has the polymeric beryllium dichloride structure (*288*). Trigonal copper(I) occurs in the dicyanocuprate(I) ion where there are helical Cu–C–N–Cu–C chains with tactic CN groups bound normal to the helices (*92*).

B. TRIHALOCUPRATE(I) COMPLEXES

In most instances further coordination of halide ion takes place and $[CuX_3]^{2-}$ complexes are formed (*23, 37, 121, 134, 141, 172, 186, 279, 334*)

$$[CuX_2]^- + X' \rightleftharpoons [CuX_2X']^{2-} \tag{4}$$

Potassium and ammonium trichloro- and tribromocuprate(I) have been prepared (*247, 350*) and their structures determined by Brink *et al.* (*44, 45*). The isomorphous anions are made up from CuX_4 tetrahedra sharing corners. Tricyanocuprate(I) salts have also been isolated (*138, 325*), and the Raman spectrum of the anion in aqueous solution indicates that it is trigonal and contains no coordinated water (*76*). Recently, $[Cu(N_3)_4]^{2-}$ has been reduced electrolytically to $[Cu(N_3)_3]^{2-}$ (*259*).

C. TETRAHALOCUPRATE(I) COMPLEXES

Only the most strongly coordinating anions are capable of forming significant quantities of $[CuX_4]^{3-}$ anions in the equilibrium (*141, 172, 203, 295*)

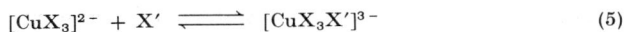

$$[CuX_3]^{2-} + X' \rightleftharpoons [CuX_3X']^{3-} \tag{5}$$

The most important ion of this stoichiometry is $[Cu(CN)_4]^{3-}$. Its infrared and Raman spectra have been obtained with increasing precision with the passing years (*40, 66, 180, 287*) and together with X-ray crystallography of the potassium salt show it to have a regular tetrahedral grouping about the copper atom (*90*).

D. Polynuclear Anions

The polynuclear anions are usually chlorocuprate(I) complexes, but compound $K[Cu_2(CN)_3] \cdot H_2O$ has been identified in the phase diagram of the $CuCN-KCN-H_2O$ system (334). The anion's structure is made up of $[Cu_2(CN)_3]$ sheets with the lattice water clathrated in $(CuCN)_6$ rings (93).

The chloro anions are all derived from linked $CuCl_4$ tetrahedral units; for instance in $Cs[Cu_2Cl_3]$ (349) there are double chains formed by $CuCl_4$ tetrahedra sharing edges (43). In bis-(N-benzoylhydrazine)-copper(II) pentachlorotricuprate(I) (18), there are cylinders of distorted $CuCl_4$ tetrahedra in which one-fifth of the chloro ligands link four Cu(I) atoms, three-fifths link two copper(I) atoms, and the remaining one-fifth link a copper(I) and copper(II) atom. Several pentacyano-tricuprate(I) salts have also been isolated (138). There is one free chloride ion in the structure of $[Co(III)(NH_3)_6][Cu_5Cl_{17}]$, and the Cu_5Cl_{16} units consist of four of the ubiquitous $CuCl_4$ tetrahedra, each bound through one corner to a tetrahedrally coordinated, central copper atom (254).

E. $[Cu(biX)_n]^{1-2n}$ Complexes

The $[Cu(biX)_n]^{1-2n}$ compounds all occur with sulfur acids. The best understood are the thiosulfato complexes. However, these have been but little investigated compared with their technically important silver analogs. They have usually been prepared by the action of alkali thiosulfates on copper(II) salts, e.g.,

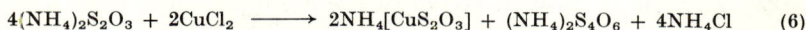

$$4(NH_4)_2S_2O_3 + 2CuCl_2 \longrightarrow 2NH_4[CuS_2O_3] + (NH_4)_2S_4O_6 + 4NH_4Cl \qquad (6)$$

It is also possible to obtain mixed Cu(II)–Cu(I) salts, such as $Na_4[Cu(NH_3)_4][Cu(S_2O_3)_2]_2 \cdot Y$ (Y = H_2O, NH_3), in other reactions due to incomplete reduction (153).

The anion most commonly obtained in the preparation of thiosulfato complexes appears to be $[Cu(S_2O_3)_3]^{5-}$ (41, 100, 300, 320), and other salts include hydrated sodium, potassium, and ammonium derivatives of $[Cu_2(S_2O_3)_3]^{4-}$ (320) and $[Cu(S_2O_3)]^-$ (300, 320). Double salts are frequently obtained and examples of these are $[NH_4]_7[Cu(S_2O_3)_4] \cdot 2NH_4X$ (X = ClO_3, NO_3), $K_7[Cu(S_2O_3)_4] \cdot 2KNO_3$, and $Na_3[Cu(S_2O_3)_2] \cdot 2NaNO_3$ (320). Mixed complexes $Na_2[CuX(S_2O_3)]$ (X = I, NCS) have also been reported (141).

Many similar sulfito complexes are known (300), e.g., $NH_4[CuSO_3]$, $[NH_4]_4[Cu_2(SO_3)_3] \cdot 3H_2O$, and $[NH_4]_5[Cu(SO_3)_3] \cdot H_2O$, but in the sodium and potassium series double-salt formation is even more prevalent than with the thiosulfatocuprate(I) salts of these cations, e.g., $K[CuSO_3]\cdot3KHSO_3$.

Treatment of aqueous solutions of copper(II) salts with ammonium polysulfide gives rise to $NH_4[CuS_4]$ (163), from which other $[CuS_4]^-$ salts can be obtained (35, 163, 283). Two thiocarbonato complexes, $M[CuCS_3]$ (M = K, NH_4), are also known (162).

III. [CuXL]$_n$ Complexes

In virtually all the (CuXL)$_n$ complexes, the anionic ligand is halide or pseudohalide. It is superficially attractive to regard them as linear copper(I) complexes—neutral analogs of $[Cu(NH_3)_2]^+$ and $[CuCl_2]^-$. However, in the majority of cases where molecular weight or X-ray structural determinations have been made the complexes have been shown to be tetrameric. Severe deviations from orthogonality make the new formulation as "inorganic cubane" molecules (268) less apt than the original description of a copper-cornered tetrahedron with a halide bound to each face (Fig. 3). In the structure of $[CuIAsEt_3]_4$ (234) the copper atoms are only 4 pm further apart than in copper itself, and because of this, one review considers copper(I) to be seven-coordinate in these complexes (253). It is worthy of note that the silver(I) analogs are also tetramers but the similar gold(I) complexes are linear monomeric molecules (235).

The majority of the complexes listed in Table I have been prepared

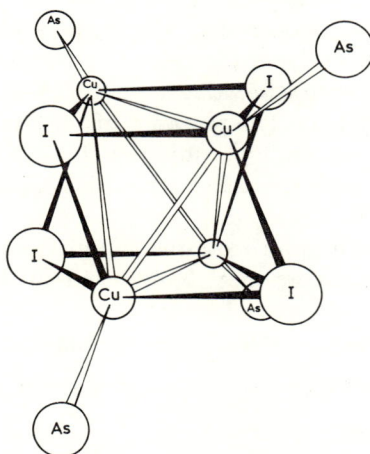

FIG. 3. Structure of $[CuIAsEt_3]_4$. [Redrawn by permission from $J.\,Chem.\,Soc.$, London, p. 1503 (1936). The ethyl groups have been omitted for clarity.]

TABLE I

TETRAMERIC CuXL COMPLEXES

Ligand	Halide	References
RNH$_2$ (R = C$_8$H$_{17}$, C$_{10}$H$_{21}$, C$_{18}$H$_{37}$)	Cl, Br, I	(354)
PHPh$_2$	Cl, Br, I	(1)
PR$_3$ (R = Et, Pr^n, Bu^n, C$_5$H$_{11}$)	I	R = Bu^n (234)
		(182)
PEt$_3$	Cl	(176)
PEt$_2$Ph	I	(62)
PPh$_3$	Cl, Br, I	(56, 85, 176)
PEt$_2$(p-C$_6$H$_4$·CF$_3$)	I	(62)
P(NEt$_2$)$_2$Ph	I	(114)
P(NMe$_2$)F$_2$	Cl	(74)
2-Phenylisophosphindoline	Cl, Br, I	(75)
AsR$_3$ (R = Me, Et, Pr^n, Bu^n)	I	(234)
AsEt$_3$	Br	(234)
AsMePh$_2$	Cl, Br, I	(267)
AsPh$_3$	Cl	(222)
AsPh$_3$	Br, I	(178)

by direct interaction of the ligand and copper(I) halide in an appropriate solvent. Recently, it has been found more convenient to prepare some of the complexes in homogeneous solution from copper(II) halides if the ligand is a tertiary phosphine or arsine (75, 176, 178).

If [CuClPPh$_3$]$_4$ is reacted with NaBH(OMe)$_3$ in deoxygenated N,N-dimethylformamide (DMF), red, hexameric [CuHPPh$_3$]$_6$ is obtained (69). In this complex the copper atoms are located at the apices of an irregular octahedron with triphenylphosphine bound to each and pointing away from the center of the octahedron. The mean Cu–Cu distance is 260 pm (the same as in [CuIAsEt$_3$]$_4$), so there is some metal–metal bonding (31). The presence of the hydrido ligands was not detected by X-ray crystallography nor by NMR or infrared spectroscopy, but the complex reacted with PhCO$_2$D to give a mixture of H$_2$ and HD but no D$_2$.

The above reaction is an example of anion substitution but, more importantly, the tetramers may be cleaved by heterocyclic nitrogen ligands (176, 234),

$$[CuI(PBu_3^n)]_4 + 4\text{bipy} \longrightarrow 4CuI(\text{bipy})(PBu_3^n) \qquad (7)$$

$$[CuBrPPh_3]_4 + 4py \longrightarrow 2[CuBr(PPh_3)py]_2 \qquad (8)$$

or by excess ligand (62),

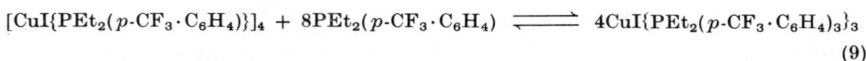

$$[CuI\{PEt_2(p\text{-CF}_3\cdot C_6H_4)\}]_4 + 8PEt_2(p\text{-CF}_3\cdot C_6H_4) \rightleftharpoons 4CuI\{PEt_2(p\text{-CF}_3\cdot C_6H_4)_3\}_3$$
$$(9)$$

Several complexes have been found to be dimeric. Infrared spectroscopy shows that the CuNCS(L) complexes (L = py, 2- or 4-picoline, 3,5-lutidine, or quinoline) contain bridging thiocyanato groups (332). The mass spectra of CuX{P(cyclohexyl)$_3$} complexes (X = Cl, Br, I) show no peaks of higher mass than those corresponding to dimeric molecules, and the far-infrared spectrum of the chloro complex indicates that they are dimeric with halo bridges (248). Triphenylphosphine reacts with copper(I) trifluoroacetate in dichloromethane to give [Cu(O$_2$C·CF$_3$)(PPh$_3$)]$_2$ which is dimeric in chloroform but partially dissociated in dichlorobenzene (107).

Polymeric complexes are formed when copper(I) cyanide reacts with ammonia or alkyl and aryl isocyanides. The ammine complex consists of Cu–C–N–Cu–C helices linked by metal–metal bonds to form sheets. The NH$_3$ ligands bond normal to the copper atoms and interlock with NH$_3$ protrusions of other sheets (94). Complex CuCN(Bu^{n}NC) is polymeric (272), and the ethyl, propyl, and 2-butyl isocyanide complexes are also known (139).

The reaction of CuCN with MeI surprisingly gives polymeric CuI·MeNC [which can be prepared more obviously from CuI and MeNC (169)]; this unusual exchange reaction (149) has been confirmed by the determination of the structure of the polymer (118). The structure is shown in Fig. 4. Other halocopper(I) isocyanide complexes have been prepared by reaction (10).

$$CuX + RNC \xrightarrow{\text{MeCN}} CuX·RNC \tag{10}$$

Compound CuClPhNC is also known. By contrast, similar acetonitrile complexes are easily air oxidized, but they can be prepared by reducing CuX$_2$ (X = Cl, Br) with copper in MeCN (250); several alkyl or aryl nitriles complex with CuCl (290).

Two complexes are known definitely to be monomeric—the triphenylphosphine sulfide complexes, CuX(Ph$_3$PS) (X = Cl, Br). They were prepared by reacting the ligand with copper(I) halide in chloroform, and molecular weight determinations in this solvent show them to be monomeric (95).

There remains a variety of unstable, insoluble, or incompletely characterized complexes whose true formulas remain open to speculation. Many unstable 1:1 complexes between copper(I) halides and ammonia (51) or the lower aliphatic amines (71, 281, 356) have been studied manometrically, but the compounds formed have high dissociation pressures, and attempts to isolate them have been unsuccessful owing to loss of ligand or to oxidation. Complexes with higher amines (348) or with pyridine (354) and its oligomers (178, 229, 230) are, in

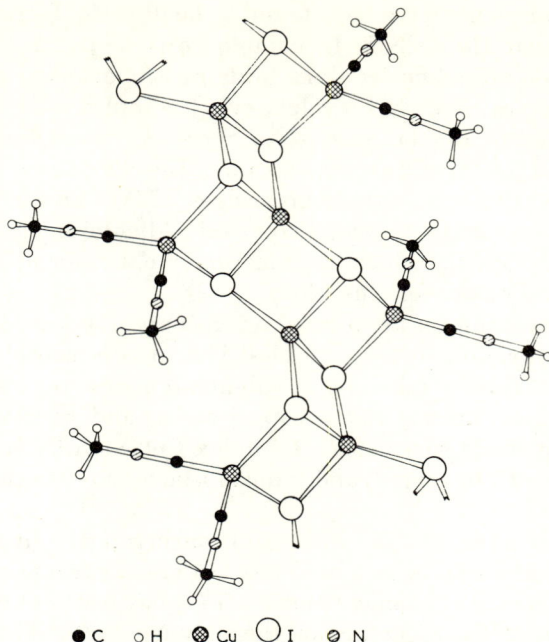

● C   o H   ⊗ Cu   ○ I   ⊘ N

Fig. 4. Structure of [CuI·MeNC]$_\infty$. [Redrawn by permission from *J. Chem. Soc., London*, p. 2303 (1960).]

general, more stable, but the solubility of compounds of the latter class is low.

Similarly, phosphine complexes (*165, 296, 309*) easily lose the ligand. No molecular weights are available for the few phosphite complexes isolated (*16, 17, 99*) or for the complex CuCl{P(CH$_2$SiMe$_3$)$_3$}. Early work on dimethylphenylarsine complexes (*50*) has been corrected (*267*), but the complexes of diphenylmethylarsine (*50*) merit reinvestigation.

Compound CuCl(Et$_2$NH) can be obtained from the bis(amine) complex, but the molecular weight has not been determined. It is even more unfortunate that no attempts have been made to determine the molecular weights of certain complexes containing sulfur ligands since these could be either monomeric (like CuClPh$_3$PS) or polymeric containing sulfur bridges. Compound CuCl{SC(NH$_2$)$_2$} has been prepared from copper(II) chloride (*292*) and so have both the bis- and tris(thiourea)copper(I) complexes. Since thiourea is a good bridging ligand, this complex is probably polymeric, as are CuX{SC(NH$_2$)(OEt)} [X = Br

(299), NCS *(101)*, Cl, I *(101, 299)*]. Thioacetamide has been shown to be S-bonded *(327)* in CuCl(thioacetamide) *(171)* since the C–S stretch frequency decreases upon coordination.

IV. $Cu_2X_2L_3$ Complexes

Recent X-ray crystallographic investigations of $Cu_2Cl_2(PPh_3)_3$ *(5)* (Fig. 5) and its benzene adduct *(217)* have shown one copper atom in the molecule to be three-coordinate and the other four-coordinate. The following complexes of this type have been isolated from reactions of the ligands with copper(I) halides: $Cu_2Cl_2\{P(p\text{-}CH_3 \cdot C_6H_4)_3\}_3$ *(252)*, Cu_2Cl_2- $(PHPh_2)_3$ *(1)*; $Cu_2X_2(PPh_3)_3$ (X = Cl, Br, I) *(85)*; and $Cu_2X_2(NH_3)_3$

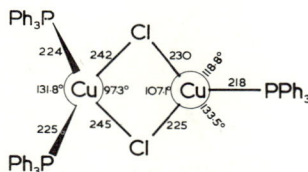

FIG. 5. Structure of $Cu_2Cl_2(PPh_3)_3$ (bond lengths in picometers). [Redrawn by permission from *J. Chem. Soc., Dalton Trans.*, p. 171 (1972).]

(X = Cl, Br, I) *(225)*. More recently, other complexes, $Cu_2Cl_2(AsPh_3)_3$ and $Cu_2X_2(SbPh_3)_3$ (X = Cl, Br, I), have been obtained by interaction of the ligands and copper(II) halides in ethanol or with $K[CuI_2]$ in aqueous acetone *(178)*. Complex $Cu_2Cl_2\{P(2\text{-pyridyl})Ph_2\}_3$ has been obtained by heating $CuCl_2\{P(2\text{-pyridyl})Ph_2\}_3 \cdot H_2O$ *in vacuo* at 413° *(14)*.

The conditions under which the complexes can be isolated are quite critical owing to the lability of the neutral ligands. Nuclear magnetic resonance studies have shown $PHPh_2$ to exchange rapidly at room temperature *(1)*, and $P(p\text{-}CH_3 \cdot C_6H_4)_3$ is significantly labile above 193° *(252)*. The lability of the ligands contributes toward the destruction of the complexes, since the free ligand produced cleaves the halo bridges, whereas loss of ligand followed by dimerization yields $[CuXL]_4$ species. Thus,

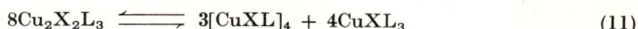

$$8Cu_2X_2L_3 \rightleftharpoons 3[CuXL]_4 + 4CuXL_3 \qquad (11)$$

In general, the complexes are not readily accessible but, para-doxically, are often obtained by metathesis from other copper(I) complexes *(56, 84, 98)*.

V. CuXL$_2$ Complexes

A. COMPLEXES CONTAINING A BIDENTATE ANION

The complexes in this class are generally well-characterized because the denticity of the polyatomic anion is often apparent from its infrared spectrum, and the molecular weights of these compounds in solution approximate their formula weights. Additionally, several complexes of this type have had their structures determined by X-ray crystallography.

Bis(triphenylphosphine)nitratocopper(I) (Fig. 6) has been investigated by all the preceding techniques. It can be prepared from

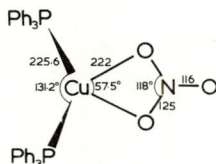

FIG. 6. Structure of CuNO$_3$(PPh$_3$)$_2$ (bond lengths in picometers). [Redrawn by permission from *Inorg. Chem.* **8**, 2750 (1969).]

Cu(NO$_3$)$_2$·3H$_2$O and PPh$_3$ in alkanolic solution (*86*) or by recrystallization of CuNO$_3$(PPh$_3$)$_3$ from methanol (*177*). The A_1 and B_2 vibrations of the nitrato group in the infrared spectrum of the complex show a separation of ca. 200 cm^{-1} characteristic of bidentate coordination; its molecular weight in chloroform has been reported as 639 (formula wt., 649) (*177*). Its crystal structure has also been determined (*243, 244*).

Similar complexes containing two molecules of tri-(*m*-tolyl)- or tri-(cyclohexyl)phosphine have been prepared, and the structure of CuNO$_3${P(C$_6$H$_{11}$)$_3$}$_2$ determined (*10*). The importance of steric crowding between the tertiary phosphine groups of the latter complex is shown by the large P–Cu–P angle and the long Cu–P bonds (Table II).

A related complex whose structure is also known is CuBH$_4$(PPh$_3$)$_2$ (*219, 220*). It has been prepared by reacting NaBH$_4$ with [CuClPPh$_3$]$_4$ in ethanolic chloroform (*56*) or with an unspecified triphenylphosphine-copper(I) sulfate complex in ethanol (*98*). Its molecular weight in CHCl$_3$ has been found to be 612 (formula wt., 603). The B–H vibrations in its infrared spectrum and those of the corresponding tetradeuteroborato complex (*222*) indicate that the tetrahydridoborato group is bidentate.

Analogous complexes containing tri-(*p*-tolyl)-, tri-(*o*-tolyl)-, and tri-(*p*-anisyl)phosphines are known (*222*).

TABLE II

P–Cu–P Angles and Mean Cu–P Bond Lengths in $CuX(PR_3)_2$
Complexes

Complex	P–Cu–P	Cu–P (pm)	Reference
$CuNO_3(PPh_3)_2$	131°	225.6	(244)
$CuNO_3\{P(C_6H_{11})_3\}_2$	140°	229	(10)
$CuBH_4(PPh_3)_2$	123°	227.6	(220)
$CuB_3H_8(PPh_3)_2$	120°	228.1	(221)
$Cu(CF_3COCHCOCH_3)(PPh_3)_2$	127°	225	(22)
$[CuN_3(PPh_3)_2]_2$	122°	226.6	(359)

Oligomers of $BH_4{}^-$ form similar copper(I) complexes, except that in $CuB_3H_8(PPh_3)_2$ (Fig. 7) the hydrogen atoms attached to copper are bonded to different boron atoms (221). White $CuB_3H_8L_2$ complexes, where L = PPh_3 (323), $AsPh_3$ (222), or $PMePh_2$ (221), have been precipitated by addition of water to mixtures of CsB_3H_8 (in acetone) and a suitable halocopper(I) complex (in ethanol). The infrared spectra of these

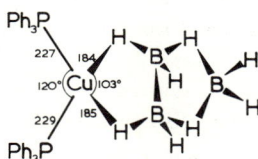

Fig. 7. Structure of $CuB_3H_8(PPh_3)_2$ (bond lengths in picometers). [Redrawn by permission from Inorg. Chem. 8, 2755 (1969).]

complexes show two bands between 2500 and 2200 cm^{-1} due to terminal B–H vibrations; other bands below the latter frequency were assigned to bridging hydrogen atom vibrations. The triphenylarsine and -phosphine complexes have molecular weights in benzene solution close to their formula weights.

Bis(triphenylphosphine)copper(I) complexes containing octahydropentaborato or nonahydrohexaborato groups have been synthesized from $CuCl(PPh_3)_3$ and KB_xH_y in dichloromethane–tetrahydrofuran mixtures at 195° (42). The latter complex is less stable to aerial oxidation than the former. Terminal B–H vibrations were observed in their infrared spectra, but no vibrations due to bridging hydrogen atoms were reported. Compound $CuB_{10}H_{13}(PPh_3)_2 \cdot CH_2Cl_2$ has been synthesized from $CuCl(PPh_3)_3$ and $NaB_{10}H_{13}$ in dichloromethane (192).

Studies using ^1H and ^{11}B NMR have proved of little value in investigating anion structure of the hydridoborato complexes, owing either to nonrigidity of the coordinated anion or to exchange between free and coordinated anion (27).

Compound $CuBH_4(PPh_3)_2$ can readily be converted into other copper(I) complexes, thus it reacts with ethanolic perchloric or tetrafluoroboric acids to give salts of the cation $[(Ph_3P)_2CuH_2BH_2Cu(PPh_3)_2]^+$ further, if the perchlorate is reacted with $NaBPh_4$ in methanol, the tetraphenylborate salt can be isolated (57). The structure of the cation was inferred from its infrared spectrum, which shows no terminal B–H vibrations. The $CuX(PPh_3)_2$ (X = acac, PhCOCHCOPh, NO_3) complexes are obtained by reacting $CuBH_4(PPh_3)_2$ with acetylacetone, dibenzoylmethane, or nitric acid (56), whereas passing HCl gas into a benzene solution of the complex yields $Cu_2Cl_2(PPh_3)_3$ (98).

Six bis(triphenylphosphine)(β-diketonato)copper(I) complexes are known. They may be prepared by one or more of reactions (12)–(14).

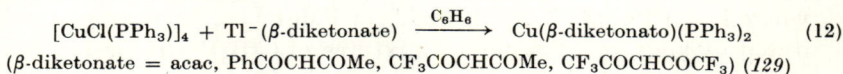

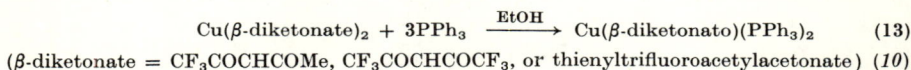

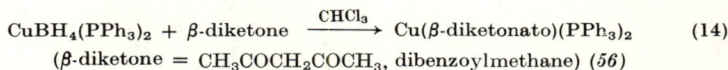

$$[CuCl(PPh_3)]_4 + Tl^-(\beta\text{-diketonate}) \xrightarrow{C_6H_6} Cu(\beta\text{-diketonato})(PPh_3)_2 \qquad (12)$$

(β-diketonate = acac, PhCOCHCOMe, $CF_3COCHCOMe$, $CF_3COCHCOCF_3$) (129)

$$Cu(\beta\text{-diketonate})_2 + 3PPh_3 \xrightarrow{EtOH} Cu(\beta\text{-diketonato})(PPh_3)_2 \qquad (13)$$

(β-diketonate = $CF_3COCHCOMe$, $CF_3COCHCOCF_3$, or thienyltrifluoroacetylacetonate) (10)

$$CuBH_4(PPh_3)_2 + \beta\text{-diketone} \xrightarrow{CHCl_3} Cu(\beta\text{-diketonato})(PPh_3)_2 \qquad (14)$$

(β-diketone = $CH_3COCH_2COCH_3$, dibenzoylmethane) (56)

Reaction (12) is probably the most useful, since, despite the apparent simplicity of reaction (13), neither $Cu(acac)_2$ nor $Cu(Bu^tCOCHCOBu^t)_2$ (175) is reduced under these conditions. The apparent molecular weights of the complexes in chloroform solution are slightly less than their formula weights. This is most probably due to the loss of a triphenylphosphine ligand. Studies using ^1H NMR show the β-diketonato groups to be symmetrically bonded (129), and C–O vibrations coupled with C=C were observed in their infrared spectra in the expected region of 1700 to 1500 cm^{-1} (10, 129). The structure of $Cu(CF_3COCHCOCH_3)$-$(PPh_3)_2$ has been determined by X-ray crystallography (22). A related, monomeric, bis(triphenylarsine) complex may be obtained by fusing the ligand and $Cu(CF_3COCHCOCF_3)_2$ at 343° (357).

Certain copper(II) carboxylates can be reduced by triphenylphosphine in alkanolic solution to $Cu(carboxylato)(PPh_3)_2$ complexes, where carboxylate = formate, acetate, benzoate, o- or m-toluate, p-nitrobenzoate, phenylacetate, $\frac{1}{2}$-maleate, $\frac{1}{2}$-fumarate, $\frac{1}{2}$-monochloromaleate, or $\frac{1}{2}$-succinate (146). The acetato and benzoato complexes have been obtained less conveniently from the unstable copper(I) salts (294). When it was possible to determine the molecular weights of these complexes in chloroform—the dibasic acid complexes were insoluble—the found values agreed closely with the formula weights. Their infrared spectra show them to contain bidentate carboxylato groups (146). Thus, it would appear that the dibasic carboxylato groups bridge two $Cu(PPh_3)_2$ units. Generally, complexes of this stoichiometry are formed only from the copper(II) salts of weak acids, but it has proved possible to prepare $CuO_2C \cdot CF_3(PPh_3)_2$ directly from $CuO_2C \cdot CF_3$ and triphenyl-phosphine (107).

Bis(triphenylphosphine)copper(I)(N,N-dialkyldithioarbamato) complexes have also been prepared. The N,N-dimethyl and N,N-diethyl complexes have been obtained by reducing their copper(II) salts with PPh_3 in chloroform, but the N,N-dipropyl and N,N-dibutyl derivatives are best prepared by fusing tetrakis-[N,N-dialkyldithiocarbamato-copper(I)] with the ligand. Complex $Cu(S_2CNEt_2)(PPh_3)_2$ has an apparent molecular weight in solution some 40% lower than the formula weight, but opinions vary as to whether this is due to ionization (198) or to loss of a PPh_3 ligand (46).

B. Dimeric Complexes

Only a few complexes have definitely been established as dimeric. The best characterized is $[CuN_3(PPh_3)_2]_2$ (Fig. 8), whose structure has been determined by X-ray crystallography (359, 360). The complex was

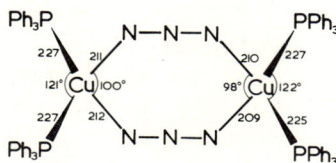

FIG. 8. Structure of $[CuN_3(PPh_3)_2]_2$ (bond lengths in picometers). [Redrawn by permission from Inorg. Chem. **10**, 1289 (1971).]

prepared by reacting copper(I) chloride, triphenylphosphine, and sodium azide in chloroform. It was long thought to be monomeric because the reaction,

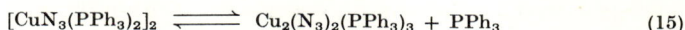

$$[CuN_3(PPh_3)_2]_2 \rightleftharpoons Cu_2(N_3)_2(PPh_3)_3 + PPh_3 \qquad (15)$$

gives rise to misleading molecular weight data from solution studies (294, 358). Reactions of this type make it possible that several "monomeric" complexes would prove to be dimeric after more searching investigation. It is believed that the similar complexes [CuNCS-(PPh$_2$Me)$_2$]$_2$ (*127*) and [CuXL$_2$]$_2$ (X = N$_3$, NCS; L = MeOPPh$_2$, EtOPPh$_2$) (*361*) are also dimeric.

The bis(triphenylphosphine)azido complex undergoes some interesting addition reactions, with CF$_3$CN to give a bis(trifluoromethyl-tetrazole) derivative, and with CS$_2$ to give a bis(thiotriazole) complex. The latter may be photolyzed in chloroform to [CuNCS(PPh$_3$)$_2$]$_2$ (*361*).

A dimeric structure has been proposed for [Cu(BH$_3$CN)(PPh$_3$)$_2$]$_2$ on the basis of its infrared spectrum and a preliminary X-ray crystallographic study. These are consistent with Cu–H–B–C–N–Cu bridges, but no molecular weight in solution was determined to support this view. The complex was prepared by precipitating it from a chloroform solution of [CuCl(PPh$_3$)]$_4$ and NaBH$_3$CN by ethanol (*224*).

The molecular weights of CuX(2-phenylisophosphindoline)$_2$ complexes (X = Cl, Br, I) and CuX(PHPh$_2$)$_2$ (X = Br, I) in chloroform or benzene led to the conclusion that they are dimeric in solution. The chloro- and bromo- 2-phenylisophosphindoline complexes were prepared by reacting the ligand with CuX$_2$ (X = Cl, Br) in aqueous acetone, and the remainder by interaction of the ligands and copper(I) halides (*1, 218*). Similarly, the molecular weight determinations made on the amine complexes [CuCl(NHEt$_2$)$_2$]$_2$ (*70*), [CuBr(RNH$_2$)$_2$]$_2$ [R = C$_8$H$_{17}$, C$_{10}$H$_{21}$, C$_{18}$H$_{37}$ (*354*), and C$_{12}$H$_{25}$ (*49*)] implied that they are also dimeric. The conductivities of solutions of the bis(dodecylamine) complex in benzene increase with time which is probably due to the rearrangement,

$$[CuXL_2]_2 \rightleftharpoons [CuL_4][CuX_2] \tag{16}$$

This reaction is also believed to take place in nitrobenzene solutions of [CuX(AsMePh$_2$)$_2$]$_2$ complexes (X = Cl, Br) (*267*).

Complex CuCl{SC(NH$_2$)(OEt)}$_2$ can be prepared by reacting the ligand with either copper(I) or copper(II) chloride in ethanol, or with CuCl{SC(NH$_2$)(OEt)} (*101*). Paradoxically, the complex was found to be dimeric in boiling benzene but monomeric in freezing acetic acid (*299*).

C. MONOMERIC COMPLEXES

Nine complexes of this stoichiometry have been established as monomers containing three-coordinate copper(I). Of these the best characterized is the complex between 2-(dimethylarsino)allylbenzene and copper(I) iodide. This complex is monomeric in chloroform and only slightly associated in concentrated acetone solutions. Further, its mull infrared spectra showed no positional change for the weakly infrared active C=C vibration found at 1640 cm^{-1} in the spectrum of the free ligand; additionally, the ^{1}H NMR spectrum of the ligand is little changed on complexation (28).

Fusion of triphenylphosphine or arsine with CuCN results in the formation of CuCN(ZPh$_3$)$_2$ (Z = P, As). Their infrared spectra indicate that, although weak cyano bridging occurs in their crystals, they are monomeric in 2-aminoethanol solution (78). Compound CuCN(PPh$_3$)$_2$ has an apparent molecular weight of 685 (formula wt. 611) in chloroform solution (294).

The bulky ligand tri(cyclohexyl)phosphine reduced copper(II) chloride or bromide in ethanolic solution to CuX{P(C$_6$H$_{11}$)$_3$}$_2$ (X = Cl, Br) complexes. The far-infrared spectrum of the chloro complex indicated that it contained terminal chloride, and the molecular weights of these complexes in solution are consistent with a monomeric formulation (248). Similarly, the molecular weight of CuCl(SbPh$_3$)$_2$—obtained by fusion of the ligand with CuCl—in benzene shows it to be monomeric (329).

Reaction of CuCl(PPh$_3$)$_3$ with NaOPh or NaSPh gives the bis(triphenylphosphine)copper(I) phenato or thiophenato complex. Their molecular weights in solution are much lower than their formula weights, as is the apparent molecular weight of CuCl(PPh$_3$)$_2$ (294). This would seem to indicate that the three complexes are monomeric in solution.

In the absence of any evidence to the contrary, the following complexes must be considered to be monomeric, but further investigation would probably show several to be dimeric with halogen bridges. Difficulties in determining their degree of polymerization arise because of the limited solubilities of some of the complexes, although molecular weight determinations were seldom undertaken in work carried out before the 1950s.

Insoluble complexes include CuI(PPhMe$_2$)$_2$, CuI(p-Me$_2$N·C$_6$H$_4$·PR$_2$)$_2$ (R = Me, Et), and CuI(p-Me$_2$N·C$_6$H$_4$·AsMe$_2$)$_2$ (61, 62); CuX(AsPhMe$_2$)$_2$ (X = Cl, Br) (50), CuBr(SbPh$_3$)$_2$ (178); and bis{tri-(2-thiophenyl)phosphine}chlorocopper(I) (170).

Complexes for which no molecular weight data have been reported despite their solubility are the unstable phosphine complexes, CuX(PH$_3$)$_2$ [X = Cl, Br (165), I (165, 309)], which readily lose one phosphine ligand; bis{tri(cyclohexyl)phosphite}bromocopper(I) (15); CuCl(Me$_2$NPF$_2$)$_2$ (74), which is probably P-bonded; and CuXpy$_2$ [X = Br (341), I (340)]. The corresponding cyano complex (340) may possibly be dimeric since copper(I) cyanide is dimeric in pyridine in contrast to copper(I) bromide (351); CuCN also combines with two molecules of EtNC in diethyl ether to give CuCN(EtNC)$_2$ (161). With the exception of $CuBr(SbPh_3)_2$, which was obtained from $CuBr_2$, all the above complexes have been obtained as a result of reacting the ligand with a copper(I) salt.

D. [CuL₂]X COMPLEXES

The $[CuL_2]X$ salts commonly occur with $L = NH_3$; three have been prepared in liquid ammonia (257, 314):

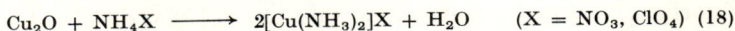

$$[Cu(NH_3)_4][NO_3]_2 + Cu \longrightarrow 2[Cu(NH_3)_2]NO_3 \qquad (17)$$

$$Cu_2O + NH_4X \longrightarrow 2[Cu(NH_3)_2]X + H_2O \qquad (X = NO_3, ClO_4) \quad (18)$$

$$CuI + LiIO_3 \longrightarrow [Cu(NH_3)_2]IO_3 + LiI \qquad (19)$$

By contrast, $[Cu(NH_3)_2]_2SO_4$ or its monohydrate are precipitated by ethanol after either hydrazine sulfate (278) or electrolytic (119) reduction, respectively, of $CuSO_4$ in aqueous ammonia. Copper metal reduces copper(II) chlorate in aqueous ammonia to a colorless solution which presumably contains the unisolated salt $[Cu(NH_3)_2]ClO_3$ (248a).

Copper metal also reduces copper(II) perchlorate to [Cu(4-cyano-pyridine)₂]ClO₄ in butan-2-ol at 353° in the presence of the ligand. The cyano group's infrared absorption at 2240 cm⁻¹ is unchanged on complexation, so the ligands are coordinated solely through the ring nitrogen (117).

Bis{tri(cyclohexyl)phosphine}copper(I) perchlorate is obtained on refluxing the tertiary phosphine with copper(II) perchlorate in ethanol; smaller tertiary phosphines give $[Cu(PR_3)_4]ClO_4$ complexes; thus, this complex is obtained because of the impossibility of coordinating more than two tri(cyclohexyl)phosphine ligands to one copper atom (248).

Compound $[Cu\{SC(NH_2)_2\}_2]Cl$ (195) is perhaps best regarded as an ionic complex since the Cu–Cl distance in its crystals is very long (322); the cation is polymeric and made up of trigonal CuS_3 units sharing corners. Nuclear quadrupole resonance (NQR) experiments also lead to the conclusion that the halogen in $[Cu\{SC(NH_2)_2\}_2]X$ and $[Cu\{SC(NH)_2-C_2H_4\}_2]X$ (X = Cl, Br) complexes is ionic (135). The bis-(2-imidazolidinethione) complexes can be prepared by reacting the ligand with CuX (X = Cl, Br, I) in water (249).

VI. CuXL₃ Complexes

A. COVALENT COMPLEXES

There are a very large number of complexes in this subgroup. The principal complexes isolated are listed in Table III. It has been widely assumed that the ligands form a distorted tetrahedron around the copper atom. Generally speaking, their physical properties are in accordance with this view. They are unionized in solution, but one neutral ligand is often dissociated in solution, e.g., the molecular weight of $CuO_2C \cdot CF_3(PPh_3)_3$ is half its formula weight in chloroform (146) or dichlorobenzene (107). The structure found for $CuNO_3(PMePh_2)_2$ (238) by X-ray crystallography confirms these assumptions. In this compound the large tertiary phosphine groups encroach upon the small monodentate nitrato group. Cursory studies of the structures of CuCl-$(PMePh_2)_3$ (273) and $CuCl(PPh_3)_3$ (6) indicate that they too have a distorted tetrahedral arrangement of ligands.

A variety of methods has been used to prepare these complexes; the following seem to be of the most general application.

From copper(I) salts (1),

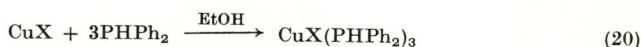

$$CuX + 3PHPh_2 \xrightarrow{EtOH} CuX(PHPh_2)_3 \qquad (20)$$

from other copper(I) complexes (133),

$$[CuCl(PPh_3)]_4 + 8PPh_3 + 4Ph_3GeLi \xrightarrow{(CH_2OMe)_2} 4Cu(GePh_3)(PPh_3)_3 + 4LiCl \qquad (21)$$

or from copper(II) salts using the ligand as reducing agent (176)

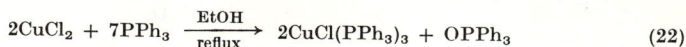

$$2CuCl_2 + 7PPh_3 \xrightarrow[\text{reflux}]{EtOH} 2CuCl(PPh_3)_3 + OPPh_3 \qquad (22)$$

B. IONIC COMPLEXES

With few exceptions, the ionic complexes are derived from sulfur ligands. In $[Cu(SPMe_3)_3]ClO_4$ (241) there is a trigonal planar arrangement of ligands around the central copper atom (111). There is a whole series of $[CuL_3]X$ complexes where $L = Me_3PS$, Et_3PS, Pr_3^iPS, Me_2PhPS, Ph_3PS, Ph_3AsS, or Ph_3PSe and $X = ClO_4$ or BF_4. These have been prepared from the ligand and appropriate copper(II) salt in ethanol, and prior reduction with SO_2 is required for the Ph_3PS, Me_2PhPS, and Ph_3PSe complexes. All the complexes were electrolytes in nitromethane or acetone, and their infrared spectra showed a reduction in the sulfur–pnictide stretch frequency upon coordination.

There also exists a considerable number of ionic tris(thiourea) complexes. Thiourea is capable of reducing copper(II) salts to copper(I) complexes in acid solution to form $[Cu(thiourea)_3]^+$ salts. The following have been isolated: chloride (194, 195), nitrate (194), oxalate (194, 195, 298), monohydrogen arsenate, and phosphate (298).

Neutral ligand	F	Cl	Br	I	NO₃	Anionic RCO₂
NH₃	—	(280)	(280)	(280)	—	—
PHPh₂	—	(1)	(1)	(1)	—	—
PPh₃	{(176)	(56) (176)	(56) (176)	(56)	(177)	(107)[a] (146)[b]
(ortho-phenylene-bis(methylene)PPh ligand)	—	—	—	(75)	—	—
PMe₂Ph	—	(218)	(218)	(218)	(10)	—
PMePh₂	—	(218)	(218)	(218)	(10)	—
PEtPh₂	—	—	—	—	(10)	—
PEt₂(p-CF₃·C₆H₄)	—	—	—	(62)	—	—
P(C≡CPh)Ph₂	—	(188)	—	—	—	—
P(CH₂SiMe₃)₃	—	(168)	—	—	—	—
AsMePh₂	—	(267)	(267)	(49, 267)	—	—
AsPh₃	—	(329)	(178)	—	(177)	—
SbPh₃	—	—	—	—	(177)	—
Me₃CNC	—	—	(272)	—	—	—

[a] R = CF₃.
[b] R = H, CH₃, CF₃, CH₂Cl, Ph, o-tolyl, o-, m-, or p-C₆H₄·NO₂, o-C₆H₄·OH, ½-succinate.

The structures of the chloride (193, 269) and perchlorate have been determined, and both contain copper atoms surrounded tetrahedrally by four sulfur atoms. The perchlorate salt is binuclear (147); four of the six thiourea ligands are monodentate and the remaining two bridge the copper atoms. The cation of the chloride is polymeric consisting of CuS₄ tetrahedra sharing corners. This polymeric structure is obviously degraded by water to thiourea(aquo)copper(I) complexes since the conductivities of [Cu(thiourea)₃]Cl solutions increase markedly with increasing dilution (298).

Tris-(2-imidazolidinethione)copper(I) acetate and sulfate have similarly been obtained by reducing the appropriate copper(II) salt with the ligand in water. They are presumably ionic since solutions of the latter gave an immediate precipitate of BaSO₄ when treated with BaCl₂ (249). Tris-(O-ethylaminothioformate)copper(I) chloride (101, 299) and bromide (299) have been prepared from the copper(I) salts.

Steric hindrance between the methyl groups of adjacent 2-picoline ligands results in the formation of [Cu(C₆H₇N)₃]ClO₄ when Cu(ClO₄)₂ is reduced by copper metal in solution of the ligand (213). Other 2-substituted pyridines, e.g., 2,5-lutidine, 2-ethyl or 2-(isopropyl)pyridine, give tris(amine) complexes, whereas pyridine or 4-picoline give [Cu(amine)₄]ClO₄.

III

COMPLEXES

ligand							
GePh$_3$	SnCl$_3$	BF$_4$	B$_3$H$_8$	B$_9$H$_{14}$	B$_{11}$H$_{14}$	BH$_3$CN	B$_9$H$_{12}$S
—	—	—	—	—	—	—	—
—	—	—	—	—	—	—	—
(133)	(106)	(56)	—	(192)	(192)	(223) (224)	(192)
—	—	—	—	—	—	—	—
—	—	—	—	—	—	—	—
—	—	—	—	—	—	—	—
—	—	—	—	—	—	—	—
—	—	—	—	—	—	—	—
—	—	—	—	—	—	—	—
—	—	—	—	—	—	—	—
—	—	—	—	—	—	—	—
—	—	—	—	—	—	(224)	—
—	—	—	(222)	—	—	(224)	—
—	—	—	—	—	—	—	—

VII. [CuL$_4$]X Complexes

The majority of the [CuL$_4$]X are either perchlorate or nitrate complexes, but some iodide and tetrafluoroborate salts are known.

The most numerous subgroup is that where L is an alkyl or aryl cyanide. A wide variety of preparative methods have been employed, e.g.,

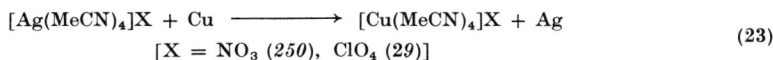

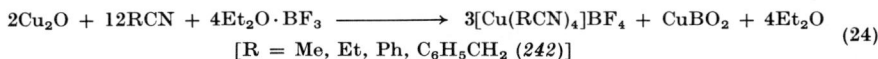

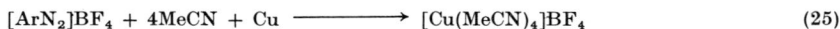

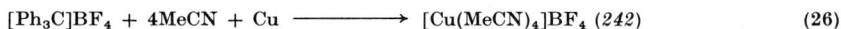

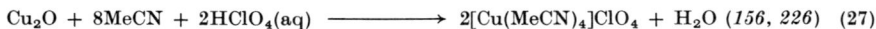

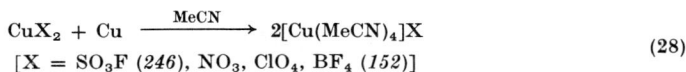

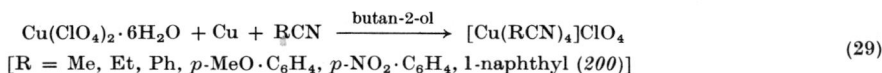

$$[Ag(MeCN)_4]X + Cu \longrightarrow [Cu(MeCN)_4]X + Ag$$
$$[X = NO_3 \ (250), \ ClO_4 \ (29)] \tag{23}$$

$$2Cu_2O + 12RCN + 4Et_2O \cdot BF_3 \longrightarrow 3[Cu(RCN)_4]BF_4 + CuBO_2 + 4Et_2O$$
$$[R = Me, Et, Ph, C_6H_5CH_2 \ (242)] \tag{24}$$

$$[ArN_2]BF_4 + 4MeCN + Cu \longrightarrow [Cu(MeCN)_4]BF_4 \tag{25}$$

$$[Ph_3C]BF_4 + 4MeCN + Cu \longrightarrow [Cu(MeCN)_4]BF_4 \ (242) \tag{26}$$

$$Cu_2O + 8MeCN + 2HClO_4(aq) \longrightarrow 2[Cu(MeCN)_4]ClO_4 + H_2O \ (156, 226) \tag{27}$$

$$CuX_2 + Cu \xrightarrow{\text{MeCN}} 2[Cu(MeCN)_4]X$$
$$[X = SO_3F \ (246), \ NO_3, \ ClO_4, \ BF_4 \ (152)] \tag{28}$$

$$Cu(ClO_4)_2 \cdot 6H_2O + Cu + RCN \xrightarrow{\text{butan-2-ol}} [Cu(RCN)_4]ClO_4$$
$$[R = Me, Et, Ph, p\text{-}MeO \cdot C_6H_4, \ p\text{-}NO_2 \cdot C_6H_4, \ 1\text{-naphthyl} \ (200)] \tag{29}$$

Of these, reactions (28) and (29) have undoubtedly the widest application and are the most convenient.

There is much evidence for the ionic nature of these complexes. Many give conducting solutions in solvents such as methyl cyanide, nitromethane, or nitrobenzene. The infrared spectra of the polyatomic anions are all typical of the free anions.

Pyridine, quinoline, or 4-picoline solutions of $Cu(ClO_4)_2$ are reduced by copper metal to $[Cu(amine)_4]ClO_4$ salts. X-Ray crystallographic investigations reveal the tetrakis(pyridine)copper(I) cation is tetrahedral (212). Compound $[Cupy_4]ClO_4$ can also be prepared by displacement of MeCN from $[Cu(MeCN)_4]ClO_4$ by pyridine under anaerobic conditions (199). The complex $[Cu(p\text{-}CH_3 \cdot C_6H_4 \cdot NC)_4]ClO_4$ is also known (302).

Tertiary phosphines (10, 56, 75, 86), arsines (178, 267, 329), and the phosphite $P(OCH_2)_3CMe$ (342) reduce copper(II) salts to $[CuL_4]X$ complexes if X is a poorly coordinating anion. Almost invariably copper(II) perchlorate yields this type of complex, but copper(II) nitrate forms ionic complexes with these ligands only if certain steric conditions are satisfied, e.g., triphenylarsine gives $CuNO_3(AsPh_3)_3$ (177), but the smaller tertiary arsine $AsMePh_2$ forms the ionic complex $[Cu(AsMePh_2)_4]NO_3$ (267). If triphenylarsine is fused with either CuCl or $CuCl_2$, air-unstable $[Cu(AsPh_3)_4]Cl$ is formed. This salt is a 1:1 electrolyte in nitrobenzene. Complexes $[CuX(AsMePh_2)_2]_2$ (X = Cl, Br, I) are believed to isomerize in nitrobenzene to $[Cu(AsMePh_2)_4][CuX_2]$ salts (49).

If copper(II) perchlorate is reacted with tris(dimethylamino)phosphine sulfide in methyl cyanide, complexes $[(Me_2N)_3PS_2P(NMe_2)_3]$-$[ClO_4]_2$ and $[Cu\{SP(NMe_2)_3\}_4]ClO_4$ are formed (313). This reaction is typical of the reducing action of sulfur ligands. Thus, copper(II) nitrate is reduced to $[Cu(2\text{-imidazolidinethione})_4]NO_3$ by the ligand, and, if this complex is heated with sodium hydroxide, the oxide $[Cu(2\text{-imidazolidinethione})_4]_2O$ is formed. This latter compound is slightly air-sensitive but dissolves in mineral acids to give other copper(I) complexes (249). The nitrate salt has been shown to be ionic by polarography (206). Thioacetamide reduces copper(II) chloride to $[Cu(thioacetamide)_4]Cl$ (90); the infrared spectrum of the cation shows the ligand to be sulfur-bonded (204) in accordance with the X-ray crystal study (333).

VIII. CuX(biL) Complexes

Although copper is nominally three-coordinate in CuX(biL) complexes, bridging by the anionic ligand has been widely proposed (despite very little evidence for this) to increase the coordination number of copper to its customary four.

Many of the iodo complexes may be prepared from copper(I) iodide by the reaction

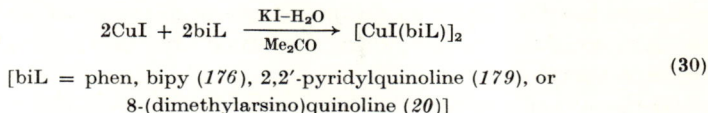

$$2CuI + 2biL \xrightarrow[\text{Me}_2\text{CO}]{\text{KI–H}_2\text{O}} [CuI(biL)]_2$$

[biL = phen, bipy (176), 2,2′-pyridylquinoline (179), or 8-(dimethylarsino)quinoline (20)]

(30)

or by displacement of monodentate ligands from other copper(I) complexes, e.g. (234),

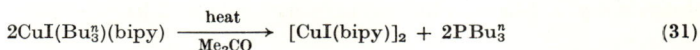

$$2CuI(Bu_3^n)(bipy) \xrightarrow[\text{Me}_2\text{CO}]{\text{heat}} [CuI(bipy)]_2 + 2PBu_3^n$$

(31)

or by reduction of a copper(II) salt in the presence of the ligand and iodide ion, e.g. (328),

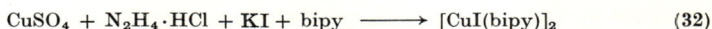

$$CuSO_4 + N_2H_4 \cdot HCl + KI + bipy \longrightarrow [CuI(bipy)]_2$$

(32)

Other iodo complexes have been isolated containing 1,2-bis(dimethylarsino)benzene (181), bis(diphenylphosphino)methane and 1,2-bis-(diphenylphosphino)ethane (237), 2,9-dimethyl-1,10-phenanthroline (145), 8-methylthioquinoline (214), 2-(diphenylphosphino)allylbenzene and 2-(diphenylphosphino)methallylbenzene (28), and 2-(dimethylarsino)-N,N-dimethylaniline (136). The structure of the last mentioned complex was investigated by X-ray crystallography and was found, indeed, to be dimeric. Molecular weight determinations have shown bis{bis(diphenylphosphino)methane}di-μ-iododicopper(I) and the corresponding $Ph_2PC_2H_4PPh_2$ complex to be dimeric in dichloromethane (237), and the halocopper(I) complexes of 2-(diphenylphosphino)allylbenzene and 2-(diphenylphosphino)methallylbenzene (28), and bis-(2,9-dimethyl-1,10-phenanthroline)di-μ-acetatodicopper(I) (146) to be dimeric in chloroform. One factor that may give rise to misleading molecular weight results is the possibility of ionization isomerism, e.g. (49, 181),

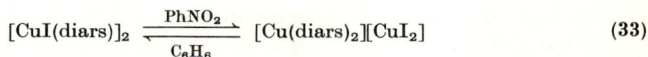

$$[CuI(diars)]_2 \underset{\text{C}_6\text{H}_6}{\overset{\text{PhNO}_2}{\rightleftharpoons}} [Cu(diars)_2][CuI_2]$$

(33)

Ionic complexes of the preceding type are included in Section IX which deals with $[Cu(biL)_2]^+$ complexes.

Only two of the phosphorus atoms in cyclic phosphine ligands $(RP)_4$ ($R = Bu^i$, C_8H_{17}, cyclohexyl) coordinate to the metal and $(Bu^iP)_4CuCl$ is dimeric in benzene. The other cyclotetraphosphines form similar complexes on reaction with copper(I) halides in carbon tetrachloride (160).

The CuCN(biL) complexes (biL = 2,2′-bipyridyl, 1,10-phenanthroline) are monomeric in 2-aminoethanol, but their infrared spectra indicate that they may be di- or polymeric in the solid state; there are no data available to elucidate the structures of the analogous thiocyanato complexes (78, 328).

If the anion is bidentate (but nonbridging), then the complexes are monomeric; this the case for $(Ph_2PC_2H_4PPh_2)CuBH_4$ (56) and probably for $(Ph_2PC_2H_4PPh_2)CuNO_3$ (120).

The tetrameric complex $(RCu)_4$ (R = 2-benzyldimethylamine) is cleaved by 1,2-bis(diphenylphosphino)ethane (diphos):

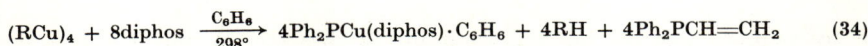

$$(RCu)_4 + 8diphos \xrightarrow[298°]{C_6H_6} 4Ph_2PCu(diphos)\cdot C_6H_6 + 4RH + 4Ph_2PCH{=}CH_2 \qquad (34)$$

It seems likely that this complex is three-coordinate with weakly coordinated benzene occupying the fourth coordination position.

IX. [Cu(biL)₂]X Complexes

The [Cu(biL)₂]X compounds are particularly well-characterized since many are potentially useful in the spectrophotometric determination of copper. The cations containing two molecules of 1,10-phenanthroline or 2,2′-bipyridyl, or their derivatives have molar absorbtivities (ε) in the range 4000–7000 M^{-1} dm² for their charge transfer bands in the visible spectrum, and have been extensively investigated (166, 173, 216, 240, 251, 284, 315–317). The charge transfer bands in the visible range are lost, however, if partial hydrogenation of the ligand disrupts its extended π-orbital system (336).

The presence of two bidentate, π-bonding ligands stabilizes copper(I) toward oxidation or disproportionation under analytical conditions. Further, it has been found that the greater the basicity of the nitrogenous ligand, the less readily is the complex oxidized at an electrode (174). Spectral and polarographic studies showed that copper-(I) was four-coordinate in these complexes, but this has not been confirmed by isolation and analysis of the complexes of the more esoteric ligands in the two series (8, 335).

The complexes may be prepared by (a) the action of reducing agents on copper(II) complexes of the ligands (144),

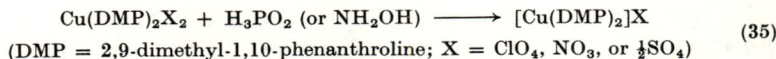

$$Cu(DMP)_2X_2 + H_3PO_2 \text{ (or NH}_2\text{OH)} \longrightarrow [Cu(DMP)_2]X$$

(DMP = 2,9-dimethyl-1,10-phenanthroline; X = ClO₄, NO₃, or ½SO₄) $\qquad (35)$

TABLE IV
ISOLATED $[Cu(biL)_2]X$ COMPLEXES

biL	X	References
	I ClO_4 NO_3 Br	(176, 328) (308) (177) (178)
	ClO_4, BF_4	(179)
	NO_3 ClO_4, BF_4	(177) (179)
	Phenylalaninate, tyrosinate, tryptophanate ClO_4, NO_3, $\frac{1}{2}SO_4$ Cl, Br, I, I_3	(208) (144) (145)
	ClO_4, NO_3, I, Br·H_2O, Cl·H_2O, $[CuX_2]$ (X = Cl, Br, I)	(143)
	NO_3 BF_4	(177) (179)
	Br, I, $[CuX_2]$ (X = Cl, Br, I)	(49, 181)
	$[CuI_2]$	(148)
	ClO_4	(20)
	R = Et; Cl, Br, I, $[CuCl_2]$, 4-bromocamphorsulfonate R = Ph; ClO_4	(73) (260)

(continued)

TABLE IV—*continued*

biL	X	References
[pyridine]–CH$_2$SMe	ClO$_4$	(65)
[pyridine]–(CH$_2$)$_2$SMe	ClO$_4$	(64)
[quinoline]–SR	R = Me; HSO$_4$, HCO$_3$, [CuCl$_2$]; R = Me, Ph; ClO$_4$	(155)
[benzene]–AsMe$_2$, SMe	Cl, Br, I, NCS, ClO$_4$	(67)
Ph$_2$PC$_2$H$_4$PPh$_2$	NO$_3$	(10)
	ClO$_4$	(120)
	CF$_3$CO$_2$	(107)
2-Aminopyridine	I, [CuX$_2$] (X = Cl, Br, I, NCS)	(326)
Me$_2$As(CH$_2$)$_3$SMe	ClO$_4$, [CuX$_2$] (X = Cl, Br, I)	(68)

(*b*) direct interaction of copper(I) halides and ligands (*181*),

$$\text{CuX} + 2\text{diars} \xrightarrow[\text{H}_2\text{O}]{\text{KX}} [\text{Cu(diars)}_2]\text{X} \qquad (36)$$

(diars = 1,2-bis(dimethylarsino)benzene; X = Br, I)

if a 1:1 ratio of reactants is used, then [CuX$_2$]$^-$ salts commonly result,

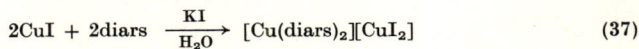

$$2\text{CuI} + 2\text{diars} \xrightarrow[\text{H}_2\text{O}]{\text{KI}} [\text{Cu(diars)}_2][\text{CuI}_2] \qquad (37)$$

(*c*) using the ligand to reduce a copper(II) salt (*20*),

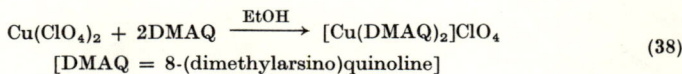

$$\text{Cu(ClO}_4)_2 + 2\text{DMAQ} \xrightarrow{\text{EtOH}} [\text{Cu(DMAQ)}_2]\text{ClO}_4 \qquad (38)$$

[DMAQ = 8-(dimethylarsino)quinoline]

and (*d*) displacement of monodentate ligands from a copper(I) complex (*177*),

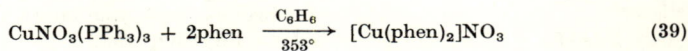

$$\text{CuNO}_3(\text{PPh}_3)_3 + 2\text{phen} \xrightarrow[353°]{\text{C}_6\text{H}_6} [\text{Cu(phen)}_2]\text{NO}_3 \qquad (39)$$

Isolated complexes of $[Cu(biL)_2]X$ are listed in Table IV. All the complexes are electrolytes in suitable ionizing solvents.

Exceptions to the simple monomeric cation structure are encountered when the bidentate ligand is an α,ω-dinitrile. For example, the complex $[Cu(CNCH_2CH_2CN)_2]NO_3$, first prepared in 1923 (*250*), has recently been shown to have a polynuclear cation made up of

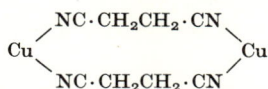

$$
\begin{array}{ccc}
 & NC \cdot CH_2CH_2 \cdot CN & \\
Cu & & Cu \\
 & NC \cdot CH_2CH_2 \cdot CN & \\
\end{array}
$$

units and free nitrate ions distributed throughout the crystal (*189*); the corresponding complexes containing $CN(CH_2)_3CN$ and $CN(CH_2)_4CN$ ligands have similar structures (*190*). The cation in $[Cu\{(CNCH_2CH_2)_2\}_2]$-ClO_4 is isostructural with that of the nitrate salt (*36*).

X. Organocopper(I) Compounds and Complexes

The widespread use of copper(I) compounds in organic syntheses (*116*) has served to draw attention to the organometallic compounds, which are themselves useful synthetic reagents in organic chemistry (*262*) (Fig. 9).

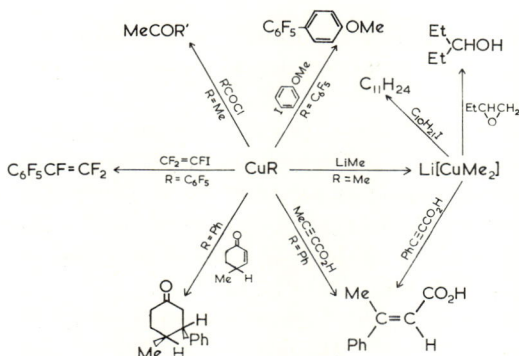

FIG. 9. Some synthetic uses of organocopper compounds. (Data from Ref. *262*)

A. ALKYL AND ARYL COMPOUNDS

The reactive and unstable alkyl and aryl compounds may be prepared at low temperatures from the alkyls and aryls of other metals as follows:

$$CuX + RLi \xrightarrow[273°]{Et_2O} CuR + LiX$$

(40)

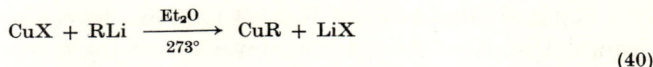

[X = I, R = Me (*167*), *cis-* or *trans*-but-2-enyl (*352*);
X = Br, R = Ph (*80*), *o*-tolyl, *p*-tolyl, *o*-anisyl (*55*)]

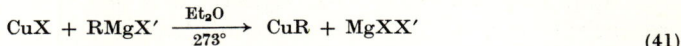

$$CuX + RMgX' \xrightarrow[273°]{Et_2O} CuR + MgXX'$$

(41)

[X = I, X' = Br, R = Ph (*150, 293*), *p*-tolyl (*150*); X = X' = I, R = Ph (*131*);
X = Cl, X' = Br, R = Et in tetrahydrofuran (THF) at 195° (*345*)]

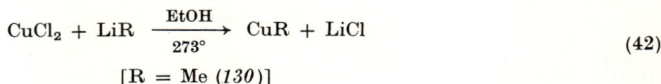

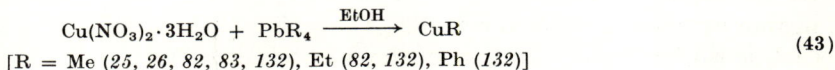

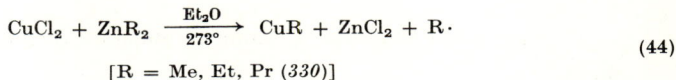

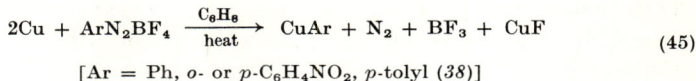

$$CuCl_2 + LiR \xrightarrow[273°]{EtOH} CuR + LiCl$$

(42)

[R = Me (*130*)]

$$Cu(NO_3)_2 \cdot 3H_2O + PbR_4 \xrightarrow{EtOH} CuR$$

[R = Me (*25, 26, 82, 83, 132*), Et (*82, 132*), Ph (*132*)]

(43)

$$CuCl_2 + ZnR_2 \xrightarrow[273°]{Et_2O} CuR + ZnCl_2 + R\cdot$$

(44)

[R = Me, Et, Pr (*330*)]

$$2Cu + ArN_2BF_4 \xrightarrow[heat]{C_6H_6} CuAr + N_2 + BF_3 + CuF$$

(45)

[Ar = Ph, *o*- or *p*-C$_6$H$_4$NO$_2$, *p*-tolyl (*38*)]

The formation of copper(I) fluoride in reaction (45) seems unlikely, but low yields of aryl copper compounds are obtained owing to pyrolysis during preparation.

None of these methods appears to be of general applicability; a disadvantage of reaction (40)—at least in the case of methylcopper—is that the precipitated alkyl dissolves in excess reagent to give Li[CuR$_2$]. Preparation from Grignard reagents requires careful choice of halides if optimum yields are to be obtained, and the preparation of ethylcopper is feasible only at low temperatures (*345*).

The decomposition of alkyl and aryl copper compounds has been the subject of much debate. Initially free radical mechanisms were advanced, such theories being supported by the reduced yields of hydrocarbons in the presence of benzoquinone or other free-radical scavengers (*24, 150*). However, under all conditions very poor yields of dimeric alkanes are obtained. These would be the likely products of a free-radical decomposition,

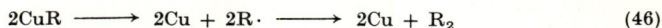

$$2CuR \longrightarrow 2Cu + 2R\cdot \longrightarrow 2Cu + R_2$$

(46)

Instead, RH and R−H alkene are formed in approximately equal

quantities (*345*), and the following scheme, in which copper(I) hydride participates, has been suggested to explain these products (*352*):

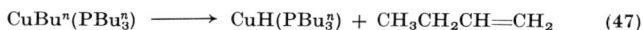

$$CuBu^n(PBu_3^n) \longrightarrow CuH(PBu_3^n) + CH_3CH_2CH{=}CH_2 \qquad (47)$$

$$CuH(PBu_3^n) + CuBu^n(PBu_3^n) \longrightarrow 2Cu + 2PBu_3^n + CH_3(CH_2)_2CH_3 \qquad (48)$$

Biaryls are, however, the main decomposition products of ArCu compounds even though free-radical reactions are not believed to be involved.

Thermally stable arylcopper(I) derivatives result if there is a good donor atom in an aryl side chain, since chelation can occur, e.g.,

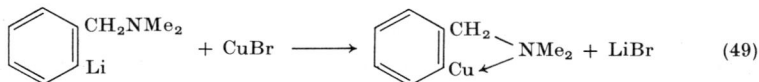

Various aryl-substituted analogs of 2-cupriobenzyldimethylamine can also be obtained by reaction (49) (*338*). The 4-methyl compound has been found to be tetrameric (*140*).

B. FLUOROALKYL AND -ARYL COMPOUNDS

The electron-withdrawing power of fluorine reduces the electron density at the copper–carbon bond and renders it less reactive. Therefore these complexes are more stable than their analogs containing hydrogen.

They can be prepared from Grignard reagents (*53*),

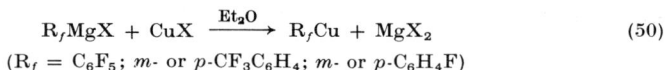

$$R_f MgX + CuX \xrightarrow{\text{Et}_2\text{O}} R_f Cu + MgX_2 \qquad (50)$$
$$(R_f = C_6F_5;\ m\text{- or } p\text{-}CF_3C_6H_4;\ m\text{- or } p\text{-}C_6H_4F)$$

or by exchange reactions,

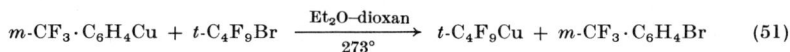

$$m\text{-}CF_3 \cdot C_6H_4Cu + t\text{-}C_4F_9Br \xrightarrow[273°]{\text{Et}_2\text{O--dioxan}} t\text{-}C_4F_9Cu + m\text{-}CF_3 \cdot C_6H_4Br \qquad (51)$$

The fluoroalkyls are obtained as 1:1 and 2:1 dioxan adducts after the magnesium halides have been precipitated by this reagent. Their enhanced thermal stability allows them to be freed from dioxan by heating them *in vacuo* for 5 hr at 403°. At higher temperatures they pyrolyze to biaryls, but the *m*-trifluoromethylphenylene compound decomposes in stages (*54*)

$$[m\text{-}CF_3C_6H_4Cu]_8 \longrightarrow (m\text{-}CF_3C_6H_4)_6Cu_8 + (CF_3C_6H_4)_2 \qquad (52)$$

$$(m\text{-}CF_3C_6H_4)_6Cu_8 \longrightarrow (m\text{-}CF_3C_6H_4)_4Cu_8 + (CF_3C_6H_4)_2 \qquad (53)$$

The intermediate compounds are mixed Cu(I)–Cu(0) complexes. Perfluorophenylcopper decomposes directly to fluorocarbon and copper

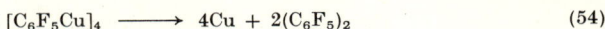

$$[C_6F_5Cu]_4 \longrightarrow 4Cu + 2(C_6F_5)_2 \tag{54}$$

The polymericity of these compounds has been determined by osmometry (54), cryoscopy (52, 54), and mass spectrometry (52). They are hydrolyzed or oxidized by air to biaryls. Despite being oxidized to biaryls by dibenzoyl peroxide, they do not give rise to free radicals in these decompositions.

C. Complex Copper(I) Alkyls and Aryls

Complexing the unstable organocopper compounds with π-bonding ligands fails to confer much additional stability, but several complexes are sufficiently stable to be isolated. It seems best to prepare them from an existing copper(I) complex, e.g. (167),

$$[CuI(PBu_3^n)]_4 + 4LiMe \xrightarrow{273°} 4CuMe(PBu_3^n) + 4LiI \tag{55}$$

rather than to react an organocopper compound with ligand, e.g. (330),

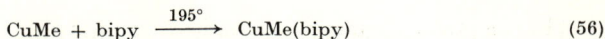

$$CuMe + bipy \xrightarrow{195°} CuMe(bipy) \tag{56}$$

Other complexes that have been isolated include $CuAr(py)_3$ [Ar = Ph, p-$NO_2 \cdot C_6H_4$ (38)], CuMeDMSO (330), $CuMe(PPh_3)_3$ (84), and $(CuPh)_n(PPh_3)$ [n = 2, 3 (81)]. Compound Cu(2-benzyldimethylamine)-(diphos) has been prepared by cleaving tetrakis-(2-cupriobenzyldi-methylamine) with $Ph_2PC_2H_4PPh_2$ (339).

D. Olefin Complexes

Olefinic complexes of copper have been reviewed in Volume 12 of this series (289). Since the appearance of this review, [Cu(cycloocta-1,5-diene)$_2$]BF$_4$ has been prepared by electrolysis of $Cu(BF_4)_2$ in methanolic diolefin at copper electrodes (231). Complexes (cycloocta-1,5-diene)$CuO_2C \cdot CF_3$, (cyclooctatetrene)$(CuO_2C \cdot CF_3)_2$, and (cyclo-octa-1,5-diene)$(CuO_2C \cdot CF_3)_2$ have been obtained as white or pale yellow solids by reacting $CuO_2C \cdot CF_3$ with the olefins in pentane or benzene, respectively (107). The structure of $Cu_2Cl_2(trans$-cyclooctene)$_3$ (Fig. 10) has been determined (124) and is reminiscent of the structure of $Cu_2Cl_2(PPh_3)_3$ (Fig. 5).

When copper(I) oxide is dissolved in a benzene solution of trifluoro-methane sulfonic acid the complex $Cu(CF_3SO_3) \cdot \frac{1}{2}C_6H_6$ results (319).

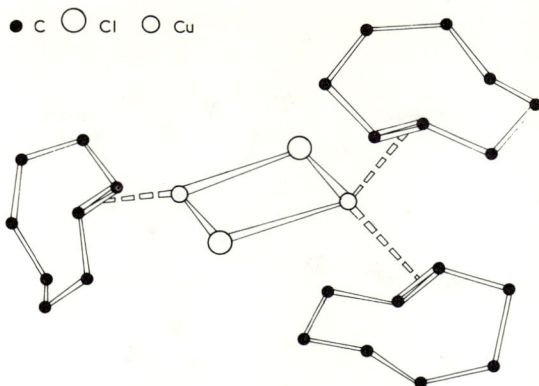

● C ◯ Cl ◯ Cu

FIG. 10. Structure of $Cu_2Cl_2(trans\text{-cyclooctene})_3$. [Redrawn by permission from *Chem. Commun.* p. 1054 (1969).]

This reacts with cycloocta-1,5-diene or *endo*-dicyclopentadiene to give $[Cu(diolefin)_2]O_3S \cdot CF_3$ complexes, and with cyclooctatetrene, cyclododeca-1,5,9-triene, cycloocta-1,3-diene, or norbornylene to give $Cu(olefin)O_3S \cdot CF_3$ complexes; cyclohepta-1,3,5-triene can replace benzene in the parent complex yielding $Cu(CF_3SO_3) \cdot \frac{1}{2}(triolefin)$.

Substituted olefins also complex with CuCl. The halide dissolves in allyl alcohol owing to complex formation (*183, 184*). Other unsaturated alcohols give complexes of the type (alkenol)CuCl and [(alkenol)Cu]$^+$ when added to solutions of CuCl in aqueous HCl (*185*). Similar complexes are formed with various unsaturated carboxylic acids, and the stability of the complexes is influenced by the stereochemistry about the double bond (*13*). Maleic acid, however, forms the anionic complex $[CuCl(O_2C \cdot CH{=}CH \cdot CO_2H)]^-$ in addition to the neutral and cationic complexes (*11, 12*).

E. CYCLOPENTADIENYL COMPLEXES

The cyclopentadienyl complex $Cu(C_5H_5)PEt_3$ was formerly believed to contain a σ-bonded cyclopentadienyl group (*355*). This erroneous conclusion was reached because of its reactivity toward acids, but it was at variance with the behavior of π-bonded complexes such as ferrocene. Careful reappraisal of its infrared spectrum (*87, 122*) led to the view that the cyclopentadienyl group was π-bonded; this was confirmed by X-ray crystallography (*103*). The structure of $Cu(\pi\text{-}C_5H_5)(PPh_3)$ (Fig. 11) has been determined to a high degree of precision (*89*).

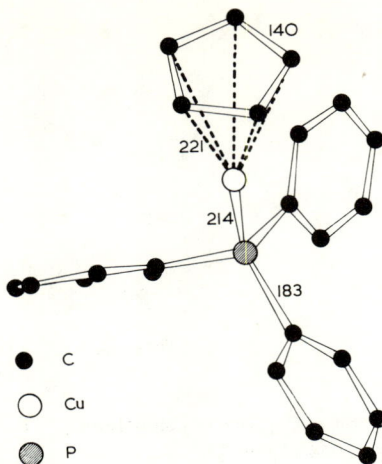

FIG. 11. Structure of $Cu(\pi\text{-}C_5H_5)(PPh_3)$ (bond lengths in picometers). [Redrawn by permission from *J. Amer. Chem. Soc.* **92**, 2353 (1970).]

The most convenient route to the complexes is from an iodocopper(I) complex (*88*),

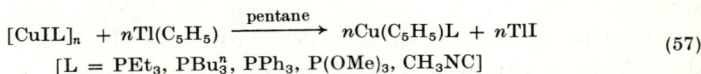

$$[CuIL]_n + nTl(C_5H_5) \xrightarrow{\text{pentane}} nCu(C_5H_5)L + nTlI \tag{57}$$
$$[L = PEt_3, PBu_3^n, PPh_3, P(OMe)_3, CH_3NC]$$

However, cyclopentadienyl and indenyl t-butyl isocyanide complexes have been obtained by the original method, that is, by heating Cu_2O with hydrocarbon and ligand. Although $Cu(\pi\text{-}C_5H_5)(Bu^tNC)$ is undoubtedly isostructural with the tertiary phosphine complexes, the indenyl complex, $Cu(\text{indenyl})(Bu^tNC)_3$, is almost certainly σ-bonded on account of its infrared and NMR spectra to say nothing of its stoichiometry (*303*).

An unstable carbonyl complex (*88*) has also been prepared,

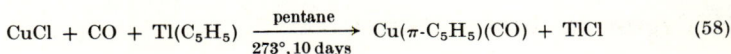

$$CuCl + CO + Tl(C_5H_5) \xrightarrow[273°, 10\,\text{days}]{\text{pentane}} Cu(\pi\text{-}C_5H_5)(CO) + TlCl \tag{58}$$

F. CARBONYL COMPLEXES

Although copper(I) compounds have been widely used as adsorbents for CO in classic methods of gas analysis (*324*), the stability of the complexes is low, and the 1:1 complexes of CO and copper(I) halides have high dissociation pressures (*30, 233, 346*). Apart from the cyclopentadienyl complex already mentioned, the only really stable complexes are those with copper(I) trifluoroacetate (*310*). These may be prepared either in trifluoroacetic acid,

$$Cu_2O + CO + CF_3CO_2H \longrightarrow CuO_2C \cdot CF_3(CO) \cdot CF_3CO_2H \qquad (59)$$

or in aqueous solution

$$Cu_2O + CO + (CF_3CO)_2O + H_2O \longrightarrow CuO_2C \cdot CF_3(CO) \qquad (60)$$

Trifluoroacetic acid may be removed from the adduct *in vacuo*, but it requires prolonged pumping to remove carbon monoxide. The C–O stretch in the infrared spectrum of $CuO_2C \cdot CF_3(CO)$ occurs at 2155 cm^{-1} a value similar to that of 2093 cm^{-1} for the cyclopentadienyl complex (*88*).

G. Acetylides and Acetylene Complexes

Acetylene reacts with copper(I) halides in ammoniacal solution to give the dangerously explosive copper(I) acetylide (*301*); a monohydrate has also been prepared (*306*). The compound can also be prepared in liquid ammonia (*256*).

Monosubstituted acetylenes give polymeric monoacetylides:

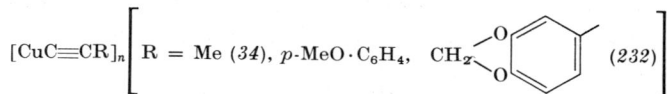

The methylacetylide is probably a tetramer such as the gold(I) compound, and there may be interaction between the metal atoms and triple bonds in addition to the acetylide linkage.

If the monosubstituted acetylides are prepared in liquid ammonia, alkynylocuprate(I) complexes can be formed if there is excess of alkali acetylide present (*225*):

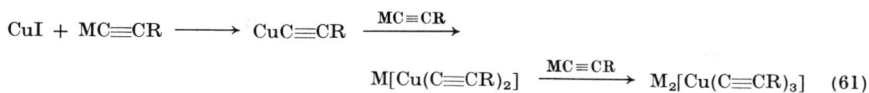

$$CuI + MC{\equiv}CR \longrightarrow CuC{\equiv}CR \xrightarrow{MC{\equiv}CR}$$

$$M[Cu(C{\equiv}CR)_2] \xrightarrow{MC{\equiv}CR} M_2[Cu(C{\equiv}CR)_3] \qquad (61)$$

$$(M = Na, K; R = H, Me, Ph)$$

Several of these complexes are precipitated as ammoniates, and they also undergo metathetical reactions:

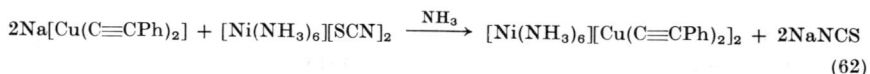

$$2Na[Cu(C{\equiv}CPh)_2] + [Ni(NH_3)_6][SCN]_2 \xrightarrow{NH_3} [Ni(NH_3)_6][Cu(C{\equiv}CPh)_2]_2 + 2NaNCS$$
$$(62)$$

The polymeric acetylides are cleaved by tertiary phosphines or aryl isocyanides, e.g.,

$$[CuC \equiv CBu^t]_n + nPMe_3 \xrightarrow{C_6H_6} n/3[Cu(C \equiv CBu^t)(PMe_3)]_3 \qquad (63)$$

This yellow-green complex is trimeric in benzene, but dimeric in nitrobenzene (72). By analogous reactions, many other tertiary phosphine complexes of monoalkylacetylides have been obtained. Among these is $[Cu(C \equiv CPh)PMe_3]_4$ (Fig. 12), in which two copper atoms are each

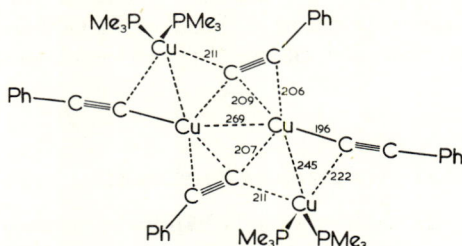

FIG. 12. Structure of $[Cu(C \equiv CPh)PMe_3]_4$ (bond lengths in picometers). [Redrawn by permission from *Acta Crystallogr.* **21**, 957 (1966).]

bound to two trimethylphosphine ligands and the tetramer is held together by metal–metal bonds and interactions with the acetylene ligands (79).

Complexes $Cu(C \equiv C \cdot C \equiv CPh)(PR_3)_n$ have also been prepared and are claimed to be more stable to aerial decomposition than the monoacetylide complexes (72).

Weak ammoniates of $[Cu(C \equiv CPh)]_3$ are formed when the complex is dissolved in liquid ammonia (34). A stronger monomeric complex is formed when the phenylacetylide is reacted with p-tolyl isocyanide in pyridine–diethyl ether at 263° K (191).

At low temperatures in acid or ethanolic solution, CuCl adds acetylene to give $CuCl \cdot C_2H_2$ (63), and $(CuCl)_3 \cdot C_2H_2$ (164), respectively. These observations are supported by a study of the slow equilibrium between the reactants (271). Addition compounds are formed between CuCl and phenylacetylene or (232),

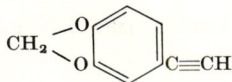

These are difficult to free from alkyne and, in ammoniacal solution, are converted to the corresponding acetylides.

The structure of CuCl(hepta-1,6-diyne) consists of Cl–Cu–Cl chains cross-linked by the diyne. Compound CuCl(but-2-yne) is tetrameric, and has a puckered Cu_4Cl_4 ring with the copper atoms on the perpendiular bisectors of the triple bonds. Unlike the C=C bond in Zeise's salt, the C≡C bond in the alkyne complex is coplanar with the copper and two adjacent chlorine atoms (58).

XI. Miscellaneous Compounds

A. COMPLEXES CONTAINING NEUTRAL BRIDGING LIGANDS

Various pyrazines (210, 211) and quinoxalines (32) react with copper(I) halides to give complexes of the general formula $(CuX)_2(biN)$. The infrared spectrum of $(CuCN)_2$quinoxaline shows the cyano groups to be terminal, and it seems most likely that the complexes have the structure XCu(biN)CuX. Complexes of similar formula have been obtained by reducing $CuCl_2$ with hypophosphorous acid or KI in the presence of azo-2-pyridine (19). However, it is believed that these contain trigonal copper and have the structure shown in Fig. 13.

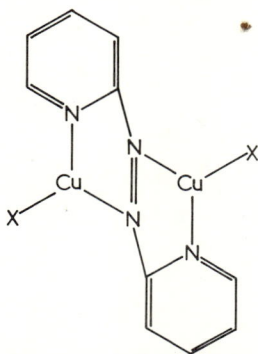

FIG. 13. Probable structure of [CuX(azo-2-pyridine)]₂ complexes.

The linear ditertiary phosphine $Ph_2PC≡CPPh_2$ (DPPA) forms six complexes of the formula $(CuX)_2(DPPA)_3$ [X = Cl, Br, I, NCS, NO_3 (60), BH_4 (59, 60)], which are believed to have the structure shown in Fig. 14. It has been assumed in the past that the $(CuX)_2(diphos)_3$ complexes [X = Cl, Br, I (237), NO_3 (10), N_3 (128)] had a similar structure. However, recent X-ray structural determinations on the azido (128) (Fig. 15) and chloro (4) complexes show them to contain only one bridging diphos ligand, and the other diphos complexes are almost certainly isostructural. The azido complex can be reacted with trifluoromethyl cyanide in chloroform to give a bis[2-{5(trifluoromethyl)-tetrazolato}] complex wherein the two copper atoms are linked by a diphos ligand (126).

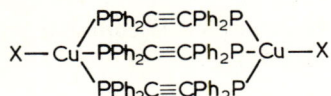

$$X—Cu \underset{\underset{PPh_2C\equiv CPh_2P}{\overset{PPh_2C\equiv CPh_2P}{-PPh_2C\equiv CPh_2P-}}}{} Cu—X$$

FIG. 14. Probable structure of the $(CuX)_2(Ph_2PC\equiv CPPh_2)_3$ complexes.

There are several other complexes containing the ligands Ph_2PC_2-H_4PPh_2 or $(Ph_2P)_2CH_2$ that may contain neutral bridging ligands. They are best isolated from reactions of copper(I) halides under specified conditions and include $Cu_4X_4(biL)_2$ (X = Cl, Br, I, biL = $(Ph_2P)_2CH_2$;

● C   ○ Cu   ◉ P   ○ N

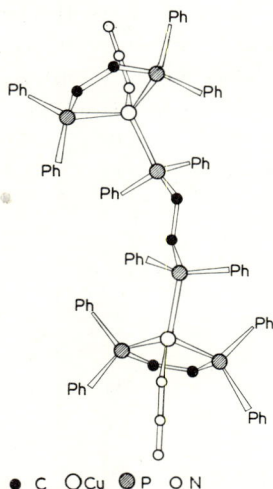

FIG. 15. Structure of $Cu_2(N_3)_2(diphos)_3$. [Redrawn by permission from *Inorg. Chem.* **10**, 2776 (1972).]

X = Cl, Br, biL = $Ph_2PC_2H_4PPh_2$). The molecular weights of the $Cu_4Cl_4(diphos)_2$ complex in both dichloroethane and chloroform are in accordance with this formula, but the solutions are unstable as their conductivity increases with time (237). Similar complexes containing one more molecule of bidentate ligand, i.e., $Cu_4X_4(biL)_3$, may be isolated from benzene solutions. Eight tricopper(I) complexes have also been prepared. The $[CuX(diphos)]_3$ (X = Cl, Br) complexes are trimeric in dilute chloroform solution but have higher apparent molecular weights in more concentrated solutions. This is also the case with the $Cu_3X_3(biL)_2$ (X = Cl, Br, I, biL = $(Ph_2P)_2CH_2$, $Ph_2PC_2H_4$-PPh_2) complexes. The diphos complexes appear to be coordinatively unsaturated since they readily retain chlorinated solvent in their crystals.

Polymeric complexes are formed when copper(I) chloride reacts with dialkylhydrazines (105) or with 3,5,5-trimethylpyrazolidine. In $Cu_2Cl_2(MeN=NMe)$ the structure consists of parallel Cl–Cu–Cl chains cross-linked by weak Cu–Cl bonds and strong Cu–N σ bonds (47). Structures of CuI(PhN=NH) and $Cu_4Cl_4(PhN=NH)$ may be similar (282, 290). Diazoaminobenzene copper(I) (110, 245) can be prepared from copper and the ligand; it is dimeric with each copper linearly coordinated to 1,1'N or 3,3'N atoms (48). The cation in $[Cu(PhN_2Ph)]$-ClO_4 may have a related structure (265).

B. COMPLEXES OF TRIDENTATE LIGANDS

6,4',6''-Triphenyl-2,2',2''-tripyridyl gives colored, complex, copper(I) ions suitable for spectrophotometric analysis (307). Complexes CuBr-(triars) [triars = bis-(2-dimethylarsinophenyl)methylarsine] (142), $CuI[MeAs\{(CH_2)_3AsMe_2\}_2]$ (21), and $CuCl\{Me(CS)CH_2(CS)CH_2(CS)Me\}$ (123) have also been prepared. These complexes are monomeric and analogs of the $CuXL_3$ complexes. It is possible that tri-(2-pyrryl)-phosphine is tridentate (via two nitrogen atoms and the phosphorus atom) in its chloro- and bromocopper complexes obtained from the copper(II) salts.

1,3,5-Trithian yields polymeric complexes when reacted with CuCl in ethanol. Complexes $(CuCl)_3(C_3H_6S_3)$ and $(CuCl)_3(C_3H_6S_3)_2$ may be isolated, depending on the conditions used. The structure of the latter complex contains chains of copper atoms alternately bridged by two chloro and two trithian ligands. The chains are cross-linked by bonding between copper atoms and the third sulfur atom of trithian (108).

C. COMPLEXES CONTAINING DITHIO LIGANDS

Complexes with dithio ligands have been isolated containing the following ligands: $S_2P(OPr^i)_2$ (104), S_2CNRR' [R = R' = Me, Et, Pr^n, Bu^n, C_5H_{11}, Pr^i, Bu^i, iso-C_5H_{11}, cyclohexylene, cyclopentylene; R = Et, R' = Pr^n; R = Me, R' = Ph (2, 3)], and S_2PR_2 (R = Et, Pr^n) (201, 202).

The molecular weights of the N,N-dialkyldithiocarbamato complexes indicate that they are tetramers in solution. This has been confirmed by X-ray crystallography of the diethyl complex (158) which has

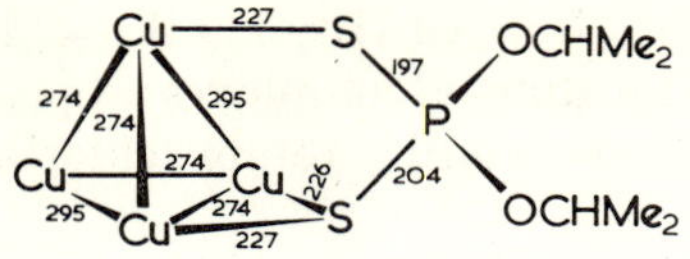

FIG. 16. Partial structure of tetrakis[O,O-di(isopropyl)phosphorodithioato-copper(I)] (bond lengths in picometers). [Redrawn by permission from *Inorg. Chem.* **11**, 612 (1972).]

a similar structure to that of tetrakis[O,O-di(isopropyl)phosphorodi-thioatocopper(I)] (*209*) (Fig. 16). By contrast, N,N-diethyldiseleno-carbamatocopper(I) is monomeric (*337*). However, N,N-dipropyl-monothicarbamatocopper(I) is hexameric and contains a central octa-hedron of copper atoms (*159*). Higher polymerization occurs in the anion of [PhMe$_3$N]$_4$[Cu$_8${S$_2$C$_2$(CN)$_2$}$_6$] (*115*) which contains a cube of copper atoms embedded in a distorted icosahedron of sulfur atoms (*239*).

D. POLYNUCLEAR COMPLEXES CONTAINING OTHER SULFUR LIGANDS

The ability of the sulfur atom in thiourea and related ligands to bridge copper(I) atoms results in a large number of complex structures. Unfortunately, many were first isolated before the advent of modern techniques of structural investigation.

One compound, [Cu$_4${SC(NH$_2$)$_2$}$_9$][NO$_3$]$_4$ (*344*), has had its structure determined by X-ray crystallography, which revealed that there were five different types of Cu–S bonds in the polymeric structure. Other thiourea salts known include [Cu$_2${SC(NH$_2$)$_2$}$_5$]X$_2$ (X = NO$_3$, $\frac{1}{2}$SO$_4$) (*298*), and the nitrate and cyanate of the similar cation [Cu$_2$(2-imidazoli-dinethione)$_5$H$_2$O]$^{2+}$ (*249*).

Aquothiourea complexes are produced by hydrolysis of thiourea-copper(I) salts, e.g., [Cu{SC(NH$_2$)$_2$}$_2$H$_2$O]X (X = NO$_3$, HSO$_4$, ClO$_4$, $\frac{1}{2}$C$_2$O$_4$) (*195*).

The phenylthiocarbamide complexes [Cu$_2$(PhCSNH$_2$)$_6$]SO$_4$ and [Cu$_2$(PhCSNH$_2$)$_3$]X$_2$ (X = Cl, $\frac{1}{2}$SO$_4$) (*305*) are also known.

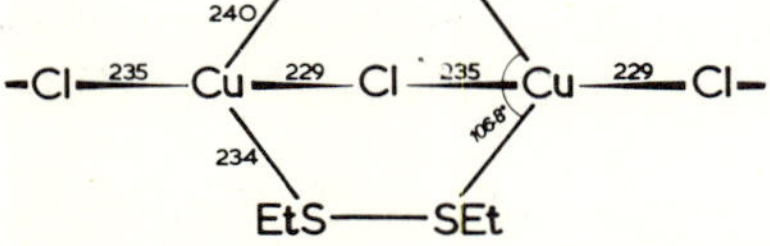

FIG. 17. Structure of [CuCl(Et$_2$S$_2$)]$_\infty$ (bond lengths in picometers). [Redrawn by permission from *Acta Chem. Scand.* **21**, 1000 (1967).]

An interesting bridged structure is found in $CuClEt_2S_2$ (Fig. 17) which contains both chloro and disulfide bridges; the complex was prepared by reacting CuCl with diethyl disulfide (39).

E. COMPLEXES CONTAINING TWO DIFFERENT NEUTRAL LIGANDS

These complexes with two different neutral ligands may be obtained by displacement or addition reactions (176):

$$[CuX(PR_3)]_4 + 4phen \longrightarrow 4CuX(PR_3)(phen) \tag{64}$$

$$CuX(PPh_3)_3 + phen \longrightarrow CuX(PPh_3)(phen) \tag{65}$$

If the anion is poorly coordinating, ionic complexes result (56), e.g., $[Cu(PPh_3)_2(phen)]ClO_4$.

The behavior of the nitrato group in $CuNO_3(PPh_3)_3$ is intermediate (177): only large bidentate ligands form $CuNO_3(PPh_3)(biN)$ complexes in boiling benzene, whereas at room temperature $[Cu(PPh_3)_2(biN)]NO_3$ complexes are formed. With $CuNO_3(ZPh_3)_3$ (Z = As, Sb) only the first type of complex is formed, owing to the weakness of Cu–As and Cu–Sb bonds.

Tertiary arsines, phosphines, and stibines can also be displaced from their $CuX(ZPh_3)_3$ or $Cu_2X_2(Z'Ph_3)_3$ (Z = As, P; Z' = As, P, Sb; X = halogen) complexes by pyridine and related amines to give $CuX(ZPh_3)(amine)$ complexes. The structures of these are uncertain since their dissociation or decomposition in solution makes molecular weight data unreliable (176, 178).

F. OTHER COMPLEXES

The following unusual complexes have been isolated: a red polymeric complex of formula $[CuClpy(C_6H_4O_2)]_\infty$ reputed to contain bridging p-benzoquinone molecules (151); the salt $[Cu(HCN)]F$, obtained by dissolving CuCN in liquid HF (109); and $(CuCN)_3 \cdot MeI$ from a sealed tube reaction of the components (149). Similarly, $(CuCl)_2 \cdot PCl_3$ results from dissolving CuCl in PCl_3 (99). A structural study of the first complex would be helpful and reinvestigation of the remainder beneficial.

There are several well-defined mixed Cu(I)–Cu(II) complexes containing cyano or thiocyanato anions. These are $Cu(II)Cu(I)_2(NH_3)_4$-$(CN)_4$, its 1,2-diaminoethane analog $Cu(II)Cu(I)_2(en)_2(CN)_4$ (77), and $Cu(II)Cu(I)(NH_3)_3(NCS)_3$ (125). The last complex is polymeric and contains tetrahedrally coordinated copper(I) atoms which are S-bound to two thiocyanato ligands and N-bound to two further thiocyanato ligands. The distorted octahedral coordination of the copper(II) atoms is made up of three ammine ligands, one N-bonded thiocyanato group, and two long bonds to the S atoms of other thiocyanato groups. All the NCS groups bridge two copper atoms.

ACKNOWLEDGMENT

The author is grateful to Dr. A. G. Vohra for his assistance with the literature survey.

REFERENCES

1. Abel, E. W., McLean, R. A. N., and Sabherwal, I. H., *J. Chem. Soc., A* p. 133 (1969).
2. Åkerström, S., *Acta Chem. Scand.* **10**, 699 (1956).
3. Åkerström, S., *Ark. Kemi* **14**, 387 (1959).
4. Albano, V. G., Bellon, P. L., and Ciani, G., *J. Chem. Soc., Dalton Trans.* p. 1938 (1972).
5. Albano, V. G., Bellon, P. L., Ciani, G., and Manassero, M., *J. Chem. Soc., Dalton Trans.* p. 171 (1972).
6. Albano, V. G., Bellon, P. L., and Sansoni, M., *J. Chem. Soc., A* p. 2420 (1971).
7. Altermatt, J. A., and Manahan, S. E., *Inorg. Nucl. Chem. Lett.* **4**, 1 (1968).
8. Anderegg, G., *Helv. Chim. Acta* **46**, 2397 (1963).
9. Anders, U., and Plambeck, J. A., *Can. J. Chem.* **47**, 3055, (1969).
10. Anderson, W. A., Carty, A. J., Palenik, G. J., and Schreiber, G., *Can. J. Chem.* **49**, 761 (1971).
11. Andrews, L. J., and Keefer, R. M., *J. Amer. Chem. Soc.* **70**, 3261 (1948).
12. Andrews, L. J., and Keefer, R. M., *J. Amer. Chem. Soc.* **71**, 2379 (1949).
13. Andrews, L. J., and Keefer, R. M., *J. Amer. Chem. Soc.* **71**, 2381 (1949).
14. Ang, H. G., Kow, W. E., and Mok, K. F., *Inorg. Nucl. Chem. Lett.* **8**, 829 (1972).
15. Arbuzov, A. E., and Valitova, F. G., *Bull. Acad. Sci. USSR, Div. Chem. Sci.* p. 723 (1952).
16. Arbuzov, A. E., and Zorastrova, V. M., *Bull. Acad. Sci. USSR, Div. Chem. Sci.* p. 697 (1952).
17. Arbuzov, A. E., and Zorastrova, V. M., *Bull. Acad. Sci. USSR, Div. Chem. Sci.* p. 713 (1952).
18. Baker, R. J., Nyburg, S. C., and Szymanski, J. T., *Inorg. Chem.* **10**, 138 (1971).
19. Baldwin, D. A., Lever, A. B. P., and Parish, R. V., *Inorg. Chem.* **8**, 107 (1969).
20. Barclay, G. A., Harris, C. M., and Kingston, J. V., *Chem. Ind. (London)* p. 227 (1965).
21. Barclay, G. A., and Nyholm, R. S., *Chem. Ind. (London)* p. 378 (1953).
22. Bartlett, M., and Palenik, G. J., *Acta Crystallogr., Sect. A* **25**, S173 (1969).
23. Basset, H., and Corbet, A. S., *J. Chem. Soc., London* **125**, 1660 (1924).
24. Bawn, C. E. H., and Johnson, R., *J. Chem. Soc., London* p. 4162 (1960).
25. Bawn, C. E. H., and Whitby, F. J., *Discuss. Faraday Soc.* **2**, 228 (1947).

26. Bawn, C. E. H., and Whitby, F. J., *J. Chem. Soc.*, *London* p. 3926 (1960).
27. Beall, H., Bushweller, C. H., Dewkett, W. J., and Grace, M., *J. Amer. Chem. Soc.* **92**, 3484 (1970).
28. Bennett, M. A., Kneen, W. R., and Nyholm, R. S., *Inorg. Chem.* **7**, 552 (1968).
29. Bergerhoff, G., *Z. Anorg. Chem.* **327**, 139 (1964).
30. Berthelot, M., *Ann. Chim. Phys.* **23**, 32 (1901).
31. Bezman, S. A., Churchill, M. R., Osborn, J. A., and Wormald, J., *J. Amer. Chem. Soc.* **93**, 2063 (1971).
32. Billing, D. E., and Underhill, A. E., *J. Chem. Soc.*, *London* p. 6639 (1965).
33. Bjerrum, J., and Nielsen, E. J., *Acta Chem. Scand.* **2**, 297 (1948).
34. Blake, D., Calvin, G., and Coates, G. E., *Proc. Chem. Soc.*, *London* p. 396 (1959).
35. Blitz, H., and Herms, P., *Ber. Deut. Chem. Ges.* **40**, 974 (1907).
36. Blount, J. F., Freeman, H. C., Hemmerich, P., and Sigwart, C., *Acta Crystallogr.*, *Sect. B* **25**, 1518 (1969).
37. Bodländer, G., and Storbeek, O., *Z. Anorg. Chem.* **31**, 1 (1902).
38. Bolth, F. A., Whalley, W. M., and Starkey, E. B., *J. Amer. Chem. Soc.* **65**, 1456 (1943).
39. Brändén, C. I., *Acta. Chem. Scand.* **21**, 1000 (1967).
40. Braune, H., and Englebrecht, G., *Z. Phys. Chem.*, *B* **11**, 409 (1930).
41. Brehmer, T., *Finska Kemistsamfundets Medd.* **57**, 67 (1948).
42. Brice, V. T., and Shore, S. G., *Chem. Commun.* p. 1312 (1970).
43. Brink, C., Binnendijk, N. F., and van de Linde, J., *Acta Crystallogr.* **7**, 176 (1954).
44. Brink, C., and MacGillavry, C. H., *Acta Crystallogr.* **2**, 158 (1949).
45. Brink, C., and van Arkel, A. E., *Acta Crystallogr.* **5**, 506, (1952).
46. Brinkhoff, H. C., Matthijessen, A. G., and Oomes, C. G., *Inorg. Nucl. Chem. Lett.* **7**, 87 (1971).
47. Brown, I. D., and Dunitz, J. D., *Acta Crystallogr.* **13**, 28 (1960).
48. Brown, I. D., and Dunitz, J. D., *Acta Crystallogr.* **14**, 480 (1961).
49. Burkin, A. R., *J. Chem. Soc.*, *London* p. 538 (1956).
50. Burrows, G. J., and Sandford, E. P., *J. Proc. Roy. Soc. N.S.W.* **69**, 182 (1935).
51. Bütz, W., and Stollenwerk, W., *Z. Anorg. Chem.* **119**, 97 (1921).
52. Cairncross, A., Omura, H., and Sheppard, W. A., *J. Amer. Chem. Soc.* **93**, 248 (1971).
53. Cairncross, A., and Sheppard, W. A., *J. Amer. Chem. Soc.* **90**, 2186 (1968).
54. Cairncross, A., and Sheppard, W. A., *J. Amer. Chem. Soc.* **93**, 247 (1971).
55. Camus, A., and Marsich, N., *J. Organometal. Chem.* **14**, 441 (1968).
56. Cariati, F., and Naldini, L., *Gazz. Chim. Ital.* **95**, 3 (1965).
57. Cariati, F., and Naldini, L., *J. Inorg. Nucl. Chem.* **28**, 2243 (1966).
58. Carter, F. L., and Hughes, E. W., *Acta Crystallogr.* **10**, 801 (1957).
59. Carty, A. J., and Efraty, A., *Can. J. Chem.* **46**, 1598 (1968).
60. Carty, A. J., and Efraty, A., *Inorg. Chem.* **8**, 543 (1969).
61. Cass, R. C., Coates, G. E., and Hayter, R. G., *Chem. Ind. (London)* p. 1485 (1954).
62. Cass, R. C., Coates, G. E., and Hayter, R. G., *J. Chem. Soc.*, *London* p. 4007 (1955).
63. Chavastelon, M., *C. R. Acad. Sci.* **126**, 1810 (1898).
64. Chia, P. S. K., Livingstone, S. E., and Lockyer, T. N., *Aust. J. Chem.* **19**, 1835 (1966).

65. Chia, P. S. K., Livingstone, S. E., and Lockyer, T. N., *Aust. J. Chem.* **20**, 239 (1967).

66. Chioboli, P., and Testa, C., *Ann. Chim.* (*Paris*) [13] **47**, 639 (1957).

67. Chiswell, B., and Livingstone, S. E., *J. Chem. Soc., London* p. 2931 (1959).

68. Chiswell, B., and Livingstone, S. E., *J. Inorg. Nucl. Chem.* **23**, 37 (1961).

69. Churchill, M. R., Bezman, S. A., Osborn, J. A., and Wormald, J., *Inorg. Chem.* **11**, 1818 (1972).

70. Clifton, J. R., and Yoke, J. T., *Inorg. Chem.* **5**, 1630 (1966).

71. Clifton, J. R., and Yoke, J. T., *Inorg. Chem.* **6**, 1258 (1967).

72. Coates, G. E., and Parkin, C., *J. Inorg. Nucl. Chem.* **22**, 59 (1961).

73. Cochran, W., Hart, F. A., and Mann, F. G., *J. Chem. Soc., London* p. 2816 (1957).

74. Cohn, K., and Parry, R. W., *Inorg. Chem.* **7**, 46 (1968).

75. Collier, J. W., Fox, A. R., Hinton, I. G., and Mann, F. G., *J. Chem. Soc., London* p. 1819 (1964).

76. Cooper, D., and Plane, R. A., *Inorg. Chem.* **5**, 16 (1966).

77. Cooper, D., and Plane, R. A., *Inorg. Chem.* **5**, 1677 (1966).

78. Cooper, D., and Plane, R. A., *Inorg. Chem.* **5**, 2209 (1966).

79. Corfield, P. W. R., and Shearer, H. M. M., *Acta Crystallogr.* **21**, 957 (1966).

80. Costa, G., Camus, A., Gatti, L., and Marisch, N., *J. Organometal. Chem.* **5**, 568 (1966).

81. Costa, G., Camus, A., Marisch, N., and Gatti, L., *J. Organometal. Chem.* **8**, 339 (1967).

82. Costa, G., Camus, A. M., and Pauluzzi, E., *Gazz. Chim. Ital.* **86**, 997 (1956).

83. Costa, G., and de Alti, G., *Gazz. Chim. Ital.* **87**, 1273 (1957).

84. Costa, G., Pellizer, G., and Rubessa, F., *J. Inorg. Nucl. Chem.* **26**, 961 (1964).

85. Costa, G., Reisenhofer, E., and Stefani, L., *J. Inorg. Nucl. Chem.* **27**, 2581 (1965).

86. Cotton, F. A., and Goodgame, D. M. L., *J. Chem. Soc., London* p. 5267 (1960).

87. Cotton, F. A., and Marks, T. J., *J. Amer. Chem. Soc.* **91**, 7281 (1969).

88. Cotton, F. A., and Marks, T. J., *J. Amer. Chem. Soc.* **92**, 5114 (1970).

89. Cotton, F. A., and Takats, J., *J. Amer. Chem. Soc.* **92**, 2353 (1970).

90. Cox, E. G., Wardlaw, W., and Webster, K. C., *J. Chem. Soc., London* p. 775 (1936)

91. Creighton, J. A., and Lippincot, E. R., *J. Chem. Soc., London* p. 5314 (1963).

92. Cromer, D. T., *J. Phys. Chem.* **61**, 1388 (1957).

93. Cromer, D. T., and Larson, A. C., *Acta Crystallogr.* **15**, 397 (1962).

94. Cromer, D. T., Larson, A. C., and Roof, R. B., *Acta Crystallogr.* **19**, 192 (1965).

95. Dalziel, J. A. W., Holding, A. F., and Watts, B. E., *J. Chem. Soc., A* p. 358 (1967).

96. Datta, R. L., and Sen, J. N., *J. Amer. Chem. Soc.* **39**, 750 (1917).

97. Datta, R. L., and Sen, J. N., *J. Amer. Chem. Soc.* **39**, 759 (1917).

98. Davidson, J. M., *Chem. Ind.* (*London*) p. 2021 (1964).

99. Davies, T. L., and Ehrlich, P., *J. Amer. Chem. Soc.* **58**, 2151 (1936).

100. Davis, D. G., *Anal. Chem.* **30**, 1729 (1958).

101. Debus, H., *Ann. Chem. Pharm.* **82**, 253 (1852).

102. Dehrain, P., *C. R. Acad. Sci.* **55**, 808 (1862).

103. Delabaere, L. T. J., McBride, D. W., and Ferguson, R. B., *Acta Crystallogr.*, *Sect. B* **26**, 515 (1970).

104. Dickert, J. J., and Rowe, C. N., *J. Org. Chem.* **32**, 647 (1967).

105. Diels, O., and Koll, W., *Justus Liebigs Ann. Chem.* **443**, 262 (1925).

106. Diltz, J. A., and Johnson, M. P., *Inorg. Chem.* **5**, 2079 (1966).

107. Dines, M. B., *Inorg. Chem.* **11**, 2949 (1972).

108. Domenicano, A., Spagna, R., and Vaciago, A., *Chem. Commun.* p. 1291 (1968).

109. Dove, M. F. A., and Hallett, J. G., *J. Chem. Soc.*, *A* p. 2781 (1969).

110. Dwyer, F. P., *J. Amer. Chem. Soc.* **63**, 78 (1941).

111. Eller, P. G., and Corfield, P. W. R., *Chem. Commun.* p. 105 (1971).

112. Endicot, J. F., and Taube, H., *J. Amer. Chem. Soc.* **86**, 1686 (1964).

113. Espenson, J. H., Shaw, K., and Parker, O. J., *J. Amer. Chem. Soc.* **89**, 5730 (1967).

114. Ewart, G., Lane, A. P., and Payne, D. S., *Proc. Int. Conf. Coord. Chem.*, *7th, 1962* Abstracts, p. 229 (1962).

115. Fackler, J. P., and Coucouvanis, D., *J. Amer. Chem. Soc.* **88**, 3913 (1966).

116. Fanta, P. E., *Chem. Rev.* **64**, 163 (1964).

117. Farha, F., and Iwamoto, R. T., *Inorg. Chem.* **4**, 844 (1965).

118. Fisher, P. J., Taylor, N. E., and Harding, M. M., *J. Chem. Soc.*, *London* p. 2303 (1960).

119. Foerster, F., and Blankenberg, F., *Ber. Deut. Chem. Ges.* **39**, 4428 (1906).

120. Fogleman, W. W., *Diss. Abstr. B* **29**, 3246-B (1969).

121. Fridman, Y. D., and Sarbaev, D. S., *Russ. J. Inorg. Chem.* **4**, 835 (1959).

122. Fritz, H. P., *Advan. Organometal. Chem.* **1**, 288 (1964).

123. Furuhashi, A., Kawano, M., Tashiro, N., and Ouchi, A., *J. Inorg. Nucl. Chem.* **34**, 2961 (1972).

124. Ganis, P., Lepore, U., and Paiaro, G., *Chem. Commun.* p. 1054 (1969).

125. Garaj, J., *Inorg. Chem.* **8**, 304 (1969).

126. Gaughan, A. P., Bowman, K. S., and Dori, Z., *Inorg. Chem.* **11**, 601 (1972).

127. Gaughan, A. P., Ziolo, R. F., and Dori, Z., *Inorg. Chim. Acta.* **4**, 640 (1970).

128. Gaughan, A. P., Ziolo, R. F., and Dori, Z., *Inorg. Chem.* **10**, 2776 (1971).

129. Gibson, D., Johnson, B. F. G., and Lewis, J., *J. Chem. Soc.*, *A* p. 367 (1970).

130. Gilman, H., Jones, R. G., and Woods, L. A., *J. Org. Chem.* **17**, 1630 (1952).

131. Gilman, H., and Straley, J. M. S., *Rec. Trav. Chim. Pays-Bas* **55**, 821 (1936).

132. Gilman, H., and Woods, L. A., *J. Amer. Chem. Soc.* **65**, 435 (1943).

133. Glocking, F., and Hooton, K. A., *J. Chem. Soc. London* p. 2658 (1962).

134. Golub, A. M., and Skopenko, V. V., *Russ. J. Inorg. Chem.* **5**, 961 (1960).

135. Graybeal, J. D., and Ing, S. D., *Inorg. Chem.* **11**, 3104 (1972).

136. Graziani, R., Bombieri, G., and Forsellini, E., *J. Chem. Soc.*, *A* p. 2331 (1971).

137. Grossmann, H., *Ber. Deut. Chem. Ges.* **36**, 1600 (1903).

138. Grossmann, H. and von der Forst, P., *Z. Anorg. Chem.* **43**, 94 (1905).

139. Guillemard, H., *Ann. Chim. Phys.* [2] **14**, 429 (1908).

140. Guss, J. M., Mason, R., Stofte, I., van Koten, G., and Noltes, J. G., *Chem. Commun.* p. 446 (1972).

141. Gyunner, E. A., and Yakhkind, N. D., *Russ. J. Inorg. Chem.* **13**, 1420 (1968).

142. Haines, R. J., Nyholm, R. S., and Stiddard, M. H. B., *J. Chem. Soc.*, *A* p. 46 (1968).

143. Hall, J. R., Litzow, M. R., and Plowman, R. A., *Aust. J. Chem.* **18**, 1339 (1965).

144. Hall, J. R., Marchant, N. K., and Plowman, R. A., *Aust. J. Chem.* **15**, 480 (1962).

145. Hall, J. R., Marchant, N. K., and Plowman, R. A., *Aust. J. Chem.* **16**, 35 (1963).

146. Hammond, B., Jardine, F. H., and Vohra, A. G., *J. Inorg. Nucl. Chem.* **33**, 1017 (1971).

147. Hanic, F., and Ďurčanská, *Inorg. Chim. Acta* **3**, 293 (1969).

148. Hart, F. A., and Mann, F. G., *J. Chem. Soc., London* p. 3939 (1957).

149. Hartley, E. G. J., *J. Chem. Soc., London* p. 780 (1928).

150. Hashimoto, H., and Nakano, T., *J. Org. Chem.* **31**, 891 (1966).

151. Hashimoto, H., Noma, T., and Kawaki, T., *Tetrahedron Lett.* No. 30, 3411 (1968).

152. Hathaway, B. J., Holah, D. G., and Postlethwaite, J. D., *J. Chem. Soc., London* p. 3215 (1961).

153. Hathaway, B. J., and Stephens, F., *J. Chem. Soc., A* p. 884 (1970).

154. Hawkins, C. J., and Perrin, D. D., *J. Chem. Soc., London* p. 2996 (1963).

155. Hein, F., and Vogt, K., *Justus Liebigs Ann. Chem.* **689**, 202 (1965).

156. Hemmerich, P., and Sigwart, C., *Experientia* **19**, 488 (1963).

157. Henry, P. M., *Inorg. Chem.* **5**, 688 (1966).

158. Hesse, R., *Ark. Kemi* **20**, 481 (1963).

159. Hesse, R., and Aava, U., *Acta Chem. Scand.* **24**, 1355 (1970).

160. Hicks, D. G., and Dean, J. A., *Chem. Commun.* p. 172 (1965).

161. Hofmann, K. A., and Bugge, G., *Ber. Deut. Chem. Ges.* **40**, 1772 (1907).

162. Hofmann, K. A., and Höchtlen, F., *Ber. Deut. Chem. Ges.* **36**, 1146 (1903).

163. Hofmann, K. A., and Höchtlen, F., *Ber. Deut. Chem. Ges.* **36**, 3090 (1903).

164. Hofmann, K. A., and Küspert, F., *Z. Anorg. Chem.* **15**, 204 (1897).

165. Höltje, R., and Schlegel H., *Z. Anorg. Chem.* **243**, 246 (1940).

166. Hoste, J., *Anal. Chim. Acta* **4**, 23 (1950).

167. House, H. O., Respess, W. L., and Whitesides, G. M., *J. Org. Chem.* **31**, 3128 (1966).

168. Hsieh, A. T. T., Ruddick, J. D., and Wilkinson, G., *J. Chem. Soc., Dalton Trans.* p. 1966 (1972).

169. Irving, H., and Jonason, M., *J. Chem. Soc., London* p. 2095 (1960).

170. Issleib, K., and Brack, A., *Z. Anorg. Chem.* **292**, 245 (1957).

171. Iyengar, R. R., Sathyanarayana, D. N., and Patel. C. C., *J. Inorg. Nucl. Chem.* **34**, 1088 (1972).

172. Izatt, R. M., Johnson, H. D., Watt, G. D., and Christiansen, J. J., *Inorg. Chem.* **6**, 132 (1967).

173. James, B. R., Parris, M., and Williams, R. J. P., *J. Chem. Soc., London* p. 4630 (1961).

174. James, B. R., and Williams, R. J. P., *J. Chem. Soc., London* p. 2007 (1961).

175. Jardine, F. H., unpublished results.

176. Jardine, F. H., Rule, L., and Vohra, A. G., *J. Chem. Soc., A* p. 238 (1970).

177. Jardine, F. H., Vohra, A. G., and Young, F. J., *J. Inorg. Nucl. Chem.* **33**, 2941 (1971).

178. Jardine, F. H., and Young, F. J., *J. Chem. Soc., A* p. 2444 (1971).

179. Jardine, F. H., and Young, F. J., unpublished results.

180. Jones, L. H., *J. Chem. Phys.* **29**, 463 (1958).

181. Kabesh, A., and Nyholm, R. S., *J. Chem. Soc., London* p. 38 (1951).

182. Kauffman, G. B., and Teter, L. A., *Inorg. Syn.* **7**, 10 (1959).
183. Keefer, R. M., and Andrews, L. J., *J. Org. Chem.*, **13**, 208 (1948).
184. Keefer, R. M., and Andrews, L. J., *J. Amer. Chem. Soc.* **71**, 1723 (1949).
185. Keefer, R. M., Andrews, L. J., and Kepner, R. E., *J. Amer. Chem. Soc.* **71**, 3906 (1949).
186. Kiang-Shu Chang and Ya-Teh Cha, *J. Chin. Chem. Soc. (Peiping)* **2**, 293 (1934).
187. Kiang-Shu Chang and Yen-Ming Liu, *J. Chin. Chem. Soc. (Peiping)* **2**, 307 (1934).
188. King, R. B., and Efraty, A., *Inorg. Chim. Acta* **4**, 319 (1970).
189. Kinoshita, Y., Matsubara, I., and Saito, Y., *Bull. Chem. Soc. Jap.* **32**, 741 (1959).
190. Kinoshita, Y., Matsubara, I., and Saito, Y., *Bull. Chem. Soc. Jap.* **32**, 1216 (1959).
191. Klages, F., and Mönkemeyer, K., *Chem. Ber.* **85**, 109 (1952).
192. Klanberg, F., Muetterties, E. L., and Guggenberger, L. J., *Inorg. Chem.* **7**, 2272 (1968).
193. Knobler, C. B., Okaya, Y., and Pepinsky, R., *Z. Kristallogr.* **111**, 385 (1959).
194. Kohlschütter, V., *Ber. Deut. Chem. Ges.* **36**, 1151 (1903).
195. Kohlschütter, V., and Brittlebank, C., *Justus Liebigs Ann. Chem.* **349**, 232 (1906).
196. Kohn, M., *Monatsh. Chem.* **33**, 919 (1912).
197. Kolthoff, I. M., and Coetzee, J. F., *J. Amer. Chem. Soc.* **79**, 1852 (1957).
198. Kowala, C., and Swan, J. M., *Aust. J. Chem.* **19**, 555 (1966).
199. Kuang-Ling Chen, and Iwamoto, R. T., *Inorg. Nucl. Chem. Lett.* **4**, 499 (1968).
200. Kubota, M., and Johnson, D. L., *J. Inorg. Nucl. Chem.* **29**, 769 (1967).
201. Küchen, W., and Hertel, H., *Angew. Chem., Int. Ed. Engl.* **8**, 89 (1969).
202. Küchen, W., and Mayatepek, H., *Chem. Ber.* **101**, 3454 (1968).
203. Kunschert, F., *Z. Anorg. Chem.* **41**, 337 (1904).
204. Kutzelnigg, W., and Mecke, R., *Spectrochim. Acta* **17**, 530 (1961).
205. Laitinen, H. A., Onstott, E. I., Bailar, J. C., and Swan, S., *J. Amer. Chem. Soc.* **71**, 1550 (1949).
206. Lane, T. J., Quagliano, J. V., and Bertin, E., *Anal. Chem.* **29**, 481 (1957).
207. Latimer, W. M., "Oxidation Potentials," Chapters 9 and 11. Prentice-Hall, Englewood Cliffs, New Jersey, 1961.
208. Laurie, S. H., *Aust. J. Chem.* **20**, 2597 (1967).
209. Lawton, S. L., Rohrbaugh, W. J., and Kokotailo, G. T., *Inorg. Chem.* **11**, 612 (1972).
210. Lever, A. B. P., Lewis, J., and Nyholm, R. S., *Nature (London)* **189**, 58 (1961).
211. Lever, A. B. P., Lewis, J., and Nyholm, R. S., *J. Chem. Soc., London* p. 3156 (1963).
212. Lewin, A. H., Michl, R. J., Ganis, P., Lepore, U., and Avitabile, G., *Chem. Commun.* p. 1400 (1971).
213. Lewin, A. H., Michl, R. J., Ganis, P., and Lepore, U., *Chem. Commun.* p. 661 (1972).
214. Lindoy, L. F., Livingstone, S. E., and Lockyer, T. N., *Aust. J. Chem.* **19**, 1391 (1966).
215. Linell, R. H., and Manfredi, D., *J. Phys. Chem.* **64**, 497 (1960).

216. Lingane, J. J., *Ind. Eng. Chem., Anal. Ed.* **15**, 583 (1943).

217. Lippard, S. J., Lewis, D. F., and Welcker, P. S., *J. Amer. Chem. Soc.* **92**, 3805 (1970).

218. Lippard, S. J., and Mayerle, J. J., *Inorg. Chem.* **11**, 753 (1972).

219. Lippard, S. J., and Melmed, K. M., *Inorg. Chem.* **6**, 2223 (1967).

220. Lippard, S. J., and Melmed, K. M., *J. Amer. Chem. Soc.* **89**, 3929 (1967).

221. Lippard, S. J., and Melmed, K. M., *Inorg. Chem.* **8**, 2755 (1969).

222. Lippard, S. J., and Ucko, D. J., *Inorg. Chem.* **7**, 1051 (1968).

223. Lippard, S. J., and Welcker, P. S., *Chem. Commun.* p. 515 (1970).

224. Lippard, S. J., and Welcker, P. S., *Inorg. Chem.* **11**, 6 (1972).

225. Lloyd, S. J., *J. Phys. Chem.* **12**, 398 (1908).

226. Lontie, R., Blaton, V., Albert, M., and Peeters, B., *Arch. Int. Physiol. Biochim.* **73**, 150 (1965).

227. Luume, P., and Junkkarinen, K., *Suom. Kemistelehti B* **41**, 122 (1968).

228. Macfarlane, A. J., and Williams, R. J. P., *J. Chem. Soc., A* p. 1517 (1969).

229. Malik, A. U., *Z. Anorg. Allg. Chem.* **344**, 107 (1966).

230. Malik, A. U., *J. Inorg. Nucl. Chem.* **29**, 2106 (1967).

231. Manahan, S. E., *Inorg. Nucl. Chem. Lett.* **3**, 383 (1967).

232. Manchot, W., *Justus Liebigs Ann. Chem.* **387**, 257 (1912).

233. Manchot, W., and Friend, J. N., *Justus Liebigs Ann. Chem.* **359**, 100 (1907).

234. Mann. F. G., Purdie, D., and Wells, A. F., *J. Chem. Soc., London* p. 1503 (1936).

235. Mann, F. G., and Wells, A. F., *Nature (London)* **140**, 502 (1937).

236. Marsh, W. C., and Trotter, J., *J. Chem. Soc., A* p. 1482 (1971).

237. Marsich, N., Camus, A., and Cebulec, E., *J. Inorg. Nucl. Chem.* **34**, 933 (1972).

238. Matthew, M., Palenik, G. J., and Carty, A. J., *Can. J. Chem.* **49**, 4119 (1971).

239. McCandlish, L. E., Bissell, E. C., Coucouvanis, D., Fackler, J. P., and Knox, K., *J. Amer. Chem. Soc.* **90**, 7357 (1968).

240. McCurdy, W. H., and Smith, G. F., *Analyst* **77**, 846 (1952).

241. Meek, D. W., and Nicpon, P., *J. Amer. Chem. Soc.* **87**, 4951 (1965).

242. Meerwein, H., Hederich, V., and Wunderlich, K., *Arch. Pharm. (Weinheim)* **291**, 541 (1958).

243. Messmer, G. G., and Palenik, G. J., *Can. J. Chem.* **47**, 1440 (1969).

244. Messmer, G. G., and Palenik, G. J., *Inorg. Chem.* **8**, 2750 (1969).

245. Meunier, L., and Rigot, M., *Bull. Soc. Chim. Fr.* **23**, 103 (1900).

246. Milne, J. B., *Can. J. Chem.* **48**, 75 (1970).

247. Mitscherlich, E., *J. Prakt. Chem.* **19**, 449 (1840).

248. Moers, F. G., and Op Het Veld, P. H., *J. Inorg. Nucl. Chem.* **32**, 3225 (1970).

248a. Möller, H., and Leschewski, K., *Z. Anorg. Allg. Chem.* **243**, 346 (1940).

249. Morgan, G. T., and Burstall, F. H., *J. Chem. Soc., London* p. 143 (1928).

250. Morgan, H. H., *J. Chem. Soc., London*, p. 2901 (1923).

251. Moss, M. L., Mellon, M. G., and Smith, G. F., *Ind. Eng. Chem., Anal. Ed.* **14**, 931 (1942).

252. Muetterties, E. L., *J. Amer. Chem. Soc.* **92**, 4115 (1970).

253. Muetterties, E. L., and Wright, C. M., *Quart. Rev., Chem. Soc.* **21**, 109 (1967).

254. Murray-Rust, P., Day, P., and Prout, C. K., *Chem. Commun.* p. 277 (1966).

255. Nast, R., and Pfab, W., *Chem. Ber.* **89**, 415 (1956).

256. Nast, R., and Pfab, W., *Z. Anorg. Allg. Chem.* **292**, 287 (1957).

257. Nast, R., and Schultze, C., *Z. Anorg. Allg. Chem.* **307**, 15 (1960).

258. Nelson, I. V., and Iwamoto, R. T., *Inorg. Chem.* **1**, 151 (1962).
259. Neves, E. F. A., and Senise, P., *J. Inorg. Nucl. Chem.* **34**, 1915 (1972).
260. Nicpon, P., and Meek, D. W., *Inorg. Chem.* **6**, 145 (1967).
261. Nord, H., *Acta Chem. Scand.* **9**, 430 (1955).
262. Normant, J., *Synthesis* p. 63 (1972).
263. Noyes, A. A., and Chow, M., *J. Amer. Chem. Soc.* **40**, 739 (1918).
264. Nunes, T. L., *Inorg. Chem.* **9**, 1325 (1970).
265. Nuttall, R. H., Roberts, E. R., and Sharp, D. W. A., *J. Chem. Soc., London* p. 2854 (1962).
266. Nyholm, R. S., *J. Chem. Soc., London* p. 1767 (1951).
267. Nyholm, R. S., *J. Chem. Soc., London* p. 1257 (1952).
268. Nyholm, R. S., Truter, M. R., and Bradford, C. W., *Nature (London)* **228**, 648 (1970).
269. Okaya, Y., and Knobler, C. B., *Acta Crystallogr.* **17**, 928 (1964).
270. Onstott, E. I., and Laitinen, H. A., *J. Amer. Chem. Soc.* **72**, 4724 (1950).
271. Osterlöf, J., *Acta Chem. Scand.* **4**, 374 (1950).
272. Otsuka, S., Mori, K., and Yamagami, K., *J. Org. Chem.* **31**, 4170 (1966).
273. Palenik, G. J., and Lippard, S. J., *Inorg. Chem.* **10**, 1322 (1971), and references therein.
274. Parker, O. J., and Espenson, J. H., *Inorg. Chem.* **8**, 185 (1969).
275. Parker, O. J., and Espenson, J. H., *Inorg. Chem.* **8**, 1523 (1969).
276. Parker, O. J., and Espenson, J. H., *J. Amer. Chem. Soc.* **91**, 1313 (1969).
277. Parker, O. J., and Espenson, J. H., *J. Amer. Chem. Soc.* **91**, 1968 (1969).
278. Péchard, E., *C. R. Acad. Sci.* **136**, 504 (1903).
279. Peters, D. G., and Caldwell, R. L., *Inorg. Chem.* **6**, 1478 (1967).
280. Peters, W., *Z. Anorg. Chem.* **77**, 137 (1912).
281. Peters, W., *Z. Anorg. Chem.* **89**, 191 (1914).
282. Petredis, D., Burke, A., and Balch, A. L., *J. Amer. Chem. Soc.* **92**, 428 (1970).
283. Peyronel, G., De Filippo, D., and Marcotrigiano, G., *Gazz. Chim. Ital.* **91**, 1190 and 1196 (1961).
284. Pflaum, R. T., and Brandt, W. W., *J. Amer. Chem. Soc.* **77**, 2019 (1955).
285. Phillips, C. S. G., and Williams, R. J. P., "Inorganic Chemistry," pp. 314–321. Oxford Univ. Press, London and New York, 1965.
286. Polly, G. W., Jackson, D. E., and Bryant, B. E., *Inorg. Syn.* **5**, 17 (1957).
287. Poulet, H., and Mathie, J. P., *Spectrochim. Acta* **15**, 932 (1959).
288. Prout, C. K., and Murray-Rust, P., *J. Chem. Soc., A* p. 1520 (1969).
289. Quin, H. W., and Tsai, J. H., *Advan. Inorg. Chem. Radiochem.* **12**, 217 (1968).
290. Rabaut, C., *Bull. Soc. Chim. Fr.* **19**, 786 (1898).
291. Randles, J. E. B., *J. Chem. Soc., London* p. 802 (1941).
292. Rathke, B., *Ber. Deut. Chem. Ges.* **14**, 1774 (1881).
293. Reich, R., *C. R. Acad. Sci.* **177**, 322 (1923).
294. Reichle, W. T., *Inorg. Chim. Acta* **5**, 325 (1971).
295. Reisfeld, M. J., and Jones, L. H., *J. Mol. Spectrosc.* **18**, 222 (1965).
296. Riban, J., *C. R. Acad. Sci.* **88**, 581 (1879).
297. Ritthausen, H., *J. Prakt. Chem.* **59**, 369 (1853).
298. Rosenheim, A., and Loewenstamm, W., *Z. Anorg. Allg. Chem.* **34**, 62 (1903).
299. Rosenheim, A., and Stadler, W., *Z. Anorg. Allg. Chem.* **49**, 1 (1906).
300. Rosenheim, A., and Steinhauser, S., *Z. Anorg. Allg. Chem.* **25**, 72 (1900), and references therein.
301. Rupe, H., *J. Prakt. Chem.* [N. S.] **88**, 79 (1913).

302. Sacco, A., *Gazz. Chim. Ital.* **85**, 989 (1955).

303. Saegusa, T., Ito, Y., and Tomita, S., *J. Amer. Chem. Soc.*, **93**, 5656 (1971).

304. Saglier, A., *C. R. Acad. Sci.* **104**, 1440 (1887).

305. Sahasrabudhey, R., and Krall, H., *J. Indian Chem. Soc.* **19**, 25 (1942).

306. Scheiber, J., and Reckleben, H., *Ber. Deut. Chem. Ges.* **44**, 210 (1911).

307. Schilt, A. A., and Smith, G. F., *Anal. Chim. Acta* **15**, 567 (1956).

308. Schilt, A. A., and Taylor, R. C., *J. Inorg. Nucl. Chem.* **9**, 211 (1959).

309. Scholder, R., and Pattock, K., *Z. Anorg. Allg. Chem.* **220**, 250 (1934).

310. Scott, A. F., Wilkening, L. L., and Rubin, B., *Inorg. Chem.* **8**, 2533 (1969).

311. Shaw, K., and Espenson, J. H., *Inorg. Chem.* **7**, 1619 (1968).

312. Shaw, K., and Espenson, J. H., *J. Amer. Chem. Soc.* **90**, 6622 (1968).

313. Slinkard, W. E., and Meek, D. W., *Inorg. Chem.* **8**, 1811 (1969).

314. Sloan, W. H., *J. Amer. Chem. Soc.* **32**, 972 (1910).

315. Smith, G. F., and McCurdy, W. H., *Anal. Chem.* **24**, 371 (1952).

316. Smith, G. F., and Schilt, A. A., *Anal. Chim. Acta* **16**, 401 (1957).

317. Smith, G. F., and Wilkins, D. H., *Anal. Chim. Acta* **10**, 139 (1954).

318. Smyth, F. H., and Roberts, H. S., *J. Amer. Chem. Soc.* **42**, 2582 (1920).

319. Solomon, R. G., and Kochi, J. K., *Chem. Commun.*, p. 559 (1972).

320. Spacu, G., and Murgulescu, J. G., *Kolloid-Z.* **91**, 294 (1940).

321. Specker, H., and Pappert, W., *Z. Anorg. Allg. Chem.* **341**, 287 (1965).

322. Spofford, W. A., and Amma, E. L., *Chem. Commun.* p. 405 (1968).

323. Stephen, J., Lippard, S. J., and Ucko, D. A., *Chem. Commun.* p. 983 (1967).

324. Stewart, R., and Evans, D. G., *Anal. Chem.* **35**, 1315 (1963).

325. Straumanis, M., and Circulis, A., *Z. Anorg. Allg. Chem.* **251**, 315 (1943).

326. Sutton, G. J., *Aust. J. Chem.* **17**, 1360 (1964).

327. Suzuki, I., *Bull. Chem. Soc. Jap.* **35**, 1449 (1962).

328. Tartarini, G., *Gazz. Chim. Ital.* **63**, 597 (1933).

329. Tayim, H. A., Bouldoukian, A., and Awad, F., *J. Inorg. Nucl. Chem.* **32**, 3799 (1970).

330. Thiele, K. H., and Köhler, J., *J. Organometal. Chem.* **12**, 225 (1968).

331. Tiethof, J. A., Hetey, A. T., Nicpon, P. E., and Meek, D. W., *Inorg. Nucl. Chem. Lett.* **8**, 841 (1972).

332. Toeniskoetter, R. H., and Solomon, S., *Inorg. Chem.* **7**, 617 (1968).

333. Truter, M. R., and Rutherford, K. W., *J. Chem. Soc., London* p. 1748 (1962).

334. Truthe, W., *Z. Anorg. Allg. Chem.* **76**, 129 (1912).

335. Uhlemann, E., and Thomas, P., *Z. Anorg. Allg. Chem.* **341**, 17 (1965).

336. Uhlemann, E., Thomas, P., and Kempter, G., *Z. Anorg. Allg. Chem.* **341**, 11 (1965).

337. van der Linden, J. G. M., and Geurts, P. J. M., *Inorg. Nucl. Chem. Lett.* **8**, 903 (1972).

338. van Koten, G., Leusink, A. J., and Noltes, J. G., *Chem. Commun.* p. 1107 (1970).

339. van Koten, G., and Noltes, J. G., *Chem. Commun.* p. 452 (1972).

340. Varet, R., *C. R. Acad. Sci.* **112**, 390 (1891).

341. Varet, R., *C. R. Acad. Sci.* **124**, 1156 (1897).

342. Verkade, J., and Piper, T. S., *Inorg. Chem.* **1**, 453 (1962).

343. von Stäckelberg, M., and von Freyhold, H., *Z. Elektrochem.* **46**, 120 (1940).

344. Vranka, R. G., and Amma, E. L., *J. Amer. Chem. Soc.* **88**, 4270 (1966).

345. Wada, K., Tamura, M., and Kochi, J., *J. Amer. Chem. Soc.* **92**, 6656 (1970).

346. Wagner, O. H., *Z. Anorg. Allg. Chem.* **196**, 364 (1934).

347. Waters, D. H., and Basak, B., *J. Chem. Soc., A* p. 2733 (1971).
348. Weiss, J. F., Tollin, G., and Yoke, J. T., *Inorg. Chem.* **3**, 1344 (1964).
349. Wells, H. L., *Z. Anorg. Allg. Chem.* **5**, 306 (1894).
350. Wells, H. L., and Hurlbert, E. B., *Z. Anorg. Allg. Chem.* **10**, 157 (1895).
351. Werner, A., *Z. Anorg. Allg. Chem.* **15**, 1 (1897).
352. Whitesides, G. M., and Casey, C. P., *J. Amer. Chem. Soc.* **88**, 4541 (1966).
353. Wieland, H., and Franke, W., *Justus Liebigs Ann. Chem.* **473**, 289 (1929).
354. Wilkins, R. G., and Burkin, A. R., *J. Chem. Soc., London* p. 127 (1950).
355. Wilkinson, G., and Piper, T. S., *J. Inorg. Nucl. Chem.* **2**, 32 (1956).
356. Yoke, J. T., Weiss, J. F., and Tollin, G., *Inorg. Chem.* **2**, 1210 (1963).
357. Zelonka, R. A., and Baird, M. C., *Can. J. Chem.* **50**, 1269 (1972).
358. Ziolo, R. F., and Dori, Z., *J. Amer. Chem. Soc.* **90**, 6560 (1968).
359. Ziolo, R. F., Gaughan, A. P., Dori, Z., Pierpoint, C. G., and Eisenberg, R., *J. Amer. Chem. Soc.* **92**, 738 (1970).
360. Ziolo, R. F., Gaughan, A. P., Dori, Z., Pierpoint, C. G., and Eisenberg, R., *Inorg. Chem.* **10**, 1289 (1971).
361. Ziolo, R. F., Thich, J. A., and Dori, Z., *Inorg. Chem.* **11**, 626 (1972).
362. Zuberbühler, A., *Helv. Chim. Acta* **50**, 466 (1967).

COMPLEXES OF OPEN-CHAIN TETRADENTATE LIGANDS CONTAINING HEAVY DONOR ATOMS

C. A. McAULIFFE*

Department of Chemistry, Auburn University, Auburn, Alabama

I. Introduction

During the past decade there has been growing interest in the synthesis of quadridentate chelating ligands containing "heavy" donor atoms, i.e., P, As, Sb, S, and Se. Previously the coordination chemistry of multidentate chelates had been restricted to those containing class *a* donors, and the chemistry of such ligands derived, for example, from Schiff base condensation reactions has been much studied and reported upon. The multidentate chelates containing class *b* donors received sparse attention mainly because of the intricate preparative routes involved in their synthesis. Subsequently, however, largely through the work of the groups led by Venanzi, Meek, and Sacconi, and their students the complex organic chemistry associated with the preparation of intricate chelates containing phosphines, arsines, thioethers, etc., was developed leading to a whole series of ligands.

The first of these quadridentate ligands produced were of the tripod

* Permanent address: Department of Chemistry, University of Manchester, Institute of Science and Technology, Manchester M60 1QD, England.

type (**I**) rather than the linear or open-chain type (**II**). The tripod ligands have been extensively studied and generally they form trigonal-bipyramidal complexes with d^7 and d^8 metal ions (*1–5*). When considering the relative crystal field stabilization energies (cfse) for low-spin d^7 or d^8 square planar or the two common pentacoordinate structures, then it is clear that on cfse grounds all quadridentate ligands should form planar

rather than pentacoordinate complexes. Presumably the formation of a further metal–ligand bond is energetically favorable enough to offset the loss in cfse resulting from stereochemical changes. When comparing the cfse for the trigonal bipyramidal (D_{3h}) and square pyramidal (C_{4v}) arrangements, it appears that the square pyramid should be the preferred structure (*4, 6*). However, the overwhelming majority of pentacoordinate complexes containing tripod tetradentate ligands have been characterized as trigonal bipyramidal. Studies (*2*) of a number of these complexes have attributed this structure to factors such as steric requirements due to the trigonal symmetry of the ligands, the rigid nature of the linkage (*o*-phenylene usually) between the donor atoms, and the steric bulk of the terminal donors. Observations that tripod tetradentate ligands with the flexible trimethylene linkage replacing the rigid *o*-phenylene linkage also form trigonal bipyramidal complexes rather than square pyramidal complexes have been interpreted as showing the importance of factors such as the repulsive interaction between bonding electron pairs and the π-acceptor ability of the ligand donor atoms in determining the structure of the complex (*7, 8*). However, mention should be made of the unusual behavior of the ligand tris-(2-diphenylphosphinoethyl)amine (tpn), which forms trigonal-bipyramidal complexes [Ni(tpn)I]$^+$ (**III**), (*9*), and [Co(tpn)Cl]$^+$ (*10*), but the square-pyramidal [Co(tpn)I]$^+$ (**IV**), (*11*). This coordination geometry

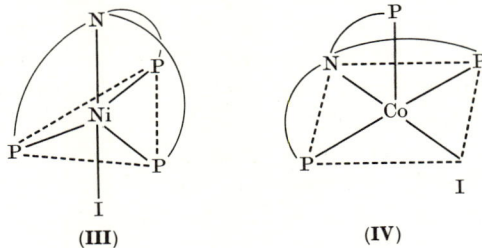

transition is probably due to the influence of increased cfse (square-pyramidal) over ligand geometry (trigonal-bipyramidal), even though, in the case of cobalt(II), complications can arise from Jahn–Teller effects. It should also be pointed out that there are only quite small bond angle movements needed to convert one idealized geometry into another:

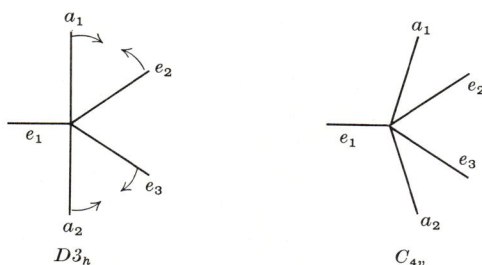

transition is probably due to the influence

That, in the absence of strong constraints such as rigid ligands, the energy difference between D_{3h} and C_{4v} structures is small has recently been demonstrated by the isolation of two distinct pentacoordinate molecules of formulation, chlorobis-(1,2-bisdiphenylphosphinoethane)-cobalt(II) trichlorostannate, [Co(diphos)Cl][SnCl$_3$] (12). X-Ray crystallographic work by Meek and co-workers (13) has shown the red isomer to have a square pyramidal structure with four basal phosphines and an equatorial chlorine, and the green isomer to be trigonal-bipyramidal with two axial phosphines, two equatorial phosphines, and an equatorial chlorine. It is thus seen that, although tripodal tetradentate ligands have contributed significantly to recent coordination chemistry, the type of complex formed by them appears to be restricted by the steric requirements of the ligand, and in order for the metal ion to exert a preference more flexible ligands are needed. This has led coordination chemists to study open-chain quadridentates, since in complexes formed from such ligands the metal atoms and donor atoms will be more free of the constraints placed upon them by the links between the donor atoms.

Although open-chain quadridentates do not have trigonal symmetry properties, results from X-ray structural determinations of tripod ligand complexes are of interest to chemists studying open-chain quadridentates. Crystallographic studies have shown that the metal ions in [Co(qp)Cl]BPh$_4$ [qp = tris-(o-diphenylphosphinophenyl)phosphine] (14), [Ni(pts)Cl]ClO$_4$ [pts = tris-(o-methythiophenyl)phosphine] (15), [Ni(tpn)I]$^+$ (9), and [Pt(qas)I]BPh$_4$ [qas = tris-(o-diphenylarsion-phenyl)arsine] (16) reside a significant distance below the plane of the three equatorial atoms. However, in the case of [Ni(ptas)CN]ClO$_4$

[ptas = tris-(3-dimethylarsinopropyl)phosphine] (*17*), the metal ion lies above the plane of the three equatorial arsenic atoms. Thus, it appears that when apical and equatorial donor atoms are linked by *o*-phenylene chelate chains the equatorial donors are below the metal atom. This is, of course, due to the inability of the two-carbon and three-carbon chains to provide for an L(apical)—M—L(equatorial) angle of 90°; for the two-carbon case this L_a—M—L_e angle will be less than 90° (usually it is about 83°), and for the three-carbon case the L_a—M—L_e angle will be greater than 90° (*18, 19*). Reflectance spectra of trigonal bipyramidal nickel(II) complexes formed by the ligands (*o*-Ph_2PC_6-$H_4)_3L$ (L = P, As, Sb) have supported this argument. The spectro-chemical series formed by changing L in [NiL-(*o*-$Ph_2PC_6H_4)_3X]^+$ complexes exhibited anomalous behavior, the order found being P > As < Sb (*20*) in contrast to the spectrochemical order R_3P > R_3As > R_3Sb found for monodentate ligands R_3L (*21*). This effect has been attributed to compression of the apical ligand to metal bond complexation (*20*). To a certain extent this compression is relieved by a displacement of the metal atom below the plane of the three equatorial ligands.

Although chelate ring strain interested Corey and Bailar (*22*) as long ago as 1959, it is a facet of chelates that has not attracted much attention from coordination chemists despite its obvious importance. There are indications that this state of affairs is changing, (*18–20, 23*), and, in studying open-chain facultative quadridentates [facultative quadridentates are flexible and are able to coordinate to a metal atom in either planar or nonplanar arrangement (*24*)], the effect of chelate chain length can be shown to have important stereochemical and spectrochemical effects. Consider a square-planar structure containing monodentate ligands, L:

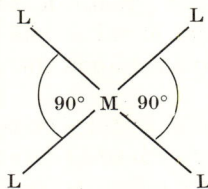

The LML is, ideally, 90°, and in this situation one might confidently expect optimum M—L orbital overlap. A planar complex containing similar donors connected by three dimethylene or *o*-phenylene linkages L—L—L—L may be altered in one of three ways (or, more likely, in a combination of these three ways): (*a*) by distorting the overall geometry of the complex (**V**); (*b*) as a result of the distortion caused by LML

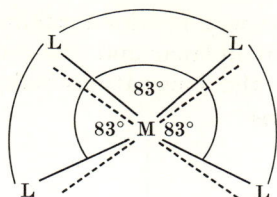

(V)

~ 83°, very poor orbital overlap between terminal L and the metal may be expected, resulting in a decrease in stability of complex (V) and the formation of a bidentate ligand complex (VI) or the bridging quadriden-

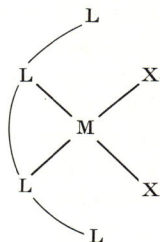

(VI)

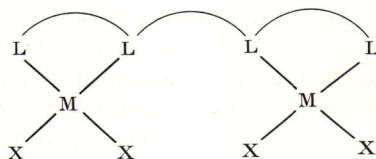

(VII)

tate complex (VII) [both (VI) and (VII) have been isolated using the S_2As_2 quadridentates of Busch and co-workers (25)]; (c) if terminal donors, L_t, do coordinate to form a mononuclear complex, then, in order to form the strongest possible L_t—M bonds, there will be a tendency for the L_t—M—L (VIII) to open up in order to facilitate better L_t—M orbital overlap. This will result in a shortening of the L—M bond due to compression, resulting also in an artificially high, ligand field strength of L.

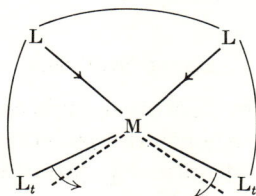

(VIII)

Such deviations from ideal behavior can be prohibited by varying the chelate backbone lengths. Thus, an L_t—C_2—L—C_3—L—C_2—L_t system or an L_t—C_3—L—C_2—L—C_3—L_t system might be expected to relieve strains a–c, and it has been found that altering chelate backbone can have quite startling effects on the resulting complex stereochemistry and spectrochemistry. One way in which strains a–c can be relieved is for the ligand to become nonplanar and adopt configuration (**IX**). The tetrahedral linkages about the donor atoms make this configuration less

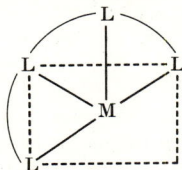

(**IX**)

strained and, although these ligands do form planar complexes, it is likely that many of the pentacoordinate derivatives have configuration (**IX**) (*26*). It will be seen later in this article that low-spin pentacoordinate complexes are frequently encountered in studies with these linear quadridentates containing heavy donor atoms. However, a number of six-coordinate species have been isolated, and it should be pointed out that a number of octahedral configurations are possible:

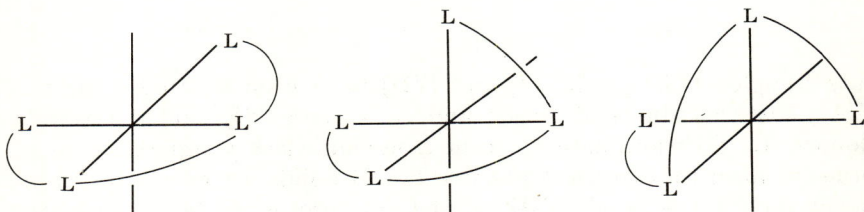

Observations that linear quadridentates form planar, trigonal-bipyramidal, square-pyramidal, and octahedral complexes testify to the flexibility of this ligand arrangement.

II. Ligand Synthesis

A number of linear quadridentate ligands have been made *in situ* and complexed to a metal ion without prior isolation of the ligand. However, most have been synthesized quite separately from complexation reactions, and these are listed in Table I. A number of ligand syntheses of general application will be outlined here. It must be emphasized that these are not complete syntheses and that many of the precursors used are themselves prepared by quite arduous routes.

1. Ligands with the $PhAs(CH_2)_nAsPh$ intermediate linkage require

two major steps in their preparation.

TABLE I

LINEAR QUADRIDENTATE LIGANDS AND THEIR ABBREVIATIONS

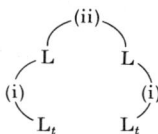

L_t	L	(i)	(ii)	Ligand[a]	Ref.
PPh_2	PPh	$-(CH_2)_2-$	$-(CH_2)_2-$	p_4	(27)
PPh_2	AsPh	$o\text{-}C_6H_4$	$-(CH_2)_3-$	p_2as_2	(28)
PPh_2	S	$-(CH_2)_3-$	$-(CH_2)_3-$	p_2s_2	(29)
PPh_2	N	$-(CH_2)_2-$	$=C(Me)C(Me)=$	p_2n_2	(30)
$AsPh_2$	PPh	$o\text{-}C_6H_4$	$-(CH_2)_3-$	as_2p_2	(28)
$AsPh_2$	$AsPh_2$	$o\text{-}C_6H_4$	$-(CH_2)_3-$	as_4ph	(28)
$AsPh_2$	S	$-(CH_2)_3-$	$-(CH_2)_3-$	as_2s_2	(31)
$AsPh_2$	S	$o\text{-}C_6H_4$	$-(CH_2)_2-$	as_2s_2e	(32)
$AsPh_2$	S	$o\text{-}C_6H_4$	$-(CH_2)_3-$	as_2s_2p	(33)
$AsPh_2$	S	$o\text{-}C_6H_4$	$-(CH_2)_4-$	as_2s_2b	(33)
$AsPh_2$	N	$-(CH_2)_2-$	$=C(Me)C(Me)=$	as_2n_2	(34)
$AsPh_2$	O	$o\text{-}C_6H_4$	$-(CH_2)_2-$	as_2o_2	(32)
$AsMe_2$	AsPh	$-(CH_2)_3-$	$-(CH_2)_3-$	qas	(35)
$AsMe_2$	AsMe	$o\text{-}C_6H_4$	$o\text{-}C_6H_4$	as_4me	(26)
$AsMe_2$	S	$-(CH_2)_3-$	$-(CH_2)_3-$	as_2s_2me	(31)
SMe	AsPh	$o\text{-}C_6H_4$	$-(CH_2)_2-$	s_2as_2e	(25)
SMe	AsPh	$o\text{-}C_6H_4$	$-(CH_2)_3-$	s_2as_2p	(25)
SMe	AsPh	$o\text{-}C_6H_4$	$-(CH_2)_4-$	s_2as_2b	(36)
SMe	S	$o\text{-}C_6H_4$	$-(CH_2)_3-$	s_4	(31)
NH_2	S	$o\text{-}C_6H_4$	$-(CH_2)_2-$	n_2s_2e	(32)
NH_2	S	$o\text{-}C_6H_4$	$-(CH_2)_3-$	n_2s_2p	(37)
NH_2	S	$o\text{-}C_6H_4$	$-(CH_2)_4-$	n_2s_2b	(37)

[a] The abbreviations may, in a small number of cases, appear to be somewhat cumbersome. However, in devising abbreviations for these complicated tetradentates attempts at consistency have led, inevitably, to a number of cumbersome symbols.

a. Preparation of the alkanebis(phenylchloroarsine):

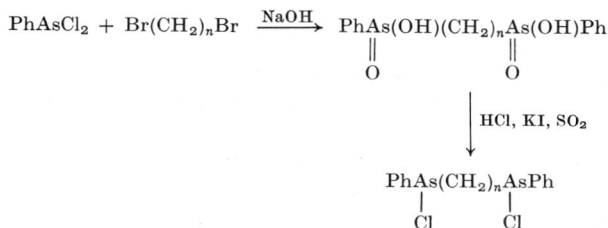

$$PhAsCl_2 + Br(CH_2)_nBr \xrightarrow{NaOH} PhAs(OH)(CH_2)_nAs(OH)Ph$$

with the two As centers each doubly bonded to O, then

$$\downarrow \text{HCl, KI, SO}_2$$

$$PhAs(CH_2)_nAsPh$$

with each As bearing a Cl substituent

b. Preparation of the terminal ligand moiety, such as

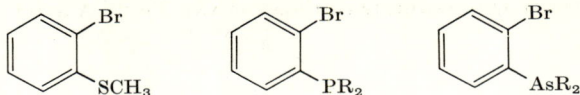

The final reaction is

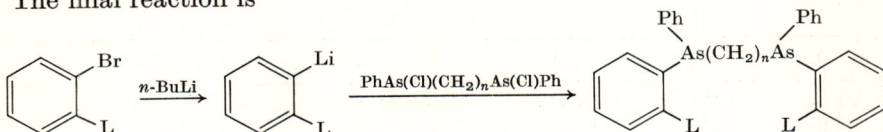

2. Ligands that have an —$S(CH_2)_nS$— intermediate linkage are frequently prepared as follows:

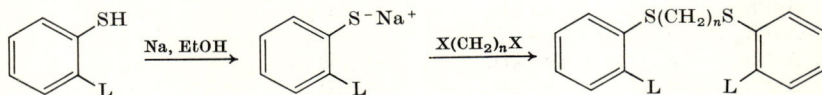

(L = NH_2, PR_2, AsR_2)

or:

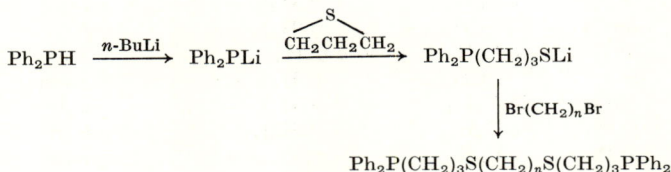

$$Ph_2P(CH_2)_3S(CH_2)_nS(CH_2)_3PPh_2$$

3. Ligands containing the —$N{=}C(CH_3)C(CH_3){=}N$— intermediate linkage are readily synthesized:

$$Ph_2LCH_2CH_2NH_2 + CH_3C(O)C(O)CH_3 \longrightarrow$$
$$Ph_2LCH_2CH_2N{=}C(CH_3)C(CH_3){=}NCH_2CH_2LPh_2$$

(L = P, As)

4. Linear tetratertiary phosphine, P_4, was obtained by King and Kapoor (27), by the addition of phosphorus–hydrogen bonds in 1,2-bisphenylphosphinoethane to the carbon–carbon double bonds in two equivalents of diphenylvinylphosphine, according to the following equation:

$$2Ph_2PCH{=}CH_2 + PhP(H)CH_2CH_2P(H)Ph \longrightarrow$$
$$Ph_2P(CH_2)_2P(Ph)(CH_2)_2P(Ph)(CH_2)_2PPh_2$$

Most of these chelates are air stable or sufficiently stable to allow complexing reactions to be carried out with a minimum of difficulty with regard to transfer of reactants.

III. Studies on Coordination Complexes

A. TETRAPHOSPHINE CHELATES

The elegant syntheses of a whole range of polytertiary phosphines devised by King and Kapoor have been reviewed (38). They prepared both the tripod tetradentate, tris-(2-diphenylphosphinoethyl)phospine (qp) (X), and the open-chain tetraphosphine, 1,1,4,7,10,10-hexaphenyl-1,4,7,10-tetraphosphadecane (p$_4$) (XI). King (38) points out that these tetratertiary phosphines can bond to one or more metal atoms in ten fundamental ways: (a) monoligate monometallic; (b) biligate mono-metallic; (c) triligate monometallic; (d) tetraligate monometallic; (e) biligate bimetallic; (f) triligate bimetallic; (g) tetraligate bimetallic; (h) triligate trimetallic; (i) tetraligate trimetallic; (j) tetraligate tetra-metallic.

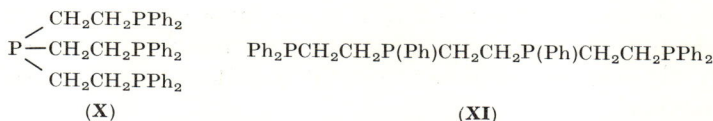

$$\begin{array}{l} \diagup CH_2CH_2PPh_2 \\ P{-}CH_2CH_2PPh_2 \\ \diagdown CH_2CH_2PPh_2 \end{array} \qquad Ph_2PCH_2CH_2P(Ph)CH_2CH_2P(Ph)CH_2CH_2PPh_2$$

(X)   (XI)

Four of the ten possible ways of bonding the tripod chelate qp have been shown (39) to exist in (i) biligate monometallic Pt(qp)Cl$_2$, M(qp)-(CO)$_4$ (M = Cr, Mo), (CH$_3$CO)Mn(qp)(CO)$_3$, [C$_5$H$_5$Mn(qp)(NO)]PF$_6$, and [C$_5$H$_5$Fe(qp)(CO)]I; (i) triligate monometallic M(qp)Cl$_3$ (M = Rh, Re), M(qp)(CO)$_3$ (M = Cr, MO), CH$_3$Mn(qp)(CO)$_2$, and Mn(qp)(CO)$_2$Br; (iii) tetraligate monometallic [M(qp)Cl]$^+$ (M = Ni, Co, Fe), and M(qp)-(CO)$_2$ (M = Mo, W); (iv) tetraligate tetrametallic [(qp){Mo(CO)$_2$-(COCH$_3$)(C$_5$H$_5$)}$_4$], and [(qp){Fe$_2$(CO)$_2$(C$_5$H$_5$)$_2$}$_2$]. Significantly, the linear tetradentate P$_4$ ligand has produced complexes in which six of the ten possible ways of bonding are involved: (i) monoligate mono-metallic CH$_3$COFe(p$_4$)(CO)(C$_5$H$_5$); (ii) biligate monometallic M(p$_4$)-(CO)$_4$ (M = Cr, Mo), CH$_3$Mn(p$_4$)(CO)$_3$, [C$_5$H$_5$Mo(p$_4$)(CO)$_2$]Cl, C$_5$H$_5$Mn-(p$_4$)(CO); (iii) triligate monometallic M(p$_4$)Cl$_3$ (M = Rh, Re), M(p$_4$)-(CO)$_3$ (M = Cr, Mo), Mn(p$_4$)(CO)$_2$Br, and [C$_5$H$_5$Fe(p$_4$)]$^+$; (iv) tetraligate monometallic [M(p$_4$)]$^{2+}$ (M = Ni, Pd, Pt), [Co(p$_4$)Cl]$^+$, and [Rh(p$_4$)]$^+$; (v) tetraligate bimetallic [(C$_5$H$_5$)$_2$Mn$_2$(p$_4$)(CO)(NO)$_2$](PF$_6$)$_2$; (vi) tetra-ligate tetrametallic [(p$_4$){Mo(CO)$_2$(COCH$_3$)(C$_5$H$_5$)}$_4$], and [(p$_4$){Fe$_2$-(CO)$_2$(C$_5$H$_5$)$_2$}$_2$].

It is somewhat hard to imagine either of these two chelates forming all the possible complex types devised by King. Nonetheless, the versatility of these ligands has been amply demonstrated by King and co-workers (39). As well as enumerating the bonding types in the preceding, it is also worth mentioning that the same transition metal reactant forms different types of complexes with qp (X) and p$_4$ (XI):

(i) the linear tetratertiary phosphine reacts with nickel(II) chloride to give, after treatment with PF_6^-, the yellow planar $[Ni(p_4)](PF_6)_2$, whereas qp produces the blue pentacoordinate $[Ni(qp)Cl]PF_6$ [this complex is reported (39) as having $\mu_{eff} = 2.23$ BM, suggesting that reinvestigation would be desirable]; (ii) the linear p_4 does not react with ferrous chloride in boiling ethanol, whereas under the same conditions qp forms the blue pentacoordinate $[Fe(qp)Cl]^+$; (iii) the open-chain p_4 reacts with $[C_5H_5Mn(CO)_2NO]PF_6$ to form the triligate bimetallic derivative, $[(C_5H_5)_2Mn_2(p_4)(CO)(NO)_2]PF_6$, whereas the tripod tetradentate, under identical conditions, yields the biligate monometallic derivative, $[C_5H_5Mn(qp)(NO)]PF_6$.

Examining these differences King (38) has been tempted to suggest that of the two tetraphosphine ligands the tripod qp is the better "chelating agent" (i.e., better at forming complexes with several phosphorus atoms all bonded to a single metal atom). However, the open-chain p_4 is more adept at forming polymetallic complexes. This is exactly the behavior predicted from the considerations discussed in the Introduction.

The current interest in open-chain tetradentates is such that p_4 is now commercially available, and Sacconi and co-workers (40) have extended the work of King et al. (39) and prepared a series of complexes of general formula $[M(p_4)X]BPh_4$ (M = Fe, Co, Ni; X = Cl, Br, I). Spectral and X-ray investigations suggest a square-pyramidal geometry for all complexes. The cobalt(II) and nickel(II) complexes are low spin, and the magnetic properties of the iron(II) complexes are consistent with a spin equilibrium between a singlet ground state and a thermally accessible low-lying triplet state. Complex $[Fe(p_4)(NCS)_2]$ is diamagnetic and probably has a trans-octahedral structure.

B. DIPHOSPHINE–DIARSINE CHELATES

Initial studies with 1,3-bis-[phenyl-(o-diphenylphosphinophenyl)-arsino]propane (p_2as_2) by McAuliffe (28) have led to the isolation of the diamagnetic planar $[Ni(p_2as_2)](ClO_4)_2$ and the purple $[Ni(p_2as_2)X]$-BPh_4 (X = Cl, Br, I) complexes. The latter have been assigned square-pyramidal coordination about the nickel atom, on the basis of electronic spectral studies.

C. DIPHOSPHINE–DITHIOETHER CHELATES

Dubois and Meek (29) synthesized 1,3-bis-(diphenylphosphino-propylthio)propane or $Ph_2P(CH_2)_3S(CH_2)_3S(CH_2)_3PPh_2$ (p_2s_2), a chelate with three trimethylene bridges between the donors. A series of planar d^8 complexes $[M(p_2s_2)](ClO_4)_2$ (M = Ni, Pd, Pt) and penta-coordinate $[M(p_2s_2)X]^+$ (M = Ni, Pd, Pt; X = halogen) have been pre-pared (29). The flexibility of this ligand is demonstrated by the fact that the five-coordinate nickel complexes are trigonal-bipyramidal, whereas those of palladium and platinum are square-pyramidal. This is a most important result, since it shows that the open-chain quadridentate can allow the central metal ion to dictate its stereochemical surroundings to a large degree. As discussed earlier, these considerations predict that the square-pyramidal geometry is more stable than trigonal-bipyramidal. This effect should increase in descending the triad, Ni, Pd, Pt, and is apparently demonstrated by this system. DuBois and Meek suggest that either of the isomers (**XII**) or (**XIII**) are compatible with their results.

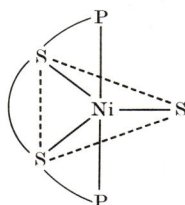
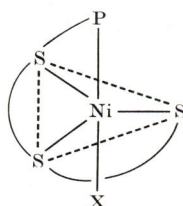

(**XII**)   (**XIII**)

Electronic spectral results provide strong evidence, especially in the case of nickel(II) complexes, for particular pentacoordinate geom-etries, which is somewhat surprising in view of the fact that the ideal geometries are interchanged by relatively small bond movements. None-theless, it has been demonstrated conclusively that low-spin trigonal-bipyramidal nickel(II) complexes have their main d-d absorptions in the region of ∼ 16 kK (7, 8, 41–44), whereas those with square-pyramidal geometry have visible absorptions in the region of ∼ 20 kK (45). With p_2s_2, DuBois and Meek (29) also prepared the maroon pentacoordinate $[Co(p_2s_2)X]^+$ (X = Cl, Br, I) derivatives, and these were tentatively assigned a square-pyramidal structure, although it should be empha-sized that this assignment is by no means as certain as that of the nickel complexes.

D. DIPHOSPHINE–DIIMINE CHELATES

Recently, DuBois (30) has prepared the phosphorus–nitrogen tetra-dentate 2,3-butanedionebis-(2-diphenylphosphinoethylimine), Ph$_2$-PCH$_2$CH$_2$N=C(CH$_3$)C(CH$_3$)=NCH$_2$CH$_2$PPh$_2$ (p_2n_2), and isolated the planar $[Ni(p_2n_2)](ClO_4)_2 \cdot 0.5EtOH$ and the square-pyramidal $[Ni-(p_2n_2)X]\ ClO_4 \cdot 0.5EtOH$ (X = Cl, Br, I). Dubois argues that, because of the rigidity of the imine linkages, these pentacoordinate complexes contain the ligand chelated in the plane. However, although anionic ligands can be added to the planar $[Ni(p_2n_2)]^{2+}$ species to yield mono-meric pentacoordinate complexes, DuBois and Smith (34) have found that planar nickel(II) derivatives of the analogous 2,3-butanedionebis-(2-diphenylarsinoethylimine) (as_2n_2), $[Ni(as_2n_2)]^{2+}$, do not add anions to produce monomeric $[Ni(as_2n_2)X]^+$ species. Pentacoordinate deriva-tives are, indeed, produced, but spectrometric titrations of $[Ni(as_2n_2)]^{2+}$ with $(n\text{-}C_7H_{15})_4NX$ (X = Cl, Br) indicate that ligand displacement from the planar cation takes place. These authors suggest structures (**XIV**) or (**XV**) as the most likely and also suggest that the difference in behavior between the p_2n_2 and as_2n_2 ligands is probably due to the ability of phenyl groups on the *cis*-arsines to interfere sterically with each other.

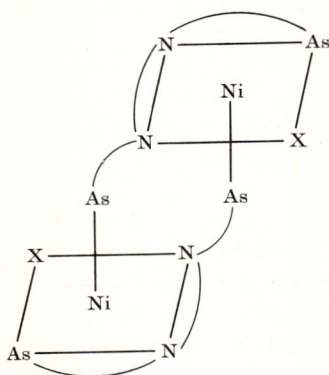

(**XIV**)

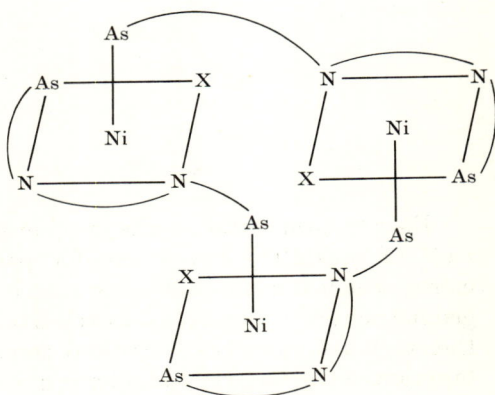

(**XV**)

E. TETRAARSINE CHELATES

The only complex of a linear quadridentate that has been studied by X-ray techniques is that of $[Pd(as_4me)Cl]ClO_4 \cdot C_6H_6$ (as_4me = *o*-phenylenebis-*o*-dimethylarsinophenylmethylarsine) (**XVI**) (26). The pentacoordinate complex cation crystallizes with perchlorate ions in an ionic lattice in which benzene molecules are included as a clathrate. The cation has a square-pyramidal coordination (**XVII**), with a chlorine atom in the basal plane in which the palladium atom is coplanar with the four attached atoms (the palladium, chlorine, and three arsenic

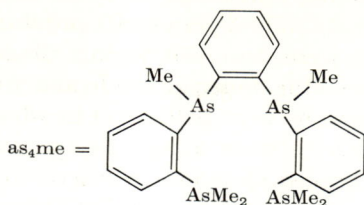

as$_4$me =

(XVI)

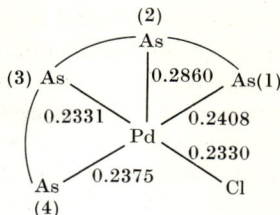

(XVII)

atoms lie within 0.06 Å of the least-squares plane through them). The apical palladium–arsenic bond is almost 0.05 nm longer than those in the basal plane, which also differ significantly from one another. The distortions in the molecule are essentially a consequence of strain due to the o-phenylene linkages between the arsenic donors; this is reflected in the bond angles, as shown in Table II. One most significant distortion is seen in the As(1)—Pd—As(2) angle of 80.1°, which results from the ideal apical Pd—As bond being pulled to one side by short o-phenylene links.

The tetradentate arsine ligand containing three trimethylene linkages, 1,3-propanebis-(3-dimethylarsinopropylphenylarsine) (qas) forms a series of square-pyramidal nickel(II) complexes of the type [Ni(qas)X]BPh$_4$ (X = Cl, Br, I, NO$_3$) and [Ni(qas)(H$_2$O)](ClO$_4$)$_2$ (35, 46). Similar results have been obtained with the tetraarsine as$_4$ph (28).

TABLE II

GEOMETRIC PARAMETERS OF
[Pd(as$_4$me)Cl]ClO$_4$. C$_6$H$_6$

Arsenic Molecule	Bond angle
As(1)—Pd—As(2)	80.1°
As(1)—Pd—As(3)	95.9°
As(1)—Pd—As(4)	178.2°
As(1)—Pd—Cl	89.8°
As(2)—Pd—As(3)	81.9°
As(2)—Pd—As(4)	99.3°
As(2)—Pd—Cl	101.8°
As(3)—Pd—As(4)	85.7°
As(3)—Pd—Cl	173.7°
As(4)—Pd—Cl	88.6°

F. DIARSINE–DITHIOETHER CHELATES

Initial studies with as$_2$s$_2$ and as$_2$s$_2$me have led to the isolation of pentacoordinate [NiL(H$_2$O)](ClO$_4$)$_2$ complexes (31). The arsenic–sulfur chelates as_2s_2e (32) and as_2s_2b (33), have been investigated by Venanzi (32) and by McAuliffe and co-workers (33, 47). A series of complexes $[Pd(as_2s_2e)X_2]$ (X = Cl, Br, I, NCS) were assigned square-planar structures in which one arsenic and one sulfur donor of the ligand was coordinated (32). In contrast to the reactions with palladium in which the ligand retains its integrity, reactions between as_2s_2e, as_2s_2p, and as_2s_2b with nickel(II) salts in acetone proceed as shown in reaction (1) (47). The trans structure has been assigned to the final complex because

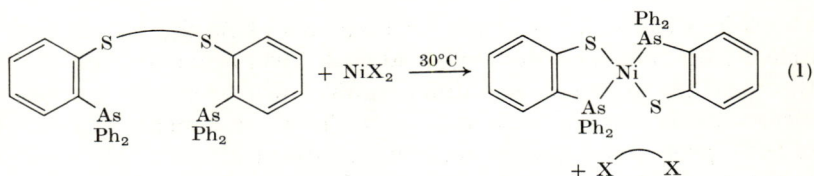

$$+ \text{NiX}_2 \xrightarrow{30°\text{C}} \qquad (1)$$

the visible and infrared spectra of the products of the reaction are identical with those of the $[NiL_2]$ complex formed in the reaction between nickel(II) salts and o-diphenylarsinophenylthiol. This suggests possible intermediate reactions of the type shown in reaction (2). The deep blue-black colors obtained when NiX_2 (X = Cl, Br, I, NO_3, ClO_4) and these ligands are mixed suggests the formation of a pentacoordinate

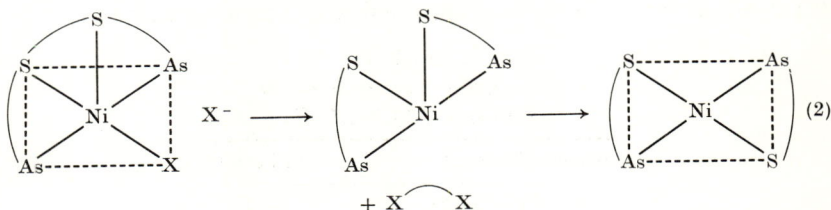

$$\xrightarrow{\text{X}^-} \qquad \longrightarrow \qquad (2)$$

intermediate. Interestingly, no reaction is observed when the reactants are mixed at temperatures below 30°C; at this temperature the deep blue-black color develops followed almost immediately by the precipitation of the green S-dealkylated complex. Although S-dealkylation is not uncommon for palladium–thioether complexes, there has only been one other report of a nickel-induced dealkylation reaction (48), and this involved boiling DMF as solvent. The facile bis-S-dealkylation of as_2s_2e, as_2s_2p, and as_2s_2b would seem to be unique. Levason et al. have also obtained bis-S-dealkylation in the reaction of cobalt(II) perchlorate with as_2s_2e, but the integrity of the ligand remained intact in the reaction between as_2s_2p and cobalt(II) bromide, leading to the eventual isolation of the $[Co(as_2s_2p)Br]BPh_4$ complex (33).

G. DIARSINE–DIOXYGEN CHELATES

Cannon *et al.* (*32*) reacted 1,2-bis-(*o*-diphenylarsinophenoxy)ethane (as_2o_2) with palladium(II) salts and obtained the complexes $[Pd(as_2o_2)X_2]$ (X = Cl, Br, I, NCS), which are nonconductors in nitrobenzene. These workers doubt that coordination of two oxygen atoms or one oxygen and one arsenic atom would occur in palladium(II) complexes of as_2o_2, as the bond between ether oxygen and palladium is expected to be very weak. Because the $[Pd(as_2o_2)Cl_2]$ complex is monomeric in methylphenyl ketone and because all the complexes of the general formula $[Pd(as_2o_2)X_2]$ have similar ultraviolet–visible spectra, therefore, there are two structural possibliities: (*a*) a bridging ligand with the two arsenic atoms in trans positions or (*b*) a bridging ligand with the two arsenic atoms in cis positions. From molecular models, Cannon *et al.* conclude that the first of these two possibilities is the most likely.

H. DITHIOETHER–DIARSINE CHELATES

Dutta, Meek, and Busch (*25, 36*) have made a thorough study of the arsenic ligands s_2as_2e, s_2as_2p, and s_2as_2b (**XVIII**) and their reactions

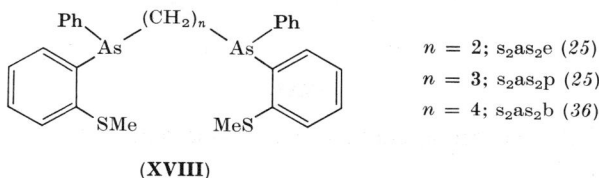

$n = 2$; s_2as_2e (*25*)
$n = 3$; s_2as_2p (*25*)
$n = 4$; s_2as_2b (*36*)

(**XVIII**)

with palladium(II) salts and, more recently, have studied the reactions of s_2as_2e and s_2as_2p with potassium tetrachloroplatinite (*49*). The following complexes were isolated.

(i) Compounds $PdLX_2$ (L = s_2as_2e, X = Cl, Br, I; L = s_2as_2p, X = Cl, Br, I, SCN), which contain coordinated arsenic groups; NMR studies indicated uncoordinated —SMe groups. However, in nitromethane and acetonitrile, the $Pd(s_2as_2p)I_2$ complex behaves as a 1:1 electrolyte.

(ii) Compounds Pd_2LX_4 (L = s_2as_2e, X = Cl, I) in which two halides and one thioether plus one arsenic atom attached to the same benzene ring are coordinated to each palladium(II) ion in a probable structure of type (**XIX**).

(iii) The $Pd_2L_2X_2^{2+}$ species formed probably from $PdLX_2$ on displacement of a coordinated halide ion by a thiomethyl group accompanied by dimerization, and having suggested structure (**XX**).

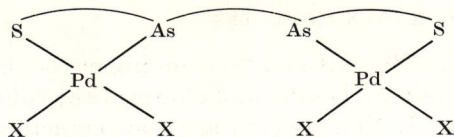

(XIX)

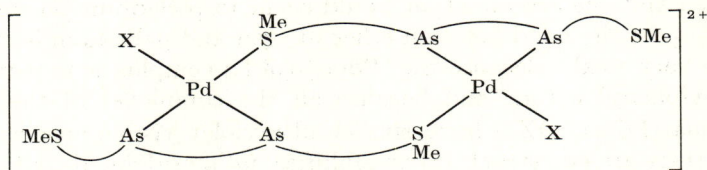

(XX)

(iv) The $[Pd_2L_2]^{4+}$ complexes (L = s_2as_2e, s_2as_2p) obtained as the perchlorate salts, having square-planar structures with the ligands functioning as bridging quadridentates (XXI).

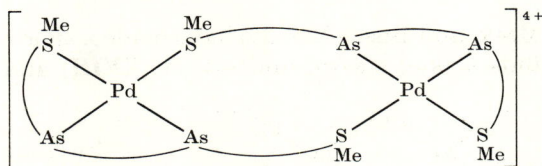

(XXI)

Some of the reactions carried out in this comprehensive study are illustrated in Fig. 1.

The platinum complexes formed (49) were (i) $[Pt_2L_2X_2]^{2+}$ (L = s_2as_2p, X = Cl, I) of which the proposed structure is (XXII), and

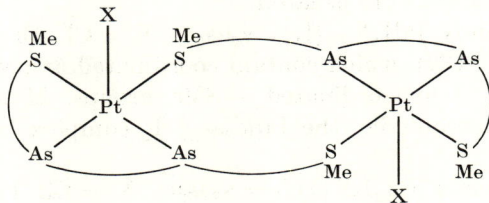

(XXII)

(ii) $[PtLI_2]$ (L = s_2as_2e, s_2as_2p) formed by the reaction of MeI with the dealkylated complexes, as for the palladium analogs. The structure suggested for the dealkylated complexes (XXIII) is dimeric, whereas the MLI_2 complexes are monomeric.

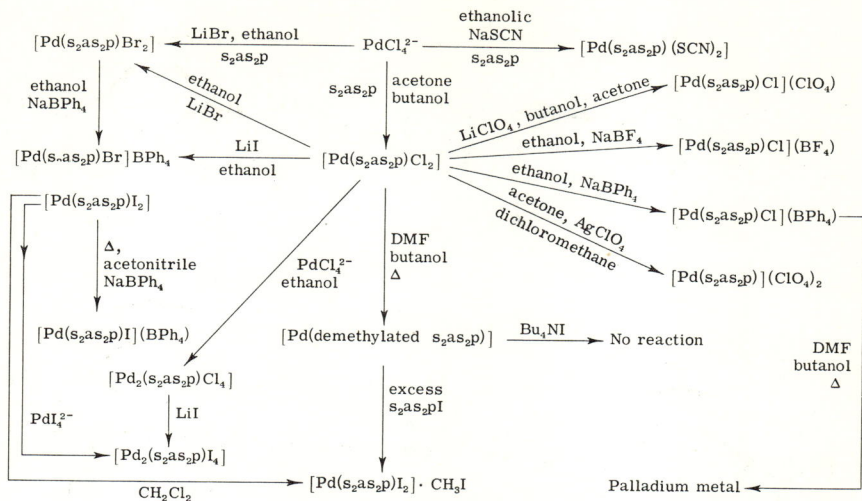

FIG. 1. Methods of preparation of the palladium(II) complexes of s_2as_2p. [From Dutta *et al.* (25).]

The complexes formed by palladium with s_2as_2b were (i) $[Pd_2LX_4]$ (X = Cl, I) having a proposed structure of type (**XIX**), and (ii) a dimeric series $[Pd_2L_2X_2](BPh_4)_2$ containing two arsenic atoms, one exchanging thiomethyl group, and one halide ion coordinated to each palladium(II) ion.

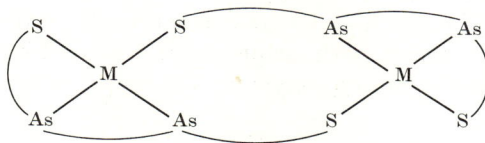

(**XXIII**)

I. TETRATHIOETHER CHELATES

A number of metal ions in biological situations are bound essentially by sulfur donor groups (50), and thus chelates of type 1,3-bis-(*o*-methylthiophenylthio)propane (s_4) (31) have added interest. Apparently this ligand does not form monomeric complexes; the only species isolated thus far are the dimeric $[Pd_2(s_4)X_4]$ (X = Cl, Br, I) and $[Pt_2(s_4)X_4]$ (X = Cl, I), and the trimeric $Rh_3(s_4)_2Cl_9$.

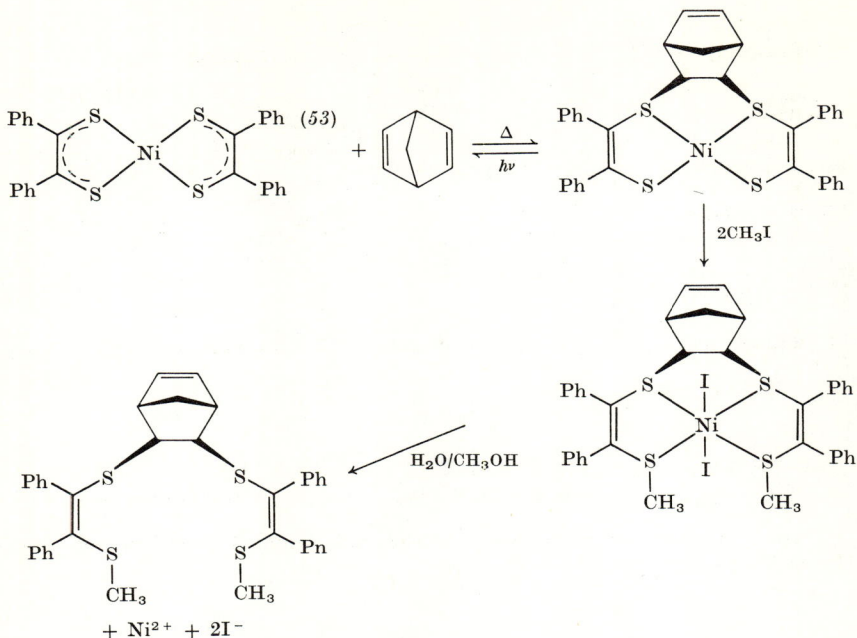

SCHEME 1

A novel method of preparing sulfur quadridentate ligands by a template type reaction has been reported (51, 52) Schrauzer et al. (52) isolated the free ligand by the sequence shown in reaction Scheme 1. However, it appears that this rather incomplete work on a very interesting system has not been pursued further.

Schrauzer (52) also found that a macrocyclic sulfur quadridentate (**XXIV**) could be formed by the sequence of Scheme I if the iodomethane is replaced by α,α'-dibromo-o-xylene. This type of ligand is of interest

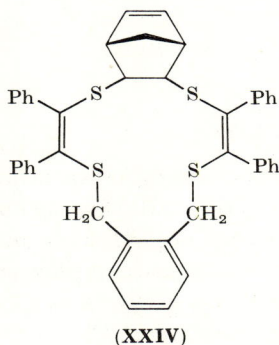

(**XXIV**)

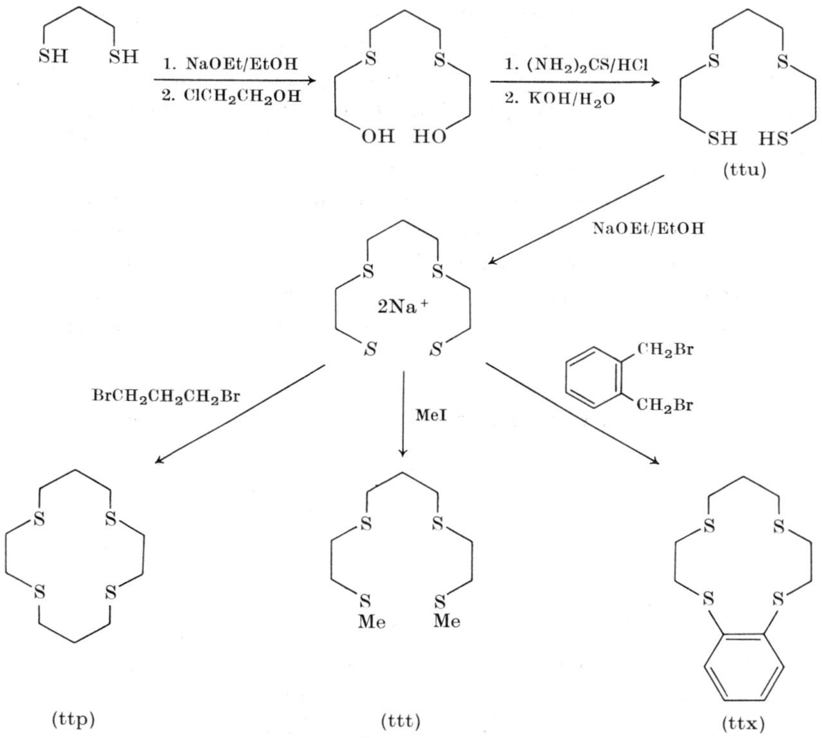

SCHEME 2

since it contains four thioether donors, and, although S^{2-} and RS^- donors have been well studied (54), comparatively little has been done on the RSR-type donors. Rosen and Busch have studied multidentate sulfur ligands and have synthesized the quadridentate macrocycles ttp, ttx, and the open-chain ttt (55) (Scheme 2). Complexation of these ligands was accomplished by reaction with the hexaacetic acid derivative of nickel(II) as the tetrafluoroborate salt in nitromethane. These complexes are red-orange and soluble in nitromethane; they react immediately with solvents of good class a-type donor ability, e.g., water, ethanol, and dimethyl sulfoxide, to give the free ligand and $Ni(solvent)_6(BF_4)_2$. This is why the preceding method of complexation was used instead of using a more standard source of nickel(II) ions.

It was shown that the tetrafluoroborate anions could be replaced in metathetical reactions to produce some tetragonal species in which the anions were coordinated. Thus, two types of complex were reported for $NiLX_2$: for $X = BF_4^-$ or ClO_4^-, the complexes were diamagnetic $2:1$

electrolytes and the electronic spectra were typical of square-planar nickel(II) complexes; for X = NCS⁻, Cl⁻, Br⁻, or I⁻, the solids were all paramagnetic and had four bands in their diffuse transmittance spectra. From the spectra of these tetragonal complexes, a value of Dq^{xy} for the ligand ttp was calculated as 1070 cm⁻¹ which is the weakest field yet observed for a symmetrical tetradentate macrocycle. The case in which X = I⁻ was anomalous in that the iodide had a conductivity in nitromethane which was concentration-dependent, whereas the complexes in which X = NCS⁻, Cl⁻, and Br⁻ were non-electrolytes at all concentrations. Equilibria of the species $Ni(L)I_2$, $Ni(L)(I)^+$, and $Ni(L)^{2+}$ was postulated to explain the existence of both 2:1 and 1:1 electrolyte types.

Rosen and Busch (56) also prepared the macrocycles ttc (**XXV**), ttd (**XXVI**), and tte (**XXVII**) by methods similar to those described in the

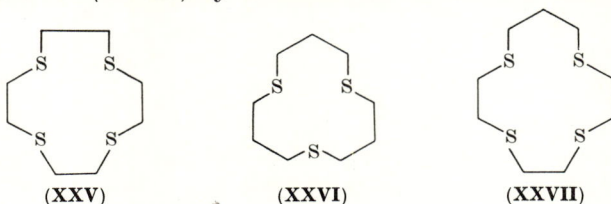

(**XXV**)   (**XXVI**)   (**XXVII**)

foregoing, and the nickel(II) complexes were prepared as before. In this study the metal-to-ligand ration of the complexes was determined by a simple procedure utilizing the reaction of these complexes with water. The macrocycle was displaced from a known amount of complex and isolated quantitatively; the metal ion was then precipitated by dimethylglyoxime and isolated quantitatively. From the quantitative amounts of ligand and metal complex, the stoichiometry of the compounds was afforded. For the terdentate ttd, the complex formed was formulated as $NiL_2(BF_4)_2$, but the complexes of the tetradentate macrocycles, ttc and tte, were formulated as $Ni_2L_3(BF_4)_4$. These latter complexes gave the remarkable conductivity value of over 300 mhos in nitromethane solution which appears reasonable for such a complex if it were a dimer having the postulated structure (**XXVIII**), by analogy to that

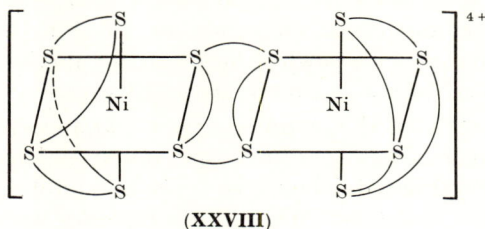

(**XXVIII**)

proposed for $Ni_2(trien)_3Cl_4 \cdot 2H_2O$ (57). The magnetic moments of these dimers were typical for octahedral high-spin nickel(II).

J. DIAMINODITHIOETHER CHELATES

With n_2s_2e, Venanzi's group (32) prepared the six-coordinate $[Co(n_2s_2e)Cl_2]Cl$ complex, and the high-spin six-coordinate $[Ni(n_2s_2e)X_2]$ (X = Cl, Br, I, NCS, NO_3, ClO_4). McAuliffe and co-workers (37) reinvestigated the nickel complexes and confirmed the earlier conclusions, but showed that the trans-octahedral complexes are tetragonally distorted. The $10Dq^{xy}$ value lay in the range 10,530–10,750 cm^{-1}, and the $10Dq^z$ values, i.e., those due to the coordinate anions, could be arranged in the order Cl$^-$ < Br$^-$ < I$^-$ < NO$_3^-$ < NCS$^-$. Chow et al. extended the work to include complexes of n_2s_2p and prepared a series $[Ni(n_2s_2p)X_2]$ (X = Cl, Br, I, NCS) (37). However, only the complex in which X = Cl showed indication of tetragonal distortion in the visible spectra. The axial field in the $[Ni(n_2s_2p)Cl_2]$ complex was determined as 10,420 cm^{-1}, showing the ligand field strengths to be in the order $n_2s_2e > n_2s_2p$, and $10Dq^z$ was found to be 872 cm^{-1}. The $10Dq^z$ values for the chloro complexes containing the n_2s_2p and n_2s_2e ligands were 872 and 609 cm^{-1}, respectively, suggesting that Drago's postulation (58) that $10Dq^z$ values for halides are transferable from one complex to another is incorrect. Chow et al. (37) also synthesized $[Cu(n_2s_2e)][CuCl_4]$, and Cannon et al. (32) prepared a series of $[Pd(n_2s_2e)X_2]$ (X = Cl, Br, NCS) in which the ligand was coordinated via both sulfur donors, and the amine groups were uncoordinated.

IV. Conclusion

It is hoped that this article has helped indicate the wide number of possibilities that exist for the synthesis of open-chain quadridentate chelates and the interesting complexes which they form. There is a great paucity of X-ray structural detail on the complexes, but electronic spectra have yielded a large amount of information. The intriguing relationship between type of chelate chain and the properties of resulting coordination compounds may well be illuminated further by future work in this area.

REFERENCES

1. Norgett, M. J., Thornley, J. H. M., and Venanzi, L. M., *Coord. Chem. Rev.* **2**, 99 (1967).
2. Venanzi, L. M., *Angew. Chem., Int. Ed. Engl.* **3**, 453 (1964).

3. Chiswell, B., *in* "Aspects of Inorganic Chemistry" (C. A. McAuliffe, ed.,) Vol. I, p. 271. Macmillan, New York, 1973.

4. Wood, J. S., *Progr. Inorg. Chem.* **16**, 227 (1972).

5. Sacconi, L., *Trans. Metal. Chem.* **4**, 199 (1968).

6. Basolo, F., and Pearson, R. G., "Mechanisms of Inorganic Reactions," 2nd ed. Wiley, New York, 1967.

7. Benner, G. S., Hatfield, W. E., and Meek, D. W., *Inorg. Chem.* **3**, 1544 (1964).

8. Benner, G. S., and Meek, D. W., *Inorg. Chem.* **6**, 1399 (1967).

9. Sacconi, L., and Dapporto, P., *Chem. Commun.*, p. 1091 (1969); *J. Chem. Soc.*, p. 1804 (1970).

10. Sacconi, L., Bianchi, A., and DiVaira, M., *J. Amer. Chem. Soc.* **92**, 4465 (1970).

11. Sacconi, L., and Orioli, P. L., *Chem. Commun.* p. 1012 (1969).

12. Dyer, G., McAuliffe, C. A., Meek, D. W., and Stalick, J. K., to be published.

13. Stalick, J. K., Corfield, P. W. R., and Meek, D. W., *Inorg. Chem.* **12**, 1668 (1973).

14. Blundell, T. L., Powell, H. M., and Venanzi, L. M., *Chem. Commun.* p. 763 (1967).

15. Haugen, L. P., and Eisenberg, R., *Inorg. Chem.* **8**, 1072, (1969).

16. Mair, G. A., Powell, H. M., and Venanzi, L. M., *Proc. Chem. Soc., London* p. 170 (1961).

17. Stephenson, D. L., and Dahl, L. F., *J. Amer. Chem. Soc.* **89**, 3424 (1967).

18. McAuliffe, C. A., Ph.D. Thesis, Oxford University (1967).

19. Dawson, J. W., Lane, B. C., Mynott, R. J., and Venanzi, L. M., *Inorg. Chim. Acta* **5**, 25 (1971).

20. Higginson, B. R., McAuliffe, C. A., and Venanzi, L. M., *Inorg. Chim. Acta* **5**, 37 (1971).

21. Goggin, P. L., Knight, R. J., Sindellari, L., and Venanzi, L. M., *Inorg. Chim. Acta* **5**, 62 (1971).

22. Corey, E. J., and Bailar, J. C., *J. Amer. Chem. Soc.* **81**, 2620 (1959).

23. DeHayes, L. J., and Busch, D. H., *Inorg. Chem.* **12**, 1505 (1973).

24. Goodwin, H. A., *in* "Chelating Agents and Metal Chelates" (F. P. Dwyer and D. P. Mellor, eds.), p. 161. Academic Press, New York, 1964.

25. Dutta, R. L., Meek, D. W., and Busch, D. H., *Inorg. Chem.* **9**, 1215 (1970).

26. Blundell, T. L., and Porsell, H. M., *J. Chem. Soc., A.* p. 1650 (1967).

27. King, R. B., and Kapoor, P. N., *J. Amer. Chem. Soc.* **93**, 4158 (1971).

28. McAuliffe, C. A., unpublished observations.

29. DuBois, T. D., and Meek, D. W., *Inorg. Chem.* **8**, 146 (1969).

30. DuBois, T. D., *Inorg. Chem.* **11**, 718 (1972).

31. McAuliffe, C. A., and Murray S. G., unpublished observations.

32. Cannon, R. D., Chiswell, B., and Venanzi, L. M., *J. Chem. Soc., A* p. 1277 (1967).

33. Levason, W., Marwood, E., and McAuliffe, C. A., unpublished observations.

34. DuBois, T. D., and Smith, F. T., *Inorg. Chem.* **12**, 735 (1973).

35. Chow, S. T., and McAuliffe, C. A., unpublished observations.

36. Dutta, R. L., Meek, D. W., and Busch, D. H., *Inorg. Chem.* **9**, 2098 (1970).

37. Chow, K. K., Tanner, J. P., and McAuliffe, C. A., unpublished observations.

38. King, R. B., *Accounts Chem. Res.* **5**, 177 (1972).

39. King, R. B., Kapoor, R. N., Saran, M. S., and Kapoor, P. N., *Inorg. Chem.* **10**, 1851 (1971).

40. Bacci, M., Midollini, S., Stoppioni, P., and Sacconi, L., *Inorg. Chem.* **12**, 1801 (1973).

41. Dyer, G., Hartley, J. G., and Venanzi, L. M., *J. Chem. Soc., London* p. 1293 (1965).

42. Dyer, G., and Venanzi, L. M., *J. Chem. Soc., London* p. 2771 (1965).

43. Dyer, G., and Meek, D. W., *Inorg. Chem.* **4**, 1398 (1965).

44. Dyer, G., and Meek, D. W., *Inorg. Chem.* **6**, 149 (1967).

45. McAuliffe, C. A., and Meek, D. W., *Inorg. Chem.* **8**, 904 (1969).

46. Chow, K. K., Dickinson, R. J., and McAuliffe, C. A., unpublished observations.

47. McAuliffe, C. A., *Inorg. Chem.* **12**, 2477 (1973).

48. Livingstone, S. E., and Lockyer, T. N., *Inorg. Nucl. Chem. Lett.* **3**, 35 (1967).

49. Dutta, R. L., Meek, D. W., and Busch, D. H., *Inorg. Chem.* **10**, 1820 (1971).

50. Hughes, M. N., "The Inorganic Chemistry of Biological Processes." Wiley (Interscience), New York, 1972.

51. Wing, R. M., Tustin, G. C., and Okamura, W. H., *J. Amer. Chem. Soc.* **92**, 1935 (1970).

52. Schrauzer, G. N., Ho, R. K. Y., and Murillo, R. P., *J. Amer. Chem. Soc.* **92**, 3508 (1970).

53. Schrauzer, G. N., and Mayweg, V. P., *J. Amer. Chem. Soc.* **87**, 1483 (1965).

54. Livingstone, S. E., *Quart. Rev., Chem. Soc.* **19**, 386 (1965).

55. Rosen, W., and Busch, D. H., *J. Amer. Chem. Soc.* **91**, 4694 (1969).

56. Rosen, W., and Busch, D. H., *Inorg. Chem.* **9**, 262 (1970).

57. Jonassen, H. B., and Douglas, B. E., *J. Amer. Chem. Soc.* **71**, 4094 (1949).

58. Rowley, D. A., and Drago, R. S., *Inorg. Chem.* **7**, 795 (1968).

THE FUNCTIONAL APPROACH TO IONIZATION PHENOMENA IN SOLUTIONS

U. MAYER and V. GUTMANN

Institut für Anorganische Chemie, Technische Hochschule Wien, Vienna, Austria

I. Introduction

Various approaches are in use to characterize properties of ions and molecules in solution.

Born (1) and later Bjerrum (2) developed a theoretical approach to ion–solvent interactions based on a rather simple electrostatic model. Ions are considered as rigid spheres of radius r and charge z in a solvent continuum of dielectric constant ε. Changes in enthalpy $\Delta H(\mathrm{sv})$ and in free energy $\Delta G(\mathrm{sv})$, respectively, associated with the transfer of the gaseous ions into the solvent are represented by the following equations:

$$\Delta G(\mathrm{sv}) = -\frac{N(ze)^2}{2r}\left(1 - \frac{1}{\varepsilon}\right) \tag{1}$$

$$\Delta H(\mathrm{sv}) = -\frac{N(ze)^2}{2r}\left\{1 - \frac{1}{\varepsilon} - \frac{T}{\varepsilon}\left(\frac{\partial \ln \varepsilon}{\partial T}\right)_p\right\} \tag{2}$$

According to this model cations and anions with the same radius and the same charge number should be equally solvated, and this is in contrast to experimental findings. Later Born's equation was modified by various authors, e.g., by using corrected ionic radii or by considering changes in the dielectric constant of the solvent in the vicinity of the ions (3–5). A more satisfactory explanation for the different behavior of cations and anions was given by Buckingham, who emphasized the

189

importance of quadrupole interaction terms between ions and solvent molecules as well as between solvent molecules of the solvation sphere (6, 7). The main difficulty in applying this model is lack of knowledge of quadrupole (or multipole) moments, solvation numbers, and the arrangement of solvent molecules in the solvation spheres, particularly in nonaqueous solvents. Furthermore absolute solvation energies are large, so that comparatively small errors in the parameters used introduce appreciable uncertainties in solvation energy differences for a given ion in different media. It may therefore be anticipated that successful calculations of solvation quantities by purely electrostatic models are not possible, both for experimental reasons and because of the inherent inadequacy of elementary electrostatic models in view of the actual charge distribution within the solvated ions.

Within the last 5 years a number of quantum mechanical calculations have been carried out, in particular for hydrogen-bonded systems such as H_2O, NH_3, and HF as solvents. Contrary to electrostatic models, MO calculations allow for the possibility of covalent bond formation and, consequently, constitute a fundamentally better approach to ion–molecule interactions. Some interesting results of recent model calculations, although qualitative in nature, are discussed in Section V.

It is well known that single-ion solvation quantities cannot be obtained by purely thermodynamic methods. Thermodynamic quantities obtained for neutral compounds may be split up into their ionic components by use of appropriate extrathermodynamic assumptions. These methods are based on the idea that changes in enthalpy or free energy associated with the transfer of a solute from one solvent to another will essentially be equal for structurally analogous ions or molecules. Examples include pairs of structural analogs such as $[As(C_6H_5)_4]^+$–$[B(C_6H_5)_4]^-$, $[As(C_6H_5)_4]^+$–$[Sn(C_6H_5)_4]$ or (less ideal) I_2–I_3^- (8–11).

A similar approach has been developed independently by Pleskov (12) and Strehlow et al. (13). They assume that redox potentials for certain redox systems such as ferrocene–ferricinium ion are essentially independent of the nature of the solvent. Recent investigations have shown that the ferrocene–ferricinium ion assumption works fairly well in nonaqueous solvents, but it cannot be applied to aqueous solutions (14).

Extrathermodynamic methods represent powerful tools for the evaluation of single-ion solvation quantities, but the available data are rather low in accuracy. Accurate knowledge of solubility and salt activity coefficients is highly desirable. Estimation of liquid–liquid junction potentials (particularly at nonaqueous–aqueous electrolyte interfaces) is another problem that has not been satisfactorily resolved. The use of extrathermodynamic data for a semiquantitative description of ionization equilibria is demonstrated in Section III.

In the following sections the functional approach (*15–17*) is used, according to which most chemical interactions in solution are considered as the result of electron pair donor (EPD)–electron pair acceptor (EPA) interactions between solute and solvent molecules. In the functional representation allowance is made for the charge transfer occurring in the course of a chemical interaction by taking into account all inductive and mesomeric effects both *between* and within the reacting molecules (*15–17*). This approach may be generally applied to all interactions involving the formation of more-or-less covalent bonds, provided the energy changes involved in covalent bonding *are high* as compared to any other energy contributions to the overall interaction energy.

II. Classification of Reactions Producing Ions in Solution

Two main groups of chemical interactions may be distinguished (*15–17*): (*1*) substitution reactions = displacement reactions = atom or group transfer reactions (EPD–EPA reactions); and (*2*) electron transfer reactions (ED–EA reactions). According to the functional approach to chemical interactions (*15–17*), substitution reactions may be initiated either by nucleophilic attack of an EPD or by electrophilic attack of an EPA. Characteristically, these reactions involve formation of a new coordinate bond between substrate and substituting agent with simultaneous cleavage of the substrate bond. Unlike substitution reactions, electron transfer reactions or ED–EA reactions proceed without mass transfer between the reacting particles. The ED–EA interactions probably involve formation of loose coordinate bonds between the reacting species which serve only as bridges for the electron transfer process, e.g.,

$$R + A\text{–}B \longrightarrow R^+ + (A\text{—}B)^-$$

Examples for displacement reactions are

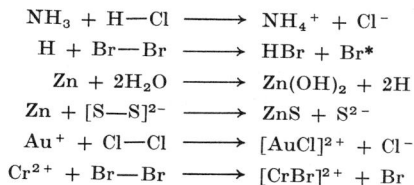

$$NH_3 + H\text{—}Cl \longrightarrow NH_4^+ + Cl^-$$
$$H + Br\text{—}Br \longrightarrow HBr + Br*$$
$$Zn + 2H_2O \longrightarrow Zn(OH)_2 + 2H$$
$$Zn + [S\text{—}S]^{2-} \longrightarrow ZnS + S^{2-}$$
$$Au^+ + Cl\text{—}Cl \longrightarrow [AuCl]^{2+} + Cl^-$$
$$Cr^{2+} + Br\text{—}Br \longrightarrow [CrBr]^{2+} + Br$$

* In solution, HBr undergoes a secondary displacement reaction, yielding H_3O^+ and Br^-.

It is apparent that both ions and radicals may be formed. Reactions between stable nonradicals will usually yield nonradical (ionic) products, whereas in substitution reactions between nonradical and radical species, radicals are produced (which will immediately recombine).

Ionization reactions in homogeneous liquid phase usually proceed as reactions between "closed-shell" molecules. According to the functional approach, these reactions are regarded as EPD–EPA reactions in the strict sense of the word. Owing to the inherently high stability of closed-shell molecules or ions, these reactions usually lead to heterolytic bond cleavage.

Heterolysis may be effected either by nucleophilic attack of an EPD or by electrophilic attack of an EPA at a given substrate A—B (18, 19):

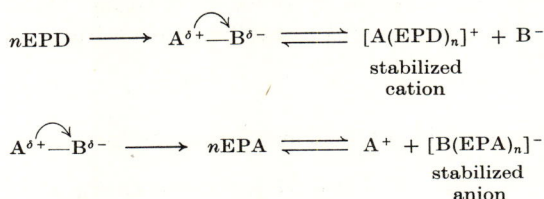

$$n\text{EPD} \longrightarrow A^{\delta+}\!\!-\!\!B^{\delta-} \rightleftharpoons [A(\text{EPD})_n]^+ + B^-$$
$$\text{stabilized cation}$$

$$A^{\delta+}\!\!-\!\!B^{\delta-} \longrightarrow n\text{EPA} \rightleftharpoons A^+ + [B(\text{EPA})_n]^-$$
$$\text{stabilized anion}$$

Examples are given in Eqs. (3) and (4)

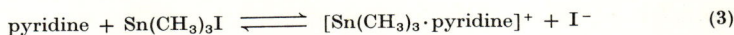

$$\text{pyridine} + \text{Sn(CH}_3)_3\text{I} \rightleftharpoons [\text{Sn(CH}_3)_3 \cdot \text{pyridine}]^+ + I^- \qquad (3)$$

$$\text{NOCl} + \text{SbCl}_5 \rightleftharpoons \text{NO}^+ + \text{SbCl}_6^- \qquad (4)$$

There are also examples of combined EPD and EPA interactions, such as

$$\text{HCl} + (n + m)\text{H}_2\text{O} \rightleftharpoons [\text{H(OH}_2)_n]^+ + [\text{Cl(H}_2\text{O)}_m]^-$$

Usually attack of the EPD occurs at the more positive, and attack of the EPA at the more negative part of the molecule. In both cases, coordination leads to an increase in polarity of the bond $A^{\delta+}\!\!-\!\!B^{\delta-}$ and finally to heterolysis with formation of stabilized cations or of stabilized anions (18, 19). It may be anticipated that a homolytic cleavage of $A^{\delta+}\!\!-\!\!B^{\delta-}$ is rather unlikely: A^+ acts as a much stronger EPA than $A^{\delta+}$, and B^- acts as a stronger EPD than $B^{\delta-}$, so that maximum stabilization is usually achieved by heterolytic cleavage of the $A^{\delta+}\!\!-\!\!B^{\delta-}$ bond.

Homolytic bond cleavage has been reported for the reaction between $CuCl_2$ and thiourea (thu) which leads to formation of a Cu(I)–thiourea complex and chlorine radicals, which are converted into free chlorine (20):

$$\text{CuCl}_2 + n(\text{thu}) \rightleftharpoons [\text{CuCl(thu)}_n] + \text{Cl}$$

This reaction has been explained by the back-bonding effect between thu and copper ion (*17*): thu acts both as σ-EPD and π-EPA; stabilization by back-bonding is particularly favorable for Cu^+ which is a stronger π-EPD than Cu^{2+} (*17*).

It has been stated that attack of an EPD usually occurs at the more positive part of the substrate and that this results in an electron shift from the more positive to the more negative part within the molecule. Exceptions to this rule are possible. In fact the nature of the reaction products formed depends on the relative thermodynamic stabilities of the individual species. An example is provided by the oxidation reaction of sulfite ions by hypochlorite ions [Eq. (5)].

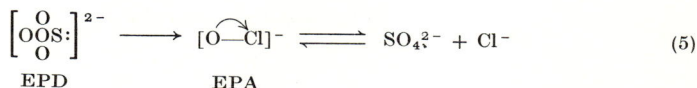

$$\left[O\underset{O}{\overset{O}{O}}S:\right]^{2-} \longrightarrow [O\frown Cl]^- \rightleftharpoons SO_4^{2-} + Cl^- \qquad (5)$$

$$\text{EPD} \qquad\qquad \text{EPA}$$

According to kinetic measurements (*21*), this reaction proceeds as a nucleophilic substitution reaction with sulfur acting as EPD and apparently involves an electron shift from the more negative to the more positive part of the hypochlorite ion. Formation of a coordinate bond between sulfur and oxygen has been demonstrated using ^{18}O-labeled OCl^-.

Traditionally reaction (5) is considered as a redox reaction. It is obvious that both from the functional point of view and from consideration of reaction mechanisms (*21, 22*), there is no basic difference between reaction (5) and "ordinary" substitution reactions such as Eq. (3).

The assignment of oxidation numbers to atoms within molecules is somewhat arbitrary and does not reflect the actual charge distribution within the molecules. For example, reaction (5) can equally well be formulated assuming an oxidation number of $-II$ for sulfur and 0 for oxygen:

$$\overset{-II\,0}{[SO_3]^{2-}} + \overset{0\,-I}{[OCl]^-} \longrightarrow \overset{-II\,0}{[SO_4]^{2-}} + \overset{-I}{Cl}$$

Then this reaction would not be considered a redox reaction since no change in oxidation numbers occurs during the reaction. Likewise the reaction

$$Cr^{2+} + Br_2 \longrightarrow [CrBr]^{2+} + Br$$

may be considered a coordinating rather than a redox reaction.

The distinction that is frequently made between redox reactions and "ordinary" displacement reactions is somewhat arbitrary and may be even misleading (see also Refs. *21–23*).

Unlike substitution reactions, electron transfer reactions proceed without mass transfer between the reacting species. Sometimes it is not possible to decide *a priori* whether a reaction is a true electron transfer reaction or rather a mass transfer (substitution) reaction (*21*). For example, evidence has been presented from kinetic studies that the reaction

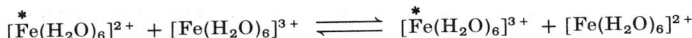

which formally is considered an electron transfer reaction, probably proceeds as a mass transfer reaction involving simultaneous transfer of a hydrogen atom and of a proton:

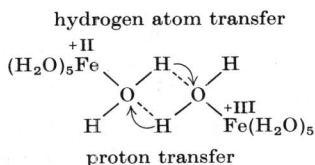

On the other hand, it has been shown that reactions between various substitution inert species, such as $[Fe(CN)_6]^{4-}–[Fe(CN)_6]^{3-}$, $[MnO_4]^{2-}–[MnO_4]^-$, $[Fe(phen)_3]^{3+}–[Fe(phen)_3]^{2+}$, clearly proceed as electron transfer reactions (*21*). Finally all electrochemical oxidation and reduction processes are to be considered as electron transfer reactions.

In numerous chemical reactions, ionic species are formed as intermediates which are subsequently converted into more stable, neutral, reaction products. Examples are provided by numerous well-known substitution reactions such as Friedel–Crafts acylations, nitration reactions, or chlorination reactions. The initial steps of these reactions are completely analogous to reactions of type (3) or (4).

For example, acetylation of benzene by use of CH_3COCl with $SbCl_5$ as a catalyst is due to the intermediate formation of the acetylium ion, which has been shown to exist in stable ionic compounds such as $[CH_3CO]^+[SbCl_6]^-$ or $[CH_3CO]^+[BF_4]^-$.

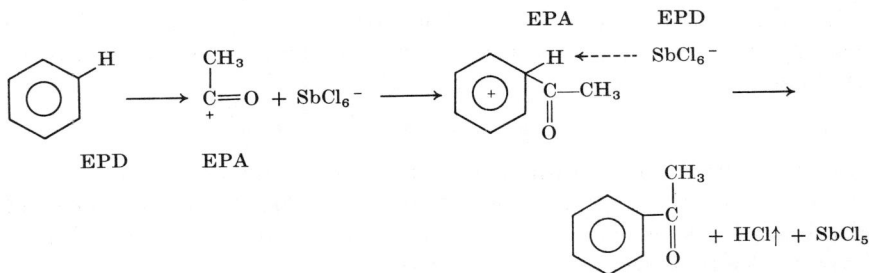

Likewise the strong nitrating action of NO_2Cl–$SbCl_5$ mixtures is due to the formation of $[NO_2]^+[SbCl_6]^-$ acting as an electrophilic agent toward the aromatic ring.

In a number of cases substitution reactions appear not to involve the formation of true ionic intermediates but rather molecular complexes with strongly polarized bonds, e.g.,

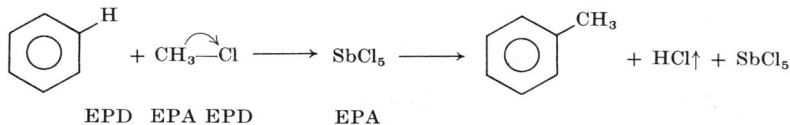

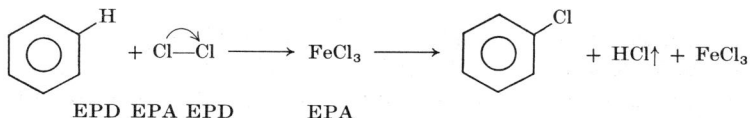

It should perhaps be mentioned that Olah and co-workers (24) have been able to prove the existence of EPA-stabilized carbonium ions in highly acidic media, for example, in mixtures of HF and SbF_5:

$$RF + SbF_5 \longrightarrow [R^+SbF_6{}^-]$$

Similar considerations apply to various nucleophilic substitution reactions.

III. Ionizing Properties of Solvents

The relationship between thermodynamic properties, for example, ionization constants or solubility products of a compound AB, and the coordination chemical properties of solvents can be readily deduced from a consideration of the energy cycles shown in Eqs. (6)–(9).

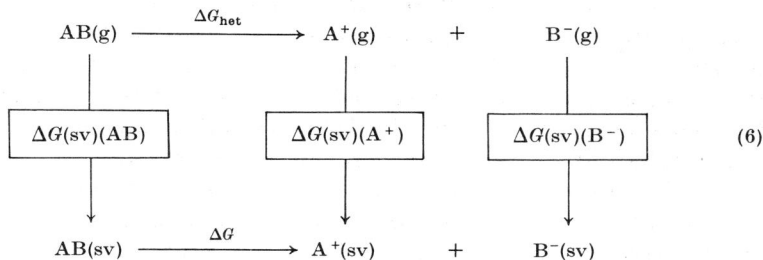

$$\Delta G = -RT \ln K \qquad K = \frac{[A^+][B^-]}{[AB]}$$

$$\Delta G = \Delta G_{het} - \Delta G(sv)(AB) + \Delta G(sv)(A^+) + \Delta G(sv)(B^-) \qquad (7)$$

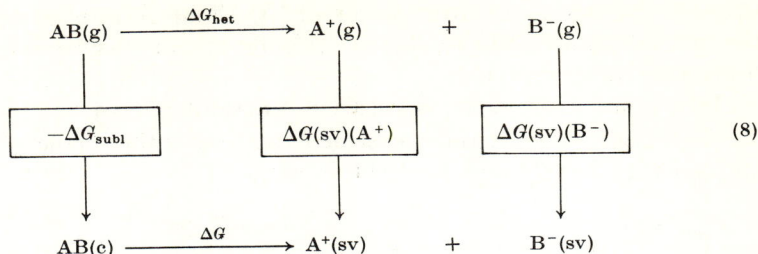

$$AB(g) \xrightarrow{\Delta G_{het}} A^+(g) \quad + \quad B^-(g)$$

$$\downarrow -\Delta G_{subl} \qquad \downarrow \Delta G(sv)(A^+) \qquad \downarrow \Delta G(sv)(B^-) \qquad (8)$$

$$AB(c) \xrightarrow{\Delta G} A^+(sv) \quad + \quad B^-(sv)$$

$$\Delta G = -RT \ln K_s \qquad K_s = [A^+][B^-]$$
$$\Delta G = \Delta G_{subl} + \Delta G_{het} + \Delta G(sv)(A^+) + \Delta G(sv)(B^-) \qquad (9)$$

Cycle (6) shows the various free-energy contributions for the ionization of a covalent compound AB; cycle (8) applies to the heterogeneous equilibrium between a solid compound AB and its saturated solution containing ions A^+ and B^- (ΔG_{het} denotes the free enthalpy change for heterolysis). From Eqs. (7) and (9), it is obvious that both ionization constant and solubility of a given compound AB in different solvents are functions of the free energies of solvation of AB, A^+, and B^-, and these represent a measure of the solvating properties of the solvents. It should be noted that according to

$$\Delta G = \Delta H - T \cdot \Delta S \qquad (10)$$

a direct relationship also exists between ionization constant or solubility and the enthalpies of solvation, $\Delta H(sv)$, of the species present in solution. The present approach considers the solvating properties of the solvents as a consequence of EPD–EPA interactions between solvent molecules and the various species present in solution, leading both to the formation of more-or-less covalent bonds between the reacting particles and to changes in bonding within the molecules (15–19, 25). Consequently, this concept provides not only a generalized and unified description of chemical interactions, but it offers also the possibility of estimating and correlating thermodynamic properties of compounds in different solvent systems, provided that suitable quantities characteristic for EPD and (or) the EPA properties of the solvents can be defined. In principle, each solvent may function as EPD as well as EPA, but for most solvents either the EPD or the EPA functions prevail; the former are denoted as EPD solvents and the latter as EPA solvents (25). Examples of typical EPD solvents are pyridine, ethers, esters, dimethyl-sulfoxide, and hexamethylphosphoricamide; examples of typical EPA solvents are the liquid hydrogen halides, liquid sulfur dioxide, sulfuric acid, and trifluoroacetic acid. Several solvents are amphoteric as they are capable of exercising both EPD and EPA functions. For example, water owes its outstanding solvent properties to the fact that toward Lewis acids such as cations it functions as a fairly strong EPD (via the oxygen atom), whereas toward certain anions it functions as a fairly strong EPA (via the hydrogen atoms). Hence water can provide stabilization for both cationic and anionic species, e.g.,

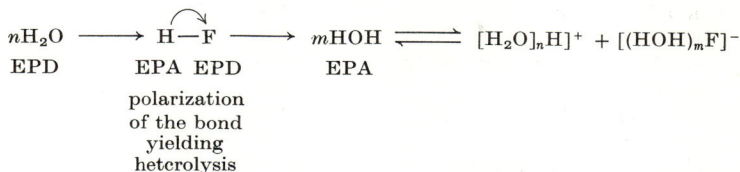

$$nH_2O \longrightarrow H\overset{\frown}{}F \longrightarrow mHOH \rightleftharpoons [H_2O]_nH]^+ + [(HOH)_mF]^-$$

EPD   EPA EPD   EPA

polarization
of the bond
yielding
heterolysis

Simultaneous stabilization of both cations and anions may also be effected by combination of an EPD solvent (with poorly developed EPA properties) and a suitable EPA. For example, nitromethane (NM) as a solvent functions as a weak EPD and hence compounds such as $CoCl_2$ remain insoluble in this solvent. Addition of a strong EPA, such as $SbCl_5$, to this suspension leads to dissolution of $CoCl_2$ with simultaneous heterolysis of the Co—Cl bonds, owing to the stabilization of both cationic and anionic species:

$$CoCl_2 + 6NM + 2SbCl_5 \longrightarrow [Co(NM)_6]^{2+} + 2[SbCl_6]^-$$

On the other hand, $CoCl_2$ is insoluble in pure liquid $SbCl_5$ since cation stabilization cannot take place. The majority of solvents that are extensively used in solution chemistry (particularly in the field of organic chemistry) are typical EPD solvents. Gutmann (25–28) has introduced the so-called donor number or donicity (DN) as a measure of the EPD properties of donor solvents. This is defined as the negative ΔH values for formation of the 1:1 adduct of the EPD with $SbCl_5$ as reference standard EPA in a dilute solution of 1,2-dichloroethane. Donicities for various solvents are listed in Table I together with their dielectric constants ε.

The basic idea in defining the donicity is based on the work of Lindqvist (29), who obtained from calorimetric measurements an order of relative, solvent, EPD strengths toward $SbCl_5$. It was then supposed that the relative EPD properties toward different EPA units (neutral and cationic) could be predicted (at least qualitatively) from the trends observed toward $SbCl_5$. This is, indeed, true for a large number of both neutral and ionic substrates (25–28). Figure 1 shows the relationship between ^{19}F chemical shifts of CF_3I dissolved in various EPD solvents as a function of solvent donicity DN (30). Nucleophilic attack of an EPD at the iodine atom causes an electron shift from iodine to the

TABLE I

DONICITIES (DN) AND DIELECTRIC CONSTANTS (ε)
FOR SEVERAL SOLVENTS

Solvent	DN	ε
1,2-Dichloroethane	—	10.1
Sulfuryl chloride	0.1	10.0
Benzene	0.1	2.3
Thionyl chloride	0.4	9.2
Acetyl chloride	0.7	15.8
Tetrachloroethylene carbonate	0.8	9.2
Benzoyl fluoride	2.0	22.7
Benzoyl chloride	2.3	23.0
Nitromethane	2.7	35.9
Dichloroethylene carbonate	3.2	31.6
Nitrobenzene	4.4	34.8
Acetic anhydride	10.5	20.7
Phosphorus oxychloride	11.7	14.0
Benzonitrile	11.9	25.2
Selenium oxychloride	12.2	46.0
Acetonitrile	14.1	38.0
Sulfolane	14.8	42.0
Propanediol-1,2-carbonate	15.1	69.0
Benzyl cyanide	15.1	18.4
Ethylene sulfite	15.3	41.0
Isobutyronitrile	15.4	20.4
Propionitrile	16.1	27.7
Ethylene carbonate	16.4	89.1
Phenylphosphonic difluoride	16.4	27.9
Methylacetate	16.5	6.7
n-Butyronitrile	16.6	20.3
Acetone	17.0	20.7
Ethyl acetate	17.1	6.0
Water	18.0[a]	81.0
Phenylphosphonic dichloride	18.5	26.0
Methanol	19.0	32.6
Diethyl ether	19.2	4.3
Tetrahydrofuran	20.0	7.6
Diphenylphosphonic chloride	22.4	—
Trimethyl phosphate	23.0	20.6
Tributyl phosphate	23.7	6.8
Dimethylformamide	26.6	36.1
N-Methyl-ε-caprolactame	27.1	—
N-Methyl-2-pyrrolidone	27.3	—
N,N-Dimethylacetamide	27.8	38.9
Dimethyl sulfoxide	29.8	45.0
N,N-Diethylformamide	30.9	—
N,N-Diethylacetamide	32.2	—
Pyridine (py)	33.1	12.3
Hexamethylphosphoric amide	38.8	30.0

[a] For water as a solvent, the bulk donicity $DN^B \sim 30$.

fluorine atoms the magnitude of which is proportional to the EPD properties (donicities) of the nucleophile:

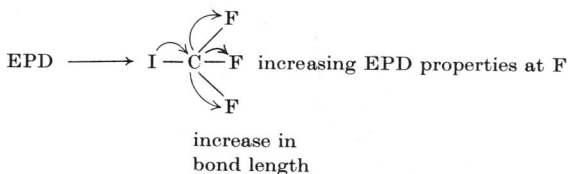

$$EPD \longrightarrow \quad I-C\overset{F}{\underset{F}{\smash{\raisebox{-1ex}{$-$}}}}F \quad \text{increasing EPD properties at F}$$

increase in
bond length

Figure 1 shows the linear correlation between ^{19}F chemical shift and solvent donicity. No relationship exists between chemical shifts and dipole moments or polarizabilities of the solvent molecules. This result is particularly significant in that it shows that the functional approach can be successfully applied even to weak chemical interactions, for example, to interactions that are usually considered as being due to Van der Waals forces or intermolecular forces in the sense of Mulliken's theory.

Essentially the same order of EPD strength as observed toward $SbCl_5$ was found for various EPA molecules such as $VO(acac)_2$ (see Table VII), $Sn(CH_3)_3Cl$, phenol, or $SbCl_3$ (25). Irregularities are

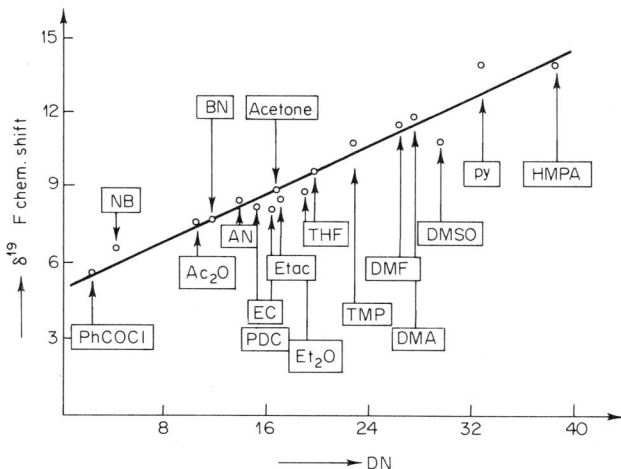

FIG. 1. Linear correlation between ^{19}F NMR chemical shifts at infinite dilution of CF_3I in various electron pair donor solvents and donicity (DN), referred to CCl_3F as external reference. NB, nitrobenzene; BN, benzonitrile; AN, acetonitrile; PDC, propanediol-1, 2-carbonate; THF, tetrahydrofuran; DMF, dimethylformamide; EC, ethylene carbonate; TMP, trimethyl phosphate; DMA, N,N-dimethylacetamide; DMSO, dimethylsulfoxide; HMPA, hexamethylphosphoric amide.

observed when relative EPD strengths for *both* hard *and* soft EPD are determined toward a hard EPA, on the one hand, and a soft reference EPA, on the other hand. For example sulfur donors (which are soft) behave as much stronger EPD toward soft EPA such as I_2, Ag^+, Cu^+ than analogous oxygen compounds (*31, 32*), whereas reverse stabilities are found toward a hard EPA such as the proton, alkali metal ions, phenol, or an EPA unit with borderline behavior (Co^{2+}, Ni^{2+}, etc.) (*32*). Similarly, nitriles that behave as rather weak EPD molecules toward $SbCl_5$ show a much higher affinity toward a soft EPA such as Ag^+ and particularly Cu^+ (*33*). This suggests that nitriles may act toward soft acids both as σ donors and π acceptors. In practice these irregularities do not seriously restrict the applicability of the donicity concept since most of the solvents and solutes are rather hard.

Use of donicity values as a measure of cation solvation and cation stabilization has been demonstrated by polarographic measurements on alkali and alkaline earth metal ions, and various transition and rare earth metal ions (*17, 16, 34*). This is illustrated in Fig. 2 which shows the variation of half-wave potentials for the reduction of $Tl^+ \rightarrow Tl^0$, $Zn^{2+} \rightarrow Zn^0$, and $Eu^{3+} \rightarrow Eu^{2+}$ as a function of solvent donicities.

From the energy cycle,

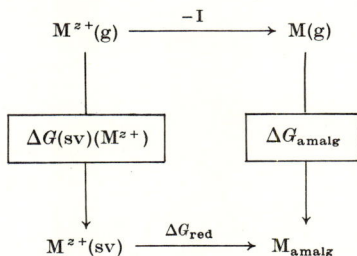

$$M^{z+}(g) \xrightarrow{\quad -I \quad} M(g)$$

$$\Delta G(sv)(M^{z+}) \qquad \Delta G_{amalg}$$

$$M^{z+}(sv) \xrightarrow{\quad \Delta G_{red} \quad} M_{amalg}$$

$$\Delta G_{red} = -zFE \qquad (E \approx E_{1/2} \text{ for reversible reductions})$$

and equation,

$$\Delta G_{red} = -I + \Delta G_{amalg} - \Delta G(sv)(M^{z+}) \qquad (11)$$

it is apparent that standard redox potential and, consequently, the half-wave potential for the reduction of a given metal cation to the metal amalgam is only a function of the free energy of solvation. Increasing EPD strength of the solvent leads to an increased stabilization of the metal cation which is, therefore, less readily reduced and this results in a shift of $E_{1/2}$ toward more negative potential values (*16, 17*). A similar consideration applies to redox reactions with both species solvated in solution: Eu^{3+} is a stronger EPA than Eu^{2+} so that with

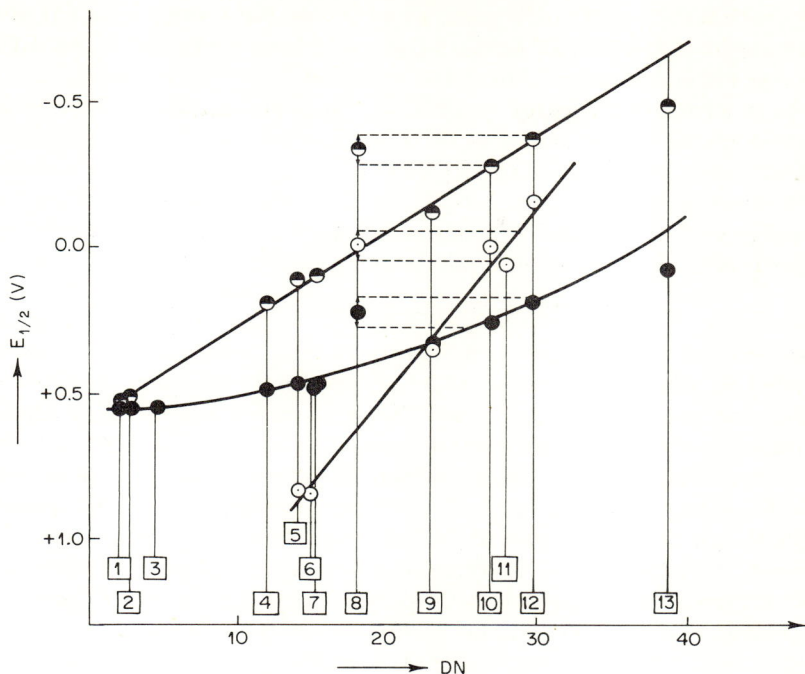

FIG. 2. Relationship between half-wave potential $E_{1/2}$ and solvent donicity (DN) for the polarographic reduction $Tl^+ \rightarrow Tl^0$ (●), $Zn^{2+} \rightarrow Zn^0$ (◖), $Eu^{3+} \rightarrow Eu^{2+}$ (○). 1—benzoyl fluoride; 2—nitromethane; 3—nitrobenzene; 4—benzonitrile; 5—acetonitrile; 6—propanediol-1, 2-carbonate; 7—ethylene sulfite; 8—Water; 9—trimethyl phosphate; 10—dimethylformamide; 11—N, N-dimethylacetamide; 12—dimethyl sulfoxide; 13—hexamethylphosphoric amide.

increasing solvent donicity the oxidized species will be more strongly stabilized. Figure 2 shows that with the exception of water, which behaves as a much stronger EPD than expected from its donicity, a good correlation between $E_{1/2}$ and DN exists for numerous redox pairs. The "anomalous" behavior of water is discussed in Section V.

In a solvent of more-or-less well-developed amphoteric properties the extent of ionization of a given compound $A^{\delta+}$—$B^{\delta-}$ [see Eq. (12)]

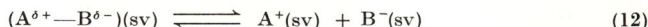

$$(A^{\delta+}\text{—}B^{\delta-})(sv) \rightleftharpoons A^+(sv) + B^-(sv) \qquad (12)$$

will depend both on the EPD and EPA properties of the solvents. Stabilization of A^+ and B^- by respective solvation will shift equilibrium (12) to the right side of the equation, whereas stabilization of the unionized substrate will shift the equilibrium to the opposite direction. Since

$A^{\delta+}$ (within $A^{\delta+}$—$B^{\delta-}$) is acting as a weaker EPA than A^+, and $B^{\delta-}$ is acting as a weaker EPD than B^-, increasing both EPD and EPA functions of a solvent will shift the equilibrium to the right side.

In order to gain a better insight into the basic relationships between ionizing power of solvents and their coordination chemical properties, it is necessary to study both the EPD effect at the cation A^+ and the EPA effect at the anion B^- separately. The role of the EPD effect has been successfully studied by the use of conductometric techniques for investigating the ionization equilibrium,

$$n\text{EPD} + \text{AB} \rightleftharpoons [A(\text{EPD})_n]^+ + B^-$$

of suitable substrates AB in a weakly coordinating medium of sufficiently high dielectric constant (35) to which an EPD solvent is added in slight excess to the solute present. Under these conditions there will be no interference from anion solvation since B^- will be solvated only by the solvent molecules that are present in large excess. Furthermore, since the ions $[A(\text{EPD})_n]^+$ and B^- are formed in a medium of invariable dielectric constant, the extent of ion pair formation (see Section IV) between $[A(\text{EPD})_n]^+$ and B^- remains essentially constant * for different EPD units so that the conductivities are a direct measure of the EPD effect at cation A^+. Media that are particularly suitable for measurements of this type are nitrobenzene (DN = 4.4; ε = 34.8), nitromethane (DN = 2.7; ε = 35.9), and eventually dichloromethane (DN < 2; ε = 9.0). Figure 3 shows the increase in molar conductivities of

TABLE II

MOLAR CONDUCTIVITIES OF $(CH_3)_3SnI$ IN NITROBENZENE IN PRESENCE OF ELECTRON PAIR DONOR[a]

EPD[b]	Λ	DN	μ	ε
Nitrobenzene	0	2.7	3.54	35.9
Acetonitrile	0.1	14.1	3.2	38.8
Tetrahydrofuran	0.2	20.0	1.75	7.6
Diphenylphosphonic dichloride	0.45	22.4	—	—
Tributyl phosphate	0.95	23.7	3.06	6.8
Dimethylformamide	1.7	26.6	3.86	36.7
Pyridine	2.7	33.1	2.3	12.3
Dimethyl sulfoxide	3.7	29.8	3.9	48.9
Hexamethylphosphoric amide	17.5	38.8	5.54	33.5

[a] Molar ratio, EPD: $(CH_3)_3SnI$ = 8 at $c \approx 7 \times 10^{-2}$ (25°C).

[b] EPD, electron pair donor; DN, donicity.

* This will only be true (see Section IV) if the coordinated cations $[A(\text{EPD})_n]^+$ are sufficiently large and coordinatively saturated.

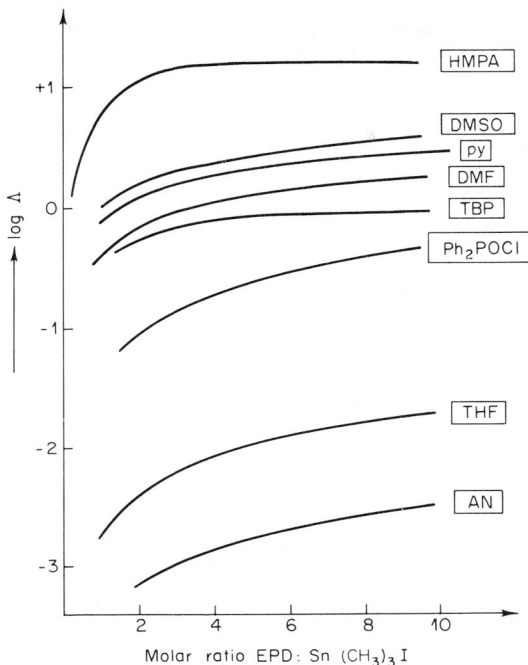

FIG. 3. Molar conductivities of $Sn(CH_3)_3I$ ($c = 7 \times 10^{-2}$ mole/liter) dissolved in nitrobenzene on addition of increasing amounts of various electron pair donor (EPD) solvents. HMPA, hexamethylphosphoric amide; DMSO, dimethyl sulfoxide; DMF, dimethylformamide; TBP, tributyl phosphate; THF, tetrahydrofuran; AN, acetonitrile.

$Sn(CH_3)_3I$ dissolved in nitrobenzene as a function of concentration of various EPD solvents added (*35*). In noncoordinating or weakly coordinating solvents, such as hexane, carbon tetrachloride, 1,2-dichloroethane, nitrobenzene, or nitromethane, $Sn(CH_3)_3I$ is present in an unionized state (tetrahedral molecules). Addition of a stronger EPD solvent to this solution provokes ionization, presumably with formation of trigonal bipyramidal cations $[Sn(CH_3)_3 \cdot (EPD)_2]^+$. Table II reveals that the molar conductivities at a given mole ratio EPD:$Sn(CH_3)_3I$ are (with the exception of pyridine) in accordance with the relative solvent donicities. No relationship appears to exist between conductivities and the dipole moments or the dielectric constants of the solvents.

The decreased EPD properties of pyridine may be due to steric hindrance between methyl groups and the α-hydrogens of the ring. Contrary to Fig. 3, conductivities measured for $Sn(CH_3)_3I$ in the pure

EPD solvents do not correctly reflect the ionizing properties of these solvents due to the differences in dielectric constants. Although $Sn(CH_3)_3I$ is considerably ionized in pure tributylphosphate, the solutions are essentially nonconducting because of the very low dielectric constant $\varepsilon = 6.8$ of this medium (see Section IV). Fuoss-Krauss analysis of conductance data for $Sn(CH_3)_3I$ in strong EPD solvents, such as dimethylformamide (DMF), dimethyl sulfoxide (DMSO), pyridine, and hexamethylphosphoric amide (HMPA), reveal that the substrate is completely ionized and consequently behaves as a 1:1 electrolyte (35).

TABLE III

COUPLING CONSTANTS J (CH_3—^{119}Sn)

Solvent	DN	J_c(Hz)
Hexamethylphosphoric amide	38.8	72.0
Dimethyl sulfoxide	29.7	69.0
Dimethylformamide	26.6	68.5
Nitrobenzene	4.4	59.0

Nuclear magnetic resonance investigations demonstrate (Table III) the existence of an empirical relationship between the ionizing power of the EPD solvents and the coupling constants $J(CH_3\text{-}^{119}Sn)$ of the ionized substrate: the coupling constants are increased by increasing donicitiy of the EPD.

Conductometric investigations in nitrobenzene (NB) have recently been extended to trialkylsilicon and trialkylgermanium compounds (36). Again the extent of ionization increases with increasing donicity. In strong EPD solvents, iodo compounds are usually completely ionized, e.g., $Si(CH_3)_3I$ is completely ionized in HMPA, DMF, N, N-dimethylacetamide (DMA), pyridine, and tetramethylurea (DN = 29.8) with formation of stable pentacoordinated "siliconium" ions, $[Si(CH_3)_3 \cdot (EPD)_2]^+$ and I^-; this is particularly interesting, since evidence for the existence of *stable* siliconium ions *in solution* is rather scarce (37, 38). All Group IV compounds exhibit hard acid behavior; consequently, sulfur compounds and phosphines behave as much weaker EPD ligands than the corresponding oxygen compounds and amines. In contrast, cacodylic iodide, $As(CH_3)_2I$, clearly behaves as a soft substrate and extensive ionization occurs on addition of phosphines and thioamides to the nonconducting solution of $As(CH_3)_2I$ in NB. On the other hand, amines and amides show only weak ionizing properties (36), the ionization increasing with increasing donicity.

In a similar manner, conductometric techniques may be applied to study the relationship between ionizing power and EPA properties of various solutes, for example, by conductometric titration of triphenyl-chloromethane dissolved in NB with EPA compounds:

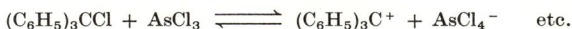

$$(C_6H_5)_3CCl + SO_2 \rightleftharpoons (C_6H_5)_3C^+ + SO_2Cl^-$$

$$(C_6H_5)_3CCl + HCl \rightleftharpoons (C_6H_5)_3C^+ + HCl_2^-$$

$$(C_6H_5)_3CCl + AsCl_3 \rightleftharpoons (C_6H_5)_3C^+ + AsCl_4^- \quad \text{etc.}$$

Since the triphenylmethyl cation exhibits a strong absorption band in the visible region, these reactions may also be studied conveniently by the spectrophotometric method. The latter method has been used to determine relative EPA strength for numerous liquid and solid EPA compounds (39).

The situation becomes more complicated if a given ionization reaction is studied in solvents that differ both in their EPD and EPA properties. This may be illustrated for the complex formation between Co^{2+} and Cl^- ions. Qualitatively, stabilities of cobalt–chloro complexes usually decrease with increasing EPD strength of the solvent (25, 26). Quantitative measurements reveal, however, a number of "irregularities" which cannot be understood by considering the differences in solvent donicities. Accurate thermodynamic data have recently been determined for the reaction

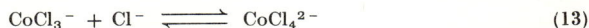

$$CoCl_3^- + Cl^- \rightleftharpoons CoCl_4^{2-} \tag{13}$$

$$K = \frac{[CoCl_4^{2-}]}{[CoCl_3^-]\cdot[Cl^-]} \quad , \quad \Delta G = -RT \ln K$$

in various nonaqueous solvents (40) and are listed in Table IV. As expected, $[CoCl_4]^{2-}$ is most stable in NM, which is a poor EPD solvent, and it is least stable in strong EPD solvents such as DMF or DMA,

TABLE IV

FREE ENERGIES OF FORMATION AND FORMATION CONSTANTS FOR THE PRODUCTION OF $[CoCl_4]^{2-}$ FROM $[CoCl_3]^-$ AND Cl^- IN NONAQUEOUS SOLVENTS AT ZERO IONIC STRENGTH AND 25°C

Solvent	ΔG(kcal/mole)	K (liters mole^{-1})
Nitromethane	-6.59 ± 0.04	$(7.0 \pm 0.5) \times 10^4$
Benzonitrile	-4.82 ± 0.01	$(3.5 \pm 0.1) \times 10^3$
Acetonitrile	-3.92 ± 0.02	$(7.7 \pm 0.2) \times 10^2$
Acetone	-4.69 ± 0.04	$(2.8 \pm 0.2) \times 10^3$
Dimethylformamide	-1.64 ± 0.07	16 ± 2
N,N-Dimethylacetamide	-2.05 ± 0.07	32 ± 4

whereas intermediate stabilities are observed in solvents of medium EPD strength, namely, benzonitrile (BN), acetone, and acetonitrile (AN). On the other hand, stabilities in acetone and AN ($K_{acetone} \gg K_{AN}$) and DMA and DMF ($K_{DMA} > K_{DMF}$), respectively, apparently do not correspond to trends predicted from the consideration of the relative solvent donicities. Water in particular (not listed in Table IV) exhibits a strongly anomalous behavior in that $[CoCl_4]^{2-}$ is even less stable than in DMF and DMA.

These discrepancies are readily explained by taking into consideration the different EPA properties of the respective solvents (42, 34, 41).

TABLE V

FREE ENERGIES OF TRANSFER OF Cl^- FROM REFERENCE SOLVENT
ACETONITRILE ($\Delta G°$) TO VARIOUS OTHER SOLVENTS

Solvent[a]	$\Delta G(sv)(Cl^-)$-$\Delta G(sv)°(Cl^-)$	Solvent[a]	$\Delta G(sv)(Cl^-)$-$\Delta G(sv)°(Cl^-)$
Water	−8.71	DMSO	0.00
Methanol	−5.98	DMF	1.22
Ethanol	−4.76	DMA	2.45
Nitro-		Acetone	3.54
methane	−3.13	HMPA	4.08
PDC	−0.14		
Acetonitrile	0.00		

[a] PDC, propanediol-1,2-carbonate; DMSO, dimethyl sulfoxide; DMF, dimethylformamide; DMA, N,N-dimethylacetamide; HMPA, hexamethylphosphoric amide.

No empirical quantity (analogous to donicity) is available to characterize relative EPA strength of solvents. Fortunately, it has been possible to obtain approximate values of free energies of solvation for individual ions by means of extrathermodynamic assumptions. Table V lists values of free energies of transfer for Cl^- (using AN as reference solvent), which have been calculated (41) from "solvent activity coefficient" data available in the literature (11). According to Table V, the following order of relative EPA strengths for EPD solvents is obtained:

$H_2O \gg$ methanol $>$ ethanol $>$ NM $\gg$ PDC $\geq$ AN $\approx$ DMSO $>$ DMF $>$

DMA $>$ acetone $>$ HMPA

The influence of anion solvation on equilibrium (13) can be derived from the energy cycle [Eq. (14)] which shows the relationship between the standard free energy ΔG of formation of $[CoCl_4]^{2-}$ in a solvent L and in the gas phase $[\Delta G(g)]$.

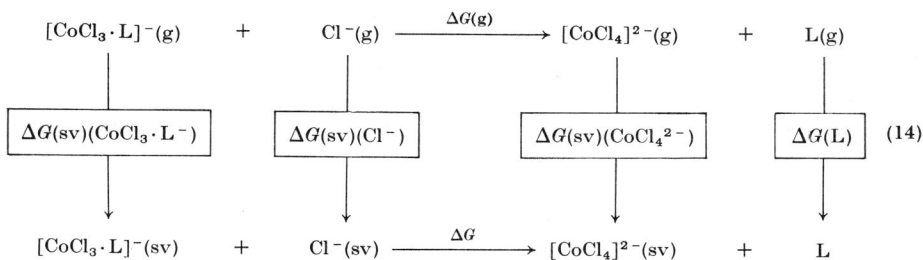

$$[CoCl_3 \cdot L]^-(g) \quad + \quad Cl^-(g) \quad \xrightarrow{\Delta G(g)} \quad [CoCl_4]^{2-}(g) \quad + \quad L(g)$$

$$\boxed{\Delta G(sv)(CoCl_3 \cdot L^-)} \quad \boxed{\Delta G(sv)(Cl^-)} \quad \boxed{\Delta G(sv)(CoCl_4{}^{2-})} \quad \boxed{\Delta G(L)} \quad (14)$$

$$[CoCl_3 \cdot L]^-(sv) \quad + \quad Cl^-(sv) \quad \xrightarrow{\Delta G} \quad [CoCl_4]^{2-}(sv) \quad + \quad L$$

$$\Delta G(g) = \Delta G + \Delta G(sv)(Cl^-) + \Delta G(sv)(CoCl_3 \cdot L^-)$$
$$- \Delta G(sv)(CoCl_4{}^{2-}) - \Delta G(L) \qquad (15)$$
$$\Delta G = -RT \ln K$$

An analogous relationship also holds for the reference solvent L^0 (L^0 = acetonitrile):

$$\Delta G^0(g) = \Delta G^0 + \Delta G^0(sv)(Cl^-) + \Delta G^0(sv)(CoCl_3 \cdot L^{0-})$$
$$- \Delta G^0(sv)(CoCl_4{}^{2-}) - \Delta G^0(L^0) \qquad (16)$$

Equations (15) and (16) reveal that a relationship between standard free energy (or stability constant) and EPD strengths of solvents can only be expected if the free energy for the gas phase reaction $\Delta G(g)$ (or the corresponding stability constant) is considered, since ΔG [in addition to $\Delta G(g)$] includes all contributions due to solvation of the reacting particles. Unfortunately, solvation energies of the complex anions are not known. Equation (15) can be simplified by assuming as a first approximation that contributions originating from groups $CoCl_3$ and L to free energies of solvation of $[CoCl_3 \cdot L]^-$, on the one hand, and $[CoCl_3 \cdot Cl]^{2-} + L$, on the other hand, will largely compensate each other and that Cl^- in $[CoCl_3 \cdot Cl]^{2-}$ will logically behave as a much weaker base than uncoordinated Cl^- By applying the same procedure to Eq. (16), one obtains by combination

$$\Delta G(g) - \Delta G^0(g) = \Delta G - \Delta G^0 + \alpha[\Delta G(sv)(Cl^-) - \Delta G^0(sv)(Cl^-)] \qquad (17)$$

where α is a constant that may take values between 0 and 1.*

Equation (17) allows calculation of standard free energies in the gas phase, [referred to the standard free energy of reaction (18)] from

---

* Since Eq. (17) applies to free-energy differences for substitution reactions of the same type, it is plausible to assume that $\Delta H(DN)-\Delta G$ relationships are not too seriously influenced by variable entropy contributions. A more rigorous treatment of the energy cycle (14) will be discussed in a subsequent paper (42).

experimentally determined free-energy data and free energies of solvation listed in Table V.

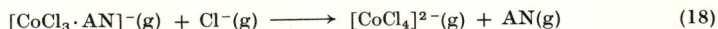

$$[CoCl_3 \cdot AN]^-(g) + Cl^-(g) \longrightarrow [CoCl_4]^{2-}(g) + AN(g) \qquad (18)$$

Values $\Delta G(g) - \Delta G^0(g)$ have been calculated for the solvents listed in Table IV assuming various values for α between 0 and 1. (Representative values, assuming $\alpha = 1$ and 0.5 are given in Table VI). Independent

TABLE VI

FREE ENERGY DIFFERENCES $\Delta G(g) - \Delta G(g)^\circ$
FOR VARIOUS ELECTRON PAIR DONORS[a]

	$\Delta G(g) - \Delta G(g)^\circ$	
EPD	$\alpha = 1$	$\alpha = 0.5$
Nitromethane	-5.80	-4.24
Acetonitrile	$+0.00$	$+0.00$
Acetone	$+2.77$	$+1.00$
Dimethylformamide	$+3.50$	$+2.89$
N,N-Dimethylacetamide	$+4.37$	$+3.095$

[a] Reference electron pair donor (EPD) is acetonitrile.

from the value of $\alpha(0.3 \leq \alpha \leq 1)$, the same relative stabilities are obtained. Contrary to stabilities observed in solution, the gas phase stabilities do correspond to stabilities predicted from the donicities of the EPD molecules:

DMA > DMF ≫ acetone > AN ≫ NM

[Benzonitrile has not been included because $\Delta G(sv)(Cl^-)$ is not available for this solvent.]

Of course, it is impossible to determine α *a priori*. If Eq. (17) is physically reasonable, it should be possible to choose α in such a way that a plot of $\Delta G(g) - \Delta G^0(g)$ versus DN can be represented by a fairly smooth curve, as shown in Fig. 4 for $\alpha \approx 0.5$. This value appears reasonable: low α values are highly improbable since the free Cl^- ion is undoubtedly a much stronger base than the coordinated Cl^- ion in $[CoCl_4]^{2-}$; on the other hand, solvation enthalpies of the complex anions will compensate only in part, so that α is necessarily < 1. Equation (17) can be used to estimate free energies of formation or stability constants of $[CoCl_4]^{2-}$ in other solvents, provided that the donicities and the values $\Delta G(sv)(Cl^-)$ (Table V) are known. Values $\Delta G(g) - \Delta G^0(g)$ required for this purpose may be interpolated or extrapolated* from Fig. 4.

* Long-range extrapolations should of course be avoided.

Application of Eq. (17) should be particularly instructive in the case of water as a solvent. According to its donicity, DN $\approx$ 18, one should expect a stability constant K_{water} for $[CoCl_4]^{2-}$ that corresponds roughly to that observed in acetone, namely $K_{water} \approx 10^{+3}$. Semi-quantitative measurements show (43) that $[CoCl_4]^{2-}$ is, in fact, extremely unstable in water—no appreciable amounts of $[CoCl_4]^{2-}$ are

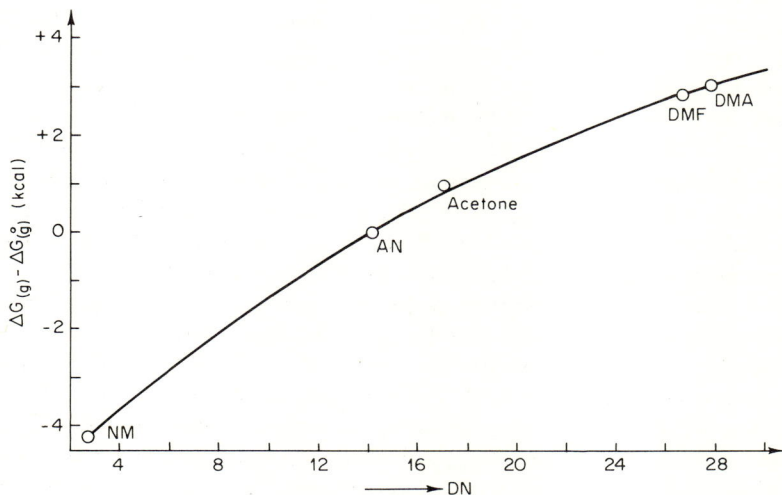

FIG. 4. Free-energy differences $\Delta G(g) - \Delta G°(g)$ for the gas phase reactions $[CoCl_3 \cdot L]^- + X^- \rightarrow CoCl_4^{2-} + L$ as a function of donicity (DN); superscript 0 refers to standard gas phase reaction with L^0 = acetonitrile (AN); $\alpha = 0.5$. DMF, dimethylformamide; DMA, N, N-dimethylacetamide.

formed even in the presence of a large excess of chloride ions ($[Cl^-]$: $[Co^{2+}] > 3000$; $(Co^{2+}) \approx 10^{-3}$ mole/liter). This apparent anomalous behavior of water is easily explained by taking into consideration the strong EPA properties of water (41). By using values $\Delta G_{H_2O}(g) - \Delta G_{AN}(g) \approx +1.1$ from Fig. 4 and $\Delta G(sv)(Cl^-) - \Delta G^0(sv)(Cl^-) = -8.71$ from Table V, one obtains ΔG_{H_2O}(solution) $\approx +1.54$, corresponding to a stability constant K_{H_2O}(solution) $\approx 7.4 \times 10^{-2}$. This predicted value is lower by more than 2 powers of ten than the stability constants measured for the strongest EPD solvents listed in Table IV and lower by more than 4 powers of ten than the expected value if anion solvation is ignored. This value is also in satisfactory agreement with experimental results.

It is highly probable that the "true" value of the stability constant in water K_{H_2O} (which has not yet been determined) is still lower, since the effective EPD strength of a highly associated liquid is considerably higher than that of the isolated molecules. This is regarded as a result of outer-sphere interactions between polarized solvent molecules of the inner and outer hydration spheres of the solvated cation (34). A more general and detailed discussion of outer-sphere effects is given in Section V.

Similar considerations may be applied to methanol as a solvent. Methanol has a donicity of DN = 19.1 so that, according to Table IV, a stability constant for $[CoCl_4]^{2-}$ of $K_{CH_3OH} \approx 5 \times 10^{+2}$ should be expected. Semiquantitative measurements carried out by Katzin (44) suggest, however, the stability of $[CoCl_4]^{2-}$ in methanol is lower than in DMF or DMA. This is readily explained by taking into account the strong anion-solvating properties of methanol: calculation of K(solution) by use of Fig. 4 ($\Delta G_{CH_3OH}(g) - \Delta G_{AN}(g) \approx +1.4$) and Table V gives values ΔG_{CH_3OH}(solution) $\approx +0.48$ and K_{CH_3OH}(solution) $\approx 4.4 \times 10^{-1}$ in agreement with semiquantitative experimental results.

The strong anion-solvating power of water, methanol, and many other hydroxylic solvents is due to hydrogen bond formation.

Calculation of the stability constant of $[CoCl_4]^{2-}$ in DMSO may serve as a final test to Eq. (17); DMSO has nearly the same donicity as DMA but its EPA functions are much stronger. Consequently, K_{DMSO} (solution) is smaller than K_{DMA}(solution): ΔG_{DMSO}(solution) $= -0.52$ and K_{DMSO} (solution) $= 2.40$ calculated from Fig. 4 and Table V. This is in excellent agreement with experimental values recently determined by Magnell and Reynolds (45) of $\Delta G_{DMSO} = -0.58$ and $K_{DMSO} = 2.66$.

It should be pointed out that the influence of anion solvation on equilibrium (13) cannot be explained by use of the Born-Bjerrum equations [Eqs. (1) and (2)]. Use of Eq. (2) generally gives negative ΔG values for the overall free-energy contribution to equilibrium (13) due to *anion* solvation. This means that formation of $[CoCl_4]^{2-}$ should be particularly favorable in solvents with high dielectric constants such as water in contrast to the experimental results.

Likewise the different behavior of $HClO_4$ and $(C_6H_5)_3COH$ in water and sulfuric acid cannot be explained by simple electrostatic models (25) as both solvents have nearly the same dielectric constant. Perchloric acid, which has pronounced EPA properties, is completely ionized in the EPD solvent, water, but it remains essentially unionized in the strong EPA solvent, sulfuric acid. Triphenylcarbinol, on the other hand, reacts quantitatively with the strong EPA solvent, H_2SO_4, but no interaction occurs with the strong EPD solvent, water.

In Section II it has been suggested that ionization reactions may be generally interpreted as substitution reactions:

$$\text{A—B} + \text{EPD} \rightleftharpoons [\text{A(EPD)}]^+ + \text{B}^-$$

In this reaction an anionic leaving group B^- is replaced by a neutral donor molecule. Consequently, if relative EPD strengths of *both* neutral *and* anionic EPD units are known, it should be possible to make at least qualitative predictions about the ionizability of substrates A—B_1, A—B_2, ... in various EPD solvents. Approximate values of relative EPD strength of halide, pseudohalide, and neutral EPD units have been determined using vanadyl(IV)acetylacetonate, VO(acac)_2, as reference EPA (46) and are listed in Table VII. These values have proved useful for the interpretation of numerous reactions involving complex formation between transition metal ions and halide or pseudohalide ions

TABLE VII

STANDARD FREE ENERGIES ΔG FOR THE 1:1 ADDUCT FORMATION OF VO(acac)_2 WITH NEUTRAL AND ANIONIC LIGANDS

Ligand[a]	ΔG (kcal/mole)	Ligand[a]	ΔG (kcal/mole)
N_3^-	-4.20	Cl^-	-0.83
CN^-	-4.20	DMF	-0.63
NCS^-	-2.69	TMP	-0.24
F^-	-2.69	Br^-	$+0.65$
HMPA	-2.50	Acetone	$+1.30$
Pyridine	-2.47	PDC	$+1.75$
$(\text{C}_6\text{H}_5)_3\text{PO}$	-2.11	I^-	$+2.00$
DMSO	-1.14		

[a] HMPA, hexamethylphosphoric amide, DMSO, dimethyl sulfoxide; DMF, dimethylformamide; TMP, trimethyl phosphate; PDC, propanediol-1,2-carbonate.

(18, 46, 28). For example, complexes with very strong EPD ligands, such as N_3^-, NCS^-, CN^-, or F^- may exist even in solvents of high DN such as HMPA or DMSO. In solvents of weak or medium EPD properties, complex formation is essentially quantitative. On the other hand, bromo and iodo complexes usually exist only in weak EPD solvents, such as NM, PDC, or AN, and are completely ionized in solvents such as DMF, DMSO, or HMPA. The stabilities of chloro complexes are somewhat higher in the respective solvents. According to Table VII the chloride ion has an EPD strength similar to that of DMF or DMSO. Consequently chloro complexes in these solvents (compare Table IV) are ionized to some extent, sometimes with autocomplex formation.

It should be emphasized that values given in Table VII as a measure of relative EPD strength are at best semiquantitative. The reasons for this situation will be analyzed briefly since they are of general importance to the understanding of ionization equilibria:

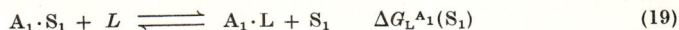

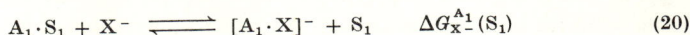

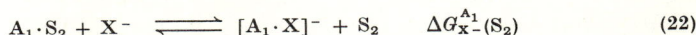

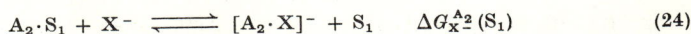

$$A_1 \cdot S_1 + L \rightleftharpoons A_1 \cdot L + S_1 \qquad \Delta G_L^{A_1}(S_1) \tag{19}$$

$$A_1 \cdot S_1 + X^- \rightleftharpoons [A_1 \cdot X]^- + S_1 \qquad \Delta G_{X^-}^{A_1}(S_1) \tag{20}$$

$$A_1 \cdot S_2 + L \rightleftharpoons A_1 \cdot L + S_2 \qquad \Delta G_L^{A_1}(S_2) \tag{21}$$

$$A_1 \cdot S_2 + X^- \rightleftharpoons [A_1 \cdot X]^- + S_2 \qquad \Delta G_{X^-}^{A_1}(S_2) \tag{22}$$

$$A_2 \cdot S_1 + L \rightleftharpoons A_2 \cdot L + S_1 \qquad \Delta G_L^{A_2}(S_1) \tag{23}$$

$$A_2 \cdot S_1 + X^- \rightleftharpoons [A_2 \cdot X]^- + S_1 \qquad \Delta G_{X^-}^{A_2}(S_1) \tag{24}$$

In Eqs. (19)–(23), L denotes a neutral and X^- an anionic EPD, for which the relative EPD strength in solvents S_1 and S_2, respectively, toward an electron pair acceptor A_1 are to be determined. Equations (23) and (24) are analogous to Eq. (19) and (20), except that A_2 is used as reference EPA instead of A_1. Using energy cycles similar to (14), the following equations can be derived, in which ΔG values refer to reactions (19)–(24) *in solution* and Δg values to the corresponding reactions *in the gas phase*. Symbols $\Delta G_{(sv)X^-}(S_1)$ and $\Delta G_{(sv)X^-}(S_2)$ are used for the free energies of solvation of X^- in the solvents S_1 and S_2, respectively, which are a measure of the EPA properties of these solvents; the constant α has the same meaning as previously.

$$\Delta G_L^{A_1}(S_1) - \Delta G_{X^-}^{A_1}(S_1) = \Delta G_L^{A_1}(S_2) - \Delta G_{X^-}^{A_1}(S_2) \\ + \alpha[\Delta G_{(sv)X^-}(S_1) - \Delta G_{(sv)X^-}(S_2)] \tag{25}$$

$$\Delta g_L^{A_1} - \Delta g_{X^-}^{A_1} - (\Delta g_L^{A_2} - \Delta g_{X^-}^{A_2}) = \\ \Delta G_L^{A_1}(S_1) - \Delta G_{X^-}^{A_1}(S_1) - [\Delta G_L^{A_2}(S_1) - \Delta G_{X^-}^{A_2}(S_1)] \tag{26}$$

From Eq. (25) and (26) the following important conclusions may be drawn (*41*).

1. The relative EPD strengths* of anionic ligands X^- as compared to neutral ligands L determined toward a *given* reference EPA (A_1) in different solvents (S_1, S_2), depend on the solvating properties of the solvents (i.e., the relative EPA properties) toward X^- and they are independent of the EPD properties of the solvents. Increasing EPA properties of the solvents causes a decrease in the EPD strength of X^-. This may lead even to a reversal in relative EPD strength between X^- and L. Assume, that in a *poorly solvating solvent* a given anion X^-

* The relative EPD strengths are characterized by the sign and magnitude of the differences $\Delta g_L - \Delta g_{X^-}$ and $\Delta G_L - \Delta G_{X^-}$, respectively.

behaves as a slightly stronger EPD than L. When the same ligands (X^- and L) are considered in *a solvent with strong EPA properties*, then X^- may certainly behave as a weaker EPD than L, because its EPD properties are much more strongly reduced than those of L by the solvent functioning as strong EPA.

2. The relative EPD strengths of anions as determined in a *given solvent* [Eq. (26)] toward *different reference EPA units* depend on the EPA strengths of the (uncoordinated) EPA molecules. If X^- is a stronger EPD than L in the gas phase, $\Delta g_L^{A_1} - \Delta g_{X^-}^{A_1}$ will be positive. If the relative EPD strengths of X^- and L are determined in a solvating solvent S_1 toward a weak reference EPA (A_1), it may be possible that X^- behaves as a weaker EPD than L [$\Delta G_L(S_1) - \Delta G_{X^-}^{A_1}(S_1)$, negative], because the gas phase EPD strength of X^- will be reduced by anion solvation. If, however, the relative EPD strengths are determined (in the same solvent S_1) toward a very strong EPA (A_2), then X^- will probably behave as a stronger EPD than L *both in solution and in the gas phase*, since the free energy of solvation of X^- will only make a comparatively small contribution to the free-energy change of the overall reaction.

This is a very important effect because it readily explains, why $[CoCl_4]^{2-}$ dissolved in NB is easily ionized on addition of a strong EPD such as HMPA, whereas $Si(CH_3)_3Cl$ is not. The argument, that the strength of the Si—Cl bond is greater than that of the Co—Cl bond is insufficient. In fact, in the gas phase Cl^- is a stronger EPD than HMPA both toward (tetrahedral) $[CoCl_3]^-$ and $[Si(CH_3)_3]^+$. This means that relative EPD strengths in the gas phase are independent of EPA strengths of the acceptor groups and therefore, independent of the strength of the Co—Cl and Si—Cl bonds, respectively*. If, however,

$$[CoCl_3 \cdot HMPA]^-(g) + Cl^-(g) \rightleftharpoons [CoCl_4]^{2-}(g) + HMPA(g) \qquad (27)$$

$$[Si(CH_3)_3 \cdot HMPA]^+(g) + Cl^-(g) \rightleftharpoons Si(CH_3)_3Cl(g) + HMPA(g) \qquad (28)$$

the systems (27) and (28) are transferred from the gas phase to a solvent, then the Co—Cl bond will be more extensively ionized, because the free energy of solvation of Cl^- will make a comparatively larger contribution to the overall free-energy change in case of the weak EPA unit $[CoCl_3]^-$.

A simple approach, which is frequently used to estimate the ionizability of compounds is based on a comparison of bond energies. In thermochemistry the term *bond energy* usually refers to the dissociation energy of the gas phase reaction (29), which may be more precisely

---

* This is valid only if both EPA units show the same [class (a) or class (b)] behavior, respectively.

termed *homolytic dissociation energy* (D_{hom}) since reaction (29) involves a homolytic cleavage of the bond A—B. It has been pointed out that,

$$A\text{—}B(g) \longrightarrow A(g) + B(g) \qquad D_{\text{hom}} \qquad (29)$$

so far as ionization reactions are concerned, bond strength should be characterized by *heterolytic dissociation energies* (D_{het}) rather than by D_{hom} (*16, 47*). Heterolytic dissociation refers to the gas phase reaction (30)

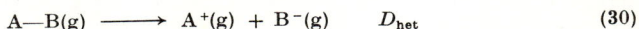

$$A\text{—}B(g) \longrightarrow A^+(g) + B^-(g) \qquad D_{\text{het}} \qquad (30)$$

which involves a heterolytic cleavage of the bond A—B.*

The relationship between D_{hom} and D_{het} is given by Eq. (31) where I_A denotes the ionization energy of A and E_B the electron affinity of B.

$$D_{\text{het}} = D_{\text{hom}} + I_A + E_B \qquad (31)$$

Although D_{het} is always greater than D_{hom},† EPD–EPA interactions in coordinating solvents usually lead to heterolytic bond cleavage. As has already been pointed out, this is owing to the fact that A^+ and B^- are considerably stronger coordinating, respectively, than the parent radicals, A and B. Consequently, stabilization of the reaction products by EPD or (and) EPA interactions with solvent molecules is more favorable for ionic reaction products than for radicals. In weakly coordinating or noncoordinating solvents, radical formation may sometimes be energetically favored, as in solutions of hexaphenylethane in benzene or in hexane.

From the foregoing considerations it is apparent that thermodynamic properties of compounds such as ionization constants or solubilities do not only depend on heterolytic dissociation energies but may be strongly influenced by free energies of solvation or free-energy contributions associated with changes in the state of aggregation.

An instructive example is the behavior of methylmercury halides, CH_3HgX. Thermochemical calculations (Table VIII) reveal that heterolytic dissociation energies for the reaction

$$CH_3HgX(g) \longrightarrow CH_3Hg^+(g) + X^-(g)$$

increase regularly in the series X = I < Br < Cl (*48*). Consequently, it may be expected that ionization constants in water increase in the order Cl < Br < I, as for the hydrogen halides (Table VIII). In fact, the

---

* Since two modes of ionization are possible ($A - B \rightarrow A^- + B^+$), D_{het} always refers to the process that requires less energy.

† The lowest value of ionization potentials of neutral atoms or groups has been measured for cesium, $I_{Cs} = +3.89$ eV, the highest value of electron affinities of neutral atoms or groups is that of chlorine, $E_{Cl} = -3.61$ eV.

TABLE VIII

Dissociation Constants (K), Solubility Products (K_s), Solubilities, and Heterolytic Dissociation Energies (D_{het}) for Various Compounds

Compound	K (moles liter^{-1})	D_{het} (kcal)	Compound	K_s (moles^2 liter^{-2})	D_{het} (kcal)
HF	6.7×10^{-4}	368	AgCl	1.6×10^{-10}	166
HCl	$> 10^{+4}$	331	AgBr	6.3×10^{-13}	163
HBr	$> 10^{+6}$	321	AgI	1.0×10^{-16}	158
CH_3HgF	3.2×10^{-2}	—	TlCl	2.2×10^{-4}	144
CH_3HgCl	5.6×10^{-6}	188	TlBr	3.9×10^{-6}	140
CH_3HgBr	2.4×10^{-7}	184	TlI	3.1×10^{-8}	136
CH_3HgI	2.5×10^{-9}	180			
	Solubility [a] (moles/1000 gm H_2O)			Solubility [a] (moles/1000 gm H_2O)	
NaF	1.00	151	KCl	4.70	116
NaCl	6.16	131	KBr	5.71	112
NaBr	9.14	127	KI	8.84	107
NaI	12.40	116			

[a] Solubility products could not be calculated due to lack of activity coefficient data for saturated solutions.

reverse stability order is found in water and this has been considered as a consequence of typical "soft acid" behavior of the CH_3Hg^+ ion (49). This behavior is explained in the following qualitative way. The D_{het} values for CH_3HgX compounds are comparatively little differentiated. The increase in bond energies in the series I < Br < Cl is therefore overcompensated by the strong increase (numerical values) in solvation energies of the halide ions. This would suggest that methylmercury halides should show normal stabilities $K_{CH_3HgI} > K_{CH_3HgBr} > K_{CH_3HgCl} > K_{CH_3HgF}$,

$$K = \frac{[CH_3Hg^+][X^-]}{[CH_3HgX]}$$

in solvents of poor solvating properties for halide ions—a behavior that was actually found in preliminary conductometric studies in HMPA (48).

Obviously soft acid behavior of methylmercury *halides* is not an inherent property of the CH_3Hg^+ ion but actually a function of the solvating properties of the solvent toward the anionic ligands; this suggests a more critical interpretation of the basic principles of the Hard and Soft Acids and Bases (HASAB) concept (50).

However, ionization constants of the hydrogen halides increase in the sequence HF < HCl < HBr < HI corresponding to decreasing values of D_{het}.

Similar considerations apply to solubility equilibria [energy cycle (8)]. The D_{het} values for Ag(I) and Tl(I) halides (Table VIII) increase in the normal sequence I < Br < Cl < F, whereas solubilities are lowest for the iodides and highest for the fluorides. On the other hand, solubilities for some alkali metal halides show the same trends as may be expected from trends in D_{het} values because, for the alkali metal halides listed in Table VIII, the contributions to ΔG [see energy cycle (8)], arising from free energies of sublimation and free energies of solvation of the halide ions, are partly compensating. Compensation is incomplete for Ag(I) and Tl(I) halides so that solubilities are strongly influenced by anion solvation.

In practice, apart from the examples listed in Table VIII, there is a large number of compounds for which D_{het} values may serve as a qualitative measure of ionizability. For reasons previously discussed this will most probably be true for compounds characterized by high values of D_{het}. Examples are provided by many inorganic and organometallic halides, for instance, SiF_4 and $SiCl_4$ give only nonionic adducts with EPD molecules, even in the case of strong EPD units such as pyridine or ammonia. In contrast, iodo compounds, such as SiI_4, SnI_4, and BI_3, are easily ionized by strong EPD molecules (*51–54*), for example:

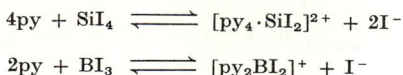

$$4py + SiI_4 \rightleftharpoons [py_4 \cdot SiI_2]^{2+} + 2I^-$$

$$2py + BI_3 \rightleftharpoons [py_2BI_2]^+ + I^-$$

IV. Ionization and Dissociation

It has been demonstrated that formation of ions from a covalent substrate is the result of EPD–EPA interactions.

So far the existence of ion pair equilibria has been excluded from our considerations. In fact, ion association is a common and characteristic phenomenon in nonaqueous solutions. Ionization of a covalent compound may be defined as the process leading to the formation of solvated ions independent of their presence as associated ions or as free entities.

Generally ionization of a covalent compound AB proceeds in two steps (*35*), which may be represented in the following way:

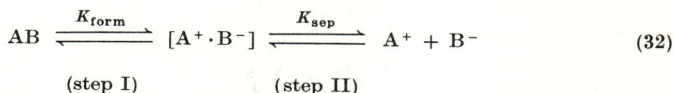

$$AB \underset{\text{(step I)}}{\overset{K_{\text{form}}}{\rightleftharpoons}} [A^+ \cdot B^-] \underset{\text{(step II)}}{\overset{K_{\text{sep}}}{\rightleftharpoons}} A^+ + B^- \tag{32}$$

Heterolysis of A—B due to EPD or EPA interactions with solvent molecules (not shown) results primarily in the formation of associated ions. This process (step I) may be characterized by the equilibrium constant K_{form} which may be termed "formation constant of the ion pair from the covalent substrate".

$$K_{\text{form}} = \frac{c_{A^+ \cdot B^-}}{c_{AB}}$$

In a medium of reasonably high dielectric constant, the ion pair undergoes electrolytic dissociation into the free ions A^+ and B^-. This process (step II) may be characterized by the equilibrium constant K_{sep} which may be termed "ion pair separation constant":

$$K_{\text{sep}} = \frac{c_{A^+} \cdot c_{B^-}}{c_{A^+ \cdot B^-}}$$

Step I is primarily dependent on the functional (coordinating) properties of the solvent; step II (according to the Coulomb law) is mainly a function of the dielectric constant of the medium (*35*)

The equilibrium constant K of the overall process.

$$\text{A—B} \rightleftharpoons \text{A}^+ + \text{B}^- \qquad K$$

is equal to the product of the equilibrium constants of the constituent equilibria.

$$K = K_{\text{form}} \cdot K_{\text{sep}}$$

Constant K is identical with the ionization or dissociation constant in the classical sense, also known as the acidity constant for Brönsted acids or the basicity constant for Brönsted bases.

In a medium of high dielectric constant, such as in water, the concentration of associated ions is negligibly small; K_{form} and K_{sep} cannot be measured separately and this is the reason why in water and in other solvents of high dielectric constant only the overall (classic) equilibrium constant K is meaningful. On the other hand, in solvents of low dielectric constant (e.g., tributylphosphate), there will be practically no electrolytic dissociation, so that the ionized substrate will be present nearly exclusively as associated ions. Consequently, the ionization process is best characterized by K_{form}.

In solvents of medium dielectric constant, K_{form} and K_{sep} may be determined separately by combination of appropriate experimental techniques. For example, spectrophotometric or NMR techniques may be applied to obtain the total concentration c_I of ionized substrate, that is,

$$c_I = c_{\text{A}^+ \cdot \text{B}^-} + c_{\text{A}^+} = c_{\text{A}^+ \cdot \text{B}^-} + c_{\text{B}^-}$$

whereas the concentration of free ions $c_{\text{A}^+} = c_{\text{B}^-}$ may be determined by conductometric measurements. With c_0 denoting the analytical concentration of the solute, one obtains:

$$c_0 = c_{\text{AB}} + c_{\text{A}^+ \cdot \text{B}^-} + c_{\text{A}^+} = c_{\text{AB}} + c_I$$

$$K_{\text{form}} = \frac{c_I - c_{\text{A}^+}}{c_0 - c_I} \qquad K_{\text{sep}} = \frac{c_{\text{A}^+}^2}{c_I - c_{\text{A}^+}}$$

If K_{form} and K_{sep} cannot be determined directly as shown in the preceding equations, approximate values of K_{sep} may be obtained by comparison with the electrolytic dissociation behavior of suitable model electrolytes such as quaternary ammonium salts.

Considering reaction (32), it is immediately realized that the ionization of a covalent substrate is a function of both the functional and the dielectric properties of the solvent. This may be illustrated by a simple model calculation as described in the following.

Consider two EPD solvents, S_1 and S_2, which have the same EPD properties and hence the same K_{form} values, e.g., $K_{form} = 1$. Solvent S_3 is assumed to be a stronger EPD solvent, characterized by $K_{form} = 3$. Solvents S_1 and S_3 may have similar dielectric constants of about 10, but S_2 is assumed to have a higher dielectric constant of about 30; this may correspond to values K_{sep} of approximately $K_{sep} \approx 10^{-4}$ (S_1, S_3) and $K_{sep} \approx 10^{-2}$ (S_2).

The extent of ionization, that is percentage of ionized substrate (ion pairs + free ions) of total solute concentration c_0 has been calculated and results are presented in Table IX. It is seen that AB is more strongly

TABLE IX

EXTENT OF IONIZATION OF A COVALENT SUBSTRATE A—B AT
$c_0 = 10^{-2}$ MOLE/LITER UNDER ARBITRARILY ASSUMED CONDITIONS

Solvent	Assumed ε	Assumed K_{sep}	Assumed K_{form}	% Ionized
S_1	10	10^{-4}	1	53
S_2	30	10^{-2}	1	75
S_3	10	10^{-4}	3	77

ionized in solvent S_2 which has the same EPD properties as S_1 but a considerably higher dielectric constant. On the other hand, AB is most strongly ionized in S_3 although its dielectric constant is low compared to S_2 and this is clearly due to its stronger EPD properties. However, solutions of AB are much better conductors in S_2 than in S_3 because in the latter solvent the ionized substrate is mainly present as ion pairs, which do not contribute to the conductivity of the solution.

Comparison of dissociation constants of acids and bases derived from electrochemical or conductivity measurements in solvents of different dielectric constants are meaningless (25, 28) since equilibrium constants determined in this way always represent the overall equilibrium constant $K = K_{form} \cdot K_{sep}$.

Although liquid ammonia is known to have considerably stronger coordinating properties than water, the acidity constant of HCl is much smaller in liquid ammonia ($K_A \approx 10^{-4}$) than in water ($K_A \gg 10^4$). Ionization of HCl is essentially complete in both solvents, hence the observed large differences in dissociation constants are mainly due to differences in dielectric constants of the media (water, $\varepsilon = 81$; ammonia, $\varepsilon = 17$).

On the other hand, the acidity constant of acetic acid is higher in liquid ammonia ($K_A \approx 10^{-4}$) than in water ($K_A \approx 10^{-5}$). Acetic acid is moderately ionized in water, but in liquid ammonia, owing to its stronger EPD properties, ionization is nearly complete. Obviously the influence of K_{sep} in liquid ammonia is overcompensated by K_{form} (25, 28).

Although comparison of classic dissociation constants for one compound in different solvents is meaningless, it remains useful to compare classic dissociation constants of different solutes in a given solvent.

The situation may be further illustrated by comparing the electrolytic dissociation behavior of lithium halides and tetrabutylammonium halides in the solvents PDC(DN = 15.1; ε = 69.0) and HMPA(DN = 38.8; ε = 29.6) (55). Lithium bromide and in particular LiCl are associated in PDC, but they are fully dissociated in HMPA. This cannot be explained by simple electrostatic considerations. The HMPA has a much lower dielectric constant, and the dipole moments of the two solvents are similar (μ_{HMPA} = 5.37; μ_{PDC} = 4.98). The behavior is, however, readily interpreted from the functional point of view. The HMPA acts as a very strong EPD and, consequently, lithium halides are completely ionized with formation of tightly solvated Li^+ ions. Due to the large effective radius of the HMPA-solvated lithium ion, the center-to-center distance in the ion pair $[Li(HMPA)_n^+ \cdot X^-]$ is so great that complete dissociation is easily effected in dilute solutions. In PDC, which is a considerably weaker EPD solvent, LiCl and LiBr exist as unionized molecules LiX ("contact" ion pairs)* in equilibrium with the free solvated ions Li^+ and X^-, the concentration of ion pairs $[Li(PDC)_n^+ \cdot X^-]$ being negligibly small due to the high dielectric constant of this solvent. The behavior of lithium halides in PDC is thus entirely analogous to the behavior of, for instance, acetic acid in water.*

The differing behavior of lithium halides in PDC and in HMPA is also in agreement with semiquantitative data for the relative EPD strengths of halide ions and EPD solvents (Table VII, Section III). The solvent HMPA is a stronger EPD than both Cl^- and Br^-, and hence these ligands are easily replaced by HMPA at the lithium ion; on the other hand, PDC is a weaker EPD than Cl^- and Br^-, and hence these ligands are only partly replaced despite the large excess of solvent molecules present.

---

* From the functional point of view, there is no basic difference between an unionized CH_3COOH molecule and unionized LiX molecules. Consequently, it has been proposed (16) to consider "contact ion pairs" as unionized species rather than ionic associates. This appears particularly true in the case of lithium compounds, as lithium ion has a strong tendency for covalent bond formation. The term "ion pair" should therefore be restricted to interactions between tightly and fully solvated ions.

Unlike lithium halides, tetrabutylammonium halides are fully dissociated in PDC, but they are associated in HMPA, and this is in agreement with simple electrostatic considerations. Similar results have been found in DMF, where alkali metal halides are fully dissociated whereas tetraalkylammonium halides are associated (*56, 57*).

It may be concluded that electrostatic models may be successfully applied only so far as interactions between weakly coordinating or noncoordinating species (such as tetraalkylammonium ions) are concerned. This is illustrated by Table X which shows that variations of association constants for tetrabutylammonium iodide as a function of dielectric constant roughly correspond to the trends predicted by the Bjerrum theory. When iodide, which is a comparatively weak base, is

TABLE X

ASSOCIATION CONSTANTS K_{ass} OF TETRABUTYLAMMONIUM IODIDE AS A FUNCTION OF DIELECTRIC CONSTANT

Solvent[a]	ε	K_{ass}	Solvent[a]	ε	K_{ass}
Pyridine	12.3	2400	Methanol	32.6	16
1-Butanol	17.5	1200	DMF	36.7	8
1-Propanol	20.1	415	Acetonitrile	36.0	3
Methylethylketone	18.5	380	DMA	37.8	0
Acetone	20.7	143	DMSO	46.7	0
Ethanol	24.3	123	N-Methyl-		
Nitrobenzene	34.8	27	acetamide	165.5	0

[a] DMF, dimethylformamide; DMA, *N,N*-dimethylacetamide; DMSO, dimethyl sulfoxide.

replaced by counterions with stronger EPD properties, such as bromide or chloride ions, anion solvation becomes increasingly effective and this is particularly true for hydrogen-bonded solvents. For example, in aprotic solvents, association constants of tetraalkylammonium halides usually increase in the series $Cl^- > Br^- > I^-$, whereas the opposite behavior is observed in hydroxylic solvents (e.g., CH_3OH and C_2H_5OH). Again, the elementary electrostatic theory cannot provide a *consistent* explanation for this behavior.

V. Outer-Sphere Interactions, Association and Self-ionization of Solvents

It has been mentioned in Section III that the polarographic reduction of metal cations occurs in water at much more negative potential values than would be expected from a donicity value of 18 for water. This has been explained in the following way (*34*). Water in the pure liquid state is a highly associated liquid. Coordination of water molecules to a metal cation (acting as EPA) causes polarization of the O—H bonds and hence an increase in acidity at the hydrogen atoms. Outer-sphere interactions between these acidic hydrogen atoms and additional water molecules lead to outer-sphere hydration. This coordination provokes at the already coordinated oxygen atoms of water molecules an increase in electron density and hence an increase in EPD strength. This transfer of negative charges toward the metal cation results in an increase in inner-sphere coordinate bond strength due to outer-sphere coordination. This has been called the *first outer-sphere effect* (*58, 59*):

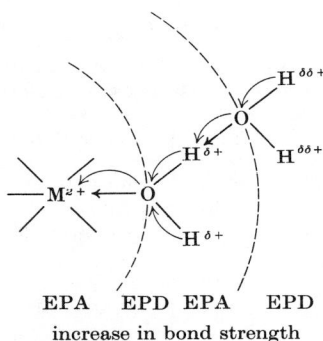

$$
\begin{array}{cccc}
\text{EPA} & \text{EPD} & \text{EPA} & \text{EPD}
\end{array}
$$

$$-M^{z+} \longleftarrow O \cdots H^{\delta+} \cdots O \cdots H^{\delta\delta+}$$

increase in bond strength

This effect may be further increased by further outer-sphere coordination, e.g., by attachment of additional hydration spheres or any other EPD ligands. As a consequence, water in the pure liquid state behaves as a stronger EPD as compared to single water molecules coordinated within *one* hydration sphere, for example, when a transition metal hydrate is dissolved in a weakly coordinating medium such as nitromethane or dichloromethane. A similar behavior may be expected for other associated liquids such as formic amide and acetic acid. In order to characterize the EPD properties of water (and other highly associated liquids), the term *bulk donicity* (DN^B) has been introduced (*34*). The bulk donicity is distinguished from donicity values (DN) determined toward (excess) $SbCl_5$ as reference acid in 1,2-dichloroethane which are a measure of the EPD strengths of the unassociated EPD molecules. Bulk donicity values of approximately 28 to 30 have been estimated for water from polarographic reductions of a number of divalent metal cations (*34*).

Outer-sphere effects occur also when an anion is acting as the coordination center: charge transfer from the anion via hydrogen bridges to the oxygen atoms leads to an increase in electron density at the oxygen atoms. Outer-sphere interactions with further water molecules (acting as EPA) in the second hydration sphere cause an increase in EPA properties of the hydrogens bonded to the anion and in this way the strength of the inner-sphere coordinate bonds is increased. This may be called the *second outer-sphere effect* because it arises from outer-sphere coordination of EPA units (*58*).

Outer-sphere interactions are by no means restricted to interactions between coordinated and uncoordinated solvent molecules (*58, 59*). Generally, all interactions involving formation of coordinate bonds between coordinated (neutral or charged) ligands and neutral or charged reactants may be termed outer-sphere interactions. For instance, it has been shown by kinetic experiments that "coordinatively saturated" complex cations, such as $[Co(NH_3)_6]^{3+}$, $[Cr(NH_3)_6]^{3+}$, or $[Co(H_2O)_6]^{3+}$, are forming stable outer-sphere complexes with various anions (*60*); this has been confirmed by conductometric, spectrophotometric, and polarographic measurements (*61, 62*). The high stability of these complexes is due to the existence of hydrogen bonds between the anions and the acidic hydrogen atoms. Similarly, it has been shown that exchange of NCS^- groups by solvent molecules S in the complex anion

$$[Cr(NH_3)_2(NCS)_4]^- + H_2O \longrightarrow Cr(NH_3)_2H_2O(NCS)_3 + NCS^-$$

proceeds much more rapidly in hydroxylic than in aprotic solvents (*63*). This is due to a weakening of the Cr—N bond by water coordinated to the NCS group of the complex ion with formation of hydrogen bridges (*59*). A similar explanation holds for the acid-catalyzed aquation of $[Cr(H_2O)_5Cl]^{2+}$ (*64*). The existence of outer-sphere interactions of this type has been clearly demonstrated by recent polarographic measurements (*65*). Half-wave potentials for the reduction of hexacyanoferrate(III) to hexacyanoferrate(II) in aprotic solvents are shifted to more positive values on addition of hydroxylic solvents. Since in the reduced state the nitrogen atoms of the coordinated ligands are more basic than in the oxidized state, the former will be more strongly stabilized by EPA outer-sphere coordination resulting in a shift to positive potentials values.

It has been mentioned in Section III that $[CoCl_4]^{2-}$ is unstable in water even in the presence of large amounts of alkali metal chlorides. In contrast $[CoCl_4]^{2-}$ is quantitatively formed in concentrated aqueous solutions of hydrochloric acid (*66*). Outer-sphere interactions between water molecules and hydronium ions will lead to the formation of highly aggregated clusters $[H(H_2O)_n]^+$. In this way the concentration of free water molecules available for outer-sphere coordination at Co^{2+} as well as for hydration of chloride ions is drastically decreased. Hence, in a concentrated HCl solution, both Co^{2+} and Cl^- are less strongly solvated than in alkali metal chloride solutions.

The existence of outer-sphere interactions in associated liquids as derived from experimental results and from functional considerations is supported by recent quantum mechanical calculations. Molecular orbital calculations (67, 68) reveal that, contrary to elementary electrostatic models, hydration of ions is associated with a considerable charge transfer from the coordination center to the water molecules and a simultaneous increase (cation solvation) and decrease (anion solvation), respectively, of the terminal O—H bond distances. Addition of further water molecules in the second hydration sphere leads to an additional charge transfer and to a decrease in potential energy of the system. The extent of charge transfer by outer-sphere hydration is smaller than that by inner-sphere hydration. Mean hydrogen bond energies ΔE between water molecules of the first and the second hydration sphere are considerably greater than in pure water. In the case of lithium ion the polarizing influence of the coordination center is noticeable even in the third hydration shell, whereas only two coordination spheres appear to be involved in the hydration of the fluoride ion. The results of model calculations are listed in Table XI. It has been emphasized that the extent of charge transfer is exaggerated by the semiempirical CNDO/2 procedure compared to the results of *ab initio* calculations, but the qualitative trends, which are found identical for both methods (67, 68), are significant.

It appears that outer-sphere effects are also responsible for the occurrence of self-ionization equilibria in pure amphoteric liquids. For instance, it has been recognized long ago that water and liquid ammonia undergo self-ionization to limited degrees. Jander (69) considered the existence of self-ionization equilibria as characteristic of so-called waterlike solvents. By analogy, self-ionization equilibria have been assumed for various nonaqueous solvents (usually because of low conductivities found in pure liquids) such as liquid sulfur dioxide or phosphorous oxychlorides (70). This concept has been widely used in a formal way on the basis of a generalized acid–base theory for the interpretation of reactions in these media. It is now generally recognized that a solvent may have good ionizing properties although there is no evidence for the existence of a self-ionization equilibrium. For example, isotopic exchange studies revealed that, in liquid sulfur dioxide, self-ionization is not detectable and that the low conductivities observed in the pure liquid are obviously due to trace impurities; on the other hand,

TABLE XI

CHARGE DISTRIBUTION FOR HYDRATED LITHIUM AND FLUORIDE IONS (SINGLE HYDRATION SHELL)

Coordination No.	Li$^+$			Coordination No.	F$^-$		
	q_{Li}	q_{H_2O}	ΔE(kcal)[a]		q_F	q_{H_2O}	ΔE(kcal)[a]
4	+0.444	+0.139	15.2	4	−0.549	−0.113	11.6
6	+0.311	+0.114	13.9	6	−0.560	−0.073	11.2
8	+0.245	+0.095	13.3	8	−0.052	−0.052	9.9

[a] Mean hydrogen bond energies ΔE for two hydration shells: each water molecule of the first hydration shell is coordinated to three water molecules of the second hydration sphere.

self-ionization is well supported for solvents such as liquid BrF_3, HF, ICl, H_2SO_4, or CH_3COOH.

A common feature of self-ionizing solvents are the amphoteric properties of the solvent molecules, since they may act both as EPD and as EPA (27). It appears therefore that outer-sphere effects, similar to those existing between polarized solvent molecules of solvated ions, occur in the pure liquids resulting in mutual polarization of solvent molecules within associated units. From the functional point of view this is described in the following way:

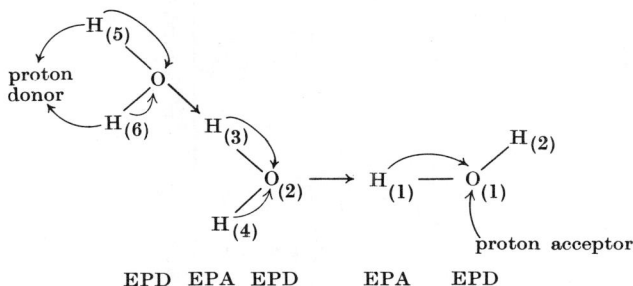

EPD EPA EPD   EPA   EPD

Nucleophilic attack of a water molecule [oxygen atom O(2)] at hydrogen atom (1) of another water molecule leads to an increase in acidity of hydrogen atoms (3) and (4) as well as to an increase in EPD strength of oxygen atom O(1). The EPD–EPA interactions between the hydrogen atoms (3) or (4) [which are more acidic than H(1)] will further increase both the acidity of the terminal hydrogen atoms H(5) and H(6) and the EPD strength of the terminal oxygen atom O(1). With increasing chain length this process may finally result in a proton transfer of an acidic hydrogen atom of one aggregate to the basic terminal oxygen atom of another aggregate. This process is known as autoprotolysis or self-ionization.

Recent quantum mechanical calculations strongly support this model. The CNDO/2 calculations on linear chains of H_2O and HF molecules reveal (Table XII) that the mean energy ΔE of the hydrogen bonds increases with increasing chain length (cooperative effect) (68). Even more significant is the energy $\Delta E_{(n-1)\to n}$ for addition of the "last" solvent molecule; it is always higher than ΔE and clearly reflects the increasing acidity (basicity) of the terminal hydrogen and oxygen atoms, respectively. Calculations for associates with tetrahedral geometry, which represent a more realistic model for the structure of liquid water, show similar trends although the effects are somewhat smaller (71). It should perhaps be mentioned that elementary electrostatic models

TABLE XII

MEAN HYDROGEN BOND ENERGIES FOR LINEAR CHAINS OF H_2O
AND HF MOLECULES AS A FUNCTION OF CHAIN LENGTH

H_2O			HF		
No. of molecules	ΔE(kcal)	$\Delta E_{(n-1)\to n}$(kcal)	No. of molecules	ΔE(kcal)	$\Delta E_{(n-1)\to n}$(kcal)
2	8.68	8.68	2	9.50	9.50
3	9.63	10.59	3	10.88	12.25
4	10.12	11.11	4	11.62	13.12
5	10.43	11.34	5	12.08	13.45
6	10.61	11.44	6	12.38	13.60
7	10.78	11.50	7	12.60	13.68
8	10.98	11.54	8	12.76	13.72

cannot account for the structure of hydrogen-bonded associates: electrostatic theory predicts a linear arrangement of H—F molecules as the most stable structure of the HF dimer, in contrast to quantum-mechanical calculations (68) from which a bent structure is predicted in agreement with spectroscopic measurements.

The existence of hydrogen bonds is not a necessary condition for association and self-ionization of solvents. Association of liquid hydrogen fluoride, for example, can be explained equally well by hydrogen bridges between fluorine atoms and fluorine bridges between hydrogen atoms (72):

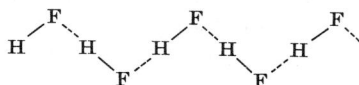

Fluorine bridges are responsible for the association of solvent molecules in liquid bromine(III)-fluoride (73):

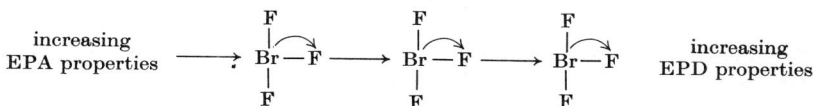

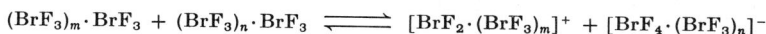

$$(BrF_3)_m \cdot BrF_3 + (BrF_3)_n \cdot BrF_3 \rightleftharpoons [BrF_2 \cdot (BrF_3)_m]^+ + [BrF_4 \cdot (BrF_3)_n]^-$$

Analogous to water, increasing chain length results in increasing EPD properties of the terminal F atom and increasing EPA properties of the terminal Br atom. In this way fluoride ion transfer is promoted between two aggregates and this means self-ionization. Association in liquid ICl, IBr, $AsCl_3$, $SbCl_3$, $HgBr_2$, etc., may be explained similarly (74–78):

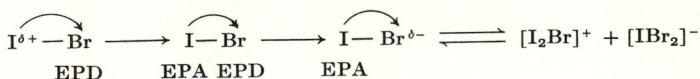

$$I^{\delta+}\!-Br \longrightarrow I\!-Br \longrightarrow I\!-Br^{\delta-} \rightleftharpoons [I_2Br]^+ + [IBr_2]^-$$
$$\quad\text{EPD}\qquad\quad\text{EPA EPD}\qquad\quad\text{EPA}$$

Little is known about the thermodynamics of self-ionization equilibria. It appears that the extent of self-ionization is primarily related to the strength of the bridge bonds. For example, self-ionization constants for HF, H_2O, and NH_3 are decreasing in the order HF ($K \approx 10^{-12}$) > H_2O ($K \approx 10^{-14}$) $\gg$ NH_3 ($K \approx 10^{-33}$) which corresponds to the order of decreasing hydrogen bond strength. However, comparison of self-ionization constants is justified only so long as the dielectric constants of the solvents are about the same.

It is apparent that association between solvent molecules and self-ionization are interrelated phenomena in that both effects are increased by increase in extent of the amphoteric properties of the solvent molecules.

VI. Conclusion

Elementary electrostatic theory cannot account for the observed ionization equilibria in nonaqueous solvents. The functional approach provides a qualitative interpretation of all ionization phenomena, and this is in agreement with quantum mechanical results. This approach considers the coordinating properties of the solvents toward neutral solutes, cations, and anions and takes into account outer-sphere coordination occurring both in the pure solvents and in the solutions. The application of the donicity and of other phenomenological properties allows a number of semiquantitative predictions.

References

1. Born, M., Z. Phys. 1, 45 (1920).
2. Bjerrum, N., and Larsson, E., Z. Phys. Chem. 127, 358 (1927).
3. Noyes, R. M., J. Amer. Chem. Soc. 84, 513 (1962).
4. Latimer, W. M., Pitzer, K. S., and Slansky, C. M., J. Chem. Phys. 7, 108 (1939).
5. Strehlow, H., Z. Elektrochem. 56, 119 (1952).
6. Buckingham, A. D., Discuss. Faraday Soc. 24, 151 (1957).
7. Buckingham, A. D., Quart. Rev. Chem. Soc. 13, 183 (1959).
8. Grunwald, E., Baughman, G., and Kohnstam, G., J. Amer. Chem. Soc. 82, 5801 (1960).

9. Arnett, E. M., and McKelvey, D. R., *J. Amer. Chem. Soc.* **88**, 2598 (1966).
10. Parker, A. J., and Alexander, R., *J. Amer. Chem. Soc.* **90**, 3313 (1968).
11. Alexander, R., Ko, E. C. F., Parker, A. J., and Broxton, T. J., *J. Amer. Chem. Soc.* **90**, 5049 (1968).
12. Pleskov, V. A., *Usp. Khim.* **16**, 254 (1947).
13. Koepp, H. M., Wendt, H., and Strehlow, H., *Z. Elektrochem.* **64**, 483 (1960).
14. Duschek, O., and Gutmann, V., *Monatsh. Chem.* **104**, 990 (1973).
15. Gutmann, V., *Monatsh. Chem.* **102**, 1 (1971); *Allg. Prakt. Chem.* **23**, 178 (1972).
16. Gutmann, V., "Chemische Funktionslehre," Springer-Verlag, Berlin and New York, 1971.
17. Gutmann, V., *Struct. Bonding (Berlin)* **15**, 141 (1973).
18. Gutmann, V., *Angew. Chem.* **82**, 858 (1970); *Angew. Chem., Int. Ed. Engl.* **9**, 843 (1970).
19. Gutmann, V., *Chem. Unserer Zeit* **4**, 90 (1970).
20. Kováčová, J., Horvath, E., and Gazo, J., *Chem. Zvesti* **23**, 15 (1969).
21. Basolo, F., and Pearson, R. G., "Mechanisms of Inorganic Reactions," 2nd ed., Wiley, New York, 1967.
22. Edwards, J. O., *Chem. Rev.* **50**, 455 (1952).
23. Ussanovich, M., *Zh. Obshch. Khim.* **9**, 182 (1939).
24. Olah, G. A., Baker, E. B., Evans, J. C., Tolgyesi, W. S., McIntyre, J. S., and Bastien, I. J., *J. Amer. Chem. Soc.* **86**, 1360 (1964); Olah, G. A., and Lukas, J., *ibid.* **89**, 4739 (1967).
25. Gutmann, V., "Coordination Chemistry in Non Aqueous Solutions," Springer-Verlag, Berlin and New York, 1968.
26. Gutmann, V., *Rec. Chem. Progr.* **30**, 169 (1969).
27. Gutmann, V., *Chem. Brit.* **7**, 102 (1971).
28. Gutmann, V., *Top. Curr. Chem.* **27**, 59 (1972).
29. Lindqvist, I., "Inorganic Adduct Molecules of Oxo—Compounds," Springer-Verlag, Berlin and New York, 1963.
30. Spaziante, P., and Gutmann, V., *Inorg. Chim. Acta* **5**, 273 (1971).
31. Tamres, M., and Searles, S., *J. Phys. Chem.* **66**, 1099 (1962).
32. Gutmann, V., Duschek, O., and Danksagmüller, K., *Z. Phys. Chem. (Frankfurt)* to be published.
33. Duschek, O., and Gutmann, V., *Z. Anorg. Allg. Chem.* **394**, 243 (1972).
34. Mayer, U., and Gutmann, V., *Struct. Bonding (Berlin)* **12**, 113 (1972).
35. Gutmann, V., and Mayer, U., *Monatsh. Chem.* **100**, 2048 (1969).
36. Mayer, U., Gutmann, V., and Weihs, P., to be published.
37. Aylett, B. J., Emeléus, H. J., and Maddock, A. G., *J. Inorg. Nucl. Chem.* **1**, 187 (1955); Campell-Ferguson, H. J., and Ebsworth, E. A. V., *J. Chem. Soc.*, A p. 705 (1967).
38. Corey, J. Y., and West, R., *J. Amer. Chem. Soc.* **85**, 4034 (1963); Beattie, I. R., and Parrett, F. W., *J. Chem. Soc.*, A p. 1784 (1966).
39. Baaz, M., Gutmann, V., and Kunze, O., *Monatsh. Chem.* **93**, 1142 (1962); Gutmann, V., and Kunze, O., *ibid.* **94**, 786 (1963).
40. Gutmann, V., and Tschebull, W., in preparation.
41. Mayer, U., and Gutmann, V., *Monatsh. Chem.* **101**, 912 (1970).
42. Mayer, U., *Monatsh. Chem.* to be published.
43. Mayer, U., unpublished data.
44. Katzin, L. I., and Gebert, E., *J. Amer. Chem. Soc.* **72**, 5464 (1950).

45. Magnell, K. R., and Reynolds, W. L., *Inorg. Chim. Acta* **6**, 571 (1972).
46. Gutmann, V., and Mayer, U., *Monatsh. Chem.* **99**, 1383 (1968).
47. Gutmann, V., *Allg. Prakt. Chem.* **23**, 2 (1972).
48. Mayer, U., unpublished data.
49. Schwarzenbach, G., and Schellenberg, M., *Helv. Chim. Acta* **48**, 28 (1965).
50. Pearson, R. G., *J. Amer. Chem. Soc.* **85**, 3533 (1963).
51. Beattie, I. R., and Leigh, G. J., *J. Inorg. Nucl. Chem.* **23**, 55 (1961).
52. Beattie, I. R., Gilson, T., Webster, M., and McQuillan, G. P., *J. Chem. Soc., London* p. 238. (1964).
53. Muetterties, E. L., *J. Inorg. Nucl. Chem.* **15**, 182 (1968).
54. Mayer, U., and Gutmann, V., *Monatsh. Chem.* **101**, 997 (1970).
55. Mayer, U., Gutmann, V., and Lodzinska, A., *Monatsh. Chem.* **104**, 1045 (1973).
56. Ames, D. P., and Sears, P. G., *J. Phys. Chem.* **59**, 16 (1955).
57. Sears, P. G., Wilhoit, E. D., and Dawson, L. R., *J. Phys. Chem.* **59**, 373 (1955).
58. Gutmann, V., to be published.
59. Gutmann, V., and Schmid, R., *Coord, Chem. Rev.* **12**, 263 (1974).
60. Taube, H., and Posey, F. A., *J. Amer. Chem. Soc.* **75**, 1463 (1953).
61. Jenkins, I. L., and Monk, C. B., *J. Chem. Soc., London* p. 68 (1951).
62. Mayer, U., and Gutmann, V., to be published.
63. Adamson, A. W., *J. Amer. Chem. Soc.* **80**, 3183 (1958).
64. Matts, T. C., and Moore, P., *J. Chem. Soc., London* p. 219 (1969); Staples, P.J., *ibid.* p. 2731 (1968).
65. Gutmann, V., Danksagmüller, K., and Duschek, O., to be published.
66. Bobtelsky, M., and Spiegler, K.S., *J. Chem. Soc., London* p. 143 (1949).
67. Russegger, P., Lischka, H., and Schuster, P., *Theor. Chim. Acta* **24**, 191 (1972).
68. Schuster, P., *Z. Chem.* **13**, 41 (1973).
69. Jander, G., "Die Chemie in wasserähnlichen Lösungsmitteln." Springer-Verlag, Berlin and New York, 1949.
70. Gutmann, V., and Baaz, M., *Angew. Chem.* **71**, 57 (1959).
71. Karpfen, A., Russegger, P., and Schuster, P., in preparation.
72. Gutmann, V., *Sv. Kem. Tidskr.* **68**, 1 (1956).
73. Banks, A. A., Eméleus, H. J., and Woolf, A. A., *J. Chem. Soc., London* p. 2861 (1949).
74. Gutmann, V., *Z. Anorg. Allg. Chem.* **264**, 151 (1951).
75. Gutmann, V., *Monatsh. Chem.* **82**, 156 (1951).
76. Davies, A. G., and Baughan, E. C., *J. Chem. Soc., London* p. 1711 (1961).
77. Gutmann, V., *Z. Anorg. Allg. Chem.* **266**, 331 (1951).
78. Jander, G., and Brodersen, K., *Z. Anorg. Allg. Chem.* **261**, 261 (1950).

COORDINATION CHEMISTRY OF THE CYANATE, THIOCYANATE, AND SELENOCYANATE IONS

A. H. NORBURY

Department of Chemistry, Loughborough University of Technology,
Loughborough, Leicestershire, England

I. Introduction

The cyanate, thiocyanate, and selenocyanate ions are pseudohalides (*106*) and have also been called *chalcogenocyanates*. They have the general formula NCX^- ($X = O, S, Se,$ or Te) and they are all potentially ambidentate, that is, they can form a coordinate bond to a Lewis acid through either N or X. Thus, the thiocyanate ion, for example, will form either N- or S-bonded complexes depending on the nature of the metal, and this preference may be modified by the presence of other ligands or by whether the complex is in the solid state or in solution. The chalcogenocyanates can also be present in a variety of bridging modes. There are many examples of such varied coordination behavior of the thiocyanate ion, and several different, sometimes conflicting, explanations have been put forward. The remaining chalcogenocyanates have not been studied to the same extent but they also show similar characteristics and have attracted a largely similar set of explanations. It has recently become apparent that kinetic factors (i.e., through which of the available atoms does the chalcogenocyanate act as a nucleophile?) as well as the thermodynamic stability of the final product need to be considered in providing a reasoned account of the coordination chemistry of these ions.

This review, therefore, is concerned with the structural information in the literature relating to complexes formed between these ions and transition metal ions, with the similar kinetic data where these relate to a known mode of coordination, and, finally, with assessing the validity of some of the explanations, initially offered for a select group of complexes, against a broader canvas. The task of reviewing the literature has been complicated by the fact that the thiocyanate ion is stable and is readily available to coordination chemists, so that, of the ions in question, it, in particular, has been widely used as a ligand without the main purpose of the investigation necessarily relating to its mode of coordination. The compilation of material has been therefore somewhat selective, concentrating more on reliable structural data than on the recording of the existence of compounds with unconfirmed bonding modes. For these reasons the extensive literature relating to solvent extraction of thiocyanate complexes has also been ignored.

Previously, reviews have appeared on the crystal chemistry of thiocyanate and selenocyanate coordination compounds (*612*), selenocyanate complexes (*334*), the stability and formation of thiocyanate complexes (*637*), the chalcogenocyanates in coordination chemistry (*474*), and the infrared spectra of thiocyanate and related complexes (*57*). Other reviews that are largely devoted to aspects of the chemistry of the chalcogenocyanates have appeared on organometallic pseudohalides of the main group elements (479, 722), nitrogen-containing pseudohalide ligands (86), ambidentate ligands (140, 141, 576), and linkage isomerism (303). Reviews on sulfur-containing ligands (421) or ligands containing Group VI donors (497) have included sections on thiocyanate and selenocyanate complexes. Intra- and intermolecular bonding and structure in NCO⁻ and NCS⁻ compounds, and some related pseudohalides, have been reviewed (398).

The reactivity of ambidentate ligands toward organic centers has been reviewed (679). The review by Beck and Fehlhammer (86) includes a most useful section on methods used for the preparation of these complexes, and this aspect will not be emphasized here. The principles of formation of selenocyanate complexes have been reviewed (682).

NOMENCLATURE

An N-bonded thiocyanate or isothiocyanate will be termed an *N*-thiocyanato complex and its formula will be written M—NCS, and an *S*-bonded or normal thiocyanate will be termed an *S*-thiocyanato complex, written M—SCN; the corresponding formalism will apply to the other ions. Where the mode of coordination is not known the complex will be termed thiocyanate and written M—CNS. Beck and Fehlhammer (86) have rightly objected to this "atom inversion, against chemical knowledge," but it, nevertheless, remains a convenient shorthand and is not likely to cause confusion in this review which contains few, if any, references to fulminates, thiofulminates, etc.

The following abbreviations for chemical names will be used throughout the text.

acacH	acetylacetone
am	ammonia (or occasionally an amine)
an	aniline
aq	aquated, H_2O
Ar	aryl or arene (ArH)
bipy	2,2′-bipyridyl
bu	butyl (prefix *n*, *i*, or *t* for normal, iso, or tertiary butyl, repectively)
can	chloroaniline
Cp	cyclopentadiene, C_5H_5
Dben	*N*,*N*′-dibenzylethylenediamine
DH_2	dimethylglyoxime
diars	*o*-phenylenebisdimethylarsine, $o\text{-}C_6H_4(AsMe_2)_2$
DMA	*N*,*N*-dimethylacetamide, $CH_3CON(CH_3)_2$
DMF	*N*,*N*-dimethylformamide, $HCONMe_2$
DMSO	dimethylsulfoxide, Me_2SO
dpm	diphenylphosphinomethane
en	ethylenediamine, $H_2NCH_2CH_2NH_2$
Et	Ethyl

fan	fluoroaniline
HMPA	hexamethylphosphoramide, $(Me_2N)_3PO$
L	ligand
lut	lutidine
M	central (usually metal) atom in compound
Mben	N-benzylethylenediamine
Me	methyl
Me_6tren	tris-(2-dimethylaminoethyl)amine, $N(CH_2CH_2NMe_2)_3$
$NTAH_3$	nitrilotriacetic acid, $N(CH_2COOH)_3$
8-oxH	8-hydroxyquinoline
Ph	phenyl, C_6H_5
phen	1,10-phenanthroline
pic	picoline
pn	propylenediamine (1,2-diaminopropane)
PNP	bis-(2-diphenylphosphinoethyl)amine, $HN(CH_2CH_2PPh_2)_2$
Pr	propyl (prefix i for isopropyl)
py	pyridine
QAS	tris-(2-diphenylarsinophenyl)arsine, $As(o\text{-}C_6H_4AsPh_2)_3$
QP	tris-(2-diphenylphosphinophenyl)phosphine, $P(o\text{-}C_6H_4PPh_2)_3$
quin	quinoline
R	alkyl or aryl group
TAN	tris-(2-diphenylarsinoethyl)amine, $N(CH_2CH_2AsPh_2)_3$
TAP	tris-(3-dimethylarsinopropyl)phosphine, $P(CH_2CH_2CH_2AsMe_2)_3$
TAS	bis-(3-dimethylarsinopropyl)methylarsine, $MeAs(CH_2CH_2CH_2AsMe_2)_2$
tet a	$trans$-1,4,8,11-tetraazacyclotetradecane
tet b	cis-1,4,8,11-tetraazacyclotetradecane
d tet	3,3-dimethyl-1,5,8,11-tetraazacyclotrideca-1-ene
t tet	2,4,4-trimethyl-1,5,8,11-tetraazacyclotrideca-1-ene
THF	tetrahydrofuran
TMED	N,N,N',N'-tetramethylethylenediamine
TMU	N,N,N',N'-tetramethylurea
tn	1,3-diaminopropane(trimethylenediamine)
tol	toluidine
TPN	tris-(2-diphenylphosphinoethyl)amine, $N(CH_2CH_2PPh_2)_3$
tren	tris-(2-aminoethyl)amine, $N(CH_2CH_2NH_2)_3$
trien	triethylenetetraamine, $(CH_2NHCH_2CH_2NH_2)_2$
TSN	tris-(2-methylthiomethyl)amine, $N(CH_2CH_2SMe)_3$
TSeP	tris-(2-methylselenophenyl)phosphine, $P(o\text{-}C_6H_4SeMe)_3$
TSP	tris-(2-methylthiophenyl)phosphine, $P(o\text{-}C_6H_4SeMe)_3$
TTA	tenoyltrifluoroacetone, $C_4H_3SCOCH_2COCF_3$
tu	thiourea
urt	urotropine
X	halogen or pseudohalogen

II. The Chalcogenocyanate Ions

A. Preparation

The ionic compounds are readily available (except for tellurocyanates) and their preparation will not be reviewed here. They can generally be prepared by the reaction of a cyanide with the appropriate Group VI element. Thus, oxidation of potassium cyanide gives potassium cyanate (323), whereas treatment with free sulfur or a polysulfide gives potassium thiocyanate (323); potassium selenocyanate is formed from the similar reaction with selenium (323). The use of large cations, such as tetraethylammonium (260) or tetraphenylarsonium (44), is apparently necessary for the reaction between cyanide and tellurium to proceed to completion: attempts to prepare potassium tellurocyanate by the aforementioned reaction were unsuccessful (347).

Potassium cyanate is readily hydrolyzed, and, for the stoichiometric reaction,

$$KNCO + 2H^+ + H_2O \longrightarrow K^+ + NH_4^+ + CO_2$$

a hydrogen ion concentration of greater than 0.06 M is required to suppress side reactions (753). The remaining salts are hydrolyzed to give varying proportions of H_2X or free X (X = S, Se, or Te) depending on X and on the conditions. Acid hydrolysis of potassium thiocyanate can give appreciable quantities of COS. The decomposition of NCSe⁻ by H^+ to give free Se is hindered in the presence of soft acids [in the Pearson sense (596, 597)] and encouraged by hard acids, and it is suggested that the soft acids compete with H^+ to attack at the selenium (74). Lodzinska (499) has noted the interdependence of pH and certain cations in affecting the stability of KNCSe. A further indication that the nature of the cation is important in determining the stability of these species is given by Songstad and Stangeland (695) who have shown that the ultraviolet spectra of NCS⁻ and NCSe⁻ in CH_3CN are very dependent on the cation employed, and that the deselenation reaction with arylphosphines to give the corresponding phosphine selenide, originally reported by Nicpon and Meek (564), also goes more readily in the presence of the more electrophilic cations. Hamada (365) observed that the standard free energies of formation of NCX⁻ (X = O, S, Se) increased linearly with the size of X.

B. MOLECULAR GEOMETRY

Details of interatomic distances and bond angles within the ions and in their hydracids are given in Table I. Similar parameters will be given for complexes of these ions in Section IV. Iqbal (398) has collated data on the lattice parameters of the ionic cyanates and thiocyanates. The ions are linear, although in many cases this is not confirmed by direct measurement but is assumed for the purposes of the subsequent calculation. Increasingly, thiocyanates of Group I and II metals are

TABLE I

INTERATOMIC DISTANCES AND BOND ANGLES OF CHALCOGENOCYANATE IONS

Compound	N—C (Å)	C—X (Å)	NCX angle	H—NC angle	Method	References
HNCO	1.19 ± 0.03	1.19 ± 0.03	$180°$	$125°$ (est.)	ED	(274, 378)
	1.207 ± 0.01	1.171 ± 0.01	$180°$	$128°5' \pm 30'$	Microwave and IR	(419)
HNCS	$1.216_4 \pm 0.007$	$1.560_5 \pm 0.003$	$180°$	$134°59' \pm 10'$	mm-wave spec.	(434)
HSCN	1.21	1.61	$180°$	$145°$	Estimated	(772)
HNCSe	1.22	1.75	$180°$	$140°$	Estimated	(772)
NCO⁻	1.17	1.23	$180°$	—	Calc. from force constants	(512)
NCSe⁻	1.16	1.79	$180°$	—	Calc. from force constants	(772)
NCTe⁻	1.16	2.03	$180°$	—	Estimated	(772)
KNCS	1.149 ± 0.014	1.689 ± 0.013	$178.3° \pm 1.2$	—	X-ray	(21)
NH₄NCS	1.15	1.63	$180°$	—	X-ray	(792)
KNCSe	1.17 ± 0.026	1.829 ± 0.025	$178.8° \pm 2.5$	—	X-ray	(710)

used in the X-ray study of complexes of cyclic ethers (e.g., *233*) but these results have not been included.

C. MOLECULAR VIBRATIONS AND FORCE CONSTANTS

The chalcogenocyanate ions are linear triatomic species belonging to the point group $C_{\infty v}$. The three normal modes of vibration shown in Fig. 1 are both infrared and Raman active. The vibrations are commonly described as though group frequencies existed unmixed in these ions,

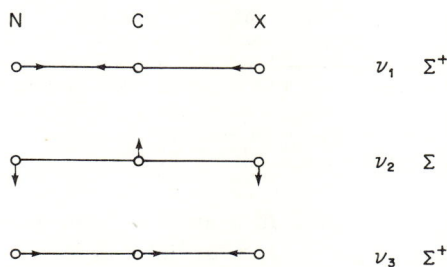

FIG. 1. Normal modes of vibration of NCX⁻.

even though this is, at best, an approximation. Thus, the pseudoantisymmetric stretching frequency (ν_1) is referred to as the CN stretching frequency (ν_{CN}); the pseudosymmetric stretching frequency (ν_3) is referred to as the CX stretching frequency (ν_{CX}); and ν_2 is the doubly degenerate deformation or bending frequency. The first overtone of ν_2 belongs to the same symmetry species at ν_3 and, in the case of the cyanate ion, these frequencies are sufficiently close for Fermi resonance to occur (*244, 512*).

The fundamental vibrations of these ions are listed in Table II. The values cited are those recorded for the potassium salts and anharmonicity corrections have not been applied. Splittings are solid state effects. The vibrational spectra of different isotopic compositions of the cyanate ion have been extensively examined in a number of different host lattices—the results of several different groups of workers are discussed in (*657, 691*).

The force constants have been calculated for the ions in a number of lattices and for different isotopic compositions for NCO⁻ (*512, 657*), NCS⁻ (*418*), and NCSe⁻ (*138, 543*). The results for the solid compounds are summarized in Table III. The effects of different potassium halide matrices on these data are not given here but are listed by Schettino and Hisatsune (*657*).

TABLE II

FUNDAMENTAL VIBRATIONS OF POTASSIUM
CHALCOGENOCYANATES

Ion	ν_{CN} (cm^{-1})	ν_{CX} (cm^{-1})	δ_{NCX} (cm^{-1})	References
NCO$^-$	2165	1254[a]	637,628	(512)
	2160	1249[a]	637,625	(347)
NCS$^-$	2053	746	486,471	(418)
	2048	747	485,470	(347)
NCSe$^-$	2070	558	424,416	(543)
	2070	561	426,417	(347)
NCTe^{-b}	2073	450	366	(267)

[a] Calculated on the basis of equal mixing; see, however, Ref. 657. Peaks were observed at 1301 and 1207 cm^{-1} (512) and at 1294 and 1205 cm^{-1} (347).

[b] (CH$_3$)$_4$N$^+$ salt.

A complete vibrational analysis has been made of the complex anions [Zn(NCX)$_4$]$^{2-}$ (X = O, S, or Se) and the force constants calculated (301). These results will be referred to later.

Infrared spectra of a number of single crystals of cyanates and thiocyanates have been recorded and correlations made with the crystal structure. However, these results are not listed here since they have been reviewed recently by Iqbal (398).

The mean square amplitude of vibrations have been calculated for these ions (550), and Bastiansen-Morino shrinkage effects reported (767, 768).

D. ELECTRONIC STRUCTURE AND ASSOCIATED PROPERTIES

1. Electronic Structure

The chalcogenocyanate ions have sixteen outer electrons. The energy levels for such linear triatomic systems have been characterized by Mulliken (546, 547) and discussed by Walsh (778); the topic has been recently reviewed (623).

The cyanate ion has the ground-state electronic configuration $1\sigma^2$, $2\sigma^2$, $3\sigma^2$, $4\sigma^2$, $5\sigma^2$, $6\sigma^2$, $1\pi^4$; the thiocyanate and subsequent ions are similar unless account is taken of the core electrons or empty d orbitals of sulfur, selenium, or tellurium which increase the complexity of the ground-state description. The most sophisticated calculations that have been carried out on these ions are those of McLean and Yoshimine (506) on NCO$^-$ and NCS$^-$. The calculated orbital energies are given in Table

TABLE III

FORCE CONSTANTS OF SOME POTASSIUM SALTS

Compound	f_{NC}	f_{CX}	f'	$f_{\alpha/l_1 l_2}$	f	References
KNCO in KBr	15.879	11.003	1.422	0.5086	0.7319	(512)
KNCS	15.95	5.18	0.9	$\left.\begin{array}{c}0.311\\0.300\end{array}\right\}$	—	(418)
KNCSe	15.97 ± 0.30	3.754 ± 0.045	0.88 ± 0.25	$\left.\begin{array}{c}0.218\\0.228\end{array}\right\}$	—	(138)

TABLE IV
CALCULATED ORBITAL ENERGIES FOR NCO⁻ AND NCS⁻

Orbital[a]	NCO⁻ [b] (eV)	NCO⁻ [c] (eV)	NCS⁻ [b] (eV)
1σ	− 552.85	− 549.06	− 2495.24
2σ	− 415.38	− 413.33	− 416.64
3σ	− 301.26	− 301.17	− 301.13
4σ	− 32.25	− 27.51	− 236.94
5σ	− 26.21	− 22.77	− 173.88
6σ	− 12.77	− 7.63	− 25.86
7σ	− 7.97	− 4.25	− 21.29
8σ	+ 16.86	—	− 10.62
9σ	—	—	− 8.32
10σ	—	—	+ 13.40
1π	− 10.00	− 6.56	− 173.80
2π	− 3.67	− 0.43	− 7.58
3π	+ 15.01	—	− 3.06
4π	—	—	+ 12.00

[a] The orbitals are not listed necessarily in order of decreasing energy.
[b] From Ref. 506.
[c] From Ref. 117.

IV together with those of comparable calculations; less meticulous calculations will be referred to as appropriate in the following sections. Electronic energy surfaces have been calculated for NCO⁻ using cuspless wave functions (789).

2. Absorption Spectra

Bands at 300 nm (4.1 eV) and 190–205 nm (6.5–6.0 eV) have been assigned to the transitions $^3\Sigma^+ \leftarrow {}^1\Sigma^+$ and $^1\Sigma^- \leftarrow {}^1\Sigma^+$, respectively, for the cyanate ion (621, 622), while the corresponding transitions for the thiocyanate ion occur at 340–360 nm (3.6–3.5 eV) and 220–240 nm (5.6–5.2 eV) (503). Higher energy transitions have been reported (503, 622) for details of which the reader is referred to the original references. Both the spin-forbidden and the allowed low-energy bands are essentially $\pi \rightarrow \pi^*$ in nature, and the assignments are justified by the authors on the bases of calculations and of analogies with comparable systems. However, Trenin and co-workers have claimed that the bands in question for the cyanate ion (487) and for the thiocyanate, selenocyanate, and tellurocyanate ions (353) are charge transfer to solvent (CTTS) in character, and support this by demonstrating appropriate solvent shifts and by measurements at variable temperatures. Other workers have argued in favor of the spin-forbidden transition just mentioned and suggest that it may become allowed on coordination (249). This point of view has been examined qualitatively by Barnes and Day (75) (see Section III, B) who have shown that the nature of the band in question depends on the metal; the same observation applied to cyanate and selenocyanate complexes (242). Thus, it appears that internal transitions have been adequately observed and assigned for NCO⁻ and NCS⁻, but that, because of the nature of these ions, the transitions are particularly susceptible to perturbations from neighboring metal ions or solvent molecules.

3. Nuclear Magnetic Resonance Spectroscopy

The ^{14}N nuclear magnetic resonances (NMRs) of the chalcogeno-cyanate ions have been measured by a consistent procedure and compared with the results of previous workers (431). The observed chemical shifts for these ions and for some related linear molecules vary linearly over a considerable range of values with π-electron densities calculated by a LCGO—MO method (431). The shift values were NCO⁻ = 288, NCS⁻ = 165, and NCSe⁻ = 135 ppm (relative to $NO_3^- = 0$ ppm) (431) and have been confirmed by later workers (192, 390). The π-electron charges on the atoms and the π-bond orders that correlate with these results are given in Table V, together with the calculated data for NCTe⁻ which had not been characterized at that time (772). More detailed calculations of the chemical shifts in the cyanate ion have been compared with those in the azide ion, and in the corresponding hydracids (191).

The ^{13}C NMRs of aqueous solutions of potassium cyanate and thiocyanate are -1.1 and -5.7 ppm (relative to benzene) (504) and comparison of these data with those from related compounds suggests that the ions have a small or zero formal charge on the carbon, in general agreement with the ^{14}N results.

TABLE V

CALCULATED π-ATOMIC CHARGES AND π-BOND ORDERS[a]

NCX⁻	Q_N	Q_C	Q_X	π_{NC}	π_{CX}
NCO⁻	-0.7712	-0.0442	-0.1846	1.5503	1.2629
NCS⁻	-0.4826	$+0.1934$	-0.7108	1.8243	0.7964
NCSe⁻	-0.3941	$+0.2345$	-0.8404	1.8943	0.5973
NCTe⁻	-0.4919	$+0.1859$	-0.6940	1.8156	0.8179

[a] Data from Ref. 772.

4. Electron Spectroscopy for Chemical Analysis (ESCA)

The N(1s), C(1s), and appropriate X (except Se) binding energies have been measured (579) for the series $Ph_4As(NCX)$ (X = O, S, Se, Te) and the results are given in Table VI, together with the previously determined N(1s) binding energies in NCO^- and NCS^- as the potassium (376) or bis-(triphenylphosphine)iminium (711) salts, and the Se(3p) binding energy in KNCSe (712).

Basch (76) has suggested that a direct correlation should exist between chemical shifts from NMR measurements and ESCA measurements. On this basis, it is not surprising that, in view of the small ^{13}C shifts observed for NCO^- and NCS^- (504) (see preceding section), the C(1s) binding energies of the ions in question should be almost identical. Similarly, the results in Table VI would predict only small changes in ^{14}N NMR measurements for these ions, whereas, in fact, considerable chemical shifts were observed and were correlated with changes in π-electron density at the nitrogen atom (431). In many cases binding energies show a linear correlation with the calculated total charge, and extended Hückel calculations give a total charge of -1.572 and -1.672 (-1.711 if d orbitals are included) for the nitrogen atoms in NCO^- and NCS^-, respectively (376).

Thus, on the one hand, the ^{14}N chemical shift data and calculated total charges indicate that changing X in NCX^- does alter the electron density on the nitrogen, and, on the other, the ESCA measurements support no such effect. It has been suggested (579) that this apparent discrepancy may arise from the importance of a direct electrostatic interaction between the cation and the nitrogen end of chalcogeno-cyanate ion, which masks any mesomeric or inductive effect of X transmitted through carbon to nitrogen. In support of this hypothesis

TABLE VI

BINDING ENERGIES FOR THE ATOMS IN THE CHALCOGENATOCYANATE IONS

Compound	N(1s) (eV)	C(1s) (eV)	X (eV)	References
Ph_4AsNCO	397.0	291.2	532.0 (1s)	(579)
KNCO	398.3	—	—	(376)
$(Ph_3P)_2NNCO$	400.5	—	—	(711)
Ph_4AsNCS	396.6	291.3	161.8 (2p)	(579)
KNCS	398.5	—	—	(376)
$(Ph_3P)_2NNCS$	400.2	—	—	(711)
$Ph_4AsNCSe$	397.0	291.3	—	(579)
KNCSe	—	—	159.3 ($3p_{3/2}$)	(712)
			165.1 ($3p_{1/2}$)	(712)
$Ph_4AsNCTe$	396.8	—	573.3 ($3d_{5/2}$)	(579)
			583.6 ($3d_{3/2}$)	(579)

it was pointed out that the N(1s) binding energies for either NCO⁻ or NCS⁻ show greater differences due to changing the cation than to changing X (*579*).

There appear to be no examples of the photoelectron spectra of appropriate inorganic molecules containing the chalcogenocyanate group (*788*).

5. Nuclear Quadrupole Resonance Spectroscopy

Ab initio calculations of the ^{14}N nuclear quadrupole coupling constant suggest that the cyanate nitrogen atom is more ionic than that of the thiocyanate ion (*118*), but measurements show the latter to be more ionic than that in the selenocyanate ion (*397*).

E. THERMODYNAMIC AND RELATED PROPERTIES

Compounds of these ions do not occur frequently in the lists of thermodynamic data; some of the few results available are given in Table VII. The dissociation constants of hydrocyanic, hydrazoic, cyanic, thiocyanic, and selenocyanic acids are in the approximate ratios $1:10^5: 10^6:10^9:10^{10}$ (*123*). Some thermodynamic functions of these acids have been obtained, but no attempt was made to confirm that the latter three acids were indeed the N-bonded compounds and that the X-bonded isomers were absent (*123*). Hydrogen bonding between these acids and various bases has been studied but will not be considered further in this review (see, e.g., Refs. *68* and *362*, and references therein). Some electronegativities are also included in Table VII; of the cases quoted, only Huheey (*394*) considered both possible atoms and he assumed -tri di di hybridization in each case.

Early work by Birckenbach and Kellermann (*106*) established the following order of increasing reducing power of the halide and pseudohalide ions: F⁻, NCO⁻, Cl⁻, N₃⁻, Br⁻, CN⁻, NCS⁻, I⁻, SeCN⁻, TeCN⁻. Few data on the pseudohalides are available to justify this sequence on other than the original chemical grounds.

The activity and osmotic coefficients of sodium thiocyanate in water have been determined (*531*). The solvation number of NCS⁻ has been reported as zero. (*588*), and the ion tends to be less structure-breaking than most anions (*136*). Its transport number and ionic conductance have been measured in formamide (*581*), and 1:1 solvates have been reported from dimethylformamide solutions of M—NCS (M = NH₄, Na, K) (*595*).

III. Physical Methods for Determining the Mode of Coordination

The determination of the mode of coordination of chalcogenocyanate ligands is fraught with difficulties. Although various techniques have

TABLE VII

SOME THERMODYNAMIC AND RELATED PROPERTIES OF CHALCOGENOCYANATES

Parameter	Units	NCO⁻	NCS⁻	NCSe⁻	Reference
ΔH_f (X_g^-)	kcal/mole	−19	−4	—	(248)
	kcal/mole	—	−24.3 to −23.0	—	(544)
ΔH for $X_{(g)}^- \rightarrow X_{(aq)}^-$	kcal/mole	−93	−74	—	(761)
Electron affinity	kcal/mole	—	49.9	—	(551)
Gaseous entropy	eu	54.4	56.9	—	(470)
	eu	53.0	55.5	—	(29)
Ionic entropy	eu	—	34.23	—	(754)
Partial molal entropy	eu	31.1	—	—	(29)
Entropy of hydration	eu	−15	—	—	(29)
Partial molal volume	ml/mole	26.7	41.0	50.3	(523)
Electronegativity	Pauling scale	3.52	—	—	(786)
	Pauling scale	3.05	2.9	2.6	(203)
	Pauling scale	4.46	4.17	—	(394)
	Pauling scale	4.66 (—OCN⁻)	3.91 (—SCN⁻)	—	(394)

been used successfully in many instances, there are a number of cases where the application of these previously satisfactory techniques gives ambiguous answers. Thus, X-ray crystallography remains the most reliable technique. A representative selection of illustrative structures is given in Table VIII; further crystal structures will be given subsequently. In addition to the tabulated types of coordination, O-cyanato complexes and various selenocyanate-bridged complexes have been characterized by spectroscopic methods.

TABLE VIII

SOME TYPES OF CHALCOGENOCYANATE COORDINATION CONFIRMED BY X-RAY CRYSTALLOGRAPHY

Compound	Type of coordination	Comments	References
$Ph_4As[Ag(NCO)_2]$	$[OCN—Ag—NCO]^-$	Linear anion	(1)
$(\pi\text{-}C_5H_5)Cr(NO)_2NCO$	$Cr—NCO$		(165)
$AgNCO$	$\begin{smallmatrix} & O & \\ & \| & \\ & C & \\ & N & \\ \diagdown & & \diagup \\ Ag & & Ag \end{smallmatrix}$		(125)
$[Ni_2tren_2(NCO)_2](BPh_4)_2$	$\begin{smallmatrix} & O—C—N & \\ Ni & & Ni \\ & N—C—O & \end{smallmatrix}$		(265)
$[Co(NH_3)_5NCS]Cl_2$	$Co—NCS$		(693)
$[Co(NH_3)_5SCN]Cl_2$	$Co—SCN$		(693)
$[Ph_2P(CH_2)_3NMe]Pd(CNS)_2$	Pd—NCS, SCN	Both types of monodentate coordination in a single molecule	(198, 199) (102)
$[Ph_2P(CH_2)_3PPh_2]Pd(CNS)_2$	Pd—NCS, SCN		
$K_2Pd(SCN)_4$	$Pd—SCN \downarrow Pd$	Weak coordinate bond to a second Pd	(524, 525)
$Co(NCS)_4Hg$	$Co—NCS—Hg$		(407)
$Pt_2Cl_2(PPr_3)_2(SCN)_2$	$\begin{smallmatrix} & SCN & \\ Pt & & Pt \\ & NCS & \end{smallmatrix}$		(348)
$Co(NCS)_6Hg_2 \cdot C_6H_6$	$Co—NCS\diagdown_{Hg}^{Hg}$		(351)
$Ni(DMF)_4(NCSe)_2$	$Ni—NCSe$		(736)
$K[Co(DH)_2(SeCN)_2]$	$Co—SeCN$		(8)

A. Infrared Spectroscopy

Infrared spectra of thiocyanate and related complexes have been thoroughly reviewed by Bailey *et al.* (*57*). Because of the importance of this technique for structural assignments, a certain amount of repetition of material included in their review is necessary here, although the retabulation of data already collated has been kept to a minimum.

Various criteria have been examined at different times with a view to correlating frequency shifts with the mode of bonding of the thio-cyanate group. Chatt *et al.* (*186, 187*) showed that the CN stretching frequency is found at higher wave numbers in bridging than in terminal complexes. Later it was shown that, for terminal thiocyanate groups, this same frequency often occurred at higher wave numbers for *S*-thio-cyanates than for N-bonded complexes; but it was also shown that other structural and electronic factors prevented this criterion from having a general application (*533, 727, 728*). The C—S stretching frequency has been considered also, when frequencies near 700 cm^{-1} have been taken to be indicative of S-bonding, whereas those between 800 and 830 cm^{-1} to suggest N-bonding (*489*). As well as being subject to the structural and electronic effects referred to in the foregoing, this frequency occurs in the same region of the spectrum as frequencies associated with other ligands or counterions so that, being only of medium or weak intensity, it is difficult to assign. A further complication is caused by the fact that this frequency can sometimes be confused with the first overtone of the bending frequency (*640*).

A single sharp band at ~ 480 cm^{-1} has been assigned to the bending mode in an N-bonded complex, in contrast to the several, low-intensity bands near 420 cm^{-1} observed in *S*-thiocyanates (*489, 639*).

Many of these frequency changes can be accounted for by taking a simple view of the alternative structures, which can be assumed to be predominantly M—N=C=S and M—S—C=N, respectively. However, it is not surprising that difficulties are encountered in making the assignments and that the range of frequencies observed for *N*-thio-cyanato complexes overlaps with that for S complexes, when the factors affecting frequency shifts are considered. Even in terms of the foregoing model, it is clear that changes in the mass, size, or charge of M can have profound consequences on the position of a given frequency, notwith-standing the further effects due to the size or electronic nature of other ligands.

The views of various authors on the origins of these frequency shifts, and on attempts to assess the relative significance of the different effects, have been summarized by Bailey *et al.* (*57*) and will not be repeated here. The complexity of the system is illustrated by the experimental results in Table IX and by the conclusions reached by Kharitonov *et al.* (*440*), who calculated theoretical changes in the vibrational frequencies with changes in the force constants of M—N and M—S bonds on the assumption that the internal force constants remained unchanged from the free ion values. They considered the systems M—NCS (with MN̂C = 180°), M—SCN (with MŜC = 180°, 120°, or 90°), and linear M—NCS—M, and varied the atomic weight of M from 50 to 200. The conclusions of Kharitonov *et al.* have been usefully summarized (*57*) as follows.

For the nitrogen-bound case:

1. The ν_{CN} and ν_{CS} bands should increase in energy with an increase of k_{MN} (metal–nitrogen bond force constant). Hence, ν_{CN} and ν_{CS} increase on coordination.

2. ν_{CN} is almost independent of the mass of the metal ion, and ν_{CS} only slightly more influenced by this; ν_{MN} is mass-dependent.

3. Changes in $k_{MN,CN}$, the interaction force constant, have little effect on ν_{CN} and almost none on ν_{CS} and ν_{MN}.

4. The ν_{CN}, ν_{CS}, and ν_{MN} bands are not pure vibrations; there is mixing in all cases.

5. These effects occur in addition to those due to electron redistribution from coordination, such as changes in the CN and CS force constants. These should be small since the effect of k_{MN} alone gives reasonable correlation with experimental observations.

For the sulfur-bound case:

1. For unchanged force constants, coordination of the SCN⁻ group through S has no influence on the frequency of ν_{CN}. However, ν_{CS} increases relative to the free ion but to a lesser extent than in the N-bound case.

2. ν_{CN} is almost independent of the mass of atom M, and ν_{CS} and ν_{MS} are only slightly affected by it.

3. ν_{CN} is almost independent of the CSM angle, but ν_{CS} decreases and ν_{MS} increases as the angle decreases from 180° to 90°.

4. ν_{CN} is almost independent of k_{MS}, but ν_{CS} and ν_{MS} increase as k_{MS} increases.

5. In order to fit experimental results, k_{CN} and k_{CS} and, hence, the bond orders must change on coordination (C—S decreasing, C—N increasing).

For the bridged case:

1. The ν_{CN} and ν_{CS} bands increase almost linearly with k_{MN} and k_{SM} (for $k_{MN} = k_{SM}$).

TABLE IX

INFRARED SPECTRA OF SOME THIOCYANATE COMPLEXES[a]

Compound	ν_{CN} (cm^{-1})	ν_{CS} (cm^{-1})	δ_{NCS} (cm^{-1})	A ($\times 10^4$ M^{-1} cm^{-2})	References
π-C$_5$H$_5$Mo(CO)$_3$NCS	2099 s[b]	—	—	9.80[b]	(690)
π-C$_5$H$_5$Mo(CO)$_3$SCN	2114 m[b]	699 w	—	2.19[b]	(690)
(π-C$_5$H$_5$)$_2$W(NCS)$_2$	{ 2199 s 2107 vs	—	—	—	(344)
(π-C$_5$H$_5$)$_2$W(NCS)(SCN)	{ 2121 s 2104 s 2094 s	700 w	423	—	(344)
Mn(CO)$_5$SCN	2160	676 w	—	—	(279)
Mn(CO)$_5$NCS (in CH$_3$CN)	2113	813	—	—	(279)
Mn(CO)$_3$(AsPh$_3$)$_2$SCN	2148 m[b]	—	—	—	(280)
Mn(CO)$_3$(AsPh$_3$)$_2$NCS	2103 m[b]	814 m	—	—	(280)
Mn(CO)$_3$(SbPh$_3$)$_2$SCN	2148 m[b]	—	—	—	(280)
Mn(CO)$_3$(SbPh$_3$)$_2$NCS	2097 m[b]	820 m	—	—	(280)
Mn(CO)$_3$(PPh$_3$)$_2$NCS	2096 m[b]	820 m	—	—	(280)
π-C$_5$H$_5$Fe(CO)$_2$NCS	2123 s[b]	830 m	—	6.70	(690)
π-C$_5$H$_5$Fe(CO)$_2$SCN	2118 m[b]	698 w	—	1.64	(690)
[Ru(NH$_3$)$_5$NCS](ClO$_4$)$_2$	2120 s	850	—	—	(490)
[Ru(NH$_3$)$_5$SCN](ClO$_4$)$_2$	2065	—	—	—	(490)
cis-Co(DH)$_2$(H$_2$O)NCS	2070	—	—	—	(3)
trans-Co(DH)$_2$(H$_2$O)SCN	{ 2180 2120	—	—	—	(3)

Complex					Ref.
trans-Co(DH)$_2$pyNCS	2128 s,sp	837 w	—	10.24	(574)
trans-Co(DH)$_2$pySCN	2118 s,sp	—	—	1.2	(574)
trans-Co(DH)$_2$4-t-bupyNCS	2110 s	—	—	—	(269)
trans-Co(DH)$_2$4-t-bupySCN	2055 m	—	—	—	(269)
[Co(NH$_3$)$_5$NCS](ClO$_4$)$_2$	2125 b	806	426 m	—	(134)
[Co(NH$_3$)$_5$SCN](ClO$_4$)$_2$	2100 sp	710 w	—	—	(134)
K$_3$[Co(CN)$_5$NCS]	2123 s	812 w	483 w	—	(359)
(n-Bu$_4$N)$_3$[Co(CN)$_5$NCS]	2137 s[b,c]	—	470 vw	—	(359)
[Co(CN)$_5$NCS]$^{3-}$	2118 s[b,c]	—	475 mw	—	(359)
K$_3$[Co(CN)$_5$SCN]	2144 ms[c]; 2134 s[c]; 2110 vs[c]	718 w	472 w; 461 vw; 448 w	—	(359)
[Co(CN)$_5$SCN]$^{3-}$	2112 s[c,d]; 2097 ms[c,d]	—	—	—	(359)
[Rh(NH$_3$)$_5$NCS]$^{2+}$	2145 s,b	815 s	—	—	(659)
[Rh(NH$_3$)$_5$SCN]$^{2+}$	2115 s,sp; 2122 sh	770 w,b	—	—	(659)
mer-Rh(PMe$_2$Ph)$_3$Cl$_2$NCS	2113 vs; 2113[b]	809	—	—	(126)
mer-Rh(PMe$_2$Ph)$_3$Cl$_2$SCN	2108 vs; 2115[b]	—	—	—	(126)
mer-Rh(PMe$_2$Ph)$_3$(SCN)$_3$	2126 sh; 2106 vs	702	473	—	(126)
Rh(PPh$_3$)$_3$NCS	2095	820–810	—	8.5	(35)
Rh(PPh$_3$)$_2$pip(NCS)	2090	815	—	6.5	(35)
Rh(PPh$_3$)$_2$MeCN (SCN)$_3$	2136	—	—	3.0	(35)
[Ir(NH$_3$)$_5$NCS]$^{2+}$	2140 s,b	825	—	—	(659)
[Ir(NH$_3$)$_5$SCN]$^{2+}$	2110 s,sp	700 m	—	—	(659)

(continued)

TABLE IX—*continued*

Compound	ν_{CN} (cm⁻¹)	ν_{CS} (cm⁻¹)	δ_{CNS} (cm⁻¹)	A (×10⁴ M^{-1} cm⁻²)	References
Pd(AsPh₃)₂(NCS)₂	2089 s,b	854 m	—	—	(143)
Pd(AsPh₃)₂(SCN)₂	2119 s,sp	—	—	—	(143)
Pd(bipy)(NCS)₂	2100 s,b	{ 849 sh 842 m	—	—	(143)
Pd(bipy)(SCN)₂	{ 2117 m,sp 2108 s,sp	—	—	—	(143)
Pd(4,7-diph-phen)(NCS)₂	2110 s,b	—	—	—	(103)
Pd(4,7-diph-phen)(SCN)₂	{ 2120 sh 2113 s,sp	—	419	—	(103)
[Pd(Et₄dien)NCS]⁺	2060	830	—	—	(78)
[Pd(Et₄dien)SCN]⁺	2125	710	—	—	(78)
Pd(4,4'-di-me-bipy)(NCS)(SCN)	2120 s,sp	—	458 mw	—	(103)
	2090 s,b	—	452 mw	—	(103)
Pd[Ph₂As(-o-C₆H₄PPh₂)](NSC)(SCN)	2117	—	—	—	(526)
	2085	—	—	—	(526)
	2118ᵈ	—	—	4.2	(526)
	2085ᵈ	—	—	8.2	(526)
Pd(Ph₂P(CH₂)₂NMe₂)(NCS)(SCN)	2126	—	—	—	(526)
	2108	—	—	—	(526)
	2126ᵈ	—	—	1.90	(526)
	2085ᵈ	—	—	11.71	(526)

Compound						Ref.
Pd(Ph₂P(CH₂)₂PPh₂)(NCS)(SCN)	{ 2118	—	—	—	—	(526)
	2095				—	
	{ 2121ᵈ	—	—	—	2.5	(526)
	2086ᵈ				10.7	
Cu(tripyam)(NCS)₂	{ 2100 s,b	—	—	—	—	(473)
	2070 sh,b					
Cu(tripyam)(NCS)(SCN)	{ 2128 s,sp	—	—	—	—	(473)
	2080 s,b					
Cu(tripyam)(SCN)₂	{ 2122 s,sp	—	—	—	—	(473)
	2100 s,sp					
Cu(dppa)(NCS)(SCN)	{ 2128 s,sp	—	—	—	—	(473)
	2081 s,b					
Cu(dppa)(SCN)₂	{ 2122 s,sp	—	—	—	—	(473)
	2100 s,sp					

ᵃ Data are recorded as mull spectra unless otherwise indicated.
ᵇ In CHCl₃.
ᶜ Includes ν_{CN} for cyano groups.
ᵈ In CH₂Cl₂.

2. Change in mass of M has almost no effect on ν_{CN} and only a small one on ν_{CS}; the effects appear as increases in k_{MN} and k_{MS}.

3. Constant k_{MS} has almost no effect on ν_{CN}.

4. ν_{MS} depends on both k_{MN} and k_{MS}.

5. The high values of ν_{CN} found experimentally are due to the combination of increase for N bonding and a change in force constants due to S bonding.

An alternative method for attempting to determine the mode of thiocyanate coordination by infrared measurements involves the intensity of the CN stretching frequency. This method was suggested by Fronaeus and Larsson (305), and developed by Pecile (599). As currently applied, it requires the measurement of the integrated intensity, A, (i.e., the area under the absorption peak) of the CN stretching frequency, and Ramsay's method of direct integration (625) is often used. The equation is

$$A = (\pi/2Cl)[\log (I_0/I)]\Delta\nu_{1/2}$$

where C = concentration in moles per liter, l = cell thickness, I/I_0 = percentage of transmitted light, and $\Delta\nu_{1/2}$ = apparent width of the absorption band at half the height of its peak.

Values of A in the region $3–5 \times 10^4 \ M^{-1}cm^{-2}$ are found for the free thiocyanate ion; integrated intensities below this are found for S-thiocyanates, whereas N-bonded complexes have values generally above $9 \times 10^4 \ M^{-1} \ cm^{-2}$. A theoretical justification for these results has been advanced (305, 482), whereby coordination through S would favor an increased contribution from $N{\equiv}C{-}S^-$ of the three resonance forms of the ions given in Table X, so that the dipole moment of the ion would be decreased; conversely, coordination through N would favor an increase in the contributions of the two other resonance forms, resulting in an increase of the dipole moment of the ion. The whole argument

TABLE X

PERCENTAGE CONTRIBUTIONS OF THE PRINCIPAL RESONANCE FORMS
OF THE CHALCOGENOCYANATE IONS

X atom	$N{\equiv}C{-}X^-$ (%)	$^-N{=}C{=}X$ (%)	$^{2-}N{-}C{\equiv}X^+$ (%)	References
O	75	1	24	(575)
S	76	5	19	(575)
Se	88	0	12	(568)
Te	90	4	6	(568)

depends on the assumption that a change in the magnitude of the dipole moment causes a corresponding change in the rate of change of the dipole during vibration and, hence, a change in the intensity of the band.

The disadvantage of the method and the need for care in interpreting the results centers around the fact that measurements are made on solutions rather than on solids. As will be discussed in detail later, the nature of the solvent can affect the mode of bonding of the thiocyanate ion, even to the extent of causing isomerization. Dissociation of the coordinated ion is a further chemical possibility, the degree of which will depend on the nature of the solvent. Solvation of the coordinated thiocyanate group itself can also occur and hydrogen bonding solvents can, for example, cause considerable broadening of the vibrational peaks. Thus a variety of factors can affect the integrated intensity and make difficult the interpretation of the results.

An attempt has been made to extend the intensity criterion to insoluble materials, using KBr disks. Satisfactory results have been obtained enabling N-bonded complexes to be distinguished from S-bonded (481). This technique has been further extended by using a suitable internal standard (60): the C—O stretching band in salicylic acid (1654 cm^{-1}) has been used, and the ratio of the intensity of ν_{CN} to the intensity of this band has been suggested as a satisfactory criterion to distinguish N- and S-bonded complexes, provided that known complexes are used for calibration.

Whatever the experimental approach, integrated intensities are cited per thiocyanate ion to normalize the results for different stoichiometries. For complexes containing only one thiocyanate group this presents no problems, but let us consider cis and trans isomers of square planar $ML_2(NCS)_2$ and assume that these complexes have the idealized microsymmetries C_{2v} and D_{2h} around the metal. Group theory predicts that both CN stretching frequencies are infrared-active in the former case, but that only one is in the latter case. In some N-cyanato complexes of palladium(II) and platinum(II), where an essentially similar problem arises, it was not possible to observe two CN stretching frequencies (575) even though subsequent work confirmed the presence of largely *cis*-platinum and *trans*-palladium compounds in the series studied (578). Nevertheless, the integrated intensities were expressed per cyanate group and chemically sensible results were obtained. The approach can perhaps be justified since the infrared-active mode in the trans complex is a complex vibration involving both CN-containing groups, so that the two groups both contribute to the intensity of the one band. Similarly, the coupling may be assumed to be proportionally less in the cis compound so that the two bands each contribute to the total intensity.

However, this analysis becomes further complicated in other symmetries, for example, in square planar or tetrahedral $[M(NCS)_4]^{n-}$ or in the corresponding octahedral series. The problem remains unresolved, and casts some doubt on the validity of the use of integrated intensities, notwithstanding the increasing amount of empirical data supporting this method.

TABLE XI

APPROXIMATE FREQUENCY RANGES FOR DIFFERENT TYPES
OF CHALCOGENOCYANATE COORDINATION

Compound type	ν_{CN} (cm^{-1})	ν_{CX} (cm^{-1})	δ_{NCX} (cm^{-1})	A ($\times 10^4$) M^{-1} cm^{-2})
NCS$^-$	2053	746	486, 471	3–5
M—NCS	2100–2050 s,b	870–820 w	485–475	7–11
M—SCN	2130–2085 s,sp	760–700 b	470–430	1–3
M—NCS—M	2165–2065	800–750	470–440	—
NCSe$^-$	2070	558	424, 416	2–3
M—NCSe$^-$	2090–2050 s,b	650–600	460–410	5–12
M—SeCN$^-$	2130–2070 s,sp	550–520	410–370	0.5–1.5
M—NCSe—M	2150–2100	640–550	410–390	—
NCO$^-$	2165	1254	637, 628	8
M—NCO	2240–2170	1350–1320	640–590	12–20
M—OCN	2240–2200	1320–1070	630–590	9–15[a]
M 　＼ 　　NCO 　／ M	2210–2150 s	1340–1300 w	660–610 m	—

[a] Organic cyanates ROCN have $A = 1$–2×10^4 M^{-1} cm^{-2} (35).

Table IX lists the positions of the thiocyanate fundamental frequencies in a representative selection of complexes, with the integrated intensities of ν_{CN}; metal–thiocyanate frequencies are also included and will be discussed later. The complexes are, in most instances, pairs of linkage isomers, and the results indicate the difficulties in drawing conclusions concerning the nature of effects causing vibrational shifts. In view of the difficulties, and because of the doubts expressed previously concerning the use of integrated intensities, great caution should be exercised in assigning the mode of bonding of the thiocyanate group on the basis of infrared data—wherever possible, supplementary measurements should also be made. Table XI summarizes the frequency ranges for different types of thiocyanate coordination.

The previous discussion has been concerned with thiocyanate complexes, but a similar situation obtains with selenocyanates (*437, 438*). Table XI includes the frequency ranges for different modes of selenocyanate coordination. Kharitonov *et al.* (*439*) obtained similar results for selenocyanate complexes as for thiocyanate complexes in their theoretical treatment outlined previously, and their conclusions are the same in both cases. Thus, frequency shifts can be due to electronic effects or to changes in the bonding mode. The relative contributions of the resonance forms of NCSe⁻ (Table X) are comparable to those of NCS⁻ so that integrated intensities can be used in a similar way, and with the same possible dangers. Table XII contains some infrared data

TABLE XII

INFRARED SPECTRA OF SOME SELENOCYANATE COMPLEXES[a]

Selenocyanate	ν_{CN} (cm⁻¹)	ν_{CSe} (cm⁻¹)	δ_{NCSe} (cm⁻¹)	A ($\times 10^4$ M^{-1} cm⁻²)	References
π-cpFe(CO)(PPh₃)NCSe	2120 m	663 mw	—	—	(*410*)
	2107 m[b]	—	—	5.3	(*410*)
π-cpFe(CO)(PPh₃)SeCN	2112 mw	532 w	—	—	(*410*)
	2117 mw[b]	—	—	1.7	(*410*)
cis-Co(DH)₂(H₂O)NCSe	2075	605	—	—	(*3*)
trans-Co(DH)₂(H₂O)SeCN	2140	—	—	—	(*3*)
Rh(PPh₃)₂CO(NCSe)	2094	—	—	7	(*150*)
Rh(PPh₃)₂MeCN(SeCN)	2135[d]	—	—	2.3	(*36*)
[Pd(Et₄dien)NCSe]⁺	2085 s,br	618	—	—	(*153*)
	2089[c]	—	—	6.6	(*153*)
[Pd(Et₄dien)SeCN]⁺	2121 s,sp	533 w	404 w	—	(*153*)
	2125[c]	—	—	0.63	(*153*)

[a] Data are recorded as mull spectra unless otherwise indicated.
[b] In CHCl₃.
[c] In acetone.
[d] KBr disc.

for a representative selection of selenocyanate complexes—fewer linkage isomers are known than have been reported for thiocyanate complexes.

It has been seen that correlations between the bonding modes of NCS⁻ and NCSe⁻ and infrared spectral parameters have been established largely on an empirical basis, with only partial theoretical justification. Further, the infrared data for a particular compound apparently may be consistent with either of the main bonding modes or, indeed, with a form of bridging. With these thoughts in mind the use of infrared measurements to determine cyanate-bonding modes becomes even more difficult for two reasons: (a) there is more coupling between the cyanate stretching frequencies than is observed for the other ions and this is also observed in complexes, e.g., the force constant calculations on $[Zn(NCX)_4]^{2-}$ (301); (b) there are very few reported O-cyanato complexes, none of which have been confirmed by X-ray crystallography, and only $(C_5H_5)_2Ti(OCN)_2$ has been confirmed by measurements other than infrared (^{14}N NMR; see Ref. 93). Apart from $Rh(PPh_3)_3NCO$ and its linkage isomer, there are therefore no sets of N- and O-bonded complexes comparable to the compounds in Tables IX (thiocyanate) and XII (selenocyanates) for which measurements can be compared, and from which bonding criteria may be deduced. Analogies with the theoretical justifications for frequency shifts for NCS– and NCSe$^-$ are also risky because of the lack of purity of the vibrational modes of NCO$^-$.

Kharitonov et al. (742) have carried out calculations, similar to those described for the thiocyanate and selenocyanate groups, to elucidate the changes in the vibrational frequencies of the cyanate group when it is N- and O-bonded. With similar models to those described previously, they reported the following findings.

For N-cyanato complexes:

1. ν_{CN} and ν_{CO} should increase with an increase of k_{MN}, that is, on coordination.

2. On increasing the mass M, ν_{CO} changes slightly, whereas ν_{CN} is hardly altered and ν_{MN} decreases steadily.

3. ν_{MN} increase almost linearly with k_{MN}.

4. ν_{CN}, ν_{CO}, and ν_{MN} are not pure vibrations. There is mixing in every case, which is most pronounced for ν_{CO}.

5. The contribution of the change in vibration mechanics is probably greater than that of any other change (such as that due to electronic effects) in the force constants k_{CN} and k_{CO}.

For O-cyanato complexes:

1. ν_{CN} is not affected, and ν_{CO} should increase in energy to a greater extent than the N-bonded case.

2. ν_{CN} is independent of the mass of M, and ν_{CO} is only slightly affected; ν_{MO} varies with M.

3. An increase in k_{MO} does not influence ν_{CN}, increases ν_{CO}, and appreciably increases ν_{MO}.

4. In contrast to the N-bonded case, a redistribution of electron density could outweigh the foregoing predictions which are based on a mechanical model.

It is useful at this stage, to discuss the arguments of Nelson and Nelson (*557*) who have also carried out some calculations on vibrational changes in the cyanate group.

The percentages of the three principal resonance hybrid structures of the cyanate ion are given in Table X. Nelson and Nelson (*557*) studied the effect on ν_{CN} and ν_{CO} as the force constants are altered by including progressively increasing contributions from, first, the resonance form $N\equiv C\!-\!O^-$ and, second, $^{2-}N\!-\!C\equiv O^+$; their results in the former case are repeated in Table XIII: The calculations were

TABLE XIII

EFFECT OF INCREASING THE CONTRIBUTIONS OF
$N\equiv C\!-\!O^-$ ON ν_{CN} AND ν_{CO} (RELATIVE TO NCO⁻)[a]

Percentage increase in contribution of resonance for $N\equiv C\!-\!O^-$	ν_{CN} (cm⁻¹)	ν_{CO} (cm⁻¹)
0 (free ion value)	2183	1254
5	2212	1248
10	2214	1239
30	2217	1191
50	2224	1133
70	2236	1063
100	2257	941

[a] Data from Ref. *557*.

carried out in order to determine the nature of the bridging cyanate group (M—NCO—M or M—N(CO)—M) and do not include the form $^-N\!=\!C\!=\!O$ which would be of importance in *N*-cyanato complexes. However, $N\equiv C\!-\!O^-$ is the resonance form most likely to be of importance in *O*-cyanato complexes, in which case the calculations show that ν_{CN} and ν_{CO} will, respectively, increase and decrease relative to the free ion values, in contrast to the prediction for M—OCN based on the mechanical model described previously.

The infrared spectra of a number of cyanato complexes are recorded in Table XIV. One of the compounds cp₂M(OCN)₂ (M = Ti,Hf) is incorrectly formulated. The infrared data are as listed, and ¹⁴N NMR (*93*) and mass spectral data (*145*) support cp₂Ti(OCN)₂. However, the dipole moment ratios of the compounds suggest that they cannot *both* have the same type of coordination (*411*). Most cyanato complexes can

TABLE XIV
Infrared Spectra of Some Cyanate Complexes[a]

Cyanate	ν_{CN} (cm^{-1})	ν_{CO} (cm^{-1})	δ_{NCO} (cm^{-1})	A ($\times 10^4$ M^{-1} cm^{-2})	Reference
$(\pi\text{-cp})_3$CeNCO	2145 s	1310	—	—	(423)
$(C_9H_7)_2$Ce(NCO)$_2$	2225 m	1320 m	—	—	(423)
$(\pi\text{-cp})_2$Ti(OCN)$_2$[b]	2235^c / 2196^c	1132 m	626 m / 593 m	13 / 18	(145)
$(\pi\text{-cp})_2$TiNCO	2216^d	1302 ms	599 m / 590 m	—	(145)
$(\pi\text{-cp})_2$Zr(OCN)$_2$[b]	2233^c / 2200^c	1257 w / 1070 sh	631 m / 607 m	12 / 16	(145)
$(\pi\text{-cp})_2$Hf(OCN)$_2$	2246^c / 2211^c	1257 w / 1071 sh	632 m / 606 m	12 / 18	(145)
[Mo(OCN)$_6$]$^{3-}$	2205 s	1296 m / 1104 m	595 m	—	(56)
[Re(OCN)$_6$]$^{2-}$	2224 s	1306 w / 1138 w	595 m	—	(56)
[Re(OCN)$_6$]$^{-}$	2220 s	—	—	—	(56)

Pd(py)₂(NCO)₂	2180S2210 s	1332 m,sp	586 m,sp	—	(575)
	2202^e	—	—	21.4	(575)
Pt(Ph₃P)₂(NCO)₂	{2230 sh {2200 s,sp	{1355 vw {1312 m,br	590	—	(575)
	2258^f	—	—	13.0	(575)
Rh(PPh₃)₂CO(NCO)	2239	—	—	—	(87)
Ph(PPh₃)₃NCO	2230^g	—	592	12.7	(37)
Rh(PPh₃)₃OCN^h	2215^g	1318	{607 {590	9.0	(37)
K[Cu(pic)₂(OCN)]	2143	1205	{630 {625	—	(321)

^{a} Data are recorded as mull spectra unless otherwise indicated.
^{b} One or other of these structures is incorrectly formulated (see text).
^{c} In CH_2Cl_2.
^{d} In acetone.
^{e} In $CHCl_3$.
^{f} In CH_3NO_2.
^{g} As KBr disk.
^{h} ν(Rh—OCN) at 332 cm⁻¹ (37).

be assumed to be N-bonded from infrared and other measurements, and it is seen that, in general, the CN stretching frequency increases and the NCO bending frequency decreases (slightly) on coordination, as might be expected from the preceding arguments; $(C_5H_5)_2Ti(OCN)_2$ shows similar changes. The CO stretching frequency does not alter very much in N-cyanato complexes. The percentage contributions of the different resonance forms (Table X) are comparable with those of the other ions, and this would appear to indicate that integrated intensity criteria can be used as before. Thus, for a series of palladium(II) and platinum(II) compounds, $ML_2(NCO)_2$, the integrated intensity per cyanate group was in the range $13–23 \times 10^{-4}$ liter mole^{-1} cm^{-2}, and larger than the free ion value of 8.4×10^{-4} liter mole^{-1} cm^{-2} (575) indicating N–cyanato complexes in every case. However, the integrated intensities of $(C_5H_5)_2M(OCN)_2$ are also larger than the free ion values, and it has been suggested that this criterion is inapplicable to cyanate complexes because of the small difference in mass between the nitrogen and oxygen atoms (145). The integrated intensities for the linkage isomers $Rh(PPh_3)_3NCO$ and $Rh(PPh_3)_3OCN$ do suggest, however, that values about equal to the free ion value are characteristic of O-cyanates (37).

The lack of compounds means that generalizations cannot be advanced for cyanates in the same way that they have been developed for the thiocyanate and selenocyanate complexes. Such data as are available suggest that ν_{CN} for O-cyanates lie within the range for N-cyanates and that the degeneracy of the deformation mode is apparently removed for some O-cyanates as well as for some N-cyanates.

Because the foregoing criteria all are found wanting, although on the basis of very few data, attention must now be turned to the remaining frequency, namely the CO stretching frequency. Fermi resonance occurs between ν_{CO} and 2δ in the free ion (see Section II, C). On coordination through nitrogen, ν_{CO} generally increases in magnitude and δ decreases. For these or for other reasons, Fermi resonances have not been observed in N-cyanato complexes. In O-cyanato complexes there is ample opportunity for the phenomenon of Fermi resonance to be maintained, and in $[Re(OCN)_6]^{2-}$ and $[Mo(OCN)_6]^{3-}$ the bands near 1300 and 1140 cm^{-1} (see Table XIV) have been assigned on this basis (56). Similarly, Burmeister has argued that the medium strong band at 1132 cm^{-1} in $(\pi–cp)_2Ti(OCN)_2$ is one component of such a pair of bands (145), but later he indicated that he prefers to formulate this compound cp$_2$Ti-(NCO)$_2$ (411). The situation is not clarified by the linkage isomers $Rh(PPh_3)_3NCO$ and $Rh(PPh_3)_3OCN$ since the only band observed in this region of the spectrum is at 1318 cm^{-1} in the latter compound, and this has been assigned only tentatively to ν_{CO} (*37*). It is possible that the Fermi resonance which has led to the extensive modification of ν_{CO} in some of the reported *O*-cyanates does not occur in this instance [nor in the case of $cp_2Ti(OCN)_2$?]. In either event, all the assignments for these compounds must be treated with caution pending further data.

The ranges observed for the different modes of coordination of the chalcogenocyanates have been summarized in Table XI. These ranges are based on the data in Bailey's review (*57*) and on the results included in Tables IX (thiocyanate), XII (selenocyanate), and XIV (cyanate). It will be clear from the preceding paragraphs and from these tables that many effects other than isomerization can cause considerable shifts in the frequencies concerned. It follows therefore that great care must be exercised in the application of the data of Table XI, which are not exclusive.

TABLE XV

PALLADIUM–LIGAND STRETCHING FREQUENCIES

Palladium compound	ν_{Pd-X} (cm⁻¹)[a]	Reference
trans-Pd(PPh₃)₂(NCO)₂	350	(*578*)
trans-Pd(AsPh₃)₂(NCO)₂	360	(*578*)
[PdL(NCO)]⁺[b]	365	(*482*)
cis-Pd(bipy)(NCO)₂	389, 374 sh	(*578*)
trans-Pd(γ-pic)₂(NCO)₂	416	(*578*)
cis-Pd(bipy)(NCS)₂	345, 332	(*336*)
[PdL(NCS)]⁺	365	(*482*)
trans-Pd(AsPh₃)₂(SCN)₂	306	(*336*)
cis-Pd(bipy)(SCN)₂	316, 304	(*336*)
[PdL(SCN)]⁺	320	(*482*)
[PdL(NCSe)]⁺	360	(*482*)
[PdL(SeCN)]⁺	318	(*482*)

[a] X = anion.
[b] L = Et₄dien.

Far-infrared spectra have also been considered for these complexes, and Bailey *et al.* (*57*) conclude the section in their review on this topic "Indeed, there is little reason to expect this frequency to offer a simple bonding mode criterion." The effect of stereochemistry on ν_{M-L} is particularly marked. Thus, Clark and Williams (*201*) have shown that $\nu_{M-Cl} > \nu_{M-NCS}$ for tetrahedral MX_2L_2 compounds, but $\nu_{M-Cl} < \nu_{M-NCS}$ for most octahedral compounds (see Ref. *57*). The few results for palladium(II) complexes recorded in Table XV show that other factors can have drastic effects also (see Refs. *57* and *578* for discussion).

B. Ultraviolet and Visible Spectroscopy

In Section II, D, 2 it was concluded that the internal electronic transitions have been adequately assigned for NCO⁻ and NCS⁻ and that they are particularly susceptible to perturbations from neighboring metal cations. It has been argued that S-thiocyanato complexes have a strong characteristic band at $\sim 30,000$ cm⁻¹, and N-thiocyanato complexes at $\sim 38,000$ cm⁻¹ (585). However, Barnes and Day (75) have discussed the origin of this intense band which lies between 30,000 and 40,000 cm⁻¹ in many thiocyanate complexes, and the distinction is not so clear cut. Such bands are not observed with nonreducible ions such as La^{3+}, Gd^{3+}, or Zn^{2+}, and it was concluded that these transitions were associated with a charge transfer to a reducible metal. By considering the nature of the orbitals concerned, Barnes and Day further concluded that the lowest-energy absorption in the free thiocyanate ion results in a net transfer of charge from nitrogen to carbon. In N-thiocyanato complexes, therefore, the position of this transition will depend on the oxidizing power of the reducible metal. Similarly, the lowest-energy bands at N-cyanato and N-selenocyanato complexes of the divalent ions of the first transition series are primarily ligand-to-metal charge transfer spectra (242). The spectra of a variety of tetra- and hexacoordinated metal complexes with NCS⁻ and NCSe⁻ have been collated and discussed by Schmidtke (663), but no simple distinction between the possible modes of coordination is possible. The spectra of some cyanate complexes have been discussed also (453).

Transition metal ions with partially filled d orbitals will show the expected d-d spectra which will depend on the degree of perturbation of these orbitals by the coordinating ligand. It has been well established that coordination through sulfur causes less perturbation than coordination through the nitrogen of the thiocyanate ion, or —SCN occurs lower in the spectrochemical series than —NCS (see, e.g., Ref. 664), and a similar distinction may be made between —SeCN and —NCSe (663).

Although the foregoing generalizations can assist in the assignment of the spectrum of a compound of known structure, they are less useful when it comes to determining the mode of coordination of a chalcogenocyanate group unless other known compounds are available for comparative purposes. Thus, the green solution of the unstable $[Cr(H_2O)_5SCN]^{2+}$ was identified by comparing it with the purple solution of $[Cr(H_2O)_5NCS]^{2+}$: the absorption spectra were similar except that the maxima in the former were shifted some 40 nm toward longer wavelengths (363). This and some other examples are given in Table XVI where some intraligand bands are also quoted.

TABLE XVI

Electronic Spectra of Some Thiocyanate Complexes

Thiocyanate	v (cm^{-1})	ε	Assignment[a]	References
[Cr(NCS)$_6$]$^{3-}$	31,700	27,000	$t_{1u} \rightarrow t_{2g}$	(663)
[Ru(NCS)$_6$]$^{3-}$	18,200	7,850	$t_{1u} \rightarrow t_{2g}$	(663)
[Os(NCS)$_6$]$^{3-}$	22,000	11,370	$t_{1u} \rightarrow t_{2g}$	(663)
[Rh(SCN)$_6$]$^{3-}$	34,800	25,300	$t_{1u} \rightarrow e_g$	(663)
[Pt(SCN)$_6$]$^{2-}$	34,600	44,000	$t_{1u} \rightarrow e_g$	(663)
[Cr(H$_2$O)$_5$SCN]$^{2+}$	38,200	8,000	C.T.	(587)
	22,300	20	—	
	16,100	26	—	
[Cr(H$_2$O)$_5$NCS]$^{2+}$	24,400	33.5	—	(364)
	17,600	31.4	—	
cis-[Cr(H$_2$O)$_4$(NCS)$_2$]$^+$	23,800	31.6	—	(388)
	17,600	28.2	—	
trans-[Cr(H$_2$O)$_4$(NCS)$_2$]$^+$	23,500	25.1	—	(388)
	16,900	25.1	—	
[Cr(H$_2$O)$_4$(NCS)(SCN)]$^+$	37,700	8,300	C.T.	(129)
	22,700	43	—	
	16,500	55	—	
[Ru(NH$_3$)$_5$SCN]$^{2+}$	45,000	13,300	$\pi \rightarrow \pi$ (NCS$^-$)	(490)
	36,000	209	$^2E_u \rightarrow {}^2T_{2g}$	
	31,100	407	$^2E_u \rightarrow {}^2T_{1g}$	
	20,200	3,620	$^2E_u \rightarrow {}^2E_u$	
[Fe(CN)$_5$NCS]$^{3-}$	38,500 sh	2,280	2π(NCS) $\rightarrow$ dπ(Fe)	(360)
	37,000 sh	2,130	LF	
	33,300 sh	1,360	σ(CN) $\rightarrow$ dπ(Fe)	
	31,000 sh	1,090	π(CN) $\rightarrow$ dπ(Fe)	
	28,600	995	LF	
	25,300 sh	520 }	π(CN) $\rightarrow$ dπ(Fe)	
	19,200	3,600		

(continued)

TABLE XVI—continued

Thiocyanate	ν (cm^{-1})	ε	Assignment [a]	References
$[Co(CN)_5SCN]^{3-}$	50,000	16,700	$(5e, 2b_2) \rightarrow 6e$	(359)
	44,000	4,300	$2\pi \rightarrow 3\pi$	
	37,700	17,100	$4e \rightarrow 3a_1$	
	26,500	191	$^1A_1 \rightarrow {}^1E \ (^1A_2?)$	
$[Co(CN)_5NCS]^{3-}$	49,500	28,100	$(5e, 2b_2) \rightarrow 6e$	(359)
	37,700	2,340	$4e \rightarrow 3a_1$	
	27,600	500	$^1A_1 \rightarrow {}^1E \ (^1A_2)$	
$[Co(DH)_2py(SCN)]$	40,000	—	—	(574)
	33,500	—	—	
$[Co(DH)_2py(NCS)]$	40,000	—	—	(574)
	33,000	—	—	
	21,000	—	—	
$[Co(NH_3)_5(NCS)]^{2+}$	32,700	1,490	$^1A_{1g} \rightarrow {}^1T_{1g}$	(134)
	20,100	179	—	
$[Co(NH_3)_5(SCN)]^{2+}$	34,700	15,600	$^1A_{1g} \rightarrow {}^1T_{1g}$ [b]	(134)
	19,500	74	—	
$[Rh(NH_3)_5NCS]^{2+}$	46,500	5,100 $\}$	Internal NCS bands	(659, 660)
	41,700	2,200 $\}$		
	31,200	460	$^1A_{1g} \rightarrow {}^1T_{1g}$	
$[Rh(NH_3)_5SCN]^{2+}$	42,900	19,200	Internal NCS band	(659, 661)
	35,700	450	—	
	31,000	220	$^1A_{1g} \rightarrow {}^1T_{1g}$	
	26,700	85	$^1A_{1g} \rightarrow {}^1T_{1g}$	
$[Ir(NH_3)_5NCS]^{2+}$	43,500	2,260	$^1A_{1g} \rightarrow {}^1T_{2g}$	(659, 661)
	38,500	560	$^1A_{1g} \rightarrow {}^1T_{1g}$	

	$\tilde{\nu}$	ε	Assignment[a]	Ref.
$[Ir(NH_3)_5SCN]^{2+}$	42,500	820	$^1A_{1g} \longrightarrow {}^1T_{2g}$	(659, 661)
	36,400	165	$^1A_{1g} \longrightarrow {}^1T_{1g}$ [b]	
$Pd(AsPh_3)_2(NCS)_2$	28,600	—	—	(143)
$Pd(AsPh_3)_2(SCN)_2$	25,300	—	—	(143)
	21,050	—	—	
$Pd(Ph_2PCH_2CH_2NMe_2)(NCS)(SCN)$	29,600	5,280	—	(526)
	24,400	615	—	
$Cu(tripyam)(SCN)_2$	26,300	—	—	(473)
	15,400	—	—	
	13,700	—	—	
$Cu(tripyam)(SCN)(NCS)$	26,300	—	—	(473)
	16,400	—	—	
	14,300	—	—	
$Cu(tripyam)(NCS)_2$	26,300	—	—	(473)
	23,500	—	—	
	15,400	—	—	

[a] The assignments are described using the nomenclature of the original papers and it may be necessary to consult these in some cases. Generally the conventional description of d—d transitions has been used and has been added in some cases where the authors did not make the assignments.

[b] Broad bands observed implying some splitting of the first crystal field transition.

Electronic spectra may not always be of prime importance in determining the mode of coordination of a monodentate chalcogenocyanate, although the chromophore MN_5S is sufficiently different from MN_5N to show some effects (see Table XVI), but they can be useful in the study of bridging systems. Thus, $Nian_2(NCS)_2$ was shown to contain octahedral nickel(II) with Ni—NCS—Ni bridges since S coordination is sufficiently different from the alternative nitrogen atoms to distort the ligand field and produce splitting of the first and second electronic absorption bands (*167*). Similar examples may be found in particular for nickel(II) and cobalt(II) complexes (see Sections IV, H and I).

Many of these spectra have been interpreted so as to distinguish between the extent of σ and π bonding in different systems, but some controversy is associated with the validity of these interpretations and they have not been included generally in this review.

The *N*-cyanato and *N*-selenocyanato groups have been shown to cause very similar splitting of the d orbitals to —NCS. Table XVII shows the spectral parameters for $[Co(NCX)_4]^{2-}$ (X = O, S, and Se) and for $[Co(NCX)_4Hg]$ (X = S and Se), and illustrates the similarity. It can be seen also that the values for Δ are in general accord with those expected for an N-bonded group, and the changes occurring as the terminal noncoordinating atom is changed suggest a small but significant transmitted effect. Other examples may be obtained from the references in Section IV.

The compound $Cu(pic)_2 \cdot KNCS$ was shown to be a double salt, but the electronic spectra suggested that the corresponding cyanate was five-coordinate: by comparing the spectra and making assumptions concerning the difference in behavior between the Cu—NCS and Cu—NCO moieties (if they had existed in this system), it was deduced that the latter compound should be formulated $K[Cu(pic)_2(OCN)]$ (*321*).

Apart from the preceding example, which depends on a rather indirect argument, electronic spectra are only useful to provide supporting evidence for the mode of coordination of cyanate and selenocyanate complexes. The arguments must necessarily be based on the magnitudes of crystal field parameters, and these can be assessed only by comparison with other related values.

C. ^{14}N NUCLEAR MAGNETIC RESONANCE SPECTROSCOPY

The values for the chemical shifts for the free ions were discussed in Section II, D, 3, and a general review of ^{14}N NMR data in inorganic molecules has appeared (*501*). The application of ^{14}N NMR to the problem of determining the mode of coordination of the thiocyanate ion was

TABLE XVII

ELECTRONIC SPECTRAL DATA AND CRYSTAL FIELD PARAMETERS FOR $[Co(NCX)_4]^{2-}$ (X = O, S, and Se)

Parameter	$[Co(NCO)_4]^{2-}$ [a]	$[Co(NCS)_4]^{2-}$ [b]	$[Co(NCS)_4Hg]$ [b]	$[Co(NCSe)_4]^{2-}$ [c]	$[Co(NCSe)_4Hg]$ [c]
ν_1 (cm^{-1})	—	—	—	—	—
ν_2 (cm^{-1})	7,150	7,780	8,300	7,840	8,400
ν_3 (cm^{-1})	16,100	16,250	16,700	16,000	16,000
Δ (cm^{-1})	4,150	4,550	4,880	4,710	4,980
B' (cm^{-1})	720	691	691	644	631
β ($B'/967$)	0.745	0.715	0.715	0.666	0.652

[a] Data from Ref. 211.
[b] Data from Ref. 213.
[c] Data from Ref. 212.

first demonstrated by Howarth, Richards, and Venanzi (*390*). Some results are listed in Table XVIII, where it can be seen that there is only a small downfield nitrogen chemical shift relative to the thiocyanate ion if coordination occurs through sulfur, whereas nitrogen coordination produces a significant high-field shift. These authors assumed that N bonding lowers the energy of the nonbonding orbital containing the pair of electrons and shifts the ^{14}N resonance upfield; S bonding however does not affect the nonbonding orbital but lowers the energy of the

TABLE XVIII

^{14}N CHEMICAL SHIFTS FOR SOME THIOCYANATE COMPLEXES

Thiocyanate	Solvent	δ (NO_3^-) (ppm)	References
$(NH_4)_2[Hg(SCN)_4]$	H_2O	+146	(*111*)
$Na_2[Pd(SCN)_4]$	H_2O	+148	(*390*)
$Na_2[Hg(SCN)_4]$	H_2O	+157	(*390*)
$K_3[Rh(SCN)_6]$	H_2O	+158	(*390*)
$Na_3[Ir(SCN)_6]$	H_2O	+163	(*390*)
$K_2[Pt(SCN)_4]$	H_2O	+166	(*390*)
NCS^-	H_2O	+166	(*390*)
NCS^-	H_2O	+170	(*111*)
$Na_2[Cd(NCS)_4]$	H_2O	+178	(*390*)
$K_4[Cd(NCS)_6]$	H_2O	+183	(*111*)
$Na_2[Cd(NCS)_4]$	MeOH	+220	(*390*)
$K_2[Zn(NCS)_4] \cdot 2Me_2CO$	H_2O	+220	(*111*)
$K_2[Zn(NCS)_4]$	H_2O	+238	(*111*)
trans-$Pt(PEt_3)_2(NCS)_2$	$CHCl_3$	+239	(*390*)
$K_2[Ru(NCS)_5NO]$	H_2O	+245	(*390*)
cis-$Pt(PPhBu_2)_2(NCS)_2$	$CHCl_3$	+249	(*390*)
$Na_2[Zn(NCS)_4]$	EtOH	+255.5	(*390*)
trans-$Ni(PPhBu_2)_2(NCS)_2$	$CHCl_3$	+291	(*390*)
trans-$Ni(PEt_3)_2(NCS)_2$	$CHCl_3$	+293	(*390*)
cis-$Pt(PBu_3)_2(NCS)_2$	$CHCl_3$	+302	(*390*)
cis-$Pt(AsBu_3)_2(NCS)_2$	$CHCl_3$	+303	(*390*)

delocalized antibonding orbital by mixing with metal d orbitals and so causes a small downfield shift relative to the free ion. Recently, a further study on similar compounds (*111*) has confirmed these results, including the observation that the large solvent effect on the ^{14}N shift for $[Cd(NCS)_4]^{2-}$ is due to the presence in solution of both N and S isomers in kinetic equilibrium (*390*).

Table XIX gives some ^{14}N shift data for cyanate compounds, where a high-field shift is seen for *N*-cyanato complexes and a downfield shift observed for EtOCN (*192*). Similar arguments to those used for thiocyanate complexes presumably may be also applied here.

TABLE XIX

^{14}N Chemical Shifts for Some Cyanate Complexes

Cyanate	Solvent	δ (NO$_3^-$) (ppm)	References
(Et$_4$N)[Ag(NCO)$_2$]	Me$_2$CO	+344	(192)
Pt(PPh$_3$)$_2$(NCO)$_2$	CH$_2$Cl$_2$	+340	(84)
(Et$_2$NH$_2$)$_2$[Zn(NCO)$_4$]	Me$_2$CO	+325	(192)
(Me$_4$N)$_2$[Hg(NCO)$_4$]	Me$_2$CO	+317	(192)
(Et$_4$N)$_2$[Hg(NCO)$_4$]	MeNO$_2$	+312	(192)
NCO⁻	H$_2$O	+300	(192)
EtOCN	—	+222	(192)
PhOCN	—	+208	(84)

Compound K[Ph$_3$Sn(NCSe)$_2$] resonates at +215 ppm, and NCSe⁻ at +136 ppm (relative to NO$_3^-$) (*111*) so that the same behavior as was described in the foregoing seems likely for selenocyanate complexes. The absence of further data precludes this tentative generalization from being more firmly based at present.

D. Chemical Methods

There are no general chemical methods for determining the mode of chalcogenocyanate bonding in complexes in the way that there are for organic compounds (*41*). This is no doubt a reflection on the ready occurrence of substitution reactions and on the lability of these groups in many cases. A pair linkage of isomers may be assigned on the basis of the easy conversion of one into the other (if the stable isomer is known), but such evidence is not sufficient by itself and needs further spectroscopic results for confirmation.

The compound *trans*-[Rhen$_2$Cl(NCS)]NCS reacts with twice the amount of mercuric nitrate necessary to determine the free thiocyanate ion, whereas only one equivalent of silver nitrate is consumed using the Volhard titration. This was explained by assuming that the mercuric ion was able to form a Rh—NCS—Hg bridge, which is less readily achieved with silver, and, under the conditions of the Volhard titration, the silver is removed as the thiocyanate ions are titrated in (*415*). Orhanovic and Sutin (*587*) also analyzed mixtures containing [Cr(H$_2$O)$_5$-NCS]$^{2+}$ and [Cr(H$_2$O)$_5$SCN]$^{2+}$ by adding mercuric ions and obtaining [Cr(H$_2$O)$_5$NCSHg]$^+$ and [Cr(H$_2$O)$_6$]$^{3+}$, respectively, which were determined spectrophotometrically. However, they had discovered that some of the bridged complex is also produced from the Cr—SCN compound and allowance was made for this. The same workers also bubbled chlorine through a solution of the mixture of isomers and obtained $[Cr(H_2O)_6]^{3+}$ and $[Cr(H_2O)_5Cl]^{2+}$. Control experiments had shown that $[Cr(H_2O)_5NCS]^{2+}$ gave the hexahydrate under these conditions, but comparison with the mercury reaction indicated that $[Cr(H_2O)_5SCN]^{2+}$ did not form $[Cr(H_2O)_5Cl]^{2+}$ exclusively, and again corrections have to be made.

The nature of the noncoordinated end of the chalcogenocyanate group may also be used, with caution, to distinguish between N- and Se-bonded selenocyanates. The compound π-cpFe(CO)(PPh$_3$)SeCN is stable but its N-bonded isomer readily deselenates at 27°C (410). Similarly, it is difficult to study Rh(I)—NCSe complexes because of rapid deselenation, and the loss of tellurium is a similar feature of tellurocyanate chemistry (36).

Although the foregoing reactions are not ideal for determining the mode of bonding, they do suggest that such methods are worth further study in the future and may yield interesting results.

E. Mass Spectroscopy

The mass spectrum of HNCO showed intense peaks corresponding to HN$^+$ and COH$^+$, and thus to a mixture of isomers, but the presence of CH$^+$, OH$^+$, and NO$^+$ indicated rather a triangular structure for the ion in the transition state (692). Such rearrangements make this an unreliable technique for these structural assignments, and, indeed, the isomeric compounds $(\pi$-C$_5$H$_5)_2$W(NCS)$_2$ and $(\pi$-C$_5$H$_5)_2$W(NCS)(SCN) have similar fragmentation patterns (344). Burmeister et al. (145) have reported characteristic TiO$^+$, ZrO$^+$, and HfO$^+$ peaks in the spectra of $(\pi$-C$_5$H$_5)_2$Ti(OCN)$_2$ and the zirconium and hafnium analogs, but there is some doubt that all these compounds are as designated.

F. ^{1}H Nuclear Magnetic Resonance Spectroscopy

The chemical shifts of the α-hydrogen atoms allow the distinction to be made between RNCS and RSCN (522). Similar measurements may be used in coordination chemistry, for example, the methyl protons in Co(DH)$_2$py(NCS) resonate at higher field to those in Co(DH)$_2$py(NCS) (574), but here the correlations are purely empirical, and the direction of the shift cannot be associated with a particular isomer. Among the many observations of such chemical shift measurements are included those on the systems Rh(PMe$_2$Ph)$_3$Cl$_2$(NCS) (126), $(\pi$-C$_5$H$_5$)Fe(CO)$_2$-NCS and $(\pi$-C$_5$H$_5$)Mo(CO)$_3$NCS (690), and $(\pi$-C$_5$H$_5)_2$W(NCS)(SCN) (344) together with their linkage isomers and other related compounds.

G. Dipole Moments

Although dipole moments have been used to distinguish between geometric isomers in, for example, square planar complexes there are no examples of the use of this technique in distinguishing between linkage isomers. Thus, Chatt and Hart (188) used dipole moments to show that the two isomers $(PPr_3)_2Pt_2Cl_2(CNS)$ contained phosphines that are trans to each other, but they were unable to draw any further structural conclusions without use of other techniques. On the other hand, dipole moments were used not only to show that the compounds $Znpy_2(NCX)_2$ (X = O and S) are tetrahedral, but also that they contain N-bonded anions (655). An indirect application of this method has been developed by Jensen et al. (411) who showed that

$$\frac{\mu \text{ for } cp_2Ti(NCS)_2}{\mu \text{ for } cp_2TiCl_2} = \frac{\mu \text{ for } cp_2Zr(NCS)_2}{\mu \text{ for } cp_2ZrCl_2}$$

whereas

$$\frac{\mu \text{ for } cp_2Ti(OCN)_2}{\mu \text{ for } cp_2TiCl_2} \neq \frac{\mu \text{ for } cp_2Zr(OCN)_2}{\mu \text{ for } cp_2ZrCl_2}$$

from which it was deduced that the cyanate bonding modes in the two compounds are not the same. The authors were unable to conclude from this evidence whether the titanium or the zirconium compound contained the O-bonded species.

H. Thermodynamic Methods

The theories for the interaction of the different donor atoms with different types of metal centers are discussed in Section V. These theories predict positive enthalpy and entropy changes for the reaction of a "hard" base with a hard or class a metal, whereas the reaction of a "soft" base with a class b metal shows a negative enthalpy of reaction (15). A study of the heats of formation of some polyamine copper(II) complexes with NCS^- shows a small positive enthalpy change for the reaction, and a larger positive entropy change (71). The analogous reaction with NCS^- and $[Ptdiars]^{2+}$ exhibits a negative enthalpy change (255).

IV. Chalcogenocyanate Complexes of the Transition Elements

A. Scandium, Yttrium, Lanthanum, and the Lanthanides

1. Cyanates

The compound $(n\text{-}Bu_4N)_3[Er(NCO)_6]$ contaminated with silver cyanate has been reported, and infrared data suggest this formulation even though the compound is impure (*160*). The same workers (*144*), without retracting the foregoing, later reported that the same reaction can result in the formation of $(n\text{-Bu}_4\text{N})[\text{Ag(NCO)}_2]$. The tricyclopenta dienyl and bisindenyl compounds of cerium(IV), $\text{Ce(C}_5\text{H}_5)_3\text{NCO}$ and $\text{Ce(C}_9\text{H}_7)_2(\text{NCO})_2$, have been characterized (423).

2. Thiocyanates

The X-ray structure of $(n\text{-Bu}_4\text{N})_3[\text{Er(NCS)}_6]$ shows the complex to have six octahedrally coordinated N-thiocyanato groups with an average Er—N distance 2.34(2) Å. The Er-NCS groups are approximately linear with Er—NC and NCS angles 174(2)° and 176(3)°, respectively. The N—C and C—S bonds are 1.10(3) and 1.61(3) Å. The cations do not penetrate significantly into the sphere of influence of the terminal sulfur atoms. Related complexes of other lanthanides (Pr, Nd, Sm, Eu, Ho, Tm, and Yb, and also Y) were also prepared and have infrared mull spectra similar to the preceding. Conductivity measurements in nitromethane and nitrobenzene suggest some dissociation of the complexes, especially for the lighter lanthanides, and this is confirmed by spectral measurements in the visible region (*518*). Burmeister *et al.* (*160*) have reported a very similar series of complexes $(n\text{-Bu}_4\text{N})_3$-$[\text{Ln(NCS)}_6]$ Ln = Pr, Nd, Sm, Eu, Gd, Tb, Dy, Ho, Er, and Yb) also with high molar conductivities in solution. Surprisingly, these solutions show very high integrated intensities for ν_{CN} (13–32 $\times$ 10^4 M^{-1} cm^{-2}), but perhaps these data reflect the lack of knowledge of their origin (see Section III, A) rather than contradicting the conductivity results; Bailey *et al.* (*60*) have recorded a lower intensity value (10.5 $\times$ 10^4 M^{-1} cm^{-2}) for $(\text{Et}_4\text{N})_3[\text{Yb(NCS)}_6]$.

Alkali metal hexa-N-thiocyanatoscandates have been prepared and characterized from infrared mull spectra (*653*). The complexes $(\text{R}_4\text{N})_3[\text{Sc(NCS)}_6]$ (R = Me, Et, n-Bu) have also been prepared and their solubilities determined: they are N-bonded from infrared data, and dissociate on dilution (*270*). The thermal stabilities of the alkali metal salts increase with M in the order Li < Na < K < Rb < Cs (*352*).

All the examples reported so far appear to contain N-bonded thiocyanate groups and to dissociate readily in solution. Thus, $[\text{ScL}_3](\text{NCS})_3$ (L = bipy or phen) has been prepared and characterized (*223*). Treatment with warm EtOH gives $[\text{ScL}_2(\text{NCS})_2](\text{NCS})$ where the formulation is supported by conductivity measurements (in CH_3CN or CH_3NO_2) and infrared mull spectra; in aqueous solutions the coordinated NCS$^-$ is displaced (*223, 225, 464*). Other mixed liquid complexes of these metals with NCS$^-$ are listed in Table XX. A 1:1 adduct is formed between 1,3,5-trinitrobenzene and several lanthanide thiocyanates. No

TABLE XX

SOME THIOCYANATE COMPLEXES OF SCANDIUM,
YTTRIUM, AND THE LANTHANIDES

Complex	Composition	References
Sc(NCS)$_3 \cdot n$py	n = 2–4	(652)
Sc(NCS)$_3 \cdot$L$_3$	L = py-NO, 2-, 3-, and 4-MepyNO, Ph$_3$MO (M = P, As), DMSO, DMF	(225)
Sc(NCS)$_3 \cdot$L$_n$	n = 2; L = isoamyl alcohol, C$_4$H$_8$O$_2$	(465)
	n = 1; L = C$_4$H$_8$O$_2$	(465)
Ln(NCS)$_3$(OPPh$_3$)$_3$	Ln = Y, Sm–Lu	(218)
Ln(NCS)$_3$(OPPh$_3$)$_4$	Ln = La–Sm (not Pm)	(218)
Ln(NCS)$_3$(OAsPh$_3$)$_3$	Ln = Ce, Pr	(218)
Ln(NCS)$_2$2H$_2$O$\cdot$4C$_4$H$_8$O$_2$	Ln = La, Ce, Pr, Nd	(769)
[Ln(phen)$_3$](NCS)$_3$	Ln = La–Lu (not Pm)	(369)
Ln(NCS)$_2$(OC$_2$H$_4$NHC$_2$H$_4$OH)MeOH	Ln = La, Ce, Pr, Nd	(330)
Ln(NCS)$_2$8–ox$\cdot$6MeOH	Ln = La, Pr, Nd, Sm, Eu, Gd	(331)
Ln(NCS)$_3 \cdot n$DMA	n = 5; Ln = Ce, Pr, Nd	(770)
	n = 4; Ln = Y, Sm, Eu, Gd, Tb, Dy, Ho, Er, Tm, Yb, Lu	(770)
MSm(NCS)$_4 \cdot$4EtOH	M = K, Rb, or Cs	(740)
Na$_2$Gd(NCS)$_5 \cdot$5C$_4$H$_8$O$_2$		(740)
KGd(NCS)$_4 \cdot$4EtOH		(740)
Sm(NCS)$_3 \cdot$4C$_4$H$_8$O$_2$		(740)
Ln(NCS)$_3 \cdot n$TMU	n = 5; Ln = La—Nd,	(601)
	n = 4; Ln = Sm—Er, Y	(601)
	n = 3; Ln = Tm—Lu	(601)

structural details are given but the order of increasing complexing facility is La < Ce < Nd < Yb < Gd (731). The species [O$_3$SO—Ce—NCS—Ce—OSO$_3$]$^{3+}$ has been suggested as an intermediate in the reduction of ceric ions by NCS⁻ on the basis of kinetic evidence only (730). Golub *et al.* have reported some further complexes of La, Pr, Nd (332), Sm (324), and Eu (325) but without structural data; these complexes are not listed specifically. The Mössbauer spectrum of Eu(NCS)$_3$ has been reported (205).

The tricyclopentadienyl and bisindenyl compounds of cerium(IV), Ce(C$_5$H$_5$)$_3$NCS and Ce(C$_9$H$_7$)$_2$(NCS)$_2$, have been characterized (423).

3. Selenocyanates

The compound (n-Bu$_4$N)$_3$[Y(NCSe)$_6$] has been prepared (162) as well as the corresponding lanthanide complexes (Ln = Pr, Nd, Sm, Dy, Ho,

Er) (*144*). Infrared spectra indicate that they are N-bonded. The compounds show a slightly greater tendency to dissociate in solution than the corresponding thiocyanate complexes. The magnitudes of the molar absorptivities of the absorption maxima in the visible spectra indicate pronounced deviations from O_h symmetry—a similar effect was noted for the thiocyanate complexes (*160*). The authors suggest this may arise from the nonlinearity of the M—NCSe (or M—NCS) moiety, but the structural data for the latter complexes (*518*) do not support this view. Coordination of solvent molecules in solution could account for the effect. Mixed-ligand selenocyanate complexes of Y, Sc, La, Ce, Pr, and Nd have been reported with various nitrogen and oxygen donors but in no case is the selenocyanate group coordinated other than as a monodentate ligand through nitrogen (*327, 328, 733*).

B. The Actinides

1. Cyanates

The preparation of uranyl cyanate, $UO_2(CNO)_2$, and of some of its corresponding anionic complexes were reported in 1914 (*592*); $(Et_4N)_2$-$[UO_2(NCO)_4H_2O]$ has been characterized from infrared data (*59*).

2. Thiocyanates

An X-ray structure determination has shown $(Et_4N)_4[U(NCS)_8]$ to have uranium at the center, and the nitrogen ligand atoms at the eight vertices, of a regular cube (*216*). The compounds $(Et_4N)_4[M(NCS)_8]$ (M = Th, Pa, U, Np, Pu) are all isostructural (*26*). Splitting of ν_{CN} occurs in $M_4[U(NCS)_8]$ (M = K, Cs) (*26*) and in $Cs_4[U(NCS)_8]$ which suggests some lowering of the symmetry of the complex anion with these Group I cations (*57*). The number of bands observed in infrared studies of $M_4[Th(NCS)_8] \cdot 2H_2O$ (M = Rb, Cs), the monohydrates of the analogous uranium compounds, and $(NH_4)_4[U(NCS)_8]$ suggests that these compounds exist with dodecahedral coordination in the solid or as Archimedean antiprisms in solution (*349*). Changes in the electronic spectrum of $(Et_4N)_4[U(NCS)_8]$ as a solid and in nitromethane have been attributed to changes in the nature of the cation–anion interaction on dissolution (*311*). Other eight-coordinate complexes include $U(NCS)_4 \cdot$ 4DMA (*50*) and $M(NCS)_4(R_3PO)_4$ (R = Me; M = U, Np, Pu or R = NMe$_2$; M = Th, U, Np, Pu or R = Ph; M = Th, U, Np) (*25, 51*). Also $Th(NCS)_4 \cdot 6Me_3PO$ has been reported (*25*) as have some uncharacterized mixed-ligand complexes of thorium (*538, 539*). Of these, $U(NCS)_4 \cdot$ 4DMA as a typical example shows a single sharp band at 2047 cm^{-1}, in a similar position to that observed in various uncharacterized compounds $M_3[UO_2(C_2O_4)_2CNS]$ (M = NH_4, K, Na, CN_3H_6, Et_3NH) (676).

Similar infrared results for other uranyl complexes have been interpreted as indicating N-bonded thiocyanate groups for M_3UO_2-(NCS)$_5$, $UO_2(NCS)_2L_3$, and $MUO_2(NCS)_3L_2$ (M = NH_4, K, Rb, Cs; L = H_2O) (435, 675). The preparation of these compounds is perhaps surprising in view of the fact that the former oxalato complexes contain only one coordinated thiocyanate group, regardless of the course of the reaction (676), and that the kinetic results of Kustin and Hurwitz (475), who used the temperature-jump method, show that only the 1:1 complex between the uranyl and thiocyanate ions is kinetically significant. However, the first three, stepwise stability constants have been determined for the reaction between UO_2^{2+} and thiocyanate ions (16). The foregoing infrared results are contrasted with those for the compounds $UO_2(NCS)_2(TBP)_2$ and $UO_2(NCS)_2H_2O$ and with the $[UO_2(NCS)_3]^-$ anion to show that in these systems the bridging thiocyanate group is present (764)—this mode has been suggested in other similar uranyl systems without any evidence (763).

The formation constants of thiocyanate complexes of Am(III), Cm(III), Bk(III), Cf(III), and Es(III) have been determined (367), and a spectrophotometric study made of the Am(III) complexes (366); no structural data are avilable.

3. Selenocyanates

The compounds $(Et_4N)_4[M(NSCe)_8]$ (M = Pa, U) have been characterized by comparing their spectral and crystallographic properties with the corresponding thiocyanate complexes (26). Some thorium(IV) complexes have been prepared containing DMF and N-bonded selenocyanate groups (326). Although various uranyl selenocyanate complexes have been reported, no structural data are available (686).

C. Titanium, Zirconium, and Hafnium

1. Cyanates

A number of metallocene cyanate complexes have been reported for these metals. Samuel (648) and Coutts and Wailes (219) reported $cp_2M(CNO)_2$ (M = Ti, Zr) independently in 1966. Samuel made no definite assignments of the structure, but Coutts and Wailes suggested the compounds were N-cyanato complexes. Later, Burmeister et al. (145, 146) reexamined the compounds in more detail and argued that they, and the hafnium analog, were O-cyanato complexes. This was subsequently confirmed for the titanium compound by Beck (85) using ^{14}N NMR. Later, however, dipole moment measurements suggested that the titanium and zirconium compounds could not both be O-bonded, in spite of their spectral similarities (411); the titanium(III) compound, $cp_2Ti(NCO)$, however, does contain the N-cyanato group (145, 220).

2. Thiocyanates

The compound $K_2[Ti(CNS)_6]$ has been reported (668, 669) and the effect of different bases on its preparation has been studied (718). The compound $(n\text{-}Bu_4N)_3[Ti(NCS)_6]$ has been shown to contain N-thiocyanato groups by comparing its crystal field parameters with those of other complexes (486); $K_2TiO(CNS)_4 \cdot 2NH_3$ has been reported (667). A series of complexes $TiL_2(NCS)_4$ with a variety of phosphine oxide-type ligands has been described, with N-thiocyanato groups in every case: the assignment is based on their infrared spectra and includes a discussion of Ti—NCS stretching frequencies (484). Equilibrium studies indicate the existence of a 1:1 adduct between NCS^- and Ti(III) (246). Orange $(NH_4)[Ti(CNS)_4]$ and violet $(NH_4)[Ti(CNS)_4(H_2O)_2]$ are 1:1 electrolytes and are oxidized to $(NH_4)[Ti(CNS)_4(OH)(H_2O)]$ (709); the low magnetic moment of the first of these compounds has been confirmed (647).

Golub et al. have shown that zirconium(IV) and hafnium(IV) form eight-coordinate complexes in solution with NCS^- alone or in the presence of DMF (329, 333). The compounds $(Et_4N)_2[M(NCS)_6]$ (M = Zr, Hf) both contain N-thiocyanato groups, as determined by infrared studies, and are isomorphous (61). Benzyl phenyl arsinic acid has been used as an extractant and, unlike most systems, hafnium complexes are extracted better than zirconium complexes in the presence of NCS^- (302).

Metallocene–thiocyanate complexes have been reported similar to the cyanate complexes described previously. There is general agreement that these are N-thiocyanato complexes, $cp_2M(NCS)_2$ (M = Ti, Zr, Hf) (145, 146, 219, 320, 648). The compound $cpTi(NCS)_3$ has been prepared, and infrared measurements show it to be as written (458); $cp_2Ti(NCS)$ is trimeric but the mode of bridging is not known (220).

3. Selenocyanates

The compound $(n\text{-}Bu_4N)_2Ti(NCSe)_6$ is known (163), and $co_2M(NSCe)_2)$ (M = Ti, Zr, Hf) are N-selenocyanato complexes (145, 146, 459); $(Et_4N)_2[M(NSCe)_6]$ (M = Zr, Hf) have been characterized (307).

D. Vanadium, Niobium, and Tantalum

1. Cyanates

Although the only cyanate compound reported for these metals is $cp_2V(NCO)_2$ (the analogous NCS and NCSe compounds are also known) which is in contrast to the O-cyanatotitanium analog (145, 261), there are numerous examples of thiocyanate complexes.

2. Thiocyanates

X-Ray structures have been determined for $(NH_4)_2VO(NCS)_4 \cdot 5H_2O$ (374) and $(Ph_4As)_2[NbO(NCS)_5]$ (424). The deviations from linearity of the NCS group in the latter case are not statistically significant, but the Nb—N bond trans to oxygen is appreciably longer than those in the cis plane (424). There are only planar NCS groups in the vanadium(IV) compound.

The compound $VO(phen)(CNS)_2$ has been formulated as an S-thiocyanato complex with $\nu_{CN} = 2040$ cm^{-1} (627). This frequency is very low and is much more characteristic of an N-thiocyanato complex, as is observed for most other thiocyanate complexes of these metals (see Table XXI in which the recorded changes in the oxidation number of the central metal and the changes in its ligational environment apparently have no effect on the thiocyanate mode of bonding). The compounds $M(NCS)_2(OR)_3bipy$ (M = Nb, Ta; R = Me, Et) are eight-coordinate, probably distorted dodecahedral, with bridging alkoxide and terminal thiocyanate groups (771).

Gutmann et al. (355–357) have investigated the effect of solvents on the ease of formation of thiocyanate complexes. In the series $[VO(NCS)_n]^{(2-n)+}$ ($n = 1$–4) complexes with low values of n are formed more readily in solvents with low donor numbers and vice versa (355, 356). The NCS⁻ is a stronger donor toward $VO(acac)_2$ than either the neutral molecules investigated or the halide ions (367).

The rate constants for the oxidation of $[V(H_2O)_6]^{2+}$ by $Fe(OH_2)^{3+}$, $FeCl^{2+}$, $FeNCS^{2+}$, and FeN_3^{2+} are 4×10^5, $(4.6 \pm 0.5) \times 10^5$, $(6.6 \pm 0.7) \times 10^5$, and $(5.2 \pm 0.6) \times 10^5$ M^{-1} sec^{1-}, respectively, at 25°C and ionic strength 1.0 M. No spectrophotometric evidence indicating that VX^{2+} species were formed in the reaction was obtained. When the oxidation is carried out with $[Cr(H_2O)_5SCN]^{2+}$ (see Section IV, E), the reaction proceeds in two stages:

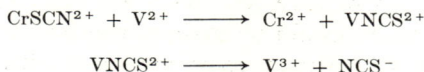

$$CrSCN^{2+} + V^{2+} \longrightarrow Cr^{2+} + VNCS^{2+}$$

$$VNCS^{2+} \longrightarrow V^{3+} + NCS^-$$

TABLE XXI

Some Thiocyanate Complexes of Vanadium, Niobium, and Tantalum

Complex	Composition	References
$VO(NCS)_2 \cdot nL$	$n = 4$ or 5; L = pyridine N-oxide	(112)
	$n = 2$; L = 4-pic N-oxide	(112)
$[VO(NCS)_5]^{3-}$		(666)
$cp_2V(NCS)_2$		(145, 261)
$[V(NCS)_6]^{3-a}$		(112, 145, 666, 774)
$[VX_a(NCS)_bL_3]^a$	X = Cl or Br; $a = 0, 1,$ or 2; $b = 3$ 2, or 1; L = CH_3CN, py, THF	(110)
$[V(CNS)_4L_2]^-$	L = py, 3-pic, 4-pic, 3,4-lut, 3,5-lut, $\frac{1}{2}$bipy, $\frac{1}{2}$phen	(361)
$[V(CNS)_3L_3]$	L = py, 3-pic, 4-pic, 3,4-lut, 3,5-lut, isoquin	(361)
$[V(CNS)_3(MeOH)_2L]$	L = Et_3N, an, quin, 2-pic, R_3P; R = Me, Et, nPr, nBu	(361)
$[V(NCS)_3(THF)_3]^-$		(109)
$[Nb(NCS)_6]^{-a}$		(114, 131, 455)
$[Nb(NCS)_5MeCN]^a$		(115)
$[Nb(NCS)_5]_2$		(115)
$[NbCl_a(NCS)_bMeCN]^a$	$a = 4–1$; $b = 1–4$	(114)
$[NbCl_a(NCS)_bEt_2O]$	$a = 4$ or 3; $b = 1$ or 2	(114)
$[NbO(NCS)_5]^-$		(424)
$[Nb(NCS)_6]^{2-}$		(455)
$[Ta(NCS)_6]^{-a}$		(113, 131, 455)
$[Ta(NCS)_5]_2$		(115)
$[Ta(NCS)_5L]^a$	L = CH_3CN, py	(113)
$[TaCl_a(NCS)_bMeCN]^a$	$a = 4–1$; $b = 1–4$	(113)
$[TaCl_a(NCS)_bEt_2O]$	$a = 4$ or 3; $b = 1$ or 2	(113)
$TaOCl_a(NCS)_b \cdot 2CH_3CN$	$a = 0–2$; $b = 3–1$	(116)
$[TaO(NCS)_4]^{-a}$		(116)
$TaSCl_a(NCS)_bC_6H_5NCCl_2$	$a = 0–2$; $b = 3–1$	(116)
$[TaS(NCS)_4 \cdot C_6H_5NCCl_2]^-$		(116)

[a] Some solvent molecules may also be present in the lattice—see the original literature.

The rates for these reactions are $7.1 \pm 0.4 \ M^{-1} \ sec^{-1}$ and $0.99 \ sec^{-1}$, respectively. At the concentrations used the rate of isomerization of $CrSCN^{2+}$ was negligible relative to these values. Sutin *et al.* (62) suggest that these results indicate that, in contrast to the V(II)–Fe(III) reactions, the reaction of $CrSCN^{2+}$ with V^{2+} proceeds via an anion-bridged intermediate. It would be entirely consistent with what is now known of vanadium–thiocyanate complexes, and with the foregoing, for this intermediate to have the structure $[(H_2O)_5V—NCS—Cr(H_2O)_5]^{4+}$.

The forward and reverse rate constants for the reaction

$$V^{2+} + NCS^- \; \rightleftharpoons \; VNCS^+$$

are $15 \pm 2\ M^{-1}\sec^{-1}$ and $1.04 \pm 0.2\sec^{-1}$ at 24°C and $\mu = 1\ M$ *(471)*. The rates for the corresponding vanadium(III) reactions have been found independently as $104 \pm 10\ M^{-1}\sec^{-1}$ and $0.7 \pm 0.1\sec^{-1}$ at 23°C and $\mu = 1\ M$ *(471)*, and $114 \pm 10\ M^{-1}\sec^{-1}$ and 1.03 ± 0.06 $\sec^{-1}$ at 25°C and $\mu = 1\ M$ *(63)*. In DMSO the forward and reverse rates for the reaction,

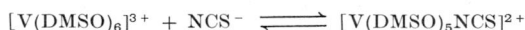

$$[V(DMSO)_6]^{3+} + NCS^- \; \rightleftharpoons \; [V(DMSO)_5NCS]^{2+}$$

are $210 \pm 10\ M^{-1}\sec^{-1}$ and $1.5\sec^{-1}$ at 25°C and $\mu = 0.15\ M$ *(478)*.

3. Selenocyanates

Only for vanadium have selenocyanate complexes been reported, and they all appear to be N-bonded. The compounds $[VO(NCSe)_4]^{2-}$ *(163, 666)* and $[V(NCSe)_6]^{3-}$ *(163, 647, 666, 685, 739)* have each been reported in the presence of various cations, and their structures confirmed by infrared and electronic spectroscopy. X-Ray data suggested that the V—Se bond distance in the latter anion lies between 5.0 and 5.1 Å, providing further evidence for this mode of bonding *(739)*. Also $cp_2V(NCSe)_2$ *(261)* has been reported and, although the authors were reluctant to define the mode of bonding in 1968, the similarity of the infrared peaks in this compound to those in other N-selenocyanato-vanadium compounds suggests that it is correctly formulated as written. Compounds $VO(NCSe)^+$, $VO(NCSe)_2$, and $VO(NCSe)_4^{2-}$ have been observed in acetone and DMF, and a number of mixed-ligand complexes with nitrogen donors (py, bipy, phen) have been reported with these species *(684, 687)*.

E. CHROMIUM, MOLYBDENUM, AND TUNGSTEN

1. Cyanates

The crystal structures of π-cpCr(NO)$_2$NCO *(165)* and π-cpMo(CO)-(PPh$_3$)$_2$NCO *(508)* have been determined. Each structure contains a linearly bound N-cyanato group with M—N, N—C, and C—O distances of 1.982(8) and 2.127(7), 1.126(4) and 1.118(14), and 1.179(5) and 1.238(16) Å, respectively. The C—O distance is particularly sensitive to vibrational motions, and its corrected value in the chromium compound is 1.27 Å. Therefore, too much should not be made of the apparent difference in these distances in the two compounds.

The anion $[Cr(NCO)_6]^{3-}$ has been characterized spectroscopically

(59) but the analogous molybdenum species is O-bonded (56). The mixed molybdenum(IV) compounds mer-MoO(NCO)$_2$L$_3$ (L = Et$_2$PhP or Me$_2$PhP) are as written (166), the assignments in each case being on the basis of infrared measurements.

The preparation of the compounds M(CO)$_5$NCO (M = Cr, Mo, W) has been achieved by a number of interesting reactions. The hexacarbonyls react with the azide ion to give the preceding products (92), for which kinetic studies suggest that nucleophilic attacks occurs at a carbon atom, followed by synchronous formation of the M—N bond and expulsion of N$_2$ (782). In other systems, it is proposed that the reaction of CO with a coordinated azide proceeds via a similar intermediate, but in the photochemical formation of π-cpMo(CO)(PPh$_3$)$_2$NCO from a molybdenum nitrosyl or azido complex, a nitrene is proposed as intermediate (508). Also Cr(CO)$_5$NCO has been formed by the reaction of NH$_2$OH or NH$_2$Cl with Cr(CO)$_6$, presumably via a hydroxamic (or N-haloamide) intermediate with subsequent loss of H$_2$O (or HCl) (88); (π-cp)$_2$W(NCO)$_2$ has been reported (344).

2. Thiocyanates

Details of X-ray structure determinations of thiocyanate complexes of these metals are given in Table XXII. Not included are the incomplete results which indicated sulfur bonding in K$_3$Cr(SCN)$_6 \cdot$4H$_2$O (793) —these findings have been superseded by the recent results on the hexa-N-thiocyanatochromate salt of a complex lanthanide cation (427).

Infrared and electronic spectra confirm the nature of the hexa-N-thiocyanatometallate anions of chromium(III) (293, 639, 647, 656, 663, 665) and molybdenum(III) (293, 489, 647, 663, 664), and molybdenum(II) (534); a variety of salts have been reported containing these anions (see the foregoing and also Refs. 30 and 528). Infrared spectra and magnetic measurements have been used to characterize [M(NCS)$_6$]$^{2-}$ (M = Mo, W) and [W(NCS)$_6$]$^-$ (386). A large number of mixed-ligand thiocyanate complexes of chromium(III) have been reported, and many of these are of the type M[Cr(CNS)$_4$L$_2$], similar to the original Reinecke's salt where L = NH$_3$.

In general, treatment of an ethanolic solution containing [Cr-(NCS)$_6$]$^{3-}$ with the appropriate ligand under reflux conditions results in the corresponding reineckate salt; some representative examples with nitrogen donors are given in Refs. 48, 308–310, 317, 633, 756, and 757. Phosphine ligands also undergo the same type of reaction (399) and with this ligand, examples of the intermediate anion [Cr(CNS)$_5$L]$^{2-}$ have also been reported (723). In view of the number of compounds that

TABLE XXII

X-Ray Data for Some Chromium and Molybdenum Thiocyanates

Compound	M—N	N—C	C—S	∠NCS	∠MNC	Reference
M[Cr(NCS)$_6$][a]	2.002	1.144	1.619	176.6	164.3	(427)
NH$_4$[Cr(NH$_3$)$_2$(NCS)$_4$]	2.05	1.14	1.80	180	180	(717)
pyH[Cr(NH$_3$)$_2$(NCS)$_4$]	1.95	1.15	1.76	180	180	(717)
C$_5$H$_{15}$NOH[Cr(NH$_3$)$_2$(NCS)$_4$][b]	1.94	1.27	1.64	180	155.5	(717)
K$_3$Mo(NCS)$_6$·CH$_3$CO$_2$H·H$_2$O	2.088(19)[e]	1.159(27)	1.630(24)	177.2[e]	169.3[e]	(271, 272)
(π-C$_3$H$_5$)Mo(bipy)(CO)$_2$NCS[c]	2.119(7)	1.163(10)	1.644(8)	175.8(5)	176.4(7)	(340)
(π-C$_4$H$_7$)Mo(phen)(CO)$_2$NCS[d]	2.146(9)	1.168(14)	1.613(12)	161.0(8)	179.2(10)	(341)

[a] M = Complex lanthanide cation.
[b] C$_5$H$_{15}$NOH = cholinium.
[c] π-C$_3$H$_5$ = π-allyl.
[d] π-C$_4$H$_7$ = π-2-Methylfallyl.
[e] These average figures obscure the fact that there are some marked differences between the coordination of the different thiocyanate groups in this compound.

have been prepared, it is perhaps surprising that so little attention has been paid to the structural aspects of these compounds. It is clear from X-ray data and spectral sources that trans isomers are formed in most circumstances, unless chelating ligands are used or when L = Ph_2EtP (*98*).

There are two reports giving full spectral data for reineckate complexes. One by Bennett *et al.* (*98*) examines the infrared data down to 70 cm^{-1} for L = NH_3, py, bipy, diars, and various phosphines, assigns the M—NCS stretching frequencies, and records the electronic transitions, although the higher-energy d-d bands are obscured by phosphine charge transfer bands. In the other, by Contreras and Schmidt (the same paper being published twice in the same journal) (*208, 209*) more limited infrared data are available but both $^4T_{2g} \leftarrow {}^4A_{2g}$ and $^4T_{2g}(F) \leftarrow {}^4A_{2g}$ transitions are observed, permitting the calculation of appropriate ligand field parameters. In the former case (*98*), it is concluded that all the complexes contain *N*-thiocyanato groups, based largely on the Cr—NCS stretching frequencies assigned at 335 to 382 cm^{-1}. In the latter case (*209*), anomalies exist in the spectrochemical series derived and these, taken with some variations in the positions of ν_{CN} and ν_{CS}, lead the authors to conclude that *N*-thiocyanato reineckate salts are formed when L = urea, 1,2-diamine cyclohexantetraacetate, thiosemicarbazide, glycinate, or alaminate and that *S*-thiocyanato complexes are obtained when L = acetamide, thiourea, ethyl xanthate, or salicylate. These two sets of results illustrate the difficulty of drawing conclusions regarding NCS coordination. The obscuring of important d-d bands in the one case or the lack of sufficient infrared data in the other mean that no firm decision can be reached at this stage, although, on chemical grounds, there seems to be no rationale for the particular ligands cited to cause the thiocyanate group to change the mode of coordination established in the starting material, $[Cr(NCS)_6]^{3-}$, nor to establish a difference in behavior from other stable chromium(III) thiocyanate complexes.

The solid state deamination of $[CrL_3](NCS)_3$ (L = en or pn) has been studied (*363*). The reaction is catalyzed by NH_4NCS, and the formation of *trans*-$[CrL—L_2(NCS)_2]NCS$ in contrast to the formation of the corresponding *cis*-chloro complex under similar conditions is tentatively ascribed to a strong trans effect for NCS$^-$ relative to a cis effect for Cl$^-$. In the former case this supposes a trans intermediate with two monodentate en or pn ligands. A large number of complexes of the type *cis*- and *trans*-$[Cren_2XY]Z_n$ have been prepared (*282*). From infrared and electronic spectral data, all the thiocyanate complexes contain *N*-thiocyanato linkages, regardless of the other ligands. The complete series of anions $[Cr(NCS)_xCN_{6-x}]^{3-}$ (x = 0–6) has been prepared and separated by gel electrophoresis (*108*): the d-d spectra show a regular variation from $x = 0$ to $x = 6$, from which it may be concluded that no drastic changes in coordination occur and that N-thiocyanato species are present throughout.

Salts of the hexa-N-thiocyanatochromate(III) anion with Cu(I), Ag(I), Cd(II), Hg(II), Tl(I), and Pb(II) contain Cr—NCS—M bridges. Infrared and electronic spectral data support this, and the latter indicate that the —NCS—M ligand forms a less covalent bond with chromium than does monodentate —NCS but that the metal-to-ligand π backbonding is increased (*779*). Similar conclusions were reached for kinetic and spectral measurements in solution (*39*) regarding the nature of the bridge in $[(H_2O)_5CrNCSHg]^{4+}$. The interactions between divalent transition metal ions and Reinecke's anion are stronger for Co(II) and Cu(II) (*583*) than for Mn(II), Ni(II), or Fe(III) (*584*) and suggest similar bridging to that described previously. For the 1:1 adduct between $[Cr(NH_3)_5NCS]^{2+}$ and Ag^+, log K is 5.11 mole^{-1} (*617*). A spectrophotometric study of the interaction of Hg^{2+} and Ag^+ ions with $[Cr(NH_3)_5NCS]^{2+}$ and cis-$[Cren_2(NCS)_2]^+$ using the method of continuous variations indicated the formation of 1:1 adducts (*773*).

Chromium(VI) compounds are reduced to chromium(III) species in molten KNCS with the formation of a variety of products (*433*). The reaction between NCS⁻ and $HCrO_4^-$ in solution indicates the formation of an intermediate $[CrO_3(CNS)]^-$ species (*562*), which is tentatively assumed to be N-bonded (*545*).

In contrast to chromium, little is known about mixed-ligand thiocyanate complexes of molybdenum and tungsten. Infrared mull spectra indicate the structures of Mo(bipy)(NCS)₄ (*130*) and *mer*-MoO(NCS)₂-(PR₃)₃ (PR₃ = Et₂PhP or Me₂PhP) (*166*) and, when coupled with magnetic measurements, are the basis for ascribing the anion in (pyH)₄[Mo₂O₄(NCS)₆] as a μ-dioxo compound with terminal N-thiocyanato groups (*534*). Other workers have reported related compounds without confirming the mode of thiocyanate coordination (*343, 644*). Some compounds MoX(CNS)₅2L (X = Cl or NCS; L = Me₂CP, MeCO·Et, or C₄H₈O₂) have been prepared (*306*).

The compounds WL₂(NCS)₄ (L = py or ½bipy) have been characterized (*130*), and a series of tungsten(V) (*750*) and tungsten(VI) (*306, 751*) compounds prepared without structural data.

The organometallic complexes of these metals with thiocyanate are important because linkage isomers have been obtained (*690*). Some of the relevant compounds are included in Table XXIII. The compound π-cpW(CO)₃SCN, which was characterized by the positions of its CN and CS frequencies and by the integrated intensity of the former, resists all attempts to convert it to the N-thiocyanato isomer, unlike

TABLE XXIII

Some Organometallic Thiocyanates of Chromium, Molybdenum and Tungsten

Formal Oxidation State	Chromium thiocyanate	References	Molybdenum thiocyanate	References	Tungsten thiocyanate	References
0	$[Cr(CO)_5NCS]^-$	(95, 491)	$[Mo(CO)_5NCS]^-$	(633)	$[W(CO)_5NCS]^-$	(633)
0 and 1	$(OC)_5Cr$—NCS—$Cr(CO)_5$	(96, 491)				
I	$Cr(CO)_5NCS$	(96, 491, 787)				
II	$(\pi\text{-cp})Cr(NO)_2NCS$	(690)	$(\pi\text{-cp})\ Mo(CO)_3NCS$	(690)	$(\pi\text{-cp})W(CO)_3SCN$	(690)
			$(\pi\text{-cp})\ Mo(CO)_3SCN$	(690)	$[W(CO)_2(NCS)_2]_n{}^a$	(206)
			$[Mo(CO)_2(NCS)_2]_n{}^a$	(206)	$W(CO)_2(PPh_3)_2(NCS)_2$	(206)
			$Mo(CO)_2(PPh_3)_2(NCS)_2$	(206)	$W(CO)_2dpm_2(NCS)_2$	(206)
			$Mo(CO)_2dpm_2(NCS)_2$	(206)		
			$\pi\text{-}C_3H_5Mo(CO)_2(CH_3CN)_2NCS$	(247)		
			$[\pi\text{-}C_3H_5Mo(CO)_2(CH_3CN)NCS]_2{}^b$	(247)		
			$[\pi\text{-}C_3H_5Mo(CO)_2(PhCN)NCS]_2{}^b$	(247)		
			$\pi\text{-}C_3H_5Mo(CO)_2bipyNCS$	(340)		
			$\pi\text{-}C_4H_7Mo(CO)_2phenNCS$	(341)		
IV	—	—	$(\pi\text{-cp})_2Mo(NCS)_2$	(344)	$(\pi\text{-cp})_2W(NCS)_2$	(344)
					$(\pi\text{-cp})_2W(NCS)(SCN)$	(344)

[a] These molecules were not isolated but their solutions show similar spectra to those of their analogs: solvent molecules may be coordinated since there is no indication of bridging NCS from the infrared data.

[b] These dimers almost certainly contain bridging NCS groups by comparison with the preceding compound.

the similarly characterized molybdenum analog that isomerizes in solution in a matter of hours (*690*). Other *S*-thiocyanato complexes of molybdenum, isomers of the compounds listed, may be able to be prepared by modifying the careful preparative techniques of Sloan and Wojcicki (*690*) and by similarly taking advantage of the kinetic factors that permitted the isolation of π-cpMo(CO)$_3$SCN. The tungsten compound on the other hand, appears to be stable, and it suggests that other similar *S*-thiocyanato complexes may be able to be prepared for that element. Thus, treatment of $(\pi$-cp$)_2$MoCl$_2$ with aqueous KNCS gave $(\pi$-cp$)_2$Mo(NCS)$_2$ but the similar reaction with $(\pi$-cp$)_2$WCl$_2$ gave a mixture of $(\pi$-cp$)_2$W(NCS) and $(\pi$-cp$)_2$W(NCS)(SCN) (*344*). No attempts were made to interconvert these isomers but their existence seems to lend support to the foregoing prediction.

The reduction of [Co(NH$_3$)$_5$NCS]$^{2+}$ with [Cr(H$_2$O)$_6$]$^{2+}$ to give [(H$_2$O)$_5$CrNCS]$^{2+}$ with extensive exchange of free NCS⁻ was studied in 1958 (*177*). Subsequently it was shown that the similar reduction of *trans*-[Coen$_2$X(NCS)]$^+$ proceeded via both remote and adjacent attacks at the coordinated NCS resulting in both CrSCN$^+$ and CrNCS$^+$ species in solution (*363, 364*). The two isomers have been characterized in solution from their electronic spectra, and kinetic studies have shown that the former isomerizes to the latter at a rate that implies either an intimate ion pair as an intermediate or, perhaps, that the chromium "slides" along the filled π orbitals of the thiocyanate. The difference in stabilities of the two isomers is ascribed mainly to the difference in their rates of aquation (*363, 364*). The isomerization of [Cr[H$_2$O)$_5$SCN]$^{2+}$ is catalyzed by Hg^{2+}, which attacks the coordinated —SCN at the sulfur to decompose in both of two ways (*586*):

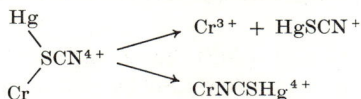

$$\begin{array}{c} \text{Hg} \\ \diagdown \\ \qquad \text{SCN}^{4+} \\ \diagup \\ \text{Cr} \end{array} \quad \longrightarrow \quad \begin{array}{l} \text{Cr}^{3+} + \text{HgSCN}^+ \\[1em] \text{CrNCSHg}^{4+} \end{array}$$

The severance of the Cr—S bond to form the stable HgSCN$^+$ species requires no further comment, and the isomerism is assumed to proceed by either of the routes just described. The stable [(H$_2$O)$_5$Cr—NCS—Hg]$^{2+}$ species has been separately characterized in solution (*39*). More recently the attack of Cr^{2+} on [Co(NH$_3$)$_5$NCS]$^{2+}$ has been shown to be remote only, whereas the attack on the linkage isomer is both adjacent and remote (*677*). The reduction of FeNCS^{2+} by Cr^{2+} in the presence of NCS⁻ leads to kinetic and spectroscopic evidence for [Cr(H$_2$O)$_4$(NCS)(SCN)]$^+$. The analysis of the results is made more difficult by the uncertainty concerning the ratio of geometric isomers present and the mechanism by which this species is formed (*129*).

There is spectroscopic evidence for the association of NCS⁻ with a number of chromium(III) complexes (603, 790), and such a stable ion pair has been identified in the substitution of H_2O in $[Cr(NH_3)_5H_2O]^{3+}$ by NCS⁻. The further kinetic results have been interpreted on the basis of a rate-limiting loss of coordinated H_2O, followed by the collapse of the solvation shell to fill the vacancy created (264, 381). Ion pairs were also indicated in the reaction of NCS⁻ with $[Cren_2Cl_2]^+$ in methanol but no further mechanistic conclusions were reached (382). The rate of substitution of H_2O in $[Cr(H_2O)_6]^{3+}$ by NCS⁻ is catalyzed by methanol due to a more favorable dissociative activation mode involving coordinated MeOH (67). The same reaction is also catalyzed by the sulfite ion, and, although evidence is offered that the formation of $[Cr(H_2O)_5OSO_2]^+$ is crucial to the subsequent substitution, the mechanism for this stage is not clear (178). Exchange between NCS⁻ and $[Cr(H_2O)_5NCS]^{2+}$ has been studied using ³⁵SCN⁻ (449), and the displacement of NCS⁻ from various complexes has also been reported (40, 795). Much of the preceding work was given direction by Adamson's earlier work in which a photochemical exchange of ligands in an ion pair was shown to take place between NCS⁻ and H_2O in either $[Cr(H_2O)_6]^{3+}$ or $[Cr(NH_3)_5H_2O]^{3+}$ (12).

3. Selenocyanates

A number of systems containing the $[Cr(NCSe)_6]^{3-}$ ion are known, and spectroscopic measurements confirm the mode of coordination (132, 530, 615, 663, 665, 738). Compounds with $[Mo(NCSe)_6]^{3-}$ have been studied similarly (647, 663, 665). Yellow and red forms of $Cr(py)_3(NCSe)_3$ have been reported (738), but Cr—NCSe linkages are present in both cases (132, 738).

Jennings and Wojcicki (409) have reported the preparation of π-cpCr(NO)$_2$NCSe and π-cpM(CO)$_3$SeCN (M = Mo, W); the mode of bonding is determined by the position of CN and CSe stretching frequencies, and by integrated intensity measurements. No indication of linkage isomerism was observed in the carbonyl compounds [cf. π-cpMo(CO)$_3$CNS], and the chromium compound readily lost Se to give the cyano complex.

F. MANGANESE, TECHNETIUM, AND RHENIUM

1. Cyanates

Tetrahedral anionic complexes $[Mn(NCO)_4]^{2-}$ have been reported and characterized by infrared and electronic spectral data (292, 296,

639). The octahedral anion $[Re(OCN)_6]^{2-}$ has been reported and its structure assigned on the basis of the positions of the infrared frequencies (*56*). Mixed-ligand complexes of manganese(II) cyanate have been reported with 2,2'-bipyridine and 1,10-phenanthroline (*335*) and with urotropine (*735*)—in all cases N-bonded cyanato complexes are formed. The compounds $MnL_2(NCO)_2$ (L = 3-cyanopyridine or 4-cyanopyridine) contain Mn—N—Mn bridges rather than bridges of the type Mn—NCO—Mn (*557*). The thermal analysis of $MnL_n(NCO)_2$ (L = py and related ligands) have been reported (*496*).

Several organometallic cyanates of these metals have been reported. In addition to substitution methods, these may be prepared by nucleophilic attack on a coordinated CO by a suitable nitrogen-containing nucleophile. Thus, $Mn(CO)_3(PMe_2Ph)_2Br$ gives $Mn(CO)_2(PMe_2Ph)_2$-$(N_2H_4)(NCO)$ when treated with hydrazine (*536*) and the $[Mn_2(CO)_6$-$(N_3)_x(NCO)_{3-x}]^-$ anion is formed by the reaction of N_3^- with $Mn(CO)_5Br$ in THF at temperatures above ambient (*520*). The analogous trisisocyanato complex is obtained when NCO$^-$ is used, and $Mn(CO)_3(PPh_3)_2NCO$ has been prepared by a nucleophilic displacement of Br$^-$ (*520*). Compounds $Re(CO)_5NCO$ (*34*) and $Re(CO)_4(NCO)_2$ (*646*) have been prepared, as have many of the rhenium analogs to the foregoing (*520*). The reaction of hydrazine with a variety of rhenium carbonyl compounds gives a similar variety of N-bonded cyanato complexes—the reaction is believed to proceed via a carbamoyl intermediate (*535*).

2. Thiocyanates

Both tetrahedral and octahedral anions are known for manganese, and $[Mn(NCS)_4]^{2-}$ and $[Mn(NCS)_6]^{4-}$ have been assigned from their electronic and infrared spectra (*293, 297, 647, 665, 666*). Reduction of an acidic solution of TcO_4^- with thiocyanate ions led to various thiocyanate–technetate complexes which, when carefully oxidized with TcO_4^-, resulted in the isolation and characterization of $(Me_4N)_2$-$[Tc(NCS)_6]$ (*672*); the bonding is confirmed by preliminary X-ray results indicating a Tc—S distance of 4.7 Å (*372*). Potentiometric titration of acidic ReO_4^- with $Hg_2(ClO_4)_2$ in the presence of NH_4NCS had indicated reduction to give an uncharged $Re(CNS)_4$ species in solution (*719*), and subsequently anionic thiocyanate complexes of rhenium(III) (*214*), (IV) (*54, 55, 214*), and (V) (*55*) have been reported but with disagreement over the mode of bonding present. Cotton *et al.* (*214*) concluded that the ions $[Re_2(NCS)_8]^{2-}$ and $[Re(NCS)_6]^{2-}$ were formed from the position and broad nature of ν_{CN} and from the position of δ_{NCS}.

Bailey and Kozak (54) prepared $[Re(CNS)_6]^{2-}$ in a fused salt and, although they also observed a strong broad CN stretching frequency, they concluded that $[Re(SCN)_6]^{2-}$ had been formed because, having no onium cations to absorb around 700 cm^{-1}, they assigned a band in that region to an S-bonded CS stretching frequency. Subsequently the latter authors reported $[Re(SCN)_6]^-$ and offered further evidence in the form of integrated intensities (55) to support their assignments for $[Re(SCN)_6]^{2-}$. When these measurements were repeated at a later stage, Bailey et al. (60) reported the compounds as $[Re(NCS)_6]^{2-}$ and $[Re(NCS)_6]^-$. Thus it was finally concluded that these rhenium thiocyanato complexes are all N-bonded, in agreement with the interpretation of their electronic spectra (663). The difficulties in reaching this conclusion illustrate the hazards of using infrared data to determine the mode of bonding of NCS$^-$; the results still await unambiguous confirmation from X-ray analysis.

The mercury(II) compounds of $[Mn(NCS)_4]^{2-}$ contain Mn—NCS—Hg bridges (293, 654). Anhydrous $Mn(NCS)_2$ probably contains manganese(II) octahedrally coordinated by two nitrogen and four sulfur atoms, with triply ligating NCS$^-$ (287). The Mn—NCS—Mn bridges are present in $Mn(py)_2(NCS)_2$, and Mn is in an octahedral environment, as in $Mn(py)_4(NCS)_2$ (201). The tetrahedral molecule MnL_2-$(NCS)_2$ (L = HMPT) have been reported and assigned from spectral data (485). Preliminary X-ray results show that $Mn(tu)_2(NCS)_2$ is isostructural (552) with the trans-octahedral nickel(II) compound (with S-bridging tu and terminal—NCS) (see Section IV, I). As part of an investigation into the magnetic properties of the MS_4N_2 system, $Mn(tu)_2(NCS)_2$ has been examined (287), and $Mn(ROH)_2(NCS)_2$ (R = Me, Et) was shown to be very similar but with —NCS— bridges (287). The compound $MnL_4(NCS)_1$ (L = N-n-butylimidazole) has been reported (631). Preliminary results suggest that $Mn(DMF)_4(NCS)_2$ contains N-thiocyanato groups (688). Although the structural data are not entirely adequate, the complexes $MnL(NCS)_2$ (L = N,N'-di-(3-aminopropyl)piperazine (319) and 2,3,5,6-tetrakis-(6-methyl-2-pyridyl)-pyrazine (337)) are each probably five-coordinate with N-thiocyanato coordination. A complex with urotropine has been reported (735). The thermal decomposition of $Mnpy_4(NCS)_2$ has been studied (94, 425, 496).

Magnetic measurements indicate that the previously discussed $[Re_2(NCS)_8]^{2-}$ has a dinuclear structure with a metal–metal bond (214). The reaction of NCS$^-$ with $Re_2Cl_8^{2-}$ in the presence of Ph_3P results in the formation of $[Re_2(NCS)_8(PPh_3)_2]^{2-}$, but here the magnetic data indicate no metal–metal bonding. The infrared spectra and other physical data are consistent with a dimeric anion in which the two rhenium atoms are bridged by two NCS groups (214). The compound $(n\text{-}Bu_4N)_3[Re_2(NCS)_8(CO)_2]$ has been prepared and, although it probably contains at least some N-thiocyanato groups, its structure remains unknown (214).

The oxygen-bridged species $Cs_3H[Re_2O(CNS)_{10}]$ and $Cs_4H_2[Re_3O_2\text{-}(CNS)_{14}]$ have been reported and formulated as S-thiocyanato complexes from ambiguous infrared data (776). Exchange of NCS$^-$ has been studied with these two species: in the former compound, one NCS$^-$ exchanges faster than the other nine (775), and in the latter, two thiocyanate groups exchange faster than the other twelve (777).

Some rhenium(V) compounds have been reported of the form $ReXY(PPh_3)_2(NCS)_2$ (X = Y = O; X = O, ·Y = OH, OEt, or Cl); $ReO(OEt)py_2(NCS)_2$ was also reported (304). Also $M[Re(CN)_5(CNS)]$ (M = K, Ag, Cu) have been prepared, but no structural data are available (651).

The anionic manganese carbonyl complexes $[Mn(CO)_4(NCS)_2]^-$ (729) and $[Mn_2(CO)_6(CNS)_4]^{2-}$ (278) are known, the latter containing both bridging —NCS— and terminal —NCS groups (278); $[Re(CO)_3(NCS)_3]^{2-}$ is also known (278, 280). However, the most interesting compounds in this category are those prepared by Wojcicki and Farona (279, 280), which are listed in Table XXIV.

Infrared spectra suggest that the solid compound prepared by the reaction of thiocyanogen chloride and sodium pentacarbonylmanganate(-I) is $[Mn(CO)_5SCN]$. The N-bonded linkage isomer is formed in acetonitrile, and an equilibrium mixture of the two isomers exists in dichloromethane, 1,2-dichloroethane, chloroform, and ethyl acetate. The relative positions of equilibrium, as indicated by the intensities of various infrared bands, change on adding acetonitrile or on altering the temperature. These changes are reversible for all the solutions and, on removal of the solvent, solid $Mn(CO)_5SCN$ may be recovered with little if any decomposition (279).

When neutral ligands are reacted with the preceding compound, a further series of N- and S-bonded products are obtained (280) (see Table XXIV). The significance of these products will be discussed later, but it is important at this stage to note the geometric isomerism associated with the linkage isomerism for $mer\text{-}Mn(CO)_3(AsPh_3)_2NCS$ and $fac\text{-}Mn(CO)_3(AsPh_3)_2SCN$, and for the corresponding SbPh$_3$ complexes. No N-bonded fac product could be detected even when the reaction was carried out in acetonitrile, a solvent already known to promote N-thiocyanato bonding in $Mn(CO)_5NCS$.

Further evidence for this solvent effect was obtained by comparing the infrared spectra of $cis\text{-}Mn(CO)_4(AsPh_3)SCN$ in chloroform and

TABLE XXIV

SOME THIOCYANATE CARBONYL COMPLEXES OF MANGANESE

Compound	References	Compound	References
Mn(CO)$_5$NCS[a]	(279)	Mn(CO)$_5$SCN	(279)
cis-[Mn(CO)$_4$(NCS)$_2$]	(729)		
		cis-Mn(CO)$_4$(PPh$_3$)SCN	(280)
		cis-Mn(CO)$_4$(AsPh$_3$)SCN	(280)
		cis-Mn(CO)$_4$(SbPh$_3$)SCN	(280)
mer-Mn(CO)$_3$(PPh$_3$)$_2$NCS[b]	(280)		
mer-Mn(CO)$_3$(AsPh$_3$)$_2$NCS[b]	(280)	fac-Mn(CO)$_3$(AsPh$_3$)$_2$SCN	(280)
mer-Mn(CO)$_3$(SbPh$_3$)$_2$NCS[b]	(280)	fac-Mn(CO)$_3$(SbPh$_3$)$_2$SCN	(280)
fac-Mn(CO)$_3$py$_2$NCS	(280)		
fac-Mn(CO)$_3$(γ-pic)$_2$NCS	(280)		
fac-Mn(CO)$_3$(p-tol)$_2$NCS	(280)		
fac-Mn(CO)$_3$(p-can)$_2$NCS	(280)		
fac-Mn(CO)$_3$(p-fan)$_2$NCS	(280)		
fac-Mn(CO)$_3$bipyNCS	(280)		
fac-Mn(CO)$_3$diphosNCS	(280)		

[a] Exists in solution only.

[b] The mer complexes can exist in two forms. The authors (280) consider the isomer with CO trans to CNS to be more probable than that with a noncarbonyl ligand trans to CNS, and cis to its auxiliary.

acetonitrile. The spectra were significantly different, and could be varied reversibly by using mixed solvents of different proportions. The S-bonded complex was recovered from the solutions without decomposition on evaporation. The most plausible explanation for these and associated spectral changes involves linkage isomerism to give cis-Mn(CO)$_4$(AsPh$_3$)NCS in acetonitrile solution. The spectra of the PPh$_3$ and SbPh$_3$ analogs in these solvents resemble those of cis-Mn(CO)$_4$-(AsPh$_3$)SCN, but additional complexities in these cases suggest that the isomerization that goes to virtual completion for the arsine is here arrested at an intermediate stage (280).

3. Selenocyanates

Tetrahedral [Mn(NCSe)$_4$]$^{2-}$ and octahedral [Mn(NCSe)$_6$]$^{4-}$ anions are known (162, 295, 647, 655, 656), and [Re$_2$(NCSe)$_8$]$^{2-}$ has been tentatively identified (377). There seem to be no reports of similar Tc compounds, nor indeed of any mixed-ligand selenocyanate complexes of either technetium or rhenium. Some mixed-ligand complexes of manganese(II) selenocyanate have been prepared, and in every case so far described the selenocyanate is N-bonded. Analogous to the thiocyanate complexes, $MnL_2(NCSe)_2$ (L = HMPT) is tetrahedral (485), $MnL(NCSe)_2$ [L = 2,3,5,6-tetrakis-(6-methyl-2-pyridyl)pyrazine] is probably five-coordinate (337), and $Mn(DMF)_4(NCSe)_2$ is octahedral (688). A complex with urotropine is isostructural with the corresponding thiocyanate (735), and a series of complexes $ML(NCSe)_2$ (L = dioxane, 3 THF, 2 phen, or 2 bipy) has been characterized (700).

The organometallic anions $[Mn_2(CO)_6Cl_2(NCSe)_2]^{2-}$ and $[Mn_2(CO)_6(NCSe)_4]^{2-}$ are each believed to contain terminal N-selenocyanato groups, and the latter also has two bridging selenocyanates (278).

G. IRON, RUTHENIUM, AND OSMIUM

1. Cyanates

The tetrahedral anions $[Fe(NCO)_4]^-$ (296) and $[Fe(NCO)_4]^{2-}$ (298) have been characterized from vibrational and electronic spectra. Mixed-ligand complexes $FeL_4(NCO)_2$ [L = py (137, 732), 4Mepy, isoquin (137), 3CN-py (557)] are octahedral with N-cyanato coordination, and the compounds containing six molecules of pyridine and five of 4-methylpyridine are correctly formulated $[Fepy_4(NCO)_2] \cdot 2py$ and $[Fe(4-Mepy)_4(NCO)_2] \cdot 4$-Mepy, respectively (137). The system $FeL_2(NCO)_2$ (L = 3-CN-py, 4-CN-py) is also octahedral with cyanate bridges of the form Fe—N—Fe (557). In methanol, there is some evidence that $[Fe(H_2O)_6](ClO_4)_2$ is oxidized to an iron(III) species by an excess of cyanate ions (618).

The compounds $[Ru(NH_3)_5NCO]^{2+}$ (490) and $Ru(NCO)(CO)(NO)(PPh_3)_2$ (476) have been prepared. Also π-cpFe(CO)$_2$NCO is formed either by nucleophilic attack of N_3^- on a coordinated CO, followed by elimination of N_2, or by nucleophilic displacement of CO by NCO⁻ (33).

Kinetic studies on the hydrolysis of $[Ru(NH_3)_5NCO]^{2+}$ to give $[Ru(NH_3)_6]^{3+}$ and CO_2 suggest that the reaction proceeds via N protonation of the coordinated NCO⁻, followed by addition of H_2O to give a carbamic acid complex, which subsequently loses H_2O and CO_2 to give the product (290).

2. Thiocyanates

Reports in the early literature of a black and a yellow isomer of $Fepy_4(NCS)_2$ have been disproved by the X-ray analysis of both forms. The black compound is crystallographically equivalent to the yellow form which is a trans-N-thiocyanato complex with Fe—N, N—C, and C—S distances of 2.088, 1.140, and 1.604 Å, respectively (696). A transpentagonal bipyramidal structure is found in $[FeL(NCS)_2]ClO_4$, where

L is the macrocyclic pentadentate ligand, 2,13-dimethyl-3,6,9,12,18-pentaazabicyclo[12.13.1]octadeca-1(18),2,12,14,16-pentaene. Here the Fe—N distance is 2.01 ± 0.02 Å (285).

The octahedral anions $[M(NCS)_6]^{3-}$ have been characterized by spectral methods for iron (293, 297, 647, 663, 665), ruthenium (647, 662, 663, 665), and osmium (647, 663, 665). The anion $[Fe(NCS)_6]^{3-}$ has the characteristic charge-transfer band associated with mixtures of iron(III) and NCS^- in analysis (297). The tetrahedral $[Fe(NCS)_4]^{2-}$ is also well-characterized (293, 297, 647, 663). The mixed complex $Fe(NCS)_4Hg$ with an extended network of Fe—NCS—Hg bridges has been studied (293, 654).

Drickamer et al. (262) have made the interesting observation from Mössbauer measurements that iron(III) compounds are reversibly reduced to iron(II) at high pressures. They record that the asymmetric spectrum of a compound they formulate as $K_3Fe(SCN)_6$ shows an increasing proportion of iron(II) with pressure and, on pressure release, a symmetric spectrum that is regenerated on repeating the experiment. The authors ascribe these changes to reversible reduction associated with linkage isomerism. However, since the original formulation of $K_3Fe(NCS)_6$ as the S-thiocyanato complex was incorrect, further experiments are necessary before this interesting interpretation can be confirmed.

A number of iron–thiocyanate complexes with different stereochemistries are listed in Table XXV; the thiocyanate group is N-bonded in every case. The anomalous magnetic behavior of cis-Fe(phen)$_2$(NCS)$_2$ (64) has been interpreted in terms of a thermal equilibrium between $^5T_2$ and $^1A_1$ ground states: the low-temperature, low-spin form shows significant changes in its infrared spectrum relative to that of the high-spin form, but, since no phase change is observed using X-ray techniques, this is not ascribed to linkage isomerism but rather to changes in the strength of bonding arising from the change in radius of the iron (456). The same phenomenon occurs for cis-Fe(bipy)$_2$(NCS)$_2$, and here the structures of the high- and low-spin forms have been confirmed by an X-ray structural determination (466). Other illustrations of this type of behavior are known for N-thiocyanato–iron(II) compounds. For example, Fe(phen)py$_2$(NCS)$_2$ behaves similarly, although Fe(py)$_4$(NCS)$_2$ shows no such variations with temperature (697). Mössbauer (698) and low-frequency infrared (716) measurements have been reported for these compounds.

The oxidation of $[Fe(CN)_5CNS]^{4-}$ with excess $[Fe(CN)_6]^{3-}$ results in a transient purple species which changes, via an isosbestic point, to a blue solution (702) characteristic of the species obtained directly from

TABLE XXV

SOME MIXED-LIGAND COMPLEXES WITH IRON THIOCYANATE

Complex	Composition	Structure	References
Iron(III)			
$FeL_3(NCS)_3$	L = R_3PO, R_2SO, or substituted pyridine N-oxides	Octahedral	(704)
$Fe(Ph_3PO)_2(NCS)_3$		Trigonal bipyramidal	(215)
$[Fe(L—L)_2(NCS)_2]$ NCS	L—L = phen, bipy	Octahedral	(236)
$Fe_2(L—L)_4O(NCS)_4$	L—L = phen, bipy	Oxygen-bridged	(236)
Iron(II)			
$Fe(phen)_2(NCS)_2$		Cis octahedral	(64, 456, 658)
$Fe(bipy)_2(NCS)_2$		Cis octahedral	(466)
$Fe(phen)py_2(NCS)_2$		Octahedral	(697)
$FeL_4(NCS)_2$	L = py, 4pic, isoquin	Trans octahedral	(137, 201, 732)
$[Fe(py)_4(NCS)_2]2py$		Octahedral	(137)
$[Fe(4pic)_4(NCS)_2]4pic$		Octahedral	(137)
$Fe(tu)_2(NCS)_2$		Tetragonal MS_4N_2	(287)
$FeL(NCS)_2$	L = pmp [a]	Trigonal bipyramidal	(231)
	L = pnp [b]	Trigonal bipyramidal	(559)
	L = L′ [c]	?	(337)

[a] pmp = 2,6-di(diphenylphosphinomethyl pyridine)
[b] pnp = 2,6-di(diphenylphosphinoethyl pyridine)
[c] L′ = 2,3,5,6-tetrakis-(6-methyl-2-pyridyl)pyrazine

$[Fe(CN)_5NH_3]^{2-}$ and NCS^- (405). Stasiw and Wilkins (702) tentatively suggested that the iron(II)–thiocyanate complex is N-bonded and that rapid oxidation gives an unstable N-thiocyanato–iron(III) complex which rearranges to the blue linkage isomer. However, orange $[Fe(CN)_5NCS]^{3-}$ has been characterized, and its solutions range in color from orange in acetone to purple in ethanol or methanol. No evidence bearing on the possibility of linkage isomers was available (360).

The thermal decomposition of $Fepy_4(NCS)_2$ has been investigated (94, 425). The stepwise formation constants of $Fe(NCS)^+$ and $Fe(NCS)_2$ in CH_3CN have been measured (469).

[14]N Nuclear magnetic resonance spectroscopy was used to characterize the bonding in $K_2[Ru(NCS)_5NO]$ (390). When heated at 70°C for

70 hrs under nitrogen, $[Ru(NH_3)_5SCN]^{2+}$ isomerizes to $[Ru(NH_3)_5-NCS]^{2+}$ as indicated by changes in the UV spectra corresponding to the different positions in spectrochemical series of —NCS and —SCN; the infrared spectra were also different (490). The polynuclear ions $[Ru_2N-(NCS)_8(H_2O)_2]^{3-}$, $[Os_2N(NH_3)_8(NCS)_2]^{2+}$, and $[Ru_2N(NH_3)_6(NCS)_3-H_2O]^{2+}$ are believed to contain a linear M—N—M unit with N-bonded thiocyanates (202). Also $[Ru(NCS)(CO)(NO)(PPh_3)_2]$ (476) and $[Ru(O_2)(NCS)(NO)(PPh_3)_2]$ (342) have been characterized by infrared measurements—the latter compound is an efficient catalyst for the homogeneous oxidation of Ph_3P (342).

Aerial oxidation of $[\pi\text{-cpFe}(CO)_2]_2$ in the presence of hexafluoro-phosphoric acid and KNCS leads to the linkage isomers, $\pi\text{-cpFe}(CO)_2$-NCS and $\pi\text{-cpFe}(CO)_2$SCN, which may be separated by column chromatography. No appreciable isomerization was detected in solution, but the S-bonded complex is converted to its N-bonded isomer by heating the solid. Sloan and Wojcicki suggest that isomerism may involve either a dimeric or polymeric intermediate with bridging NCS groups (690). This suggestion gains some support from more recent work (413) which showed that the most convenient conditions for the oxida-tion of $[\pi\text{-cpFe}(CO)_2]_2$ required iron(III) in acetonitrile or acetone, and that the reaction proceeded via an intermediate $[\pi\text{-cpFe}(CO)_2S]^+$ (S = solvent); addition of KNCS gave the two linkage isomers, which were separated as described previously. The solvent molecule is, therefore, incorporated on rupturing the thiocyanate bridge in a random fashion and prevents further isomerization. Also $[Fe(CO)_5NCS]^-$ has been prepared by a photolytic reaction, and no evidence was found for linkage isomerism (638).

The reaction of $Os_3(CO)_{12}$ with Ph_3PAuX (X = Cl, Br, I, or SCN) gives $Os_3(CO)_{19}(AuPPh_3)X$. The basic framework of the molecule is believed to be an Os_3 triangle with $AuPPh_3$ and X each bridging two $Os(CO)_3$ units. When X = SCN, ν_{NC} indicates S-bonding. If the com-pound is left on an alumina column for 1 hr, decomposition occurs to give $[Os_3(CO)_{10}(AuPPh_3)_2S_2]$ (200). The nature of the SCN-bridging is not discussed by the authors, but the foregoing would be consistent with an Os—S—Os structure which hitherto has only been suggested when N is also involved with a third center (see, for example, Ref. 287).

The rate of formation of $[Fe(H_2O)_5NCS]^{2+}$ by displacement of water from $[Fe(H_2O)_6]^{3+}$ by NCS^- has been reported as $127 \pm 10\ M^{-1}$ sec^{-1} (25°C, $\mu = 0.4$) (97) or as $122 \pm 6\ M^{-1}\ sec^{-1}$ (25°C, $\mu = 3.0$) (207). In solutions of high $[NCS^-]$, the exchange of NCS^- with $[Fe(NCS)_4]^-$, which is assumed to predominate, has been studied by

[14]N NMR line broadening: the degree of broadening is interpreted as showing Fe—N bonding (254). The cation $[Fe(H_2O)_6]^{2+}$ exchanges separately with both $FeNCS^{2+}$ and $Fe(NCS)_2^+$ (232). The reaction of NCS^- is faster with $[Fe(H_2O)_6]^{3+}$ than with $[Fe(H_2O)_6]_2^+$ (207), but the former is catalyzed by iron(II) $(207, 483)$. It is not possible to assess the relative contributions of adjacent and remote attack but the kinetic data do suggest that linkage isomerism does not provide a major pathway to $[Fe(H_2O)_5NCS]^{2+}$ (207). The kinetics of the reduction of $[Co(NH_3)_5SCN]^{2+}$ (281) and of $[Co(NH_3)_5NCS]^{2+}$ (273) by Fe^{2+} have been compared, and show rate constants of 0.12 and $< 3 \times 10^{-6}$ $M^{-1}sec^{-1}$, respectively. The possibility of adjacent and remote attacks was not considered specifically, but no evidence is offered for the formation of $[Fe(H_2O)_5SCN]^{2+}$, so that the fast reduction of $[Co(NH_3)_5SCN]^{2+}$ perhaps proceeds via remote attack, and the slower reduction of $[Co(NH_3)_5NCS]^{2+}$ by adjacent attack. In DMSO the second-order rate constant for the formation of $[Fe(DMSO)_5CNS]^{2+}$ is 670 ± 10 M^{-1} sec^{-1} (477).

The substitution of pyridine by NCS^- in optically active cis-$[Ru(phen)_2py_2]^{2+}$ in dry acetone proceeds by a first-order reaction with retention of configuration; the mode of coordination of NCS^- in the product was not reported $(120–122)$. In aqueous solution, H_2O is displaced by NCS^- from $[Ru(H_2O)(bipy)terpy]^{2+}$ and $[Ru(H_2O)bipy_2]^{2+}$ in second-order reaction with k_2 equal to 1.36×10^{-2} and 6.57×10^{-2} 1 mole^{-1} min^{-1}, respectively. Again, the mode of coordination was not reported (237).

3. Selenocyanates

The anions $[Fe(NCSe)_4]^{2-}$ $(162, 295, 647, 663, 665)$, $[Fe(NCSe)_6]^{4-}$ $(295, 663)$, and $[Fe(NCSe)_6]^{3-}$ $(162, 663, 665)$ have been characterized by spectral methods. No anionic selenocyanate complexes of ruthenium or osmium appear to have been reported.

Since —NCSe and —NCS are close to each other in the spectrochemical series, it is not surprising that a number of selenocyanate complexes of iron(II) have been prepared, similar to the thiocyanate complexes already described, with a view to studying the magnetic crossover from ${}^5T_2$ to ${}^1A_1$. These and some other mixed-ligand selenocyanate complexes are listed in Table XXVI; invariably they contain N-bonded selenocyanate groups. One compound deserves special mention, namely, the product obtained by the thermal decomposition of $[Fe(bipy)_3]$-$(NCSe)_2$. This product is formulated as $[Fe^{sp}(bipy)_2 (NCSe)_2]_2[Fe^{sf}(bipy)_2$

$(NCSe)_2](bipy)$, where Fe^{sp} and Fe^{sf} denote spin-paired and spin-free iron(II), respectively (457).

The only selenocyanate-containing complex of ruthenium or osmium is believed to be $[Ru(NH_3)_5SeCN](ClO_4)_2$ which was characterized by comparing the UV data with expected band positions from the spectrochemical series (490).

The reaction of KSeCN with π-cpFe(CO)$_2$Cl or of Se(SeCN)$_2$ with π-cpFe(CO)$_2$CH$_2$C$_6$H$_5$ results in each case in π-cpFe(CO)$_2$SeCN, characterized by infrared data (409). However, the reaction of Se(SeCN)$_2$

TABLE XXVI

SOME MIXED-LIGAND COMPLEXES WITH IRON SELENOCYANATE

Complex	Composition	Structure	References
Iron(III)			
$(n\text{-Bu}_4N)_3[Fe(CN)_5NCSe]$		Octahedral	(360)
Iron(II)			
$(Et_4N)_3[Fe(NCSe)_5H_2O]$		Octahedral	(616)
$Fe(phen)_2(NCSe)_2$		Cis octahedral	(64, 456, 683)
$[Fe(bipy)py_2NCSe](NCSe)$		Octahedral	(683)
$Fe(bipy)_2(NCSe)_2$		Octahedral	(683)
$Fe(DMF)_4(NCSe)_2$		Octahedral	(683)
$FeL_4(NCSe)_2$	L = py, 4pic, isoquin	Octahedral	(137, 732)
$[Fe(py)_4(NCSe)_2]2py$		Octahedral	(137)
$[Fe(4pic)_4(NCSe)_2]4pic$		Octahedral	(137)
$FeL(NCSe)_2$	L = 2,3,5,6-tetrakis-(6-methyl-2-pyridyl)pyrazine		(137)

with π-cpFe(CO)(PPh$_3$)CH$_2$C$_6$H$_5$ affords the linkage isomers π-cpFe-(CO)(PPh$_3$)(NCSe) and π-cpFe(CO)(PPh$_3$)(SeCN) (410). Both isomers, as solids or in solution, are stable with respect to interconversion at room temperature, but, at higher temperatures, deselenation rather than isomerization occurs; the deselenation is faster for the N-bonded isomer and is enhanced by the presence of Ph$_3$P. The interaction of [π-cpFe-(CO)$_2$SeCN] with PPh$_3$, P(C$_6$H$_{11}$)$_3$, or P(OPh)$_3$ yielded the compounds [π-cpFe(CO)$_2$(PR$_3$)] NCSe (R = Ph or C$_6$H$_{11}$) and [π-cpFe(CO)(P-(OPh)$_3$)SeCN], respectively. The compound [π-cpFe(CO)$_2$(PPh$_3$)] NCSe lost CO on heating to give [π-cpFe(CO)(PPh$_3$)SeCN], but the P(C$_6$H$_{11}$)$_3$ compound gave a number of products on similar treatment (410).

H. Cobalt, Rhodium, and Iridium

1. Cyanates

The tetrahedral complex anion $[Co(NCO)_4]^{2-}$ has been characterized by infrared and electronic spectra (292, 296, 432, 639) in the presence of a variety of cations. The infrared spectrum of the potassium salt is considerably more complex than that of the tetraethylammonium salt, indicating some loss of symmetry in the former compound. It is suggested that there is a strong interaction between K^+ and O of the cyanate group which leads to the distortion (296, 743).

A number of mixed-ligand cyanates of cobalt(III) and (II) are given in Table XXVII, which lists only a representative sample of the large number of such complexes available. In many cases the results were discussed in terms of variations in σ or π bonding or of steric effects. These discussions will not be rehearsed in this section but, where relevant to this review, will be referred to later. With the exception of

TABLE XXVII

Some Mixed-Ligand Cyanate Complexes of Cobalt

Complex	Ligand	References
Cobalt(III)		
Octahedral		
$[Co(DH)_2L(NCO)]$	L = H_2O, OH^-, NCO^-	(2)
	L = Cl^-, Br^-, I^-, NO_2^-	(10)
	L = Me	(253)
$[Co(NH_3)_5NCO]^{2+}$		(65, 149)
$[Co(CN)_5NCO]^{3-}$		(204)
Cobalt(II)		
Octahedral		
$Copy_4(NCO)_2$		(240, 290, 450, 451, 558)
$Co(2pic)_4(NCO)_2$		(451)
$CoL_4(NCO)_2$	L = 3-CN-py, 4-CN-py	(557)
$Co(urt)(H_2O)_2(NCO)_2$		(735)
$CoL_2(NCO)_2$	L = bipy, phen	(335)
Trigonal bipyramidal		
$CoEt_4dien(NCO)_2$		(159)
Tetrahedral		
$Co(py)_2(NCO)_2$		(450, 451, 558)
$Co(2pic)_2(NCO)_2$		(451)
$CoL_2(NCO)_2$	L = quin, isoquin, 2,6-DMP, py, 2-, 3-, or 4R-py	(488)
$Co(HMPA)_2(NCO)_2$		(655)

$CoL_2(NCO)_2$ (L = 3- or 4-cyanopyridine) which contain Co—N—Co bridges (557), the cyanate groups are N-bonded and monodentate. Measurements have been made of various thermodynamic parameters associated with the formation and the thermal decomposition of $Copy_4(NCO)_2$ (241, 451, 496, 500).

Compounds $[M(CO)(PPh_3)_2NCO]$ (M = Rh or Ir) have been prepared by a number of routes (87, 150, 155, 762) and show no evidence of isomerization in solution (155). In addition to $[Rh(PPh_3)_3NCO]$, however, the linkage isomer has been obtained—this is at present the only example where both cyanate isomers have been isolated (37). The compound $[Rh(CO)_2(NCO)]_2$ probably contains single-atom bridges of the type found in $CoL_2(NCO)_2$ (L = 3- or 4-cyanopyridine) (557) (see preceding text) and not the Rh—NCO—Rh bridges indicated (164). Oxidative addition reactions have been reported with $Ir(CO)(PPh_3)_2$-(NCO) to give the 1:1 adduct with tetracyanoethylene (47) and the products $[IrCO(PPh_3)_2(NCO)Y_2]$ (Y = Cl, Br, NO_3) (181). Also trans-$[Ir(pip)_4H(NCO)]^+$ has been reported (107).

Both $[Co(NH_3)_5NCO]^{2+}$ (65, 66) and $[Rh(NH_3)_5NCO]^{2+}$ (290) are hydrolyzed under acid conditions to the corresponding hexamine, and the rates of these reactions have been studied. Initial protonation of the nitrogen followed by reaction with H_2O to give a carbamic acid intermediate seems to provide a reasonable mechanism in each case. Such an intermediate has been observed in the reaction of $[Co(NH_3)_5H_2O]^{3+}$ with NCO^- (650). Hydrolysis of $K[Co(DH)_2L(NCO)]$ gives $Co(DH)_2L$-(NH_3) for L = H_2O, Cl, and Br and for L = I or NO_2 at pH 5 to 7, but at pH < 5 $[Co(DH)_2L(H_2O)]$ is formed in the last two cases (10). Similarly, $K[Co(DH)_2(NCO)_2]$ hydrolyzes to $[Co(DH)_2(NH_3)_2][Co-(DH)_2(NCO)_2]\cdot 3H_2O$ at pH 2 and to $[Co(DH)_2(NCO)NH_3]$ at pH 5 (2).

The apparent association and dissociation rate constants for the reaction of NCO^- with aquocobalamin have been determined over a range of pH, which is limited at the lower end due to hydrolysis of NCO^- (626).

A number of reactions on the coordinated cyanate group in $[M(CO)(PPh_3)_2NCO]$ (M = Rh, Ir) have been discussed (91, 783).

2. Thiocyanates

a. Cobalt. The X-ray structures of a number of cobalt thiocyanates have been determined. Some examples are listed in Table XXVIII and illustrate the interatomic distances associated with N—, S—, and bridging modes of coordination.

Tetrahedral $[Co(NCS)_4]^{2-}$ with a variety of cations is known (184,

TABLE XXVIII

INTERATOMIC DISTANCES AND BOND ANGLES OF SOME COBALT THIOCYANATES

Compound[a]	M—N or M—S (Å)	N—C (Å)	C—S (Å)	∠NCS	∠MNC or ∠MSC	References
I. $(C_{20}H_{17}N_4)_2[Co(NCS)_4]$[b]	1.945(3)	1.166(5)	1.601(4)	178.3(4)°	—	(184)
II. $Co(N_3S)(NCS)_2$,[c]						(235)
apex	1.986(12)	1.141(18)	1.641(14)	176.7(1.3)°	172.9(1.1)°	
plane	1.971(11)	1.147(18)	1.636(15)	176.9(1.2)°	161.1(1.1)°	
III. $Co(N_2O)(NCS)_2$,[d]						(316)
apex	2.01(1)	1.14(2)	1.61(2)	179.6(1.5)°	166.9(1.3)°	
plane	1.99(1)	1.21(2)	1.57(2)	177.2(1.8)°	156.9(1.4)°	
IV. $Co(NCS)_4Hg$,						(406, 407)
Co	1.926(16)	1.207(21)	1.638(14)	178.0(2)°	179(1)°	
Hg	2.559(4)	—	—	—	97.3(5)°	
V. $Co(NCS)_6Hg_2 \cdot C_6H_6$,						(351)
Co—N(1)	2.17	1.20	1.60	179°	160°	
Co—N(2)	2.08	1.11	1.76	176°	171°	
Co—N(3)	2.09	1.16	1.64	177°	167°	
Hg—S(1)	2.455	—	—	—	96°	
Hg—S(2)	2.424	—	—	—	98°	
Hg—S(3)	2.855	—	—	—	105°	
VI. $trans$-$[Co(en)_2(NCS)(SO_3)] \cdot H_2O$	1.974(18)	1.174(29)	1.629(18)	177.5(1.2)°	170.7(1.0)°	(49)
VII. $[Co(NH_3)_5(NCS)]Cl_2$	1.90(2)	1.43(11)[e]	1.32(13)[e]	—	—	(693)
VIII. $[Co(NH_3)_5(SCN)]Cl_2$	2.272(7)	1.14(4)	1.64(3)	175.0(3.0)°	104.9(1.1)°	(693)
IX. $(NH_4)[Co(DH)_2(SCN)_2] \cdot H_2O$	2.316	1.25	1.68	174°	105°	(649)

[a] The coordination around the central metal is as follows: I, tetrahedral CoN_4; II and III, distorted trigonal bipyramidal CoN_5 and $CoON_4$; IV, tetrahedral CoN_4 linked by thiocyanate to tetrahedral HgS_4; V, octahedral CoN_6 linked by thiocyanate to distorted tetrahedral HgS_4; VI, trans-octahedral CoN_5S; VII, octahedral CoN_6; VIII, octahedral CoN_5S; IX, trans-octahedral CoN_4S_2.

[b] $C_{20}H_{17}N_4$: nitron.

[c] N_3S: N,N-bis-(2-diethylaminoethyl)-2-methylthioethylamine.

[d] N_2O: 2-[2-(diethylamino)ethyl]amino]ethyl]diphenylphosphine oxide.

[e] These values are less accurate because of difficulty in placing C precisely between N and S in this disordered structure. The overall N—S distance of Å 2.73(6) is in good agreement with the other results.

293, 294, 432, 599, 647, 663). Cobalt(II) forms Co—NCS—M bridges in
the presence of Hg(II) (*293, 407, 654*), Cd(II) (*734*), Pt(IV) (*728*), or with
the reineckate anion (*583*). The compound Co(NCS)$_4$Hg contains the
tetrahedral cobalt ion, but further ligands can be added to give L$_2$Co-
(NCS)$_4$Hg (L = THF, dioxane, py, an, PPh$_3$) in which similar NCS
bridging occurs and the cobalt is in an octahedral environment (*511*).

The relative donor properties of various solvents toward Co(II) have
been compared with that of NCS$^-$ (*354, 515*).

The combination of cobalt, a popular metal for crystal field theory,
and thiocyanate, an accessible and stable ligand, has led to a very
large number of mixed-ligand cobalt thiocyanate complexes. Of the
Co(II) complexes studied there has been only one reported to contain
the *S*-thiocyanato group. A careful comparison was made of the electronic spectra and magnetic moments of Co(PPh$_3$)$_2$X$_2$ (X = Cl, Br, I,
and NCS) and of some Ph$_3$PO analogs, and from the position of NCS in
the spectrochemical series for these compounds it was concluded that
Co(PPh$_3$)$_2$(SCN)$_2$ was tetrahedral with *S*-thiocyanato groups (*213*).
Infrared measurements, on the other hand, indicated that the compound should be formulated Co(PPh$_3$)$_2$(NCS)$_2$ (*599*). Both infrared and
electronic spectral data indicate *N*-thiocyanato bonding in [Co(PEt$_3$)$_2$-
(NCS)$_2$] and [Co(P(C$_6$H$_{11}$)$_3$)$_2$(NCS)$_2$] with the former compound exhibit-
ing an equilibrium between high-spin, tetrahedral and low-spin,
pentacoordinate structures; the latter compound is tetrahedral (*563*). It
is perhaps possible that high spin–low spin equilibria, or some such
phenomenon, could account for the contradictory results obtained for
Co(PPh$_3$)$_2$(CNS)$_2$. No doubts have arisen about other cobalt(II) thiocyanates, which are N-bonded. Some examples are given in Table
XXIX, of which many of the pentacoordinate species are close to the
magnetic crossover point.

Cobalt(III) thiocyanates, isolated as solids, are listed in Table XXX
where the coordination behavior is seen to be more varied. The compound K$_3$[Co(CN)$_5$(SCN)] was characterized by Burmeister (*139*), and its
linkage isomer subsequently prepared by Stotz *et al.* (*708*). Later it was
shown that the preferred mode of bonding was influenced by the nature
of the cation in the ionic lattice. The two most stable forms are [*n*-Bu$_4$N]$_3$-
[Co(CN)$_5$NCS] and K$_3$[Co(CN)$_5$SCN] (*358, 359*). A consequence of this
work has led to the novel linkage isomers [(NH$_3$)$_5$CO—NCS—Co(CN)$_5$]
and [(NH$_3$)$_5$Co—SCN—Co(CN)$_5$] (*135*). One of the most extensively
studied systems is that containing bis(dimethylglyoximato)cobalt(III)
moiety with NCS and another ligand in the trans positions. The mode of
coordination of the thiocyanate ion is dependent on the ligand trans to
it, and both *N*- and *S*-thiocyanatobis-(dimethylglyoximato)cobalt(III)

complexes have been isolated with different ligands (see Table XXX). Linkage isomers have been isolated, for example with L = py (574) or 4t-bu-py (269). A careful study of changes in the positions and intensities of infrared frequencies, combined with NMR measurements of the methyl protons established that an equilibrium existed between the two isomers in solution,

$$\text{Co(DH)}_2\text{L(SCN)} \rightleftharpoons \text{Co(DH)}_2\text{L(NCS)}$$

in which the equilibrium position is affected both by the nature of the ligand and that of the solvent (571). This view was challenged by Hassel and Burmeister (370) who ascribed the solvent effect to kinetic factors. However, using as an additional probe the NMR signals of the t-butyl group in the t-bu-py complex, Epps and Marzilli (269) confirmed that the equilibrium existed. Both N- and S-thiocyanato groups have also been observed in solutions containing the anion $[\text{MeCo(DH)}_2\text{NCS}]^-$, and there is evidence for thiocyanate bridging (253).

A number of these trans compounds have been prepared by Ablov and his co-workers (4–6, 9, 11) who have recently described cis-[Co-(DH)$_2$H$_2$O(NCS)] (3), whereas the previous discussion shows that S-bonding predominates in the trans series unless modified by solvent effects. Although no evidence was observed for the N-bonded isomer in the study of trans-[Co(DH)$_2$(H$_2$O)(SCN)] (9, 572), this isomer has been claimed (193) from the position of ν_{CS}, a frequency which is difficult to observe in these compounds due to ligand vibrations. The difficulties in determining the coordination of NCS in bis(dimethylglyoximato)-cobalt(III) compounds are illustrated by the report of hydrolysis studies of twenty-six new salts of the bisthiocyanate anion in which the results were discussed on the basis of N-bonding (284) in spite of the available X-ray evidence to the contrary (649).

The [Cobipy$_2$(CNS)$_2$]$^+$ species has been formulated as both N- (590) and S-bonded (513, 570) compounds; the former designation may have arisen due to the presence of $[\text{Co(bipy)}_2\text{(SCN)}_2]_2[\text{Co(NCS)}_4]$ in the reaction product (570).

Whereas the previous discussion has been concerned with cobalt(II) and (III) compounds, the coordination of NCS toward cobalt in a formally lower oxidation state is illustrated by the isolation of [Co(NO)$_2$-L(NCS)] [L = Ph$_3$P, Ph$_3$As, (C$_6$H$_{11}$)$_3$P] with terminal N-thiocyanato groups and of [Co(NO)$_2$(NCS)]$_2$ with —SCN— bridges (89).

The oxidation of coordinated NCS⁻ in [Co(NH$_3$)$_5$NCS]$^{2+}$ generally leads to both the NH$_3$- and CN-substituted products: for a given oxidant the proportion of these products is dependent on the concentrations of both oxidant and acid (168, 580, 670).

TABLE XXIX

SOME COBALT(II) THIOCYANATE AND SELENOCYANATE COMPLEXES[a]

Complex	Composition	References
Octahedral[b]		
Co(py)₄(NCS)₂*		(201, 291,* 436,* 450,* 451,* 610, 671)
		(339)
Co(4pic)₄(NCS)₂		(300, 602)
Co(py)₄(NCS)₂·2I₂		(658, 683*)
Co(phen)₂(NCS)₂*		(473)
Co(tripyam)₂(NCS)₂	tripyam = tri-2-pyridylamine	(593*)
Co(Dben)₂(NCS)₂*	Dben = N,N′-dibenzylethylene diamine	(631)
CoL₄(NCS)₂	L = N-n-butyl imidazole	(393)
CoL₄(NCS)₂	L = thiazole	(735*)
Co(H₂O)₃(NCS)₂L₂	L = hexamethylenetetramine	(286)
CoL₂(NCS)₂	L = 2,5-dithiahexane or 1,2-di(isopropylthio)ethane	(688*)
Co(DMF)₄(NCS)₂*		
Cotu₂(NCS)₂	S-bridging by tu	(256, 287, 552)
Octahedral with —NCS— bridges		
CoL₂(NCS)₂*	L = py, 4pic	(339, 558*, 582, 611)
CoL₂(NCS)₂	L = 2,6-dimethylpyridine N-oxide, 2,4,6-trimethyl-pyridine N-oxide	(624)
CoL₂(NCS)₂	L = selenourea	(256)
Co(ROH)₂(NCS)₂	R = Me or Et	(288)
Tetrahedral		
CoL₂(NCS)₂*	L = py, 2Me-py, 3Me-py, 4Me-py, quin, isoquin, 2,6-dimethylpyrazine	(201, 339, 429, 450,* 451,* 488*)

Co(HMPA)₂(NCS)₂ — HMPA = hexamethylphosphoramide — (655)

Co(HMPT)₂(NCS)₂* — HMPT = hexamethylphosphoric triamide — (485*)

Co(NIPA)₂(NCS)₂ — NIPA = nonamethylimidodiphosphoramide — (243)

CoL₂(NCS)₂ — L = thioacetamide — (553)

Pentacoordinate

$$\left[\left(\underset{CH_2CH_2D}{\overset{CH_2CH_2B}{A{-}E}}\right)Co(NCS)_n\right]^{(2-n)+}$$

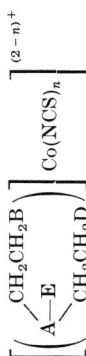

A	B	D	E	n	
N	NH₂	NH₂	CH₂·CH₂NH₂	1	(196)
N	NH₂	NH₂	H	2*	(159,* 257)
N	OMe	PPh₂	CH₂CH₂PPh₂	1	(541, 642)
N	OMe	NEt₂	CH₂CH₂PPh₂	1	(542)
N	SMe	NEt₂	CH₂CH₂NEt₂	2	(235)
N	SMe	NEt₂	CH₂CH₂PPh₂	1	(542)
N	OMe	PPh₂	CH₂CH₂PPh₂	2	(541)
N	CH₃	PPh₂	CH₂CH₂PPh₂	1	(541)
N	CH₂OMe	PPh₂	CH₂CH₂PPh₂	2	(541)
N	NEt₂	PPh₂	H	2	(540)
N	NEt₂	PPh₂	CH₂CH₂Me	2	(540)

CoL(NCS)₂

L = 1,1,1-tris-(diphenylphosphinomethyl)ethane — (238)

L = 2,6-di(diphenylphosphinomethyl)pyridine — (231)

L = 2,6-di(diphenylphosphinoethyl)pyridine — (559)

L = 2,3,5,6-tetrakis-(6-methyl-2-pyridyl)pyrazine* — (337*)

ᵃ An asterisk indicates that the corresponding selenocyanate complexes are also discussed.

ᵇ Octahedral Co(bipy)₂(NCSe)₂ is also known (683).

TABLE XXX

SOME COBALT(III) THIOCYANATE COMPLEXES

Co(III)—NCS compounds	References	Co(III)—SCN compounds	References
K₃[Co(CN)₅(NCS)]	(708)	K₃[Co(CN)₅(SCN)]ᵃ	(139)
(nBu₄N)₃[Co(CN)₅(NCS)]ᵃ	(358)	n(Bu₄N)₃[Co(CN)₅(SCN)]	(358, 359)
		Cobalamin—SCN	(380)
trans-[Co(DH)₂L(NCS)]; L = py	(574)	trans-[Co(DH)₂L(SCN)]; L = py	(574)
4t-bu-py	(269)	4t-bu-py	(269)
Meᵇ	(253)	SCN, Cl, Br, I, NH₃, H₂O, PPh₃, pip, 3Me-py,ᵇ 4Me-py,ᵇ 3Cl-py,ᵇ 3B4-py,ᵇ 4NH₂-py,ᵇ 4CN-py,ᵇ an,ᵇ pMe-an,ᵇ mNO₂-an,ᵇ pCl-an,ᵇ NO₂ᵇ	(572)
cis-[Co(DH)₂(H₂O)(NCS)]	(3)	trans-[Co(MH)₂(SCN)]⁻	(715)
[Co(bipy)₂(NCS)₂]⁺ (see text)	(590)	[Co(bipy)₂(SCN)₂]⁺	(513, 570)
		[Co(phen)₂(SCN)₂]⁺	(513)

$[CoTAAB(NCS)_2]^{+\,c}$	(227)		
$[Co(Me_2AsCH{=}CHAsMe_2)_2(NCS)_2]^+$	(100)		
$[Co(C_{16}H_{32}N_4)(NCS)_2]^{+\,d}$	(643)		
$cis\text{-}[Co(cyclam)(NCS)_2]^{+\,e}$	(608)		
$[CoL(NCS)_2]^+$; L = tet a or tet b^f	(784)		
$cis\text{-}$ and $trans\text{-}[Co(NH_2(CH_2)_3NH_2)_2(NCS)_2]^+$	(426)		
$cis\text{-}$ and $trans\text{-}[Co(en)_2(NCS)_2]^+$	(185, 498)		
$cis\text{-}$ and $trans\text{-}[Co(en)_2X(NCS)]^+$; X = NO$_2$	(498)		
= Cl	(185, 703)		
$cis\text{-}[CoNH_3en_2(NCS)]^{2+\,a}$	(133)	$trans\text{-}[CoNH_3en_2(SCN)]^{2+}$	(133)
$[Co(NH_3)_5(NCS)]^{2+\,a}$	(134)	$[Co(NH_3)_5(SCN)]^{2+}$	(134)
$trans\text{-}[Co(NH_3)_4(NCS)_2]^+$	(185)		
$\{[Co(NH_3)_4(NCS)]_2NH_2\}^{3+}$	(289)		
$\{[Co(en)_2(NCS)]_2O_2\}^{3+}$	(680)		

^{a} Stable isomer.

^{b} Linkage isomer formed in solution but not isolated as solid.

^{c} TAAB = Tetrabenzo[b,f,j,n][1,5,9,13] tetraazacyclohexadecine.

^{d} $C_{16}H_{32}N_4$ = hexamethyl-1,4,8,11-tetraazacyclotetradeca-4,11-diene

^{e} cyclam = 1,4,8,11-tetraazacyclotetradecane.

^{f} tet a and tet b = $trans\text{-}$ and $cis\text{-}$hexamethyl-1,4,8,11-tetraazacyclotetradecane.

The association of NCS$^-$ and ClO$_4$$^-$ with *trans*-[Coen$_2$(NCS)$_2$]$^+$ has been studied (*480*) and the equilibrium constants measured for the reactions of H$^+$ (*701*), Ag$^+$ (*617*), and Hg^{2+} (*276*) with various cobalt(III) amine–thiocyanate complexes. Thermodynamic data have also been obtained by various methods on a number of cobalt(II) complexes in different environments (*94, 291, 387, 425, 451, 500, 607, 645*).

Although the kinetics of hydrolysis, anation, and other substitution reactions involving cobalt(III) thiocyanates have been extensively studied, in most cases only the overall rates have been measured and only rarely has account been taken of the possibility of attack by nitrogen or sulfur of NCS$^-$, although the reality of this possibility is illustrated by the isolation of both [Co(NH$_3$)$_5$(NCS)]$^{2+}$ and [Co(NH$_3$)$_5$-(SCN)]$^{2+}$ species from the reaction of NaNCS with [Co(NH$_3$)$_5$Cl]-(ClO$_4$)$_2$ (*134*). After speculation concerning the mode of isomerization of [Co(NH$_3$)$_5$SCN]$^{2+}$ the same workers observed that *trans*-[Co(NH)$_3$en$_2$-SCN]$^{2+}$ gave *cis*-[Co(NH$_3$)en$_2$NCS]$^{2+}$ and concluded that the isomerization involves rupture of the Co—SCN bond (*133*). However, in this review, kinetic data will not generally be included when there is doubt regarding the mode of coordination.

b. Rhodium and Iridium. Octahedral [M(SCN)$_6$]$^{3-}$ (M = Rh or Ir) have been characterized by infrared (*665*), electronic (*663*), and ^{14}N NMR (*390*) spectroscopy. The *trans*-[Men$_2$Cl(NCS)]$^+$ cations have been characterized for rhodium (*415*) and iridium (*80*), as have *cis*-[Rh(cyclam)(NCS)$_2$]$^+$ and *trans*-[Rh(cyclam)(X)(NCS)]$^+$ (X = NCS, Cl, Br, I; cyclam = 1,4,8,11-tetraazacyclotetradecane) (*124*). The kinetics of the formation of [RhCl$_5$(CNS)]$^{3-}$ have been studied but the product was not isolated (*634*). The linkage isomers have been isolated for [Rh-(PMe$_2$Ph)$_3$Cl$_2$(CNS)] (*126*), [Rh(NH$_3$)$_5$(CNS)]$^{2+}$ (*659, 660*), and [Ir(NH$_3$)$_5$(CNS)]$^{2+}$ (*659, 661*), whereas *trans*-[Ir(pip)$_4$(H)NCS]$^+$ isomerizes in solution (*107*). Luminescence from [Rh(NH$_3$)$_5$(NCS)]$^{2+}$ has been measured (*724*).

Compounds [M(CO)(PPh$_3$)$_2$NCS] (M = Rh, Ir) have been characterized (*150, 155, 762*) and show no evidence of isomerization in solution (*155*). The compounds RhL$_2$(CO)(NCS) [L = P(C$_6$H$_{11}$)$_3$, PEt$_3$, PMe$_2$Ph, P(p-ClC$_6$H$_{11}$)$_3$, EPh$_3$, E(p-ClC$_6$H$_{11}$)$_3$, (E = As or Sb)] are N-bonded in the solid and solution. The arsine complexes and Rh[P(OPh)$_3$]$_3$(NCS) dissociate in solution, and the bridged compound Rh$_2$[P(OPh)$_3$]$_4$(CNS)$_2$ was isolated. The anionic species [Rh(CO)$_2$(NCS)$_2$]$^-$ was also characterized as the tetra-n-butyl ammonium salt (*408*). Compound Rh(PPh$_3$)$_3$-NCS is superficially similar to the previously mentioned species in that it remains N-bonded in the solid, but in solution in the absence of oxygen, one Ph$_3$P molecule is labile, and compounds Rh(PPh$_3$)$_2$L(SCN)

[L = $(CH_3)_2CO$, CH_3CN, Et_2O, C_6H_6] have been characterized. The $Rh(PPh_3)_2(C_6H_6)(SCN)$ reverts to the N-bonded starting material on treatment with excess PPh_3, and the C_6H_6 can be displaced to give $Rh(PPh_3)_2pip(NCS)$; the N-bonded species have been detected in pyridine or aniline solutions but have not been isolated (35).

Oxidative addition reactions of $Ir(CO)(PPh_3)_2NCS$ give the 1:1 adduct with tetracyanoethylene (47, 760) and the products [IrCO-$(PPh_3)_2(NCS)Y_2$] (Y = Cl, Br, NO_3) (181). By reacting $M(CO)(PPh_3)_2X$ (M = Rh, Ir; X = Cl, NCS, NCO) with $(SCN)_2$ the corresponding S-thiocyanato complexes have been obtained in every case (161).

3. Selenocyanates

The only selenocyanate of this group of metals to have its structure determined by X-ray crystallography is $NH_4[Co(DH)_2(SeCN)_2]$; the information available suggests that this is a trans-octahedral complex with a Co—Se distance of 2.4 Å (8). The anionic complexes $[Co(NCSe)_4]^{2-}$ (212, 295, 647, 663, 746, 747), $[Co(NCSe)_6]^{4-}$ (295, 616, 647, 663), and $[Rh(SeCN)_6]^{3-}$ (162, 662, 663, 665) have been reported and characterized by infrared and electronic spectroscopy. The bridged species $Co(NCSe)_4$-Hg has a comparable structure to its thiocyanate analog, with cobalt(II) tetrahedrally surrounded by four nitrogen atoms and mercury(II), similarly, by four selenium atoms (436, 600, 714, 747). The compound $Co(NCSe)_4Hg$ is blue, but a pink form has been reported (714), suggesting the presence of octahedrally coordinated cobalt(II); also $Co[(NCSe)_3$-$Hg]_2$ has been reported (714). Other cobalt(II) selenocyanates are listed in Table XXIX.

Fewer cobalt(II) selenocyanates than thiocyanates are known, and the bonding is not so varied. The species $[Co(CN)_5(NCS3)]^{3-}$ shows no signs of forming the Se-bonded isomer (142, 359), nor do $[Co(NH_3)_5$-$(NCSe)]^{2+}$ and $[Co(NH_3)_4(CN)(NCSe)]^+$ (142). Ablov and Samus (7) have reported $[Co(DH)_2L(SeCN)]$ (L = SeCN, NO_2, py,an) in which the mode of coordination is presumed by analogy with trans-$[Co(DH)_2$-$(SeCN)_2]^-$ (4) (see earlier). A number of salts of this anion have been prepared and its aquation studied (683, 759). Similar compounds with different oximes have been examined (755, 758). The cis-$[Co(DH)_2H_2O$-$(NCSe)]$ shows a similar reversal in the mode of selenocyanate coordination to that observed in the corresponding thiocyanate complexes (3).

The compound trans-$[Ir(pip)_4H(NCSe)]$ has been characterized (107). Also $[M(CO)(PPh_3)_2(NCSe)]$ (M = Rh, Ir) have been prepared by a number of different methods (148); the compounds show no tendency to isomerize in solution (155). The compound $Rh(CO)(PPh_3)_2NCSe$

deselenates easily in the presence of excess Ph_3P, and this type of reaction may also account for the difficulty found in attempting to prepare $Rh(PPh_3)_3NCSe$. The compound $Rh(PPh_3)_2(CH_3CN)(SeCN)$ has been isolated and characterized (*36*).

I. Nickel, Palladium, and Platinum

It is convenient to treat each metal, nickel, palladium, and platinum, in turn, because of the large number of compounds involved.

1. Nickel

a. Cyanates. The X-ray structure (*265*) of $[Ni_2tren_2(NCO)_2](BPh_4)_2$ reveals that it has bridged cyanate groups of the type:

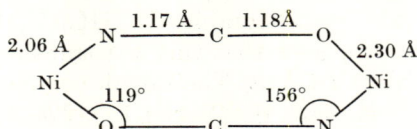

This is the first example of this type of bridging confirmed for the cyanate group.

The tetrahedral complex anion $[Ni(NCO)_4]^{2-}$ has been characterized by infrared spectroscopy (*275, 296, 639*) and by electronic spectral and magnetic (*275, 292, 647*) measurements. With the exception of $[NiEt_4dien(NCO)]BPh_4$, which has square planar coordination around the nickel (*159*), other mixed-ligand nickel(II) cyanates, for example, the compounds $NiL(H_2O)(NCO)_2$ (L = hexamethylenetetramine) (*735*), $NiL_2(NCO)_2$ (L = bipy, phen) (*335*), and $NiL_4(NCO)_2$ (L = py, 3pic, 4pic, isoquin) (*560*) are all octahedral. The compound $Ni(py)_6(NCO)_2$ is correctly formulated $[Ni(py)_4(NCO)_2]2py$ (*569*). Attempts to decompose thermally the foregoing tetrakis complexes to give the bis-ligand compounds were unsuccessful (*560*). However, with 3- and 4-cyano-pyridines, Nelson and Nelson were able to prepare $NiL_2(NCO)_2$ in which the nickel was coordinated octahedrally, with Ni—N(CO)—Ni bridges (*557*). Other workers tentatively suggested Ni—NCO—Ni bridges in the similar $NiL_2(NCO)_2$ (L = 4CN-py, 4MeO, 4-MeO_2, C-py, $\frac{1}{2}$phen (*158*), but, in spite of the preceding structure, their arguments are not as convincing as those of Nelson and Nelson (*557*). Proton NMR measurements have been made on $Nipy_4(NCO)_2$ (*291*), and its thermal decomposition has been studied (*241, 496, 500*).

b. Thiocyanates. The structures of a number of nickel(II) thio-cyanate complexes have been determined by X-ray methods, and are included in Table XXXI as are some other complexes whose structures have been elucidated by spectroscopic techniques. Results of X-ray structure determinations are not listed specifically in this case.

The hexacoordinate anion $[Ni(NCS)_6]^{4-}$ has been characterized (293, 294, 639, 647, 663) as well as $[Ni(NCS)_4]^{2-}$ (293, 294, 639, 647, 663). The latter exists either as a discrete, distorted tetrahedral anion, which is the case in the presence of large cations, or as part of a structure containing bridging thiocyanate groups with tetragonal hexacoordination around the nickel. Small cations seem to favor the tetragonal species, and both forms have been observed with the tetraphenylarsonium ion (294); such a polymeric structure is found for $[NiL_2][Ni(NCS)_4]$ (L = 2,2',2''-terpyridyl (368). Tetragonal coordination is also observed in $NiHg(SCN)_4$ (294), and presumably in $NiCd(SCN)_4$ (734). Bridging and terminal N-thiocyanato groups are present in $MNi(NCS)_3 \cdot nH_2O$ (M = alkali metal) (441, 447). Compounds $Ni[Hg(SCN)_3]_2 \cdot H_2O$ and $Ni[Hg(SCN)_4] \cdot H_2O$ also contain bridging and terminal thiocyanate groups, and there is slight evidence for two types of bridging thiocyanate, as observed in $Co[Hg(SCN)_3]_2 \cdot C_6H_6$ (351); the compounds are thermochromic (322). Thiocyanate bridges are present in nickel(II) reineckate (584) and in $[Ni(NO)NCS]_x$, although $Ni(NO)L_2$-(NCS) [L = $(PhO)_3P$, Ph_3P, $(C_6H_{11})_3P$] are monomeric (89). Other examples occur below.

Many mixed-ligand nickel thiocyanates are known and some are listed in Table XXXI. All the compounds appear to be N-thiocyanates in spite of the different ligational environments and stereochemistries observed. However, in a discussion of the electronic spectra of [Ni-$(diars)_2(NCS)]ClO_4$, Preer and Gray (614) state that N- and S-bonded isomers of this compound are in equilibrium in CH_3CN, CH_2Cl_2, and DMSO at room temperature. Further, it has been reported that some anionic complexes of the form $[NiL(CNS)_2LH(H_2O)_2]^-$ contain Ni—SCN linkages for L = glycinate or alaninate, and Ni—NCS bonding in the dianion when L = EDTA. Monomeric $Ni(DMSO)_2(SCN)_2$ was also reported, together with $NiL_2(NCS)_2$ (L = urea, α- or β-naphthylamine) which have thiocyanate bridges (173). However, the analytical data are not always good enough to be certain about the number of molecules of water present which could modify the interpretation of the visible spectra, and the infrared data and the assignments are unconvincing.

Of particular interest in Table XXXI is the variation in the steric hindrance of the ligands and also in their π-bonding propensities. Reference will be made to these results in the later discussion, but here changes in the stereochemistry of the nickel provide the main feature of interest.

TABLE XXXI
Some Nickel Thiocyanates

Complex	Composition	References
Octahedral		
trans-Ni(NH$_3$)$_4$(NCS)$_2$		(791)
trans-Ni(en)$_2$(NCS)$_2$		(127, 277, 495)
trans-NiL$_2$(NCS)$_2$	L = 2Me-pn, bn, N,N′-dimeen, N,N′-dieten	(277)
	= N,N′-dibenen	(593)
trans-Ni(pn)(NCS)$_2$		(228, 277)
NiL(NCS)$_2$	L = trien, tet a, tet b	(228)
trans-Ni(dtet)(NCS)$_2$	dtet = 3,3-dimethyl-1,5,8,11-tetra-aza-cyclotrideca-1-ene	(229)
cis-NiL(NCS)$_2$	L = tren	(628, 629)
trans-Ni(py)$_4$(NCS)$_2$		(38, 201, 560, 561, 671)
trans-Ni(py)$_4$(NCS)$_2$·2I$_2$		(300, 602)
trans-Ni(py)$_4$(NCS)$_2$·2py		(569)
NiL$_4$(NCS)$_2$	L = 3pic, 4pic, isoquin	(560, 561)
	= quin	(460, 699)
	= 2Me-quin	(532)
NiL$_2$(NCS)$_2$	L = tri-2-pyridylamine	(473)
NiL$_2$(NCS)$_2$	L = tri-(6-methyl-2-pyridylmethyl)amine	(234)
Ni(phen)$_2$(NCS)$_2$		(658)
NiL(NCS)$_2$	L = R-dionebis(3-aminopropylimine); R = 2,3-butane-, 2,3-pentane-, 1,2-cyclohexane	(721)
Ni(DMF)$_4$(NCS)$_2$		(688)
Ni(NTPA)$_2$(NCS)$_2$		(243)
NiL$_4$(NCS)$_2$	L = o-toluidine, o-anisidine, o-phenetidine, 3,4-xylidine	(167)
NiL$_2$(NCS)$_2$	L = o-phenylenediamine, 4Me-L	(516)
NiL$_4$(NCS)$_2$	L = thiazole	(393)
	L = N-n-butyl-imidazole	

Compound	Ref.
$Niquin_2(H_2O)_2(NCS)_2$	(460)
$NiL_2(H_2O)_2(NCS)_2$	(735)
$NiL_2(NCS)_2$	(286)
$NiL_2(NCS)_2 \cdot 2L_2$	(256, 287, 552, 554)
$Ni(tu)_2(NCS)_2$	

L = hexamethylenetetramine
L = 2,5-dithiahexane, 1,2-di(isopropylthio)ethane

Compound	Ref.
$Ni(SeC(NH_2)_2)_2(NCS)_2$	(256)
$NiL_2(NCS)_2$	(263)
$Ni(SN)_2(NCS)_2$	(170)
$Ni(NS)_2(NCS)_2$	(313)
$Ni(PS)_2(NCS)_2$	(681)

L = diphenyl-(o-diphenylarsinophenyl)phosphine
SN = 1,2-bis-(o-aminophenylthio)ethane
NS = thiosemicarbazide
PS = 2-diethylphosphinoethyl-1-thioethane

Five-coordinate

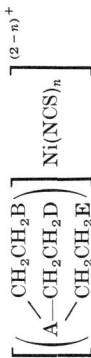

$$\left[\left(A \overset{\displaystyle CH_2CH_2B}{\underset{\displaystyle CH_2CH_2E}{-CH_2CH_2D}} \right) Ni(NCS)_n \right]^{(2-n)+}$$

A	B	D	E	n	Ref.
N	OMe	PPh_2	PPh_2	1	(104, 541, 642, 251, 252)
N	NEt_2	NEt_2	$AsPh_2$	2	
N	NEt_2	OMe	PPh_2	1	(542)
N	NEt_2	SMe	PPh_2	1	(542)
N	Me	PPh_2	PPh_2	1	(541)
N	CH_2OMe	PPh_2	PPh_2	1	(541)

Compound	Ref.
$NiL_3(NCS)_2$	(27)
	(529)

L = 9-alkyl-9-phosphafluorene; alkyl = Me, Et
= PMe_3

Compound	Ref.
$[NiL_2NCS]^+$	(614)
	(263)
	(781)

L = diars
= diphenyl[(o-diphenylarsinophenyl)phosphine
= 1,2-ethylmercaptocyclohexylphenylphosphine

Compound	Ref.
$NiL(NCS)_2$	(337)
$[NiL(NCS)]^+$	(231, 559)

L = 2,3,5,6-tetrakis-(6-methyl-2-pyridyl)pyrazine
L = 2,6-di(diphenylphosphinomethyl)pyridine

Square planar

Compound	Ref.
$[NiEt_4dienNCS]BPh_4$	(159)
$[Ninas\ NCS]BPh_4$	(250)

nas = N,N-bis-[2-(diethylamino)ethyl]-2-(diphenylarsino)ethyl-amine

(continued)

TABLE XXXI—*continued*

Complex	Composition	References
Ni(quin)$_2$(NCS)$_2$	L = 1-benzyl-2-phenylbenzimidazole	(460, 699)
NiL$_2$(NCS)$_2$		(119)
Ni(PR$_3$)$_2$(NCS)$_2$	R = n-Pr, i-Pr, s-but, cyclohexyl	(318)
	= i-Pr	(599)
	= Ph	(599, 765)
	= Me	(412)
NiL$_2$(NCS)$_2$	L = 9-phenyl-9-phosphafluorene	(27)
NiL$_2$(NCS)$_2$	L = 1,2-bis-(diphenylphosphino)ethane	(392)
	= 1,4-bis-(diphenylphosphino)butane	(641)
	= 2,6-bis-(2-diphenylphosphinoethyl)pyridine	(559)
	= 2,2′-biphenylenebisdiethylphosphine	(28)
NiL$_2$(NCS)$_2$	L = 2-diethylphosphinoethyl-1-thioethane	(681)
[NiQas(NCS)]NCS	Qas = o-Me$_2$As(C$_6$H$_4$)$_3$As	(375)
NiL(NCS)$_2$	L = diphenyl-(o-diphenylaminophenyl)phosphine	(263)
NiL(NCS)$_2$	L = 1,2-bis-(isopropylseleno)ethane	(346)
Complexes with bridging —NCS—		
NiL(NCS)$_2$	L = dien, dpt	(228)
[NiL$_2$(NCS)]ClO$_4$	L = en, ½trien	(228)
NiL$_2$(NCS)$_2$	L = various anilines	(167)
NiL(NCS)$_2$	L = quin	(460)
NiL(NCS)$_2$[a]	L = methyl-1-(6-methyl-2-pyridylmethyl)(2-pyridylmethyl)amine	(234)
NiL(NCS)$_2$[a]	L = tri-2-pyridylamine	(473)
NiL$_2$(NCS)$_2$	L = hexamethylphosphoric triamide	(485)
Ni(ROH)$_2$(NCS)$_2$	R = Me, Et	(288)
NiL$_2$(NCS)$_2$	L = 2-thioimidazolidine	(555)
	= thioacetamide	(172, 553)

[a] One terminal —NCS.

Anhydrous $Ni(NCS)_2$ contains the triply ligating NCS group with an octahedral NiN_2S_4 environment (288). Such an environment is also found in a number of other complexes (see Table XXXI), but it is of interest that it is achieved by bridging tu in $Nitu_2(NSC)_2$ (554) and bridging NCS in $NiL_2(NCS)_2$ for L = 2-thioimidazolidine (555) and thioacetamide (172). The magnetic properties of these and related compounds have been studied (268, 287, 288). In $NiL_2(NCS)_2$ (L = thiourea, thioimidazolidinone, 2-thiopyrrolidone, thioacetamide) it is suggested that ferromagnetic spin coupling occurs within the structural chains caused by the bridging sulfur atoms, with a superimposed antiferromagnetic interaction between the chains; the latter does not occur in $NiL_2(NCS)_2$ (L = N,N'-dimethylthiourea) since the chains are further apart owing to the substituents (268).

Some thermodynamic parameters have been measured for $Nipy_4$-$(NCS)_2$ (94, 291, 425, 496, 500) and other complexes (83, 416, 417, 645).

c. Selenocyanates. The compound $Ni(DMF)_4(NCSe)_2$ was prepared and characterized as a trans-octahedral complex containing N-selenocyanato groups (668). This has been confirmed by X-ray analysis which gives the following distances: Ni—N 2.05(2), N—C 1.21(8), and C—Se 1.71(3) Å; the NiNC and NCSe bond angles are 174.5° and 177°, respectively (736).

The anionic octahedral complex $[Ni(NCSe)_6]^{4-}$ has been reported (162, 295, 647, 663), as has $[Ni(NCSe)_4]^{2-}$. Unlike its thiocyanate analog, the latter ion only exists in the polymeric octahedral from with —NCSe— bridging (295, 647). Bridging also occurs in $Ni(NCSe)_4Hg$ (714).

Relatively few mixed-ligand complexes have been reported, and in these the selenocyanate group is N-bonded. The compound $NL_2(NCSe)_2$ [L = en, pn, bn, N,N'-dimeen, N,N'-dieten (277), Dben (593)] contains octahedral and $[NiEt_4dienNCSe]BPh_4$ (159) square planar nickel. Octahedrally coordinated nickel is also found in $Ni(urt)_2(H_2O)_4(NCSe)_2$ (735), $NiL_2(NCSe)_2$ (L = bipy, phen) (683), and $NiL_4(NCSe)_2$ (L = py, 3pic, 4pic, isoquin) (560); $Nipy_6(NCSe)_2$ is, correctly formulated $Nipy_4(NCSe)_2 \cdot 2py$ (569). Five coordination is believed to occur in $NiL(NCSe)_2$ [L = 2,3,5,6,-tetrakis-(6-methyl-2-pyridyl)pyrazine] (337).

2. Palladium

a. Cyanates. Palladium(II) forms a square planar tetra-N-cyanato anion with the tetramethylammonium (299, 573, 575) cation. A number of mixed-ligand compounds, $PdL_2(NCO)_2$ (L = NH_3, py, 2pic, 4pic, ½bipy, ½phen, Ph_3P), have been prepared by substitution reactions and shown to be N-bonded (573, 575); the Pd—N stretching frequencies have been studied in these compounds (566, 578). Also $Pd(PPh_3)_2(NCO)_2$ has been obtained by the reaction of CO with the corresponding azide (85), and the complexes, $PdL_2(NCO)$, [L = $C_5H_{11}N$, PBu_3, $P(C_6H_{11})_3$], have been prepared similarly (87). The compound $[PdEt_4dien(NCO)]$-BPh_4 has also been reported (147). Treatment of $Pd(PPh_3)_2(NCO)_2$ with ROH (R = Me, Et) and CO gives $Pd(PPh_3)_2(NCO)(CO_2R)$ (90), via $[(Ph_3P)_2Pd(NH_2CO_2Et)(NCO)]BF_4$ (783). Further, addition of Et_3O^+-BF_4^- to $Pd(PPh_3)_2(NCO)_2$ yields the N-cyanato bridged cation (90):

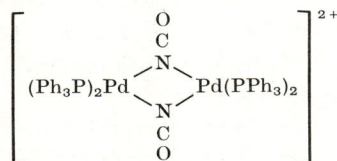

$$\left[\begin{array}{c} O \\ C \\ N \\ (Ph_3P)_2Pd \diagdown \quad \diagup Pd(PPh_3)_2 \\ N \\ C \\ O \end{array} \right]^{2+}$$

b. Thiocyanates. Complete X-ray structure analyses have been carried out for the compounds $K_2Pd(SCN)_4$ (524, 525), $Pd[Ph_2PCH_2$-$CH_2CH_2NMe_2](SCN)(NCS)$ (198, 199,) $Pd[Ph_2PCH_2CH_2PPh_2](SCN)$-(NCS) (102), and trans-$Pd(Ph_2PC{\equiv}CBu^t)_2(SCN)_2$ (101). These structures contain features of interest which have been delineated previously (Section III) and will be referred to again in the discussion. The mode of thiocyanate bonding in $K_2Pd(SCN)_4$ had been established previously by infrared (293, 639), ultraviolet (663), and ^{14}N NMR (390) spectroscopy.

A large number of mixed-ligand complexes of palladium(II) thiocyanate have been prepared. The mode of bonding of the thiocyanate group to the metal is markedly dependent on the nature of the other ligands present. Some examples are given in Table XXXII where the compounds are grouped either as N- or S-thiocyanates for each class of different ligand.

In addition to the compounds reported in Table XXXII, using bidentate ligands in PdL(CNS)$_2$, it has been possible to obtain complexes containing both N- and S-thiocyanato groups with L = 4,4'-diMe-bipy (103), $Ph_2As(-o-C_6H_4PPh_2)$ (526, 565), $Ph_2PCH_2CH_2NMe_2$ (526), $Ph_2P(CH_2)_3NMe_2$ (198, 199, 526), and $Ph_2PCH_2CH_2PPh_2$ (102, 526). The last result, confirmed by X-ray crystallography (102), contradicts the conclusions from the infrared spectrum of this compound, and thus casts some doubt on the correctness of the assignment, on infrared grounds, of Pd(diars)(SCN)$_2$ (392). Either similar mixed bonding or a mixture of linkage isomers occurs with L = 2,9-diMe-phen (606) and in solution for L = $Ph_2AsCH_2CH_2AsPh_2$, although the latter forms the S-thiocyanato complex only in the solid state (526). Similarly it

TABLE XXXII

Some Mixed-Ligand Complex Thiocyanates of Palladium(II) and Platinum(II)

Compound	References	Compound	References
L = Monodentate nitrogen donor		L = Monodentate phosphine	
trans-Pd(NH$_3$)$_2$(SCN)$_2$	(745)	Pd(Et$_3$P)$_2$(NCS)$_2$	(745)
Pd py$_2$(SCN)$_2$	(103, 640)	Pd(i-Pr$_3$P)$_2$(NCS)$_2$	(599, 745)
Pd(4-pic)$_2$(SCN)$_2$	(143)	Pd(Bu$_3$P)$_2$(NCS)$_2$	(143, 640)
Pd(4-n-am-py)$_2$(SCN)$_2$	(143)	Pd(Ph$_3$P)$_2$(NCS)$_2$	(143, 599, 640)
Pd(4-NO$_2$-py)$_2$(SCN)$_2$	(103)	Pd(o-allyl-C$_6$H$_4$PPh$_2$)$_2$(NCS)$_2$	(99)
Pd(4-CN-py)$_2$(SCN)$_2$	(103)	Pd[(C$_6$H$_{11}$)$_3$P]$_2$(NCS)$_2$	(599)
Pd(i-nicotin)$_2$(SCN)$_2$[a]	(103)	trans-PdH[(C$_6$H$_{11}$)$_3$P]$_2$(NCS)	(345)
Pd(2-Cl-py)$_2$(SCN)$_2$	(222)	trans-Pd(C$_6$F$_5$)(MePh$_2$P)$_2$(NCS)	(150)
Pd(3-Cl-py)$_2$(SCN)$_2$	(222)		
Pd(4-Cl-py)$_2$(SCN)$_2$	(222)	trans-Pt(Et$_3$P)$_2$(NCS)$_2$	(599)
Pd(4-MeO-py)$_2$(SCN)$_2$	(222)	cis-Pt(Et$_3$P)$_2$(NCS)$_2$	(599, 745)
Pd(2-F-py)$_2$(SCN)$_2$	(222)	Pt(i-Pr$_3$P)$_2$(NCS)$_2$	(745)
Pd(3-NH$_2$-py)$_2$(SCN)$_2$	(222)	Pt(Bu$_3$P)$_2$(NCS)$_2$	(390)
Pd(4-NH$_2$-py)$_2$(SCN)$_2$	(222)	Pt(Bu$_2$PhP)$_2$(NCS)$_2$	(390)
Pd(2,6-di-MeO-py)$_2$(SCN)$_2$	(222)	Pt(Ph$_3$P)$_2$(NCS)$_2$	(143)
		PtH(Et$_3$P)$_2$(NCS)	(190, 390)
cis-Pt(NH$_3$)$_2$(SCN)$_2$	(350, 745)	PtH(Ph$_3$P)$_2$(NCS)	(176)
cis-Pt(EtNH$_2$)$_2$(SCN)$_2$	(350)		
trans-Pt(py)$_2$(SCN)$_2$	(640)	L = Monodentate arsine	
cis-Pt(py)$_2$(SCN)$_2$	(350, 640)	Pd(Bu$_3$As)$_2$(SCN)$_2$[b]	(640)
		Pd(Ph$_3$As)$_2$(NCS)$_2$[c]	(79, 143, 640)
		Pd(o-allyl-C$_6$H$_4$AsPh$_2$)$_2$(6CS)$_2$	(99)
L = Miscellaneous monodentate			
Pd(CNMe)$_2$(SCN)$_2$	(599)	trans-Pt(Et$_3$As)$_2$(NCS)$_2$	(599)
		Pt(Bu$_3$As)$_2$(NCS)$_2$	(390)

(continued)

TABLE XXXII—continued

Compound	References	Compound	References
$Pd(CN\text{-}i\text{-}Pr)_2(SCN)_2$	(599)	$Pt(Ph_3As)_2(NCS)_2$	(143)
$Pd(tu)_2(SCN)_2$	(143)	$Pt(o\text{-allyl-}C_6H_4AsPh_2)_2(NCS)_2$	(99)
$Pdetu_2(SCN)_2$	(143)		
$Pd(Ph_3Sb)_2(SCN)_2$	(143, 640)		
$Pt(Ph_3Sb)_2(SCN)_2$	(143)		
Palladium (L_2 = chelate)		Platinum (L_2 = chelate)	
$Pd(bipy)(NCS)_2{}^c$	(79, 103, 143, 640)	$Pt(en)(SCN)_2$	(548, 549)
$Pd(phen)(SCN)_2$	(103, 143, 640)	$Pt(N\text{-Me-en})(SCN)_2$	(548, 549)
$Pd(5\text{-Me-phen})(SCN)_2$	(103)	$Pt(N\text{-Et-en})(SCN)_2$	(548, 549)
$Pd(5,6\text{-diMe-phen})(SCN)_2$	(103)	$Pt(1\text{-Et-en})(SCN)_2$	(548, 549)
$Pd(4,7\text{-diMe-phen})(SCN)_2$	(103)	$Pt(N\text{-}i\text{-Pr-en})(SCN)_2$	(548, 549)
$Pd(5\text{-Me-6-}NO_2\text{-phen})(SCN)_2$	(103)	$Pt(N,N\text{-diMe-en})(SCN)_2$	(548, 549)
$Pd(5\text{-}NO_2\text{-phen})(NCS)_2$	(103, 640)	$Pt(N,N'\text{-diMe-en})(SCN)_2$	(548, 549)
$Pd(5\text{-Cl-phen})(SCN)_2$	(103)	$Pt(N,N\text{-diEt-en})(SCN)_2$	(548, 549)
$Pd(4,7\text{-di-Ph-phen})(SCN)_2{}^c$	(103)	$Pt(N,N'\text{-diEt-en})(SCN)_2$	(548, 549)
		$Pt(N,N\text{-diMe-}N'\text{-Me-en})(SCN)_2$	(548, 549)
$Pd[Ph_2P(\text{-}o\text{-}C_6H_4SMe)](SCN)_2$	(526)	$Pt(N,N\text{-diMe-}N',N'\text{-tetraMe-en})(SCN)_2$	(548, 549)
$Pd[Ph_2P(\text{-}o\text{-}C_6F_4SMe)](SCN)_2$	(526)		
$Pd[Ph_2P(\text{-}o\text{-}C_6H_4SeMe)](SCN)_2$	(526)		
$Pd[Ph_2As(\text{-}o\text{-}C_6H_4P(S)Ph_2)](SCN)_2$	(526, 565)		
$Pd[Ph_2AsCH_2CH_2AsPh_2](SCN)_2$	(526)		

^{a} i-Nicotin = methyl ester of isonicotinic acid.

^{b} S-Bonded isomer as a solid but partially isomerizes on melting.

^{c} Stable linkage isomer.

has not been possible to distinguish between these two possibilities in the following dimeric compound (*179*):

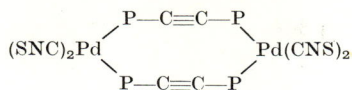

$$(SNC)_2Pd \underset{P-C\equiv C-P}{\overset{P-C\equiv C-P}{\diagdown\diagup}} Pd(CNS)_2$$

(two phenyl groups are omitted from each P for clarity)

A monomeric compound is obtained with L = $Ph_2PCH{=}CHPPh_2$, in which only *S*-thiocyanato groups are present (*194*). Using other bidentate ligands with two different atoms, the compounds Pd[*o*-X—C_6H_4—YCH_2]$_2(CNS)_2$ (X = Ph_2As, Y = O; X = Ph_2As, Y = S; X = NH_2, Y = S) have been reported (*170*). It is not always clear which are the donor atoms of the bidentate ligand and, in view of some of the results already given, the authors' tentative assignment of the complexes as *N*-thiocyanates on the basis of ν_{CN} only, seems worthy of reexamination. The bonding of the thiocyanate group has not been specified in the complex PdL(CNS)$_2$ (L = 8-dimethylarsinoquinoline) (*73*).

Basolo and his co-workers first prepared the ion [PdEt$_4$dien(CNS)]$^+$ (*77*). Compound [PdEt$_4$dienSCN]NCS is prepared by the reaction of Et$_4$dien with $K_2Pd(SCN)_4$, and the solid product isomerizes completely to the *N*-thiocyanate in 3 days. If, however, the original salt is treated with NH_4PF_6, then [PdEt$_4$dienSCN]PF$_6$ is formed which undergoes no detectable solid state isomerization. The *S*-thiocyanato cation isomerizes in solution, and kinetic studies indicate that the mechanism involves an intermolecular process (*78*). Solid [PdEt$_4$dienNCS]BPh$_4$, prepared by an analogous method, reisomerizes to the S-bonded form at room temperature (*153, 156*). The kinetics of the displacement of both N- and S-bonded thiocyanate groups from this system by the bromide ion have been studied (*78, 157*), and kinetic and thermodynamic parameters obtained for the equilibrium (*379, 414*):

$$[PdLBr]^+ + NCS^- \rightleftharpoons PdLCNS]^+ + Br^- \quad (L = dien, Et_4dien)$$

Dimeric compounds of the type $Pd_2L_2X_2(CNS)_2$ [L = As(iso-Pr)$_3$, AsBu$_3$, PBu$_3$; X = Cl, NCS] have been isolated and shown to contain NCS bridges (*186*), although the assignments of the mode of bonding of the terminal thiocyanate groups in these compounds and in the corresponding PdL$_2(CNS)_2$ were based, in 1956, on insufficient data by current standards. Such bridging is believed to occur in the low-temperature form of 2-methylallyl palladium thiocyanate (see Fig. 2) whereas the coalescence of the NMR spectrum at higher temperatures is believed to

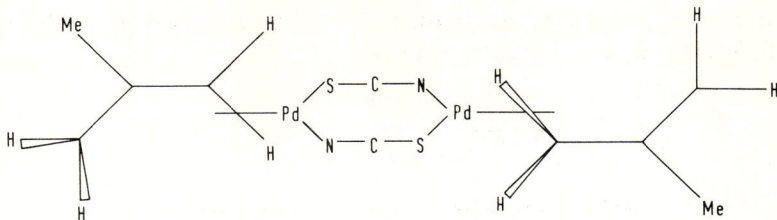

FIG. 2. Dimeric form of 2-methyl allyl palladium thiocyanate. [From Tibbetts and Brown (725).]

be due to the formation of a tetramer which is tentatively assigned the structure in Fig. 3 (725).

Some five-coordinate palladium(II) thiocyanate complexes have been reported. Both [Pd(2,9-diMe-phen)$_2$SCN]ClO$_4$ (606) and [Pd(o-Me$_2$As-C$_6$H$_4$)$_3$As(NCS)]NCS (375) have been characterized, but the mode of coordination is unspecified in [PdL$_2$CNS]NCS [L = 1,8-naphthylenebis-(dimethylarsine)] (636); in the latter system, six coordination seems to occur in solution (636).

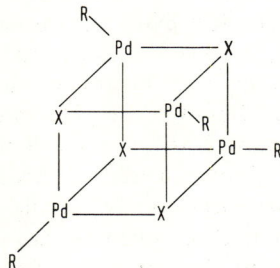

FIG. 3. Tetrametric form of 2-methyl allyl palladium thiocyanate (X = NCS with nature of bridge unspecified). [From Tibbetts and Brown (725).]

The effect of different solvents on the type of thiocyanate coordination has been studied by dissolving the two linkage isomers (Pd-(AsPh$_3$)$_2$(CNS)$_2$ in various solvents. It was found that Pd—SCN bonding was promoted by solvents with high dielectric constants and a mixture of Pd—NCS, Pd—SCN, and Pd—SCN—Pd bonding by solvents having low dielectric constants (154, 155). These and related results will be discussed in Section V.

c. Selenocyanates. The square planar anions [Pd(SeCN)$_4$]$^{2-}$ have been characterized (162, 662, 663, 665) and some mixed-ligand complexes reported. The compound Pd(n-Bu$_3$P)$_2$(SeCN)$_2$ on standing shows changes in its infrared spectrum which have been tentatively ascribed to isomerization, but some decomposition also occurs, and the N-bonded isomer was not isolated (*152*). The other reported compounds of the type PdL$_2$(SeCN)$_2$ are as written [L = NH$_3$, py, 4pic, 4NH$_2$-py, 4Ac-py, *N*,*N*-diph-tu, *N*,*N*'-diph-tu, PPh$_3$, AsPh$_3$, P(*o*-Me)$_3$] (*152*). Selenium-bonded compounds have been reported only for the bidentate ligands bipy (*142, 152*), en, phen, 5-NO$_2$-phen, 1,2-(NH$_2$)$_2$C$_6$H$_4$, tripy (*152*) (where tripy = 2,2',2"-tripyridine acting, in the solid compound, as a bidentate ligand), Ph$_2$As(-*o*-C$_6$H$_4$PPh$_2$), Ph$_2$As(-*o*-C$_6$H$_4$PS·Ph$_2$) (*526, 565*), and Ph$_2$AsCH$_2$CH$_2$AsPh$_2$ (*526*).

The system [PdEt$_4$dienCNSe]X is very similar to the corresponding thiocyanate compounds. The compound [PdEt$_4$dienSeCN]BPh$_4$ is the kinetic product of the reaction of Et$_4$dien with K$_2$Pd(SeCN)$_4$, but when dissolved in a variety of solvents it isomerizes and solid [PdEt$_4$dien-NCSe]BPh$_4$ may be isolated. This latter compound reisomerizes in the solid state to give again the Se-bonded compound. With dien, only the Se-bonded compound [Pd(dien)SeCN]BPh$_4$ is observed (*151, 153, 156*) The kinetics of displacement of —NCSe and —SeCN from [PdEt$_4$dien-CNSe]BPh$_4$ by the bromide ion have been studied, and by comparing the rates with the corresponding thiocyanate displacements it was originally concluded that displacement of —SeCN involved the opening of one or more of the chelate rings, whereas, for —SCN, a dissociative of S_N1 or solvent-assisted ligand exchange mechanism occurred (*157*). Later, the same authors extended their measurements and rejected the chelate ring opening mechanism in favor of a direct substitution of [PdEt$_4$dienSeCN]$^+$ by Br$^-$: it was proposed that the steric interaction caused by the outgoing group was responsible for the reagent-dependent path (*414*).

3. Platinum

a. Cyanates. The tetraphenylarsonium and tetraethylammonium salts of square planar [Pt(NCO)$_4$]$^{2-}$ have been prepared and characterized (*575*). A number of mixed ligand complexes PtL$_2$(NCO) (L = $\frac{1}{2}$bipy, Ph$_3$P, Ph$_3$As, Ph$_3$Sb) have also been reported (*575*) and their far-infrared spectra studied (*578*). Two groups of absorptions due to hydrido resonances were observed in the NMR spectrum of *trans*-PtH-(PEt$_3$)$_2$NCO which were ascribed to the *N*- and *O*-cyanato isomers (*613*). This view has been challenged and the effects ascribed to phosphine exchange (*13, 14*), but Pidcock (*604*) has demonstrated that such exchange cannot account for the observed broadening and has supported the idea of linkage isomers. The related compounds, *trans*-PtHL$_2$(NCO) (L = PPh$_3$, PEtPh$_2$, PBu$_3$), were also reported (*13, 14*).

Similarly to the analogous palladium compounds, $Pt(PPh_3)_2X_2$ ($X = N_3$, NCO) reacts with ROH ($R = Me$, Et, Pr, i-Pr) and CO to give $Pt(PPh_3)_2(NCO) CO_2R$ (*90, 749*). The bridged species $[(Ph_3P)_2PtX_2Pt(PPh_3)_2](BF_4)$ ($X = N_3$, NCO) reacts with CO to give $[Pt(PPh_3)_2(CO)(NCO)]BF_4$ which further reacts with CO and ROH (Me, Et) to give $[Pt(PPh_3)_2(CO)(CO_2R)]BF_4$ (*90*). The reactions appear to proceed via protonation of the cyanate group to species such as $[(Ph_3P)_2Pt(NH_2CO_2Et)(NCO)]BF_4$, which has been characterized (*783*). The compound $Pt(PPh_3)_2C_2H_4$ reacts with $RCON_3$ ($R = OBu^t$, OEt) to give $Pt(PPh_3)_2(NCO)N_3$ (*749*).

b. Thiocyanates. The X-ray structures of the α- and β- forms of $Pt_2Cl_2(PPr_3)_2(CNS)_2$, originally prepared by Chatt *et al.* (*187, 188*), have further refined (*348*) as follows:

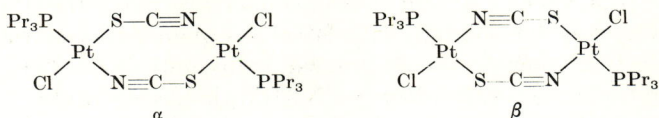

and the appropriate bond distances are listed in Table XXXIII (*348*).

Platinum(IV) forms an octahedral complex anion shown to be $[Pt(SCN)_6]^{2-}$ by infrared (*639, 665*) and electronic (*663, 713*) spectroscopy. Square planar $[Pt(SCN)_4]^{2-}$ has been similarly characterized (*293, 599, 639, 663, 665, 745*) and additionally by ^{14}N NMR spectroscopy (*390*).

Few mixed-ligand complexes of platinum(IV) thiocyanates appear to have been reported. The *cis*- and *trans*-$Pt(NH_3)_2(NO_2)_2(SCN)_2$ (*46*) and *trans*-$[Pt(NH_3)_4(SCN)_2]^{2+}$ and -$[Pt(NH_3)_4Cl(SCN)]^{2+}$ have been characterized (*521*) as have the bridged compounds, $Pt(SCN)_6M$ (M = Mn, Fe, Co, Ni, Cu, Zn, Cd, Pb) (*58, 728*). The easy reduction of $K_2Pt(SCN)_6$

TABLE XXXIII

SELECTED BOND DISTANCES AND BOND ANGLES IN THE
α AND β FORMS OF $Pt_2Cl_2(PPr_3)_2(CNS)_2$[a]

Bonds	α	β
Pt—S	2.327(5) Å	2.408(4) Å
Pt—N	2.078(13) Å	1.965(13) Å
S—C	1.643(14) Å	1.641(15) Å
C—N	1.124(19) Å	1.168(16) Å
∠Pt—S—C	103.6(6)°	102.9(5)°
∠Pt—N—C	164.9(14)°	167.3(11)°
∠S—C—N	179.3(16)°	178.6(13)°

[a] Data from Ref. *348*.

in aqueous solutions of nitrogen ligands results in platinum(II) complexes (350). Tetrameric [Pt(NCS)Me₃]₄ has been reported and is believed to break down on addition of pyridine to give an N-bonded product according to the scheme in Fig. 4 (385).

Mixed-ligand platinum(II) thiocyanates are quite numerous and many have been included in Table XXXII with some similar palladium-(II) compounds. Linkage isomerism has been observed in solution for

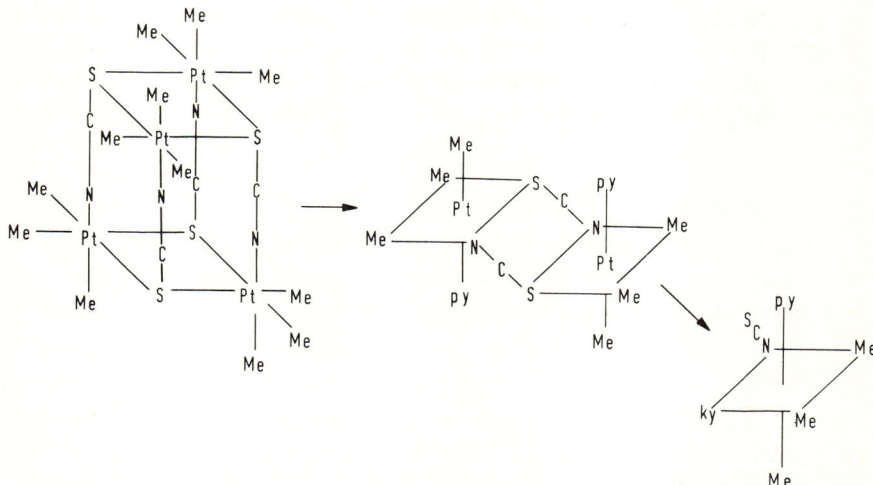

FIG. 4. Reaction scheme for the addition of pyridine to [Pt(NCS)Me₃]₄. [From Headley et al. (375).]

the complexes PtHL₂(NCS) where the —NCS to —SCN ratio increased in the order Et₃P ~ Bu₃P < Et₃As < Ph₂EtP < Ph₃P; the S-bonded complex appears to be favored by polar solvents (13, 14). Dimeric Pt₂(DPPA)₂(CNS)₄ [DPPA = bis(diphenylphosphino) acetylene], like the palladium(II) analog (see Section IV, I, 2, b) has an infrared spectrum which indicates both N- and S-bonded thiocyanates, but the authors were unable to distinguish whether these arose from a compound containing both types of coordination or from an unresolved mixture of linkage isomers (179). The thiocyanate bonding was not characterized in PtL(CNS)₂ (L = 8-dimethylarsinoquinoline) (73). However, in PtAP-(SCN)₂ (AP = o-allylphenyldiphenylphosphine), the double bond is coordinated to the metal and S coordination occurs [cf. Pd(AP)₂(NCS)₂ and the corresponding N-thiocyanato complexes in Table XXXII with the related arsine] (99). A Pt—SCN bond is reported in [PtL₃SCN]NCS (L = 9-R-9-phosphafluorene; R = Me, Et) (27).

Dimeric compounds of the type $Pt(PPr_3)_2X_2 CNS)_2$ (X = Cl, NCS) have been shown to contain —NCS— bridges (186)—the X-ray structure of the isomers when X = Cl have been mentioned previously in this section.

The five-coordinate compounds $[PtL_2(CNS)]NCS$ [L = 1,8-naphthyl-enebis(dimethylarsino) (636), 8-dimethylarsinoquinoline (73)] have been reported but the mode of bonding was not specified; in the latter, six coordination appears to occur in solution (73).

c. *Selenocyanates.* Both $[Pt(SeCN)_6]^{2-}$ (639, 663, 665) and $[Pt-(SeCN)_4]^{2-}$ (162, 662, 663, 665, 713) have been characterized. The mixed-ligand complex $Ptbipy(SeCN)_2$ has been reported (142), and the compounds trans-$PtHL_2SeCN$ (L = Ph_3P, Bu_3P) (14).

J. COPPER, SILVER, AND GOLD

Because of the different relative stabilities of the oxidation states of these elements, it is convenient to treat each element individually as was done with nickel, palladium, and platinum.

1. Copper

a. *Cyanates.* The preparation of some anionic cyanate complexes of copper(I) has been reported (694), but no structural data are available.

Copper(II) forms the anionic species $[Cu(NCO)_4]^{2-}$ (292, 296, 693). Electronic spectra suggest that the ion has a distorted tetrahedral structure (639), and the infrared spectra indicate that the distortion involves the CuN_4 tetrahedron rather than kinking of the Cu—NCO group (296). It has been suggested that, in the presence of suitable cations, this distortion can extend to give pseudo-octahedral geometry around the copper (195). In $K[Cu(NCO)_3]$ both bridging (Cu—N—Cu) and isocyanate groups occur (743). The addition of LiNCO to a solution of copper(II) in acetone has been followed spectrophotometrically and indicates the formation of a 1:1.5 complex (620, 752).

Several mixed-ligand complexes, $CuL_2(NCO)_2$ (L = ½bipy, ½phen, isoquin, quin, 6NO₂-quin, 4R-py; R = H, Me, NH₂, MeCO, CN, MeO₂C, CH·NOH), have been reported (158). In most cases these are essentially *trans*-square planar complexes (except when L = bipy or phen), but when L = isonicotinamide or 4-pyridine aldoxime there is evidence for a bridged octahedral structure; in all these cases the cyanate group acts as a monodentate nitrogen donor (158). A similar series of compounds, $CuL_2(NCO)_2$ [L = quin, isoquin, 2Me-isoquin, 3Me-isoquin, 4Me-isoquin (461), 2-pic, 3-pic, 4-pic, 2,4-lut, 2,6-lut, 2,4,6-coll (461, 462)], with various 2-substituted pyridines and related ligands show increasing tetragonal distortion but maintain the same overall stereochemistry (461, 462). The compounds $CuL(NCO)_2$ [L = quin, 2Me-isoquin, 3Me-isoquin, 4Me-isoquin (461), 2-pic, 2,4-lut (461, 462)] have also been prepared, and seem to have a deformed tetragonal pyramidal structure with extensive Cu—N—Cu bridging (461) which is believed to give rise to a ferromagnetic interaction and would account for the observed magnetic properties of the compounds (462). With substituted anilines as ligands, the complexes $CuL_2(NCO)_2$ (L = an, pCl-an, pI-an, o-tol, m-tol, p-tol) are formed, with the neutral ligands occupying the trans positions to an octahedrally coordinated copper which has the nitrogens from four bridging cyanate groups forming the plane (463, 619). The compound $[CuEt_4dienNCO]BPh_4$ contains four-coordinated copper (159). Bispicolinatocopper(II) is square planar and forms a 1:1 adduct with KNCO in which electronic spectral changes are attributed to cyanate coordination: since no thiocyanate coordination occurs in the corresponding KSCN adduct (705) (see Table XXXIV) as would have been expected if N coordination occurs, the cyanate adduct is formulated $K[Cu(pic)_2(OCN)]$ (321).

b. *Thiocyanates*. The structures of several of these complexes have been determined crystallographically, and some of the resulting data are summarized in Table XXXIV. The variations are particularly interesting for this metal. The mode of coordination of thiocyanate changes apparently with the different coordination geometries around copper(II) (structures I, IV, VI, and IX of Table XXXIV), or with the nature of the ligand (structures II–IV), or both (structures I–IX). Structures VII and VIII are the two most commonly obtained crystalline species of the eight which have been observed for this compound (53); structures X and XI are two structural modifications (422), whereas crystals of $Cu(en)(CNS)_2$ contain copper(II) in two different environments (314). The compound $Cu_2(NH_3)_3(NCS)_3$ has a polymeric structure with sheets of copper(I) tetrahedra cross-linked by pairs of copper(II) octahedra. Each copper(II) has an approximately square-planar array of four nitrogen atoms (3NH₃ and —NCS) with two sulfur atoms completing a distorted octahedron in the trans positions. One of these sulfur atoms is also coordinated directly to copper(I), whereas the other allows the existence of a conventional Cu(II)—SCN—Cu(I) bridge. The tetrahedron surrounding copper(I) consists of two N- and two S-bonded thiocyanate groups (312).

With large cations, $[Cu(NCS)_4]^{2-}$ exists as a discrete but distorted tetrahedral entity; with smaller cations, it adopts a six-coordinate tetragonal environment (293, 294, 647, 663).

Complex	C—N (Å)	C—S (Å)	∠NCS	Cu—NCS (Å)
I. $Cu(NH_3)_4(SCN)_2$	1.17	1.64	—	—
II. $Cuen_2(SCN)_2$	1.16(3)	1.62(2)	176.9°	—
III. $[Cu(N\text{-meen})_2NCS]NCS$	1.132(18)	1.620(13)	178.5(13)°	2.238(14)
	1.193(20)[c]	1.598(15)[c]	178.3(14)°	—
IV. $Cu(N,N'\text{-dimeen})_2(NCS)_2$	1.164(10)	1.636(8)	176.8(7)°	2.517(7)
V. $[Cu(aebg)NCS]NCS$[a]	1.16	1.62	178.2°	1.99
VI. $[Cu(trien) SCN]NCS$	1.164(10)	1.646(8)	178.0(6)°	—
VII. $[\alpha\text{-CuL}(NCS)]NCS$[b]	1.56(18)	1.613(13)	179.2(13)°	2.119(12)
VIII. $[\gamma\text{-CuL}(NCS)]NCS$[b]	1.111(17)	1.616(13)	179.0(13)°	2.162(14)
IX. $[Cu(tren)(NCS)]NCS$	1.142(7)	1.612(7)	177.4(5)°	1.959(5)
	1.168(7)[c]	1.624(5)[c]	178.6(5)°[c]	—
X. $\alpha\text{-}Cu(NH_3)_2(NCS)_2$	1.20(4)	1.62(3)	178 (11.4)°	1.96(3)
	1.21(4)	1.64(3)	162 (7.9)°	1.91(3)
XI. $\beta\text{-}Cu(NH_3)_2(NCS)_2$	1.20(9)	1.63(9)	159 (7.0)°	1.96(7)
	1.37(9)	1.67(7)	160 (11.8)°	1.83(7)
XII. $Cu(en) (NCS)_2$	1.25(4)	1.68(3)	172.3(1.9)°	2.01(2)
	1.24(4)	1.71(3)	165.0(4.0)°	1.99(2)
	1.29(3)	1.69(2)	161.2(4.0)°	2.01(3)
	1.32(5)	1.65(4)	154.7(4.9)°	2.05(2)
XIII. $Cu(py)_2(NCS)_2$	—	—	—	2.10
XIV. $CuHg(SCN)_4$	1.11	1.66	—	—
XV. $Cu(en)_2Hg(SCN)_4$	1.33	1.57	163°	2.58
XVI. $Cu_2(NH_3)_3(NCS)_3$	1.181(35)	1.638(28)	179.3(2.5)°	1.987(23)[e]
	1.128(35)	1.710(27)	175.1(2.5)°	1.996(24)[d]
	1.127(38)	1.671(29)	176.5(2.7)°	2.006(8)[d]
XVII. $[Cu(PPh_2Me)_2(NCS)]_2$	1.14(2)	1.64(2)	177.2°	2.02(2)
XVIII. $Cu(pic)_2 \cdot KSCN$	1.153[c]	1.628[c]	—	—

[a] aebg = 1-(2-aminoethyl)biguanide.
[b] L = 1,7-bis-(2-pyridyl)-2,6-diazaheptane.
[c] Free thiocyanate ion.
[d] See comments column.
[e] See comments column.

Cu—SCN (Å)	∠ Cu—NCS	∠ Cu—SCN	Comments	References
> 3.0	—	90°	Trans-octahedral CuN_4S_2	(609)
3.27(1)	—	79.9°	CuN_4S_2. Amino-N at corners of rectangle; trans-S completing distorted octahedron.	(128)
3.348(4)	178.2(12)°	114.2(5)°	Amino-N at corners of square with bridging NCS completing octahedral CuN_4NS.	(589)
—	—	—		
—	128.6(6)°	—	Amino-N at corners of square with N-thiocyanates completing distorted octahedral CuN_4N_2	(467)
—	158.1°	—	Three amino-N and N-thiocyanato group square with two long Cu—NCS distances (3.37 and 2.95 Å) formally completing distorted octahedron.	(31)
2.607(2)	—	89.5°	Trien forms an approximately square-based plane around Cu with —SCN at apex completing five coordination.	(517)
—	156.6(11)°	—	Approximately trigonal bipyramidal CuN_5.	(52, 53)
—	159.8(11)°	—	Approximately trigonal bipyramidal CuN_5.	(52, 53)
—	163.3°	—	Approximately trigonal bipyramidal CuN_5, with —NCS in axial position.	(400, 401)
—	—	—		
$\begin{cases}3.11(4)\\2.93(4)\end{cases}$	174 (5.5)°	85 (0.2)°	Four nitrogens ($2NH_3$ and 2-NCS) form a plane around Cu with S from one NCS bridging two coppers; the other NCS is monodentate. The structures differ in the degree of linearity of NCS.	(422)
—	169 (9.5)°	—		
$\begin{cases}3.05(6)\\2.99(6)\end{cases}$	168 (7.2)°	84.5(0.3)°		
—	165 (11.8)°	—		
$\begin{cases}3.01(2)\\3.10(2)\end{cases}$	175.0(6.0)°	$\begin{cases}85.9(3.0)°\\92.0(3.0)°\end{cases}$	As in VII and VIII, Cu has 4N in plane, and two bridging S both of which arise from one molecule which alternates throughout the unit cell with a molecule containing only terminal —NCS.	(314)
$\begin{cases}2.99(2)\\3.10(2)\end{cases}$	169.0(3.6)°	$\begin{cases}86.0(1.5)°\\97.4(1.5)°\end{cases}$		
—	165.4(4.7)°	—		
—	149.8(3.2)°	—		
3.0	—	—	NCS bridging.	(610)
—	172–180°	—	NCS bridging resulting in octahedral CuN_4S_2.	(609)
—	—	—	NCS bridging resulting in trans-octahedral CuN_4N_2.	(673)
3.286(9)[e]	172.3(2.2)°[e]	85.2°[e]	A compound containing Cu(I) and Cu(II) (designated d and e, respectively) with three different bridging NCS (see text).	(312)
$\begin{cases}2.946(7)^{e}\\2.470(7)^{d}\end{cases}$	161.0(2.3)°[od]	$\begin{cases}88.7(0.9)°^{oe}\\98.3(0.9)°^{od}\end{cases}$		
2.376(8)[d]	168.2(2.4)°[od]	96.3°[od]		
2.46(1)	158.0(2)°	99.1(6)°	Bridging NCS with each copper in a distorted tetrahedral configuration CuP_2NS.	(315)
—	—	—	1:1 adduct.	(705)

The X-ray structure of $Cu(en)_2(SCN)_2$ shows long Cu—S bonds of 3.27 Å (127) (see Table XXXIV), the infrared results indicate ionic or weakly S-bonded thiocyanates (69, 277), and electronic spectral measurements suggest a square structure for the solid that is solvated in solution (69). Similarly, infrared studies suggested $Cu(N,N'$-dimeen$)(SCN)_3$ (277), whereas X-ray studies showed the compound to be an N-thiocyanate (467). Thus, it is sometimes difficult to decide from spectral evidence whether or not the thiocyanate group is even bonded to copper(II), let alone the nature of that bond. Similarly, in $Cu(py)_2(NCS)_2$ the existence of the thiocyanate bridge (610) is not readily detected by spectroscopic measurements (201). In $Cu(NH_3)_2(NCS)_2$ (422) and $Cu(en)(NCS)_2$ (314) both bridging and N-bonded terminal thiocyanates occur. These conclusions and the effects of changing the sterochemistry make copper(II) thiocyanate complexes particularly difficult to review accurately. For example, in Table XXXV which collates known copper(II) thiocyanate complexes, although some of the compounds reported in Table XXXIV are formulated as $CuL_4(SCN)_2$, the Cu—SCN bond may be very weak and the compounds may be essentially square planar. A similar situation arises in the bridging compounds of the type $CuL_2(NCS)$, where a strong Cu—NCS bond may predominate and the bridge may be so weak as to be scarcely detectable or where more than one type of thiocyanate coordination may occur.

Notwithstanding the foregoing, a number of interesting changes can be detected in the thiocyanate coordination with different amines, especially with the polyamines. Thus, $Cupn_2(SCN)_2$ is a six-coordinate CuN_4S_2 solid which forms a 1:1 electrolyte in solution, and the presence of a weakly coordinating group, such as the solvent molecule or ClO_4^-, causes isomerization (72). Further, when L = tren in $[CuL(CNS)]^+$, an N-thiocyanato complex ion occurs, but for L = trien, S-bonding is present which is reversed by methyl substitution in L = $trienMe_6$ (72). The compound $Cuen_2(NCS)(ClO_4)$ contains a square planar $Cu(en)_2$ moiety with the nitrogen atom from NCS^- in the tetragonal positions (Cu—N = 2.73 Å), but not formally coordinated (169). It has been suggested that the presence of the perchlorate ion favors M—NCS for some stereochemistries in polyamine–copper(II) complexes (226). The heats of formation of some of these thiocyanate complexes have been measured (71).

The complexity of thiocyanate behavior toward copper(II) has also been exploited by McWhinnie with complexes of tri-2-pyridylamine (472, 473, 509). The results are shown in Fig. 5.

The differences in the powder photographs between the copper(II) and cobalt(II) salts of the reineckate anion have been attributed to the

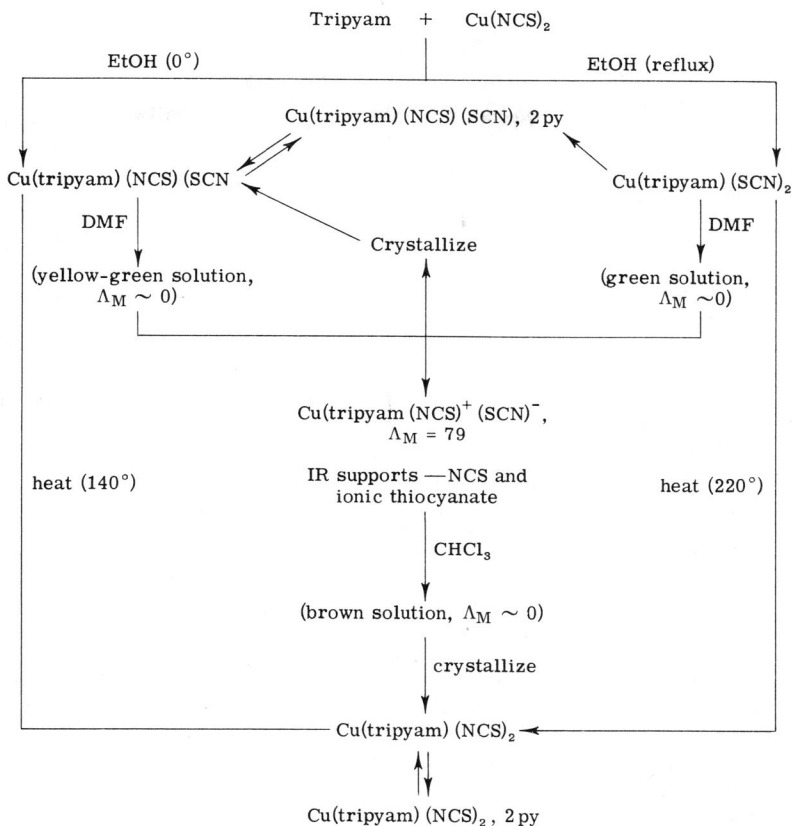

Tripyam + Cu(NCS)₂

EtOH (0°)   EtOH (reflux)

Cu(tripyam) (NCS) (SCN), 2 py

Cu(tripyam) (NCS) (SCN   Cu(tripyam) (SCN)₂

DMF   Crystallize   DMF

(yellow-green solution,   (green solution,
$\Lambda_M \sim 0$) $\Lambda_M \sim 0$)

Cu(tripyam (NCS)⁺ (SCN)⁻,
$\Lambda_M = 79$

heat (140°)   IR supports —NCS and   heat (220°)
                   ionic thiocyanate

CHCl₃

(brown solution, $\Lambda_M \sim 0$)

crystallize

Cu(tripyam) (NCS)₂

Cu(tripyam) (NCS)₂, 2 py

Fɪɢ. 5. The Cu(tripyam)(CNS)₂ system. [From Kulasingam (473).]

formation of a conventional Cr—NCS—Co bridge in the latter case, and
in the former to the more compact arrangement involving donation of
π electrons from NCS to copper(II) (583). Thus,

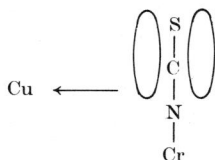

$$Cu \longleftarrow \overset{\displaystyle S}{\underset{\displaystyle \underset{Cr}{N}}{\overset{|}{\underset{|}{C}}}}$$

Bridging thiocyanate has been confirmed by X-ray studies in the
case of $Cu(NH_3)_2Ag(SCN)_3$ which contains a trigonal-bipyramidal CuN_5
with axial NH_3 (371).

The X-ray structure of $[Cu(PPh_2Me)_2NCS]_2$, which can be prepared

TABLE XXXV

SOME COPPER(II) THIOCYANATE COMPLEXES

Complex	Composition	References
Coordination number = 6		
Monodentate thiocyanate		
$Cu(NH_3)_4(SCN)_2$		*(609)*
$Cu(en)_2(SCN)_2$		*(69, 127, 277)*
$Cu(pn)_2(SCN)_2$		*(72, 277)*
$CuL_2(SCN)_2$	L = N,N-dimeen, N,N'-dieten	*(277)*
$Cu(MeBen)(SCN)_2$		*(593)*
$Cu(py)_4(NCS)_2$		*(201)*
$Cu(phen)_2(NCS)_2$		*(658)*
$Cu(bipy)_2(NCS)_2$		*(726)*
Bridging thiocyanate		
$Cu(py)_2(NCS)_2$		*(201, 502, 610, 726)*
$CuL_2(NCS)_2$	L = 2-pic, 4-pic	*(726)*
$CuL_2(NCS)_2$	L = 2,4-lut, 2,6-lut, 3,4-lut, 3,5-lut	*(537)*
$CuL_2(NCS)_2$	L = an, benzidine, o,o'-diaminobi-phenyl	*(502)*
$CuL(NCS)_2$	L = benzidine, m,m'-diaminobiphenyl-2,7-diaminofluorene	*(502)*
$CuL(NCS)_2$	L = piperazine, pyrazine	*(726)*
$CuL(NCS)_2$	L = bipy, phen, 5-NO_2-phen	*(266)*
$CuL(NCS)_2$	L = en, pn, 2Me-pn, N,N-dimeen, N,N'-dimeen, N,N'-dieten	*(277)*
$Cu(en)(NCS)_2$		*(314)*
$Cu(NH_3)_2(NCS)_2$		*(422)*
$CuHg(SCN)_4$		*(609)*
$[Cu(dpt)(NCS)(ClO_4)]$	dpt = 3,3′-diaminodipropylamine	*(72)*
Coordination number = 5		
$[Cu(tren)NCS]NCS$		*(69, 196, 630)*
$[Cu(trien)SCN]X$	X = NCS, ClO_4	*(69, 72)*
$[Cu(tpt)NCS]NCS$	tpt = 3,3′,3″-triaminotripropylamine	*(69)*
$[Cu(pn)_2NCS]ClO_4$		*(72)*
$[Cu(trienMe_2)NCS]X$	X = NCS, ClO_4	*(72)*
$[Cu(dpt)(NCS)_2]$	dpt = 3,3′-diaminodipropylamine	*(72)*
$[CuLNCS]NCS$	L = 2,3,5,6-tetrakis-(6-methyl-2-pyridyl)pyrazine	*(337)*
Coordination number = 4		
$[CuEt_4dienNCS]BPh_4$		*(159)*

by the reaction of CS_2 with the corresponding azide (*794*), has been described (*315*) (see Table XXXIV), and contains two bridging NCS groups. When Cu(CNS) is dissolved in molten $(n\text{-}C_2H_9)_4N \cdot NCS$, it is postulated that the species $[Cu_2(CNS)_6]^{4-}$ is formed with two bridging thiocyanates and two terminal S-thiocyanates on each copper (*468*). Thus, in compounds in which copper(I) has an apparent coordination number of two or three, bridging thiocyanates seem very likely due to the ability of the metal to achieve readily four coordination. Terminal N-thiocyanates have been reported for $[Cu(NCS)]_2(DPPA)_3$ (*180*):

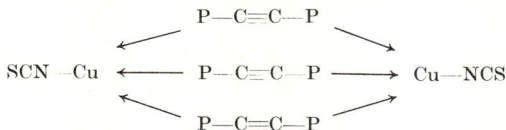

(two phenyl groups are omitted from each P for clarity)

In view of both the preceding and that bridging has been suggested in Cu(py)(SCN) (*502*), the compounds designated CuL(SCN) (L = py, 2-pic, 4-pic, 3,5-lut, quin), $(CuSCN)_4L_3$ (L = piperidine, pyrazine), $(CuSCN)_2L$ (L = piperazine), and $(CuSCN)_6bipy_5$ (*726*) need reexamining—they probably contain bridging groups in most cases as is found in $Cu_3[Cr(NCS)_6]$ (*779*).

Black copper(II) thiocyanate turns red-brown on heating, and it is postulated that the product is derived from copper(I) thiocyanate distorted by the inclusion of thiocyanogen in the lattice (*395*). A zigzag chain structure analogous to AgSCN (*492*) has been suggested for CuNCS on the basis of their similar infrared spectra (*502*).

The kinetics of the thermal decomposition of $Cupy_4(NCS)_2$ have been studied (*425*), as have the thermodynamics of the reaction of NCS with various polyamine complexes of copper(III) leading to the formation of bonds (*70*).

c. Selenocyanates. The reducing power of the selenocyanate ion has precluded all attempts to prepare anionic selenocyanate complexes of copper(II) (*295*), although some mixed-ligand complexes are known. Preliminary X-ray investigation suggests that $Cu(en)_2(SeCN)_2$ is probably isostructural with $Cu(en)_2(SCN)_2$ (*497*) so that all the problems associated with determining whether or not SCN was coordinated, described in the previous section, apply here also, and in particular to the compounds $CuL_2(SeCN)_2$ (L = en, pn, N,N'-dimeen, N,N'-dieten) which contain either ionic or Se-bonded selenocyanate (*217*). The compound $Cu(NH_3)_4(CNSe)_2$ and its monohydrate have been reported without structural data (*688*). Five-coordinate [CuLNCSe]NCSe [L = 2,3,5,6-tetrakis-(6-methyl-2-pyridyl)pyrazine] (*658*) and square planar

[Cu(Et$_4$dien)NCSe]BPh$_4$ (*159*) have been characterized. Bridging seleno-cyanate occurs in [CuMeben$_2$(NCSe)]NO$_3$ (*593*) and in the copper(II) salt of [Hg(SeCN)$_4$]$^{2-}$ (*714*).

2. Silver

a. Cyanates. The compound AgNCO contains infinite chains of —Ag—N—Ag—N—Ag zigzagging through the lattice with the angles N—Ag—N and Ag—N—Ag at 180° and 97.7°, respectively (*125*). The compound Bu$_4$N[Ag(NCO)$_2$] contains discrete approximately linear [Ag(NCO)$_2$]$^-$ (*1*), and comparative distances for the two structures are given in Table XXXVI. The [Ag(NCO)$_2$]$^-$ ion had been previously characterized by infrared (*45, 577*) and ^{14}N NMR spectroscopy (*192*). The AgNCO forms a high-temperature modification at 120°C before decomposing at 335°C to give CO, N$_2$, CO$_2$, (CN)$_2$ and NC·NCO [the last is conveniently prepared by this method (*338*)].

b. Thiocyanates. X-Ray measurements show the structure of AgNCS to consist of endless zigzag chains of —Ag—NCS—Ag—NCS—Ag—

TABLE XXXVI

COMPARATIVE STRUCTURAL DATA FOR AgNCO AND AgNCS
AND THEIR ANIONIC COMPLEXES

	AgNCO (*125*)	[Ag(NCO)$_2$]$^-$ (*1*)	
Distances			
Ag—N	2.115(8) Å	2.068(12) Å	2.015(13) Å
N—C	1.195(11) Å	1.111(18) Å	1.076(19) Å
C—O	1.180(11) Å	1.129(18) Å	1.200(17) Å
Angles			
N—Ag—N	—	177.2(5)°	
Ag—N—C	128.2°	170.0(13)°	172.1(12)°
N—C—O	178.2°	178.2(16)°	179.8(17)°
Ag—N—Ag	97.7°	—	—
	AgNCS (*493*)	(AgSCN)(SCN)$^-$ (*494*)	
Distances			
Ag—N	2.223(28) Å	—	—
N—C	1.186(68) Å	1.241(197) Å ^{a}	1.095(107) Å
C—S	1.636(29) Å	1.599(110) Å ^{a}	1.707(86) Å
Ag—S	2.428(11) Å	⎰2.474(20) Å ^{a}	2.654(19) Å
		⎱2.630(27) Å	2.742(29) Å
Angles			
N—Ag—S	164.50°	—	—
Ag—N—C	180°	—	—
N—C—S	180°	180° ^{a}	180°
Ag—S—C	103.79°	110° ^{a}	—

^{a} In molecular AgSCN.

with the sulfur atoms coordinating also to silver in neighboring chains (492). However, $NH_4[Ag(SCN)_2]$ has discrete molecules of AgSCN in a lattice containing NH_4^+ and SCN^- ions: each silver is surrounded by a distorted tetrahedron of sulfur (493). Zigzag —Ag—SCN—Ag—SCN— Ag— chains with cross-linking by sulfur also occur in $n\text{-}Pr_3PAgSCN$ (591, 744), $AgL_2(SCN)$ (L = thiosemicarbazide) (171), and $Agtu_2SCN$ (748). Under these circumstances the formulation of the compounds as S-thiocyanates is a formalism, as is reflected by the "normal" Ag—N and "long" Ag—S distances of 2.24(3) and 2.99(1) Å (shorter of two) relative to those in AgSCN (see Table XXXVI).

Compounds $AgL(CNS)$ [= PMe_3, $P(n\text{-}Bu)_3$, PEt_2Ph], $AgL_2(CNS)$ (L = PPh_3, $AsPh_3$, $SbPh_3$), and $Ag(PPh_2Et)_3(CNS)$ have been reported. The 1:1 complexes are assumed to have double-chain structures, the 1:2 complexes probably have dimeric structures similar to that found for $Ag(PPh_3)_2(CNS)$ (391), and the 1:3 compound appears to be mono-meric and contain an N-thiocyanate group (221).

Conductance studies in molten $(n\text{-}C_5H_{11})_4N \cdot NCS$ indicate the formation of $[Ag(SCN)_2]^-$ (430), but in propylene carbonate there is evidence for $[Ag(CNS)_3]^{2-}$ in addition to the former anion and AgSCN (213). The compound $Cu(NH_3)_2Ag(SCN)_3$ has $Ag(SCN)_3$ units but the nitrogen atoms are also coordinated to the copper(II) (371). The $[Ag(SCN)_2]^-$ anion has been reported with a number of different cations (514).

The infrared and Raman spectra of AgSCN and $[Ag(SCN_2]^-$ have been studied and suggest that the latter should indeed be formulated $(AgSCN)(NCS)^-$ in agreement with X-ray data—there were no spectro-scopically detectable units having strong bonds between SCN groups and silver ions (452).

The complexes $AgL(CNS)$ (L = 2-pic, 4-pic) have also been reported (594).

3. Gold

a. Cyanates. The ·compounds $Ph_4As[Au(NCO)_2]$ and $Ph_3PAuNCO$ have each been prepared by the action of CO on the corresponding azide (87); also $Ph_3AsAuNCO$ and $diphos(AuNCO)_2$ have been reported (150). Oxidation of these compounds with Br_2 gives $LAuBr_2NCO$ (L = Ph_3P, Ph_3As) and $diphosAu_2Br_4(NCO)_2$ (150). The compound $(Me_2AuNCO)_2$ has Au—N—Au bridges that are cleaved with L = Ph_3P or Ph_3As to give $cis\text{-}Me_2AuNCO \cdot L$ (706).

b. Thiocyanates. A preliminary report (507) has appeared of the X-ray structure of the gold cluster compound $Au_{11}(PPh_3)_7(SCN)_3$.

Although only the heavy atoms (Au, P, and S) have been located, the compound has a central gold atom surrounded by ten remaining gold atoms, each of which has one ligand attached to it; three of these ligands are, thus, *S*-thiocyanato groups. Some isomorphous compounds with other phosphines have been reported (*175*).

The square planar anion $[Au(SCN)_4]^-$ has been characterized by infrared (*428, 639, 663*) and electronic (*663*) spectral measurements. In addition to the normal routes to it, the anion is formed by an apparently two-stage reaction, in which the amine is also displaced, between $[Au(Et_2dien—H)X]^+$ (X = Cl, Br) and NCS^- (*780*). The $AuBr_4^-$ anion is reduced by NCS^- to an unspecified gold(I) compound (*428*). The anion $[Au(CNS)_2]^-$ has been identified in solution and the effect of temperature on its stability has been studied (*245*), but the parent compound has not been characterized structurally.

Mixed-ligand gold(I) thiocyanates have been reported. The compounds AuL(SCN) [L = PPh_3, $P(p-MeC_6H_4)_3$, $\frac{1}{2}$diphos, $PEtPh_2$, $P(p-ClC_6H_4)_3$], $AuL_2(SCN)$ (L = PPh_3, $PEtPh_2$), $Au[P(p-MeC_6H_4)_3]_3$-SCN and $Audiphos_2SCN$ have been prepared and their infrared spectra studied (*174*). This work has been confirmed for AuL(SCN) (L = PPh_3, diphos) and extended with the compound $AuAsPh_3(SCN)$ (*150*). The compounds $(CNS)Au \leftarrow PPh_2 \cdot C \vdots C \cdot PPh_2 \rightarrow Au(CNS)$ and $(AuCNS)_2$-$(DPPA)_3$ have been described, and the authors suggest S-bonding in the former and N-bonding in the latter, which is believed to have a structure similar to the analogous copper(I) compound (see Section IV, J, 1, b), without being able to obtain confirmatory evidence in either case (*180*).

Oxidation of the appropriate gold(I) compounds with Br_2 or $(SCN)_2$ yields $Au_2diphosBr_4(SCN)_2$ and $Au(PPh_3)Cl(SCN)_2$ (*150*). The thiocyanate-bridge compounds,

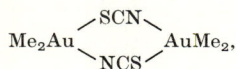

have been reported (*674*), and the bridge may be cleaved by various ligands to give *cis*-$Me_2AuL(SCN)$ (L = py, PPh_3, $AsPh_3$) (*707*).

The report (*556*) of the linkage isomers for $[Au(CN)_2(CNS)_2]^-$ has been shown to be erroneous, and $(Me_4N)[trans-Au(CN)_2(SCN)_2]$ has been characterized (*527*).

c. Selenocyanates. It has been possible to characterize $[Au(SeCN)_4]^-$ by infrared (*665*) and electronic (*663*) spectroscopy. The gold(I) compounds, AuL(SeCN) (L = PPh_3, $\frac{1}{2}$diphos), have been reported, but

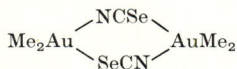

oxidation with Br_2 gives $AuLBr_3$ (150). The bridged gold(III) compound is stable and, although the bridge is cleaved quantitatively by Ph_3P in solution, no pure product was isolated (706).

K. ZINC, CADMIUM, AND MERCURY

1. Cyanates

Tetrahedral $[Zn(NCO)_4]^{2-}$ and $[Cd(NCO)_4]^{2-}$ have been reported as their Me_4N^+ salts by preparation in nonaqueous media (292, 296); force constants have been determined for the former and compared with the corresponding thio- and selenocyanate complexes (301). Even in the presence of excess of KNCO (158), $K[Cd(NCO)_3]$ precipitates from water and has been shown to contain

bridges (743). A similar type of bridging occurs in the dimeric mixed-ligand complexes, $ZnL_2(NCO)_2$ (L = 3-CNpy, 4-CNpy) (557) but $Zn(py)_2$-$(NCO)_2$ and $Zn(HMPA)_2(NCO)_2$ are monomeric tetrahedral molecules with monodentate N-bonded cyanate groups (665). Octahedral ZnL_4-$(NCO)_2$ occurs with L = 3-CNpy, 4-CNpy (557), $\frac{1}{2}$bipy, $\frac{1}{2}$phen (335). $Cd(bipy)(NCO)_2$ gas been reported (335). The ^1H NMR and infrared spectra indicate a donor-acceptor interaction between Me_4NNCO and $CdMe_2$ in solution (635). The compound $K_2[Hg(OCN_4]$ has been claimed on the basis of infrared measurements (742), but this formulation is not supported by ^{14}N NMR data (192), which do not, however, definitely eliminate an equilibrium mixture of N and O forms.

2. Thiocyanates

X-Ray structures of $[Zn(tren)NCS]NCS$ (32, 402) and $Zn(N_2H_4)_2$-$(NCS)_2$ (283) have been determined: the former contains the trigonal bipyramidal cation in an ionic lattice, and the latter is a trans-octahedral molecule. Preliminary results show $Zn(an)_2(NCS)_2$ to be tetrahedral (678). Only N-thiocyanato bonds are formed with zinc, whereas X-ray crystallography shows Cd—NCS—Cd bridges in $Cd(etu)_2(NCS)_2$ (182, 183) and in bis(thiocyanato)-N,N-diethylnicotinamide cadmium(II) (105), whereas $K_2[Cd(SCN)_4]\cdot 2H_2O$ contains cadmium octahedrally coordinated by 4S and 2N (796). Similarly, $Hg(SCN)_2$ contains mercury with a trans-octahedral HgS_2N_4 coordination but here the Hg—N bonds are very long at 2.81 Å (81). Similar Hg—N bond lengths (2.80 Å) are found in

ClHgSCN and BrHgSCN, in which octahedral mercury is achieved with bridging halides and thiocyanate to give $Hg(SX)(X_2N_2)$ coordination (797). The compound $K_2[Hg(SCN)_4]$ contains a tetrahedral anion (796), and $Hg(phen)_2(SCN)_2$ has a cis-octahedral structure (82). The compound $Hg(AsPPh_3)(SCN)_2$ has an apparent coordination number of three, but two nitrogen atoms complete an elongated trigonal bipyramid (510).

The tetrahedral anions $[Zn(NCS)_4]^{2-}$ (293, 599, 663) and $[Hg(SCN)_4]^{2-}$ (293, 663) have been characterized as their tetraalkyl ammonium salts and their bonding determined by infrared spectroscopy; force constant calculations have been carried out in the former (301). A Raman study of these ions and the corresponding cadmium complex in water confirms the preceding structural conclusions and indicates that $[Cd(NCS)_2(SCN)_2]^{2-}$ exists in solution (720). Similarly ^{14}N NMR measurements agree that $[Zn(NCS)_4]^{2-}$ exists, whereas the facts that the chemical shift for the cadmium complex was independent of concentration but showed a large solvent dependence indicated a lack of lability in addition to the existence of N and S isomers in kinetic equilibrium (390).

Although the infrared spectra show some evidence of distortion in the crystal lattice as the cation is changed, the mode of thiocyanate coordination remains the same in $[Zn(NCS)_4]^{2-}$ with Na^+, K^+, Cs^+, or NH_4^+ as cations (446) and with Mg^{2+}, Ca^{2+}, Sr^{2+} or Ba^{2+} (some thiocyanate bridges are possible with Mg^{2+}) (443). The cadmium complex anion shows a greater dependence on the nature of the cation, since the stability of the Cd—NCS—Cd bridge decreases from the situation in $K_2[Cd(SCN)_4] \cdot 2H_2O$ to the corresponding cesium salt, which is anhydrous, and, if bridges occur, they are very weak (741). Bridging thiocyanates are detectable in $M[Cd(SCN)_4]$ (M = Mg, Ca, Sr, Ba), but the bridges are broken when the salts are dissolved in water (443). A number of compounds containing the $Hg(SCN)_4$ fragment are known and have been discussed under the appropriate transition metal throughout this review—details are not repeated here. Both $ZnHg(SCN)_4$ and $CdHg(SCN)_4$ have been reported and they also appear to have structures dominated by Hg—S bonds, although some distortion of the coordination polyhedron of mercury is observed (741). Preliminary X-ray studies on $CdHg(SCN)_4$ show that it belongs to the $I\bar{4}$ space group with $a = 11.48$ and $c = 4.33$ Å (396) [cf. $CoHg(SCN)_4$: $I\bar{4}$; $a = 11.11, c = 4.38$ Å (259)]. Also $ZnPb(CNS)_4$ has been reported (259).

In view of the readiness with which these metals increase their coordination number and with which thiocyanate bridging occurs, it is not surprising that $M_4[Zn(NCS)_6]$ (M = Na, K, Cs, NH_4) have been reported (445) as well as $M_4[Cd(SCN)_6]n H_2O$ (M = Na, K, Cs, NH_4, $\frac{1}{2}Ba$), although when M = Cs or largely Cs in mixed cation compounds, some N-thiocyanato bonding is also indicated (448). Further, both bridging and N-thiocyanato groups occur in $M[Zn(NCS)_3]n \cdot H_2O$ (M = Na, K, Cs, NH_4) (441, 442), whereas in $M[Cd(SCN)_3] \cdot x H_2O$ (M = Na, K, Cs, NH_4, $\frac{1}{2}Ba$), S-thiocyanato bonding only is found for the cesium salt and in water, but the remainder as solids show bridging groups in addition (444).

Conductance studies have been interpreted to show the existence of $[Cd(CNS)_5]^{3-}$ in melts (430), whereas potentiometric and solubility measurements are explained by assuming the formation of $[Hg(CNS)_n]^{(2-n)}$ ($n = 1$–4) (197). The pH at which HgO is precipitated from $M[Hg(SCN)_3]$ varies from 7.71 (Na) to 9.98 (K) and 10.29 (NH_4): hydrogen bonding between NH_4 and NCS groups is assumed to weaken the Hg—S bond (230).

The compound $Hg_2(SCN)_2$ forms monoclinic crystals of space group $C2/c$ (258). The Raman spectra of $Hg(SCN)_2$ as a solid and in diglyme have been measured and compared with that of MeHgSCN (210); $[MeHg(SCN)_2]^-$ has also been characterized (632).

The electric moments of $Znpy_2(NCS)_2$ (and of the corresponding HMPA complex) show it to be a tetrahedral N-thiocyanato complex (655). This result conflicts with conclusions drawn from infrared studies which assign the compound as $Znpy_2(SCN)_2$ with the possibility of some bridging (42). Compounds $ZnL_2(CNS)_2$ (L = 2-pic, 3-pic) are reported to contain bridging —NCS— groups, and $ZnL_2(NCS)_2$ (L = 4-pic, quin, isoquin) as formulated (42). These results must be treated with caution since they include the only reported S-thiocyanatozinc(II) complex, since they show a thiocyanate sensitivity to a similar group of ligands, the like of which has not been observed elsewhere, and since the infrared spectral data tabulated appear very similar throughout the series of compounds. Other workers confirm the formulation $ZnL_2(NCS)_2$ (L = py, 2-pic, 3-pic, 4-pic, 2Et-py, 3Et-py, 4Et-py, 2,4-dime-py) and find no evidence for any NCS bridging (18). Other mixed-ligand thiocyanate complexes are listed in Table XXXVII.

The ^1H NMR and infrared spectra indicate a donor–acceptor interaction between Me_4NNCS and $CdMe_2$ in solution (635). Thermogravimetric studies have been made on $ZnL_2(CNS)_2$ and $CdL_4(CNS)_2$ (L = piperidine) (645), and the thermal decomposition of $Zn(py)_4(NCS)_2$ has been studied (94, 425). The stepwise stability constants of Zn^{2+}, Cd^{2+}, and Hg^{2+} with NCS⁻ have been determined (17).

TABLE XXXVII: Some Mixed-Ligand Thiocyanate Complexes of Zinc, Cadmium, and Mercury

Compound	Characteristics	References
Zinc		
$Zn(MeCS \cdot NH_2)_2(NCS)_2$	Tetrahedral	(553)
$Zn(tu)_2(NCS)_2$	Octahedral with bridging S (from tu)	(552)
$Zn(NIPA)_{3/2}(NCS)_2$	NIPA = nonamethylimidodiphosphoramide; bridging NCS	(243)
$ZnL_2(NCS)_2$	L = morpholine; octahedral with bridging NCS	(19)
Cadmium		
$Cd(MeCS \cdot NH_2)_2(NCS)_2$	Tetrahedral	(553)
$CdL_4(NCS)_2$	L = N-n-butyl imidazole; trans-octahedral	(631)
$K_4[Cd(SeCN)_4(SCN)_2] \cdot 0.5Me_2CO$	Possibly a double salt; NCSe bridging	(734)
$K_2[Cd(SeCN)_2(SCN)_2]$	Octahedral; NCSe bridging	(734)
$Cd[XC(NH_2)_2]_2(NCS)_2$	X = S, Se; trans-octahedral; X-bridging	(256, 552)
$CdL(NCS)_2$	L = py-NO, 2-pic-NO, 3-pic-NO, 4-pic-NO, 2,6-lut-NO; bridging NCS	(20)
$CdL_2(NCS)_2$	L = 4-CN-py-NO; bridging NCS	(20)
$Cd(NIPA)(NCS)_2$	NIPA = nonamethylimidodiphosphoramide; bridging NCS	(243)
$CdL_2(NCS)_2$	L = morpholine; octahedral with bridging NCS	(19)
$CdL_2(NCS)_2$	L = py, 2-pic, 3-pic, 4-pic, 3Et-py, 4Et-py, an, 2Me-an, 3Me-an, 4Me-an, 3Cl-an; dimeric octahedral with bridging NCS	(18)
$CdL(NCS)_2$	L = 2Et-py, 2,4-dimepy, 2,6-dimepy; polymeric with bridging NCS	(18)
Mercury		
Tetrahedral structures:		
$HgL(SCN)_2$	L = phen, bipy	(404)
$Hg(Ph_3P)_2(SCN)_2$		(239, 405)
$Hg(Ph_3As)_2(SCN)_2$		(239)
$HgTmen(SCN)_2$	Tmen = N,N,N',N'-tetramethylethylenediamine	(403)
$HgL_2(SCN)_2$	L = 3-CN-py, DMSO	(403)
$Hg(2\text{-}Etpy)_2(SCN)_2$		(18)
With bridging and terminal —SCN:		
$[Hg(PPh_3)(SCN)_2]_2$		(239, 404)
$[Hg(AsPh_3)(SCN)_2]_2$		(239)
$HgL(SCN)_2$	L = py, 2Me-py, 2,6-dime-py, 3Et-py	(18)
Unassigned:		
$Hg(phen)_2(CNS)_2$		(404)

3. Selenocyanates

The tetrahedral anions $[Zn(NCSe)_4]^{2-}$ (*162*, *295*), $[Cd(SeCN)_4]^{2-}$ (*24*), and $[Hg(SeCN)^4]^{2-}$ (*22*) have been characterized as well as the octahedral $[Cd(SeCN)_6]^{4-}$ (*24*). The compound $K[Ce(SeCN)_3]$ has been formulated as a double salt (*24*) containing the $[Cd_2(SeCN)_6]^{2-}$ anion with two bridging selenocyanate groups (*162*). Selenocyanate bridges are common and occur in $Cd(SeCN)_2$ (*24*) and $Hg(SeCN)_2$ (*22*) and in the just mentioned possible double salt (*24*). The compounds $KSeCN \cdot Hg(SeCN)_2$ (*734*), $M[Hg(SeCN)_3]$, and $M[Hg(SeCN)_4]$ (M = Cu, Pb) all contain bridging selenocyanate groups (*22*) which are also found in $M[Hg(SeCN)_4]$ [M = Zn, Co (*436*, *714*), Cd (*714*)]. The nature of the bridge is not always clear but if M—NCSe—M bridging occurs as expected, then cadmium and some of the other lighter elements are coordinated to nitrogen in some of these compounds. Preliminary X-ray data on $CdHg(SeCN)_4$ show electron density maxima at distances corresponding to both long and short Hg—Se distances (*737*).

A few mixed-ligand selenocyanate complexes have been reported for these metals. Tetrahedral $ZnL_2(SeCN)_2$ (L = py, 3-pic, quin, isoquin) and $Zn(4-pic)_2(SeCN)_2$ with selenocyanate bridges have been reported (*42*) but are subject to the same reservations as the corresponding thiocyanates. The compounds $ZnL_2(NCSe)_2$ (L = bipy, phen, antipyrene) (*23*) and $Zn(Me_2SO)_4(NCSe)_2$ (*23*, *689*) are trans-octahedral, whereas $Zn(quin)_2(NCSe)_2$ is tetrahedral (*23*). The hexamethylenetetramine (L) complex, $ZnL_2(NCSe)_4 \cdot 4H_2O$, has been reported (*735*).

The compound $Cd(py)_4(NCSe)_2$ appears to contain N-selenocyanato groups, but bridging occurs in $Cdpy_2(SeCN)_2$ (*24*), $Cd(Me_2SO)_2(SeCN)_2$ (*689*), and $K_2[Cd(SeCN)_2X_2]$ (X = Cl, Br, SCN) (734).

V. Factors Affecting the Mode of Chalcogenocyanate Coordination

The previous section has shown that chalcogenocyanate ions can coordinate using different donor atoms toward different metals and that a particular mode of coordination may be modified by a number of different circumstances. A variety of reasons have been offered to account for this behavior, and these explanations will now be examined. The thiocyanate group has been most widely studied and has furnished illustrations for many of the effects supposed to be important in determining the modes of coordination of these ions. Therefore, it will be discussed first followed by the selenocyanate group, which is similar in behavior in many respects; the cyanate group and how it differs from the first two groups will follow. Finally, the coordination behavior of all three groups will be compared.

It is appropriate first to make a few general comments on the nature of the interactions to be discussed and to emphasize that, unlike the problem of dual reactivity of ambidentate organic anions toward organic centers, recently reviewed by Shevelev (679), where kinetic factors predominate in importance over thermodynamic factors, in this case the reverse is true and thermodynamic factors are more important. Section IV listed the structures reported for the different complexes and indicated which of two linkage isomers, where they had been reported, was the thermodynamically more stable. Specific evidence that a given compound is thermodynamically stable and is not being prevented from isomerizing by a high activation energy is at best sparse and often lacking completely. However, in those cases where there is evidence for kinetic control [e.g., in the formation not only of $[Co(NH_3)_5NCS]^{2+}$ but also of $[Co(NH_3)_5SCN]^{2+}$ and their subsequent separation using an ion-exchange column (134)],* both linkage isomers have been detected and the relative thermodynamic instability of the latter demonstrated. Thus, available evidence indicates that any thermodynamically unstable isomers that are formed by kinetic control will isomerize either so rapidly as to be undetectable or at an appreciable rate at room temperature as is the case with those chromium(III)–SCN complexes that have been reported. This situation is entirely in accord with known features of coordination chemistry where kinetic control of products is generally less important than in organic chemistry where relatively fewer mechanistic pathways exist.

There is one property of the chalcogenocyanate complexes which may temporarily outweigh strictly kinetic or thermodynamic factors, namely, solubility. Precipitation of the less soluble of a pair of linkage isomers may give a false impression as to which is the stable compound. However, such is the wealth of compounds prepared, the majority of the structures may be regarded as the thermodynamically stable forms. This assumption is often supported by evidence concerning the relative stabilities of linkage isomers.

A. THIOCYANATES

1. Homogeneous Anionic Thiocyanate Complexes

The mode of thiocyanate coordination in known homogeneous anionic complexes is listed in Table XXXVIII. Although further

---

* The significance of this example is that, whereas cobalt(III) forms several S-thiocyanato complexes in the presence of ligands capable of participating in back-bonding, this is an example where a conventional σ-bonding ligand, such as NH_3, coexists with the $Co—SCN^{2+}$ moiety.

compounds will no doubt be prepared, the pattern appears to be well-established. Class b metals form S-thiocyanato complexes, whereas N-thiocyanato complexes are formed by class a metals. The distinction in the behavior of different types of metal toward the thiocyanate group was observed originally by Lindqvist and Strandberg (*493*), and this behavior parallels the division of metals into class a and b acceptors.

Not all ambidentate ligands containing nitrogen and sulfur as donor atoms behave like the thiocyanate group, so that the particular distribution of electrons in the ligand is in part responsible for its varied coordination behavior (*576*). Lewis et al. (*489*) made the assumption that (*a*) the lone pairs on the sulfur atom in NCS⁻ are more easily polarized than those on the nitrogen atom; (*b*) the permanent lone-pair dipole on the nitrogen atom is larger than that on the sulfur atom; and (*c*) the mode of coordination of the thiocyanate group will be decided by the relative bond energies of a covalent M—S bond and of a more ionic M—N bond. They then computed the ratio $R = IP(ne/r)$, where IP is the ionization potential in volts, ne is the formal charge on the ion, and r is the radius in angströms, and obtained a number between 5 and 20. They assumed that higher values of R indicated covalent bonding and hence S-thiocyanato complexes, and they obtained numbers for which the values for class b metals were generally larger than those for class a metals. A similar line of argument was developed by Williams and Hale (*785*) and extended to provide the most detailed description so far available of the factors governing classes a and b metal behavior, although without any particular reference to the thiocyanate group. This work also emphasized that the principle of hard and soft acids and bases developed by Pearson (*596, 597*) contained language that tended to obscure the individual effects giving rise to the overall behavior. Nevertheless, in spite of the carefully argued strictures by Williams and Hale (*785*), a commonplace description of thiocyanate coordination is that soft acceptors form S-thiocyanato complexes whereas the nitrogen end coordinates to hard acceptors. More examples of this will occur shortly.

A number of mixed metal complexes containing bridging thiocyanate groups have been described in the previous section. In general, the sulfur atom coordinates to the class b or softer metal and the nitrogen end of the bridge bonds to the class a or harder metal.

2. Mixed-Ligand Thiocyanate Complexes

Much of the work on mixed-ligand complexes containing the thiocyanate group stems from the discovery by Turco and Pecile in 1961

TABLE

Homogeneous Anionic

$Sc(NCS)_6^{3-}$	$Ti(NCS)_6^{3-}$	$VO(NCS)_5^{3-}$ $VO(NCS)_4^{2-}$ $V(NCS)_6^{3-}$	$Cr(NCS)_6^{3-}$	$Mn(NCS)_4^{2-}$ $Mn(NCS)_6^{4-}$
$Y(NCS)_6^{3-}$	$Zr(NCS)_6^{2-}$	$NbO(NCS)_5^{2-}$ $Nb(NCS)_6^{-}$ $Nb(NCS)_6^{2-}$	$Mo(NCS)_6^{3-}$ $Mo(NCS)_6^{2-}$ $Mo_2O_4(NCS)_6^{4-}$	$Tc(NCS)_6^{2-}$
	$Hf(NCS)_6^{2-}$	$Ta(NCS)_6^{-}$	$W(NCS)_6^{-}$ $W(NCS)_6^{2-}$	$Re(NCS)_6^{-}$ $Re(NCS)_6^{2-}$ $Re_2(NCS)_8^{2-}$

	$Pr(NCS)_6^{3-}$	$Nd(NCS)_6^{3-}$		$Sm(NCS)_6^{3-}$	$Eu(NCS)_6^{3-}$	
$Th(NCS)_8^{4-}$	$Pa(NCS)_8^{4-}$	$U(NCS)_8^{4-}$	$Np(NCS)_8^{4-}$	$Pu(NCS)_8^{4-}$		

(745) that the mode of coordination toward palladium(II) or platinum-(II) depends on the nature of the other ligands present. This observation was used by Basolo et al. (79, 143) in the rational synthesis of Pd(bipy)-(NCS)$_2$ and Pd(Ph$_3$As)$_2$(NCS)$_2$ and their linkage isomers. There are now many examples where a cooperative ligand effect acts to modify the mode of thiocyanate coordination. Table XXXIX summarizes the data contained in Section IV and indicates which of the thiocyanate atoms donates to the metals in their various oxidation states. Only terminal thiocyanates have been recorded, so that the complexes of copper(I), silver(I), and gold(I) have been largely ignored since the direct evidence is ambiguous and circumstantial evidence suggests that bridging thiocyanato groups are present in most of the complexes. The entries in Table XXXIX for these cases and for some others are based on very few actual compounds so that this table, unlike Table XXXVIII, may change considerably as further experimental data accumulate. In many cases where a metal bonds to either the nitrogen or the sulfur atom, depending on the ligational environment, linkage isomers have been prepared when factors favoring N coordination are about equally offset by those favoring S coordination. It is also anticipated that the number of these isomers will increase considerably.

3. Nature of Cooperative Effects in Mixed-Ligand Thiocyanate Complexes

Various explanations have been put forward to account for the effects of neutral ligands on the nature of the metal–thiocyanate bond: π-bonding, symbiosis, and antisymbiosis of hard and soft acids and bases as well as steric effects have all been discussed. In addition, directional

XXXVIII

THIOCYANATE COMPLEXES

Fe(NCS)$_6$$^{3-}$ Fe(NCS)$_4$$^{2-}$	Co(NCS)$_4$$^{2-}$	Ni(NCS)$_6$$^{4-}$ Ni(NCS)$_4$$^{2-}$	Cu(NCS)$_4$$^{2-}$	Zn(NCS)$_4$$^{2-}$ Zn(NCS)$_6$$^{4-}$
Ru(NCS)$_6$$^{3-}$	Rh(SCN)$_6$$^{3-}$	Pd(SCN)$_4$$^{2-}$	(AgSCN)(SCN)$^-$	Cd(NCS)$_2$(SCN)$_2$$^{2-}$ Cd(NCS)$_6$$^{4-}$
Os(NCS)$_6$$^{3-}$	Ir(SCN)$_6$$^{3-}$	Pt(SCN)$_6$$^{2-}$ Pt(SCN)$_4$$^{2-}$	Au(SCN)$_4$$^-$	Hg(SCN)$_4$$^{2-}$

Gd(NCS)$_6$$^{3-}$	Tb(NCS)$_6$$^{3-}$	Dy(NCS)$_6$$^{3-}$	Ho(NCS)$_6$$^{3-}$	Er(NCS)$_6$$^{3-}$	Tm(NCS)$_6$$^{3-}$	Yb(NCS)$_6$$^{3-}$

effects have been ascribed to the nature of the counterion in charged complexes and to the nature of the solvent in solution work. To some extent these effects overlap, but here an attempt will be made to discuss each effect separately.

a. Steric Effects. From the many structural details given in previous tables, it is apparent that the M—SCN bond angle is usually bent, and thus the *S*-thiocyanato group makes greater steric demands than the *N*-thiocyanato group which is usually linear. In some cases values as low as 160° have been recorded for M—NCS angles, but M—SCN angles in the region of 100° are relatively common. It is, of course, impossible to separate steric from electronic factors completely. Nevertheless, the formation of Pd(SbPh$_3$)$_2$(SCN)$_2$ in contrast to Pd(PPh$_3$)$_2$(NCS)$_2$ has been attributed to the larger antimony atom reducing the steric effects of the phenyl groups, whereas with the smaller phosphorus atom the phenyl groups are brought in and increase the overcrowding around the metal; this overcrowding is reduced by the thiocyanate group adopting a linear Pd—NCS bond (*78*). A clearer indication of such an effect is given by cations [Pd(dien)(SCN)]$^+$ and [PdEt$_4$dien(NCS)]$^+$ in which the introduction of four ethyl groups around the vacant coordination position causes the only example of the linear Pd—NCS grouping in a saturated amine complex of palladium(II) (*77*). Both kinetic and thermodynamic evidence support the steric effect in Et$_4$dien complexes (*379*). The formation of complexes containing both *N*- and *S*-thiocyanato groups in the same molecule in the presence of bidentate ligands containing two different donor atoms, such as Pd[Ph$_2$P(CH$_2$)$_3$NMe$_2$](NCS)-(SCN) (*198, 199*), has been ascribed to electronic effects, although the

TABLE XXXIX

Coordination of the Thiocyanate Group in Mixed-Ligand Complexes[a]

Sc(III)–N	Ti(IV)–N Ti(III)–N	V(IV)–N V(III)–N V(II)–N	Cr(III)–N/S Cr(II)–N/S Cr(O)–N	Mn(II)–N Mn(I)–N/S	Fe(III)–N Fe(II)–N/S Fe(O)–N	Co(III)–N/S Co(II)–N	Ni(II)–N	Cu(II)–N/S Cu(I)–N	Zn(II)–N/S
Y(III)–N	Zr(IV)–N	Nb(V)–N Nb(IV)–N	Mo(VI)–N Mo(IV)–N Mo(III)–N Mo(II)–N/S Mo(O)–N	Tc(IV)–N	Ru(IV)–N Ru(III)–N/S Ru(II)–N Ru(I)–N	Rh(III)–N/S Rh(I)–N/S	Pd(II)–N/S	Ag(I)–S	Cd(II)–N/S
La(III)–N	Hf(IV)–N	Ta(V)–N	W(IV)–N/S W(II)–N/S W(O)–N	Re(V)–N Re(IV)–N Re(I)–N	Os(III)–N	Ir(III)–N/S Ir(I)–N	Pt(IV)–S Pt(II)–N/S	Au(III)–S	Hg(II)–S

[a] M–N implies N-thiocyanato coordination. M–S implies S-thiocyanato coordination. M–N/S implies ambidentate behavior of NCS^- to the metal in that oxidation state.

bending of the S-thiocyanato group trans to the amine, away from the adjacent PhP grouping, was thought to be a steric effect. The importance of these steric effects becomes apparent in the closely similar $Pd(Ph_2-PCH_2CH_2PPh_2)(NCS)(SCN)$ where different types of thiocyanate coordination are observed trans to two equivalent phosphorus atoms and the nonlinear Pd—SCN is bent away from one of the neighboring Ph_2P groups (102). Carty $et\ al.$ (101) refer to unpublished results by Meek $et\ al.$ ($590a$) whose X-ray studies show that in $Pd(Ph_2PCH_2PPh_2)(SCN)_2$ the P—Pd—P angle is $73.24(6)°$, whereas it is $85.1(1)°$ in $Pd(Ph_2PCH_2\cdot-CH_2PPh_2)(NCS)(SCN)$ (102), and $89.31(4)°$ in $Pd(Ph_2CH_2CH_2CH_2PPh_2)$ (NCS_2). As the diphenylphosphine groups are allowed to occupy more space by the decreasing constraints of the increasing methylene chain, so the need for N-bonding becomes greater.

Interesting behavior is observed with $Pd(AsBu_3n)_2(SCN)_2$. As a solid it is S-bonded, on melting it partially isomerizes to give a mixed species, and on cooling it reverts to the S-bonded form (640). It is possible that steric effects increase on melting and that they account for the isomerization. Alternatively, lattice effects or packing considerations may tilt the balance in the solid toward S bonding, and, on removal of these constraints, the compound is free to adopt its preferred mixed structure.

Steric effects are possibly important in the formation of cis-$Co(DH)_2$-$(H_2O)NCS$, whereas the corresponding trans-compound is S-bonded, but they cannot be separated from the electronic effects of destroying the H bonding which maintains the planar $Co(DH)_2$ moiety (3). Steric effects are, however, apparent in the series of compounds $MnL_2(CO)_3$-(CNS) (280). The compounds mer-$MnL_2(CO)_3(NCS)$ (L = Ph_3P, Ph_3As, Ph_3Sb) have been characterized as have fac-$MnL_2(CO)_3(SCN)$ (L = Ph_3Sb). It was not possible to isolate fac-bistriphenylphosphine, but fac-Mn diphos$(CO)_3(NCS)$ was isolated together with a number of other N-thiocyanato complexes. Although it is not known whether CO or L is trans to the thiocyanate in these complexes, models do show that the fac-arsine and stibine complexes are considerably sterically strained. Since this strain is on one side of the molecule only, it is reduced by the formation of a bent Mn—SCN bond, whereas a linear Mn—NCS would exacerbate an already crowded situation. Further, the absence of the fac-bis(triphenylphosphine) complex is also consistent with these conclusions since the smaller phosphorus atom brings the phenyl group even closer to the metal and prevents the two neutral ligands from occupying positions mutually cis to each other.

   $b.\ Electronic\ Effects.$ Turco and Pecile (745) originally pointed out the cooperative effects of ligands with reference to the compounds Pd-$(NH_3)_2(SCN)_2$ and $Pd(PEt_3)_2(NCS)_2$. They explained these findings by assuming that the Pd—NH_3 interaction was σ in character and resulted in a small further increase of electron density on the metal, which enhanced the covalency of the Pd—SCN bond. On the other hand, backbonding occurred between the filled metal and the empty phosphorous d orbitals, and this strengthened the Pd—P bond while at the same time reducing the electron density on the metal. The metal was thus assumed to be converted to a state where it had more ionic character and, hence, preferred to form the more ionic Pd—NCS linkage. The discussion on steric effects in the preceding section suggests that these effects could provide as valid an explanation for the behavior of the two compounds just cited as does the π-bonding hypothesis. It has been observed (*194*) that the recent preparation of *cis*-Pd($Ph_2PC = CPPh_2$)$(SCN)_2$ involves a phosphine with better π-acceptor properties than that in *cis*-Pd(Ph_2-$PCH_2CH_2PPh_2$)(NCS)(SCN) (*102*), which, in turn, is contrasted with Pd$(PPh_3)_2(NCS)_2$ (*640*). Ignoring the electronic arguments and using the steric arguments cited previously, it is apparent that the P—Pd—P angle will decrease in the series $(Ph_3P)_2Pd$, $(Ph_2PCH_2CH_2PPh_2)Pd$, $(Ph_2PCH:CHPPh_2)Pd$, so that Pd—SCN bonding will become progressively more easy. The recent characterization of *trans*-Pd($Ph_2PC:CBu^t)_2$-$(SCN)_2$ and the careful study of the phosphine–thiocyanate interactions indicates that the N-bonded form would experience very similar steric interactions, so that the formation of the *S*-thiocyanate suggests that the electronic effects of $Ph_2PC:CBu^t$ on Pd(II) are insufficient to affect the mode of thiocyanate coordination (*101*). Such a conclusion is in agreement with the results of Venanzi (*605*, *766*), who has argued against the idea of appreciable double-bonding between platinum(II) and phosphine ligands.

Whatever the final conclusions concerning the reasons for many phosphine complexes of palladium(II) forming *N*-thiocyanates, there are a number of amine complexes where electronic effects are apparent and the most satisfactory explanation does involve the π-bonding hypothesis. The presence of electron-withdrawing substituents such as the nitro group in phenanthroline increases the ability of the ligand to form π bonds, and this would account for the compounds Pd(phen)$(SCN)_2$ and Pd(5-NO_2phen)(NCS_2). The same explanation serves for Pd(4,7-diph-phen)$(NCS)_2$ at low temperatures, while its isomerism at higher temperatures has been attributed to the increased thermal motion of the phenyl rings decreasing the intraligand conjugation (*103*). Similarly, the introduction of electron-donating substituents should decrease the ability of the ligand to form π bonds with the metal, and thus favor *S*-thiocyanato compounds. This is observed with the complexes Pd(bipy)-$(NCS)_2$ (*143*) and Pd(4,4'-dimebipy)(NCS)(SCN) *103*). The phenanthroline and bipyridyl series of compounds each provide some evidence for the π-bonding hypothesis, whereas the inability of 4-CNpy to form a complex other than Pd(4-CNpy)$_2$(SCN) has been attributed to its probable trans structure. It is not clear however, why the two parent chelate compounds should have different structures. Bertini and Sabatini (*103*) suggested that the σ-donor properties, as measured by the pK_a values, of the ligands were different and that this difference explained the different compounds formed. They therefore used 5Cl-phen which has a similar pK_a to bipy, but the former formed an *S*-thiocyanato complex so that pK_a alone does not provide a sufficient answer.

Electronic and steric effects have been compared and used to explain the mixed thiocyanate coordination when chelates with two different donor atoms are used (*526*), but it is now apparent that steric effects could be largely or even solely responsible.

It has been assumed that M(PPh$_3$)$_2$CO(NCS) (M = Rh, Ir) have a trans configuration, and that the N bonding is a consequence of the strong π-bonding effect of the *trans*-CO group (*155*). In the similar Rh(PPh$_3$)$_3$NCS (*35*), steric arguments could be used to explain the mode of coordination but there is no obvious reason why Rh(PPh$_3$)$_2$CO(NCS) (*155*) or Rh(PPh$_3$)$_2$L(NCS) (L = py, an, pip) should have different steric requirements from Rh(PPh$_3$)$_2$L(SCN) (L = C$_6$H$_6$, MeCN, Me$_2$CO, Et$_2$O) (*35*). It has been widely assumed but rarely established that these compounds have trans geometry. Unless the formation of cis compounds could somehow provide an explanation for the foregoing results, it seems that the rhodium(I) complexes taken together provide a clear example of a cooperative ligand effect that is electronic in origin, but which is not easily explained in terms of π bonding: Rh(PPh$_3$)$_2$pip(NCS) does not have the same π-bonding opportunities as Rh(PPh$_3$)$_2$CO(NCS) even if the thiocyanate coordination is the same.

In octahedral cobalt(III) complexes there is a very clear illustration of cooperative ligand effect as exemplified by the ions [Co(NH$_3$)$_5$NCS]$^{2+}$ (*134*) and [Co(CN)$_5$SCN]$^{3-}$ (*139*). Each of these has a linkage isomer but the preceding are the stable forms. In contrast to the previous discussion on class b metals, cobalt is class a and the ligands with strong π-bonding ability modify the expected N bonding to give the preceding *S*-thiocyanato complex. Table XXX illustrates the point most emphatically and, although [Co(bipy)$_2$(SCN)$_2$]$^+$ (*513*, *570*) and [Co(phen)$_2$(SCN)$_2$]$^+$ (*513*) probably have a cis configuration (*505*), in which case steric effects may in some way influence S coordination, such an explanation does not apply to [Co(CN)$_5$SCN]$^{3-}$ (*139*) nor to the *trans*-Co(DH)$_2$L(SCN) series of compounds (*572*). Thus, in these cobalt(III) complexes, there is clear evidence that ligands, such as CN^- and DH^-, which might be expected to take part readily in π bonding with the metal are the very ligands that encourage the formation of Co—SCN bonds. The effect is there, but the consequences are directly in opposition to that suggested first by Turco and Pecile.

The thiocyanate carbonyl compounds of manganese listed in Table XXIV follow a similar pattern to the cobalt(III) complexes. S Bonding occurs in $Mn(CO)_5SCN$ and in some tetracarbonyl derivatives (providing the ligand other than CO or SCN can play some part, which a second thiocyanate group apparently cannot) but is only present in those tricarbonyl compounds with the lopsided steric effect described in the previous section. Farona and Wojcicki (280) who first reported most of these compounds preferred to put the emphasis on Mn—CO π bonding and suggested that in $Mn(CO)_5SCN$ such π bonding is weak as a result of being spread over five CO groups so that the metal prefers the more polarizable sulfur atom which can, in turn, increase the negative charge and thus enhance π bonding between CO and Mn. As the extent of the participation of the metal in π bonding with its carbonyl groups increases (i.e., in the tricarbonyl compounds) the CO groups make less demand on the metal for negative charge, and the preference of the latter changes from the sulfur to the nitrogen atom.

The foregoing explanation, in which the extent of M—CO π bonding controls the polarizing ability of the metal and thus the preference of the metal for N or S coordination, can also be applied to cobalt(III) compounds, where ligands other than CO modify the polarizing power of the metal. These arguments, however, do not apply to the square planar palladium(II) case, just as the π-bonding arguments that appeared valid there do not apply for the class a metals.

There still remains the question, which is unanswered by either the π-bonding hypothesis or by the concept of metal polarizing power, as to why cobalt(III) and manganese(I) experience a cooperative ligand effect, whereas there is no evidence as yet for such an effect with chromium in various oxidation states [any Cr(III)—SCN compounds are purely kinetic in origin and thermodynamically unstable with respect to Cr(III)—NCS] with iron(II) or iron(III), cobalt(II), or nickel-(II), even though the stereochemistries of these last two elements may be drastically modified by ligational effects (see Tables XXIX and XXXI).

An alternative approach was initiated by Jörgensen (420), who drew attention to the fact that $[Co(NH_3)_5X]^{2+}$ is better bound for X = F than I and, thus, showed class a characteristics, whereas $[Co(CN)_5X]^{3-}$ is most stable with X = I, not known for X = F, and shows class b characteristics. The formation of $[Co(CN)_5SCN]^{3-}$ and $[Co(NH_3)_5\text{-}NCS]^{2+}$, in which the soft cyanide and S coordinate together, and the hard NH_3 and N do likewise, may be regarded as a further illustration of this effect of the ligands, which Jörgensen had termed *symbiosis* or a flocking together of like ligands. [It has been pointed out (*150*) that symbiosis, as applied to biological systems, refers to the living together of *dissimilar* organisms, in contrast to its use to describe the behavior of similar ligands in inorganic chemistry. However, it seems probable that Jörgensen's definition has been assimilated into the vocabulary of chemists, and this review follows that definition.]

Symbiosis can be used to account for the general coordination behavior of the thiocyanate group toward cobalt(III) and manganese(I) compounds described in the previous paragraphs. The compound *cis*-Co-$(DH)_2H_2O(NCS)$ (*3*) is an apparent exception, but it can be argued, notwithstanding steric effects, that destroying the $Co(DH)_2$ plane lessens the symbiotic effect of the dimethylglyoximato anions relative to that in *trans*-$Co(DH)_2(H_2O)(SCN)$. Similarly, $[Rh(NH_3)_5NCS]^{2+}$ (*659, 660*) is accountably the more stable isomer because of the hardening effect of the amine ligands. However, when palladium(II) complexes are considered, as has been seen for the previous explanations, the arguments predict the wrong effects: soft phosphine ligands encourage interaction with the hard nitrogen atom. Also, $Rh(PPh_3)_2CO(NCS)$ (*155*) and $Rh(PPh_3)_2(MeCN)(SCN)$ (*35*) are not explicable, and pK_a changes of the ligands in the phenanthroline and bipyridyl complexes (*103*) have the opposite effects to those expected if symbiosis were to apply.

Pearson (*598*) has introduced the further idea of *antisymbiosis* by applying the original concept of Chatt and Heaton (*189*), "that groups of high trans-effect, as ethylene in $Pt(C_2H_4)Cl_3^-$, render the position in mutual trans-position more susceptible to bonding by what are now known as hard bases," to his principle of hard and soft acids and bases. Pearson put the idea in the form: "two soft ligands in mutual trans positions will have a destabilising effect on each other when attached to class 'b' metal atoms." This extension, which, like the original concept of symbiosis, has a wider chemical application than that of explaining thiocyanate coordination, would thus predict that *cis*-palladium–phosphine complexes would contain *N*-thiocyanato groups, whereas the corresponding trans compounds would be S-bonded. With Rh-$(PPh_3)_2L(CNS)$ the same arguments are valid if it is assumed that all the compounds have trans geometry, in which case, for the stronger trans-directing ligands CO or piperidine, an *N*-thiocyanate is formed, whereas, for the weaker ligands MeCN and Et_2O, no antisymbiotic effect occurs and $Rh(PPh_3)_2L(SCN)$ is formed. It is not clear how this phenomenological approach should be applied to the bipyridyl and phenanthroline complexes, but it breaks down for the manganese carbonyl compounds. All the compounds in Table XXIV are assumed to have CO trans to the thiocyanate group, yet a variety of N- and S-bonded complexes are formed, so that, if antisymbiosis were to apply, a special explanation would be needed to account for the effects of the remaining four planar ligands: these range from four carbonyls in $Mn(CO)_5SCN$ to two carbonyls and a diphosphine in fac-$Mn(CO)_3$-diphosNCS via fac-$Mn(CO)_3py_2NCS$. The planar ligands in the dimethylglyoximatocobalt(III) compounds remain constant, but the ligand trans to the thiocyanate group varies considerably. Compounds $Co(DH)_2L(SCN)$ (L = —SCN, Ph_3P, H_2O, NH_3) are all unaffected by the nature of the solvent (572), whereas both linkage isomers have been isolated for L = py (574) and detected in certain solvents for L = Me (253), substituted pyridines and anilines, and for L = NO_2 (572). Such behavior is not consistent with antisymbiosis following approximately the same sequence as trans-directing ligands.

A further, semiquantitative approach has been made to the problem of providing a consistent explanation for the behavior of mixed-ligand thiocyanate complexes for both classes a and b metals (567). The distribution of electrons in the two most energetically available orbitals of the thiocyanate ion is unequal, being concentrated more on the nitrogen atom in σ_4 and on the sulfur atom in σ_3 (249). It was argued that both these orbitals would play an important part in any thiocyanate complex, and the hardness or softness of each donor atom in each orbital was calculated (567) according to the polyelectron perturbation treatment developed by Klopman (454). The treatment takes account of the varying dielectric constant of the solvent in which the reaction takes place, and the results are presented in Fig. 6 which shows how these hardness or softness parameters vary with 1-$(1/\epsilon)$ (Klopman's equations contain the dielectric constant in this form and it is retained in order to give a convenient linear plot). Inspection of Fig. 6 reveals that the hardest center available is the nitrogen atom of σ_4, so that this is the most likely donor to the hardest Lewis acid, namely, a class a metal surrounded by hard ligands (e.g., $[Co(NH_3)_5NCS]^{2+}$). The softest center is also a nitrogen atom but this time of σ_3, and it will bond to the softest Lewis acid available which this time would be a class b metal surrounded by soft ligands [e.g., $Rh(PPh_3)_2CO$-(NCS) or Pd(II) surrounded by phosphine or arsine ligands]. Intermediate situations involving class a metals with soft ligands or class b metals with hard ligands would each prefer to coordinate with the sulfur atom, as has often been observed in the previous pages. The arguments can be extended to discuss solvent effects also, as will be described in the next section.

The advantage of this explanation of thiocyanate coordination is that it does allow a description that fits both classes a and b metal complexes, although there are some examples that do not fit. It is not clear why in $Rh(PPh_3)_2L(CNS)$ the N isomer should be obtained for $L = pip$, and the S isomer for $L = MeCN$ (35). Manganese(I) has to be treated as a class a acid, the same approach as is used for it to fit with

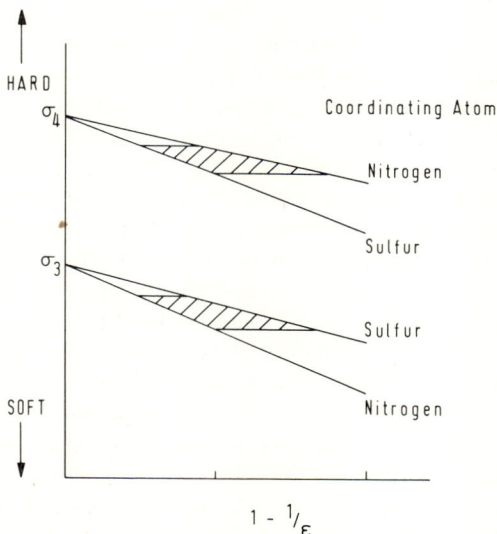

FIG. 6. Variation in softness character of donor centers on CNS⁻ with dielectric constant.

cobalt(III) in the discussion of π-bonding effects. Since neither of these metals exist as the hydrated species, they cannot be classified in the same way as the other metals, but only in their commonly ligated forms, such as $Mn(CO)_5^-$ or $[Co(NH_3)_5\text{-}]^{3+}$, where a different situation prevails. The treatment is not supported by recent, X-ray, electron spectroscopy measurements (384) on the $Pd(3d_{3/2})$ electrons of $Pd(AsPh_3)_2(NCS)_2$ and $Pd(AsPh_3)_2(SCN)_2$, with binding energies of 6.5 and 5.8 eV, respectively. These results imply that the electrons on the metal are less tightly held for $Pd(AsPh_3)_2(SCN)_2$, but care must be excercised in the use of such data, and the fact that the valence electrons $Pd(4d_{3/2})$ have identical binding energies in the two compounds may also be significant.

The major disadvantage of this semiquantitative approach, of the π-bonding treatment of Turco and Pecile, and the qualitative soft and hard acids and bases treatment is that any π bonding between the thiocyanate group and the metal is neglected. Gutterman and Gray (*359*) have assigned the electronic absorption spectra of the series $[Co(CN)_5-X]^{3-}$ ($X = $ —SCN^-, NCS^-, $NCSe^-$, and N_3^-) in terms of a molecular orbital scheme for the complex anion. They obtain no evidence for the participation of thiocyanate π-acceptor orbitals in the metal–ligand bonding, but they do find that the highest filled orbital, the 2π, contributes significantly to the absorption spectrum. The distribution of electrons in this orbital is particularly sensitive to bond distances. There is an approximately equitable distribution in the free ion, but when the group is N-bonded the 2π electrons are concentrated more on the sulfur atom. From the position of —SCN at the low field end of the spectrochemical series, and from the preceding results, the authors conclude that —NCS behaves essentially as a σ donor, whereas the often bent M—SCN bond results from the overlap of these 2π electrons with an empty metal orbital, in addition to the normal σ overlap.

c. Solvent Effects. Burmeister *et al.* (*154*) originally observed that the nature of the solvent may modify the mode of thiocyanate coordination of certain complexes in solution. They studied the behavior of Pd(As-Ph$_3$)$_2$(NCS)$_2$ and its linkage isomer in a number of different solvents, and concluded that Pd—SCN bonding was promoted by solvents with high dielectric constants, whereas solvents with low dielectric constants resulted in a mixture of Pd—NCS, Pd—SCN, and Pd—SCN—Pd bonding modes. An extension of the work (*155*) revealed that the compounds M(PPh$_3$)$_2$CO(NCS) (M = Rh, Ir) were not susceptible to these solvent effects but that PdL$_2$(NCS)$_2$ (L = Ph$_3$P, Ph$_3$As, $\frac{1}{2}$bipy) isomerized to PdL$_2$(SCN)$_2$ completely in DMF, DMSO, py, Me$_2$CO, MeCN, PhCH$_2$CN, or adiponitrile (NC(CH$_2$)$_4$CN)—solvents that were referred to as group A. In group B solvents (C$_6$H$_6$, CCl$_4$, CHCl$_3$, CH$_2$Cl$_2$, cyclopentanone, cyclohexanone, PhNO$_2$, 2-butanone, 3-pentanone), the experimental behavior was rationalized in terms of equilibria involving the bridged species as follows:

$$2Pd(AsPh_3)_2(SCN)_2 \;\rightleftharpoons\; [Pd_2(AsPh_3)_2(SCN)_4] + 2Ph_3As \;\rightleftharpoons\; 2Pd(AsPh_3)_2(NCS)_2$$

The possible existence of the mixed compound Pd(AsPh$_3$)$_2$(NCS)(SCN) was also considered, but could not be confirmed in solution. Similar behavior was observed for the analogous Ph$_3$Sb complex, but no bridged intermediate was observed for the Ph$_3$P complex. A third group of solvents (MeOH and EtOH) showed intermediate behavior for Pd-(AsPh$_3$)$_2$(SCN)$_2$, and solubility difficulties prevented further studies of this system. The induction period before changes occurred, which was observed only for this group of solvents, was attributed to the necessity of breaking hydrogen bonds.

The authors drew attention to the fact that group A solvents had generally larger dielectric constants, dipole moments, and internal pressures than those in group B, although exceptions did occur: acetone and py belong to group A, whereas 2-butanone and $PhNO_2$ belong to group B. They then argued that the results were consistent with the conclusions of Klopman (454) that solvents with high dielectric constants tend to enhance frontier-controlled (largely covalent) reactions, whereas those with low dielectric constants favor charge-controlled (largely ionic) reactions. Thus, group A and B solvents should encourage Pd—SCN and Pd—NCS bonding, respectively, as is observed, and the implication of this interpretation is that a similar effect should be observed in other systems.

However, the behavior of $Co(DH)_2py(SCN)$ shows the opposite effect to that observed for the palladium complexes and the solvent dependence of this compound may be represented (571):

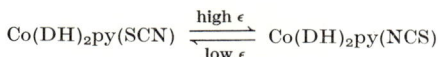

$$Co(DH)_2py(SCN) \xrightleftharpoons[\text{low } \epsilon]{\text{high } \epsilon} Co(DH)_2py(NCS)$$

Burmeister and co-workers (155, 370) challenged these results and suggested that they were due to kinetic rather than thermodynamic effects—the activation energy for the isomerization being lowered by high dielectric solvents. However, Marzilli et al. (269, 519) studied the behavior of the compound $Co(DH)_2(t\text{-Bupy})(SCN)$ in solution, using the proton NMR spectrum of the t-butyl group as a probe. They found that an equilibrium to type reported did indeed exist, and that Burmeister's results and, hence, conclusions were invalidated by the presence of traces of cobalt(II) that catalyze the isomerization.

The observation (571) that class a metals apparently show a different solvent dependence of thiocyanate coordination to class b metals was discounted by Marzilli (519) on the grounds that cobalt(III) in these systems has become sufficiently soft for it to be regarded as similar to palladium(II). The problem of the relative hardness or softness of metals was discussed in the previous section of this review from which it is clear that the association of cobalt(III) with palladium(II) in these terms must be viewed with extreme caution.

Further experimental evidence for a solvent effect is limited. The compound $Mn(CO)_5SCN$ is stable as a solid, but forms the N-isomer in acetonitrile (277). This, the isomerization of $Pd(AsBu_3)_2(SCN)_2$ on melting (640) and the partial isomerization of $Pd(Ph_2AsCH_2CH_3AsPh_2)$-$(SCN)_2$ in dichloromethane (526) all involve changes of state and are not comparable with the previous results where an equilibrium exists between the two forms in solution. The stability of $[Co(CN)_5SCN]^{3-}$ in water and the N-bonded isomer in CH_2Cl_2 (368, 369) is, however, a further example of an equilibrium situation, but it is not yet clear whether arguments that apply for aprotic solvents may be used when hydrogen bonding is also present. It does seem reasonable that, if the factors modifying the coordination behavior of the thiocyanate group are in approximate balance and linkage isomerism is observed, then the fine balance of ligational effects may be perturbed by the nature of the solvent. Although too few compounds have been studied for a firmly based generalization, it begins to appear that, for linkage isomers in aprotic solvents, M—NCS bonding is favored for class a metals in solvents with high dielectric constants and for class b metals for solvents with low dielectric constants. In addition, M—SCN bonding is promoted under the opposite conditions (class a and b metals in the presence of solvents with low and high dielectric constants, respectively).

TABLE XL

SUMMARY OF COOPERATIVE LIGAND AND SOLVENT EFFECTS
ON THIOCYANATE COORDINATION

Metal	σ-Donor ligand	Solvent		π-Acceptor ligand
		High ϵ	Low ϵ	
Class a	—NCS	—NCS	—SCN	—SCN
Class b	—SCN	—SCN	—NCS	—NCS

There is some parallel in the effects caused by ligands, where soft or π-bonding ligands favor M—NCS bonds for class b metals and M—SCN bonds for class a metals, and hard ligands have the opposite effect. The situation may be summarized tentatively using Table XL where the solvent effect is apparent only for intermediate ligand situations, and the use of the terms σ donor and π acceptor is in a descriptive sense rather than necessarily implying an understanding of the nature of the effect. It is also assumed that, under certain conditions, steric effects could override the solvent or even the ligand effects (see Section, V, A, 3, a).

Figure 6 may be used (567) to provide a self-consistent explanation for the behavior summarized in Table XL if it is assumed that the metal is unaffected by changes in the solvent, but that the hardness and softness of the electrons in σ_4 and σ_3 on nitrogen and sulfur do change. As the dielectric constant of the solvent increases, the class a metal, which had previously preferred to form an M—SCN bond, gradually changes its preference until an M—NCS bond is preferred (see areas with diagonal lines on Fig. 6). If, however, a class b metal is considered, then the opposite effect is observed in agreement with the observed experimental situation.

d. Counterion Effects. Solid [PtEt₄dien(SCN)]NCS isomerizes completely in 3 days, but under the same conditions there is no isomerization of [PdEt₄dien(SCN)]PF₆ (78). The steric effects that favor formation of the [PdEt₄dien(NCS)]⁺ cation have been detailed in Section V, A, 3, so that the stability of the hexafluorophosphate salt just mentioned is of interest. This apparent reversal of stability is illustrated more strikingly by the solid state isomerization of [PdEt₄dien(NCS)]BPh₄ to give [PdEt₄dien(SCN)]BPh₄ in which the interaction of the sulfur atom with the phenyl groups of the anion in the crystal overrides its interaction with the ethyl groups of the coordinated amine (156).

A similar type of effect has been observed with cations: K₃[Co(CN)₅-SCN)] is the stable form in the solid state (139) although its linkage isomer is also known (708). When the cation was replaced by (n-Bu₄N)⁺ it was apparently impossible to isolate the pure, solid, S-bonded compound, and the mixed solid product isomerized to give (n-Bu₄N)₃-[Co(CN)₅(NCS)] in 3 days (358). It was suggested (358) that the stabilization of the N-bonded isomer is due to an electronic effect by which the polarizable end of the thiocyanate group is better accommodated by the soft nonpolar environment of the (n-Bu₄N)⁺ cation, whereas the hard K⁺ undergoes a more favorable interaction with the hard nitrogen atom. This explanation would predict the wrong result if applied to the palladium example in the previous paragraph.

By way of contrast, the study of a series of compounds $CuL_n(CNS)_x$ $(ClO_4)_{2-x}$ $(x = 0, 1, 2; n = 1, 2; L = $ bi-, ter-, and quadridentate amines) suggests that, whatever the coordination in the bisthiocyanato complexes, the substitution of a perchlorate ion for a thiocyanate ion favors the coordination of the remaining thiocyanate by the nitrogen atom when the resulting complexes are five-coordinate (i.e., when $L_n = $ bisdiamine or a tetramine); for tridentate ligands, five coordination is achieved by Cu—NCS—Cu bridging (72, 124).

There are too few examples of the effect of the counterion on the coordination of the thiocyanate group for any generalization to be apparent. Detailed crystal structure determinations are required in order to reveal whether any effects other than those associated with crystal packing are occurring.

B. Selenocyanates

1. Homogeneous Anionic Selenocyanate Complexes

Table XLI summarizes the structures of known, homogeneous, anionic selenocyanate complexes. Although fewer examples exist than for the corresponding thiocyanate complexes, it is apparent that a similar pattern exists for the two sets of complexes. Class a metals are coordinated by the nitrogen atom, whereas selenium is the donor for the class b metals. In mixed metal complexes containing bridging selenocyanate groups the nitrogen atom coordinates to the harder or class a metal, and the selenium atom bonds to the softer or class b metal, as was observed with the corresponding thiocyanate complexes.

2. Mixed-Ligand Selenocyanate Complexes

Table XLII summarizes the different types of coordination behavior for the selenocyanate group. Relatively few examples were cited in Section IV so that, like Table XXXIX, which contains the corresponding data for thiocyanate complexes, Table XLII is likely to alter considerably in the ensuing years.

3. Nature of Cooperative Effects in Mixed-Ligand Selenocyanate Complexes

Of the data leading to Table XLII, the results for the copper(II) complexes are complicated by long and short or weak and strong bonds as indicated in Section IV, J, 1, c and will not be discussed further; similarly, the iron(II) examples are the linkage isomers $cpFe(CO)_2NCSe$ and $cpFe(CO)_2SeCN$ (410). Because the remaining compounds are few

TABLE

Homogeneous Anionic

	$Ti(NCSe)_6^{2-}$	$VO(NCSe)_4^{2-}$ $V(NCSe)_6^{3-}$	$Cr(NCSe)_6^{3-}$	$Mn(NCSe)_6^{4-}$ $Mn(NCSe)_4^{2-}$
$Y(NCSe)_6^{3-}$	$Zr(NCSe)_6^{2-}$		$Mo(NCSe)_6^{3-}$	
	$Hf(NCSe)_6^{2-}$			$Re_2(NCSe)_8^{2-}$
$Pr(NCSe)_6^{3-}$	$Nd(NCSe)_6^{3-}$		$Sm(NCSe)_6^{3-}$	
$Pa(NCSe)_8^{4-}$	$U(NCSe)_8^{4-}$			

and because the nature of the ligational effect is even less clear than for thiocyanates, no attempt has been made to divide the following discussion into subsections.

Most palladium(II) selenocyanate complexes have a Pd—Se bond, but the compound $[Pd(Et_4dien)(SeCN)]BPh_4$, isolated at low temperatures, isomerizes in a number of different solvents via a dissociative process, whereas $[Pd(dien)(SeCN)]BPh_4$ shows no signs of such isomerization. Further, if $[Pd(Et_4dien)NCSe]BPh_4$ is isolated it reisomerizes to the Se-bonded form at room temperature in the solid state (153, 156). This behavior parallels that of the thiocyanate group under similar circumstances and provides evidence for a steric effect modified by the nature of the anion.

Electronic effects account for the variations in selenocyanate coordination toward cobalt(III) recorded in Section IV, H, 3, and again the pattern is similar to that for the corresponding thiocyanate complexes. The pentamine cobalt(III) moiety bonds with nitrogen, but *trans*-cobaloxime complexes are S- or Se-coordinated: *cis*-$[Co(DH)_2(H_2O)(NCSe)]$ having a structure affected by steric hindrance (3). The explanations that have been put forward for cobalt(III) thiocyanates could, therefore, apply here where complications due to palladium(II) complexes do not exist. There is, however, one example of cooperative ligand effects with a class b metal in the form of the compounds $Rh(PPh_3)_2CO(NCSe)$ (150) and $Rh(PPh_3)_2MeCN(SeCN)$ (36). Thus, there are indications that a very similar set of arguments to those given for the thiocyanate group may be rehearsed for the selenocyanate group also, including a similar lack of definite conclusions.

XLI

SELENOCYANATE COMPLEXES

$Fe(NCSe)_6^{3-}$	$Co(NCSe)_6^{4-}$	$Ni(NCSe)_6^{4-}$		$Zn(NCSe)_4^{2-}$
$Fe(NCSe)_6^{4-}$	$Co(NCSe)_4^{2-}$			
$Fe(NCSe)_4^{2-}$				
	$Rh(SeCN)_6^{4-}$	$Pd(SeCN)_4^{2-}$		$Cd(SeCN)_4^{2-}$
				$Cd(SeCN)_6^{4-}$
	$Pt(SeCN)_6^{2-}$		$Au(SeCN)_4^{-}$	$Hg(SeCN)_4^{2-}$
	$Pt(SeCN)_4^{2-}$			
	$Dy(NCSe)_6^{3-}$	$Ho(NCSe)_6^{3-}$		$Er(NCSe)_6^{3-}$

TABLE XLII

Coordination of the Selenocyanate Group in Mixed-Ligand Complexes[a]

Ti(IV)–N	V(IV)–N V(III)–N	Cr(III)–N Cr(II)–N	Mn(II)–N Mn(I)–N	Fe(III)–N Fe(II)–N/Se	Co(III)–N/Se Co(II)–N	Ni(II)–N	Cu(II)–N/Se	Zn(II)–N
Zr(IV)–N		Mo(III)–N Mo(II)–Se		Ru(III)–Se	Rh(I)–N/Se	Pd(II)–Se/N		Cd(II)–Se
Hf(IV)–N		W(II)–Se	Re(IV)–N		Ir(III)–N Ir(I)–N	Pt(IV)–Se Pt(II)–Se	Au(III)–Se	Hg(II)–Se

[a] M–N implies N-selenocyanato coordination. M–Se implies Se-selenocyanato coordination. M–N/Se implies ambidentate behavior of NCSe⁻ to the metal in that oxidation state.

C. Cyanates

1. Homogeneous Anionic Cyanate Complexes

The homogeneous cyanato anions are presented in Table XLIII. Although the majority are N-cyanato complexes, $[Mo(OCN)_6]^{3-}$, $[Re(OCN)_6]^{2-}$, $[Re(OCN)_6]^-$ (56), and $[Hg(OCN)_4]^{2-}$ (724) have been characterized by the positions of the fundamental cyanate vibrations; only for the last compound has any confirmatory measurement been attempted, in which case ¹⁴N NMR showed a relatively small upfield shift, characteristic for an N-bonded compound (192).

2. Mixed-Ligand Complexes

Only one pair of linkage isomers has been reported in the solid state, and here the compounds $Rh(PPh_3)_3NCO$ and $Rh(PPh_3)_3OCN$ are characterized by differences in the positions and intensities of the infrared peaks (37). The possibility is that trans-$PtH(PEt_3)_2NCO$ has a linkage isomer in solution, but only ¹H NMR shifts are available for this compound (see Section IV, I, 3, a).

No mixed-ligand cyanate complexes of molybdenum(III), rhenium-(IV), rhenium(V), or mercury(II) are known, but a large number of mixed ligand N-cyanato complexes involving a variety of metals have been reported, including the compounds mer-$MoO(NCO)_2L_3$ (L = Et_2PhP, Me_2PhP) (166). These suggest that an investigation of cooperative ligand effects in molybdenum cyanate complexes would be interesting. The few reported mixed ligand O-cyanates are $K[Cupic_2-(OCN)]$ (321) and $cp_2M(OCN)_2$ (M = Ti, Zr, Hg) (145, 146), and the situation regarding the latter series well illustrates the complexity and difficulty of determining whether or not such compounds do exist. The series was reported in 1969 on the basis of mass spectroscopic and infrared shifts even though $cp_2Ti(OCN)_2$ had a rather different infrared spectrum to the others. Subsequent ¹⁴N NMR shifts were in accord with the proposed structure for the titanium compound which was the only one measured (93). However, Burmeister et al. (411), who originally reported the compounds, later found that dipole moment measurements on the titanium and zirconium compounds (the hafnium compound being insufficiently soluble for useful measurements) were incompatible with both being O-cyanates. These workers favored $cp_2Zr(OCN)_2$ which would be in agreement with the homogeneous anionic complexes but casts doubt on the use of ¹⁴N NMR for diagnostic purposes and also on the claim of linkage isomers. However, the alternative choice of $cp_2Ti(OCN)_2$ also presents its own problems. All that can be said at present is that there

TABLE XLIII

Homogeneous Anionic Cyanate Complexes

$Cr(NCO)_6^{3-}$	$Mn(NCO)_4^{2-}$	$Fe(NCO)_4^{-}$ $Fe(NCO)_4^{2-}$	$Co(NCO)_4^{2-}$	$Ni(NCO)_4^{2-}$	$Cu(NCO)_4^{2-}$	$Zn(NCO)_4^{2-}$
$Mo(OCN)_6^{3-}$				$Pd(NCO)_4^{2-}$	$Ag(NCO)_2^{-}$	$Cd(NCO)_4^{2-}$
$Re(OCN)_6^{2-}$				$Pt(NCO)_4^{2-}$	$Au(NCO)_2^{-}$	$Hg(CNO)_4^{2-}$ [a]

[a] See Text.

is an urgent need for some X-ray structure determination of O-cyanato complexes. (See Appendix, p. 382.)

Two types of bridging cyanate have been confirmed by X-ray crystallography. In AgNCO only the nitrogen atom is involved (125) but in [Ni$_2$tren$_2$(NCO)$_2$](BPh$_4$)$_2$ the bridge is of the type,

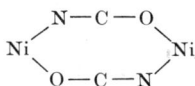

thus establishing unequivocally the involvement of the cyanate oxygen in coordination for the first time (265).

In view of the foregoing situation and the uncertainties of whether or not O-cyanates do indeed exist or whether all the many N-cyanato complexes have been correctly characterized, any explanations concerning the behavior of the cyanate group would be premature. INDO calculations showed that almost equal electron densities existed at nitrogen and oxygen (383), so there is no apparent reason why both ends should not be involved in coordination. Attention has been drawn (86) to the fact that many of the metals found to form O-cyanates have vacant or only partly filled $d\pi$ orbitals to interact with filled π orbitals of the cyanate group. Whether these are centered on the oxygen or whether that atom forms the better σ bond with these mostly class a metals remains to be seen.

VI. Comparative Coordination Chemistry of the Chalcogenocyanate Groups

The preceding sections have shown that the behavior of the thiocyanate and selenocyanate groups in homogeneous complex anions is identical: in both cases the nitrogen atom coordinates to the class a metal, and the sulfur or selenium atoms coordinate to the class b metal. Although fewer comparable cyanate complexes are known, it is quite clear that the coordination of the cyanate group is not governed by the same factors.

Similarly, with mixed-ligand complexes it is apparent that selenocyanate complexes do follow similar patterns of behavior to thiocyanate complexes [e.g., with cobalt(III); see Section IV, B, 3] although ligational changes have a smaller effect. Thus, the pentacyanocobaltate-(II) moiety forms linkage isomers with the thiocyanate group, but the N-bonded [Co(CN)$_5$NCSe]$^{3-}$ is the only selenocyanate species. Similarly the palladium complexes listed in Table XLIV show the sensitivity of the thiocyanate group to ligational change to be much greater than that of the cyanate and selenocyanate groups. At the time of writing there are no reported tellurocyanate complexes for further comparisons.

When the bridging behavior of these groups is compared, another difference emerges. Generally, thiocyanate and selenocyanate bridges have the form M—NCX—M, whereas cyanate forms M—N(CO)—M bridges in most cases, although the three-atom bridge is known. Further, the cyanate group does not form analogous compounds to the extensive series of M(NCX)Hg reported for X = S, Se.

These facts, together with the remarkably few O-cyanato complexes reported, suggest that the oxygen atom of the cyanate group is involved in coordination only with reluctance. Although there are obvious

TABLE XLIV

CHALCOGENOCYANATE COORDINATION IN SOME PdL₂(CNX)₂ COMPLEXES

L	NCO⁻ (575)	NCS⁻ (143)	NCSe⁻ (152)
NCX	N	S	Se
NH₃	N	S	Se
py	N	S (640)	Se
γ-pic	N	S	Se
½bipy	N	N	Se
½phen	—	S	Se
½5NO₂-phen	—	N (640)	Se
Ph₃P	N	N	Se
Ph₃As	N	N	—
Ph₃Sb	—	S	—

differences in the systematic chemistry of oxygen, sulfur, and selenium which will contribute to the differences in the coordination behavior of the chalcogenocyanate ions, it is not clear whether any overall explanation can, or should, be offered. Indeed, it has been pointed out that, when the donor properties of sulfur and oxygen in ambidentate ligands containing both these elements are compared, the preferred donor atom depends on the particular ligand. Thus, dimethyl sulfoxide coordinates through sulfur or oxygen, depending on the metal, $S_2O_3^{2-}$, SO_3^{2-}, SO_2, and RSO_2^- all prefer to coordinate through sulfur, and the sulfinate group shows a strong tendency to form M—O bonds (see Ref. 576 for further details and references).

It may be concluded, therefore, that attempts to provide too detailed a comparison of the coordination behavior of the chalcogeno-cyanate ions are not useful until a better understanding is obtained of the differing effects of the cyano group on the atom X in NCX⁻. Such an interaction may, and probably does, modify the comparative chemistry that might be expected for the oxygen, sulfur, or selenium atoms in these ions. Similarly, the coordination properties of the nitrogen atom will also depend, to some extent, on the particular Group VI atom in the ion; the nucleophilicity of NCO^- toward organic substrates is appreciably different from that of NCS^- and of $NCSe^-$ (43). If the gross coordination behavior of these ions is difficult to understand, it is not surprising that the accurate description of the processes modifying the behavior of any one of the ions also provides difficulties. These difficulties reach a maximum for any two linkage isomers that do not owe their existence to kinetic factors since, by definition, the energy difference between the isomers will be very small, and, thus, very difficult to account for in detail.

REFERENCES

1. Aarflot, K., and Åse, K., *Acta Chem. Scand. A* **28**, 137 (1974)
2. Ablov, A. V., Popova, A. A., and Samus, N. M., *Zh. Neorg. Khim.* **14**, 994 (1969).
3. Ablov, A. V., Proskina, N. N., Bologa, O. A., and Samus, N. M., *Russ. J. Inorg. Chem.* **15**, 1245 (1970).
4. Ablov, A. V., and Samus, N. M., *Dokl. Akad. Nauk SSSR* **113**, 1265 (1957).
5. Ablov, A. V., and Samus, N. M., *Russ. J. Inorg. Chem.* **3**, 137 (1958).
6. Ablov, A. V., and Samus, N. M., *Russ. J. Inorg. Chem.* **4**, 790 (1959).
7. Ablov, A. V., and Samus, N. M., *Dokl. Akad. Nauk. SSSR* **133**, 1327 (1960).
8. Ablov, A. V., and Samus, I. D., *Dokl. Akad. Nauk. SSSR* **146**, 1071 (1962).
9. Ablov, A. V., Samus, N. M., and Popov, M. S., *Dokl. Akad. Nauk. SSSR* **106**, 665 (1956).
10. Ablov, A. V., Samus, N. M., and Popova, A. A., *Russ. J. Inorg. Chem.* **16**, 215 (1971).
11. Ablov, A. V., and Syrtsova, G. P., *Zh. Obshch. Khim.* **25**, 1304 (1955).
12. Adamson, J. W., *J. Inorg. Nucl. Chem.* **13**, 275, (1960).
13. Adlard, M. W., and Socrates, G., *J. Chem. Soc., Chem. Commun.* p. 17 (1972).
14. Adlard, M. W., and Socrates, G., *J. Chem. Soc., Dalton Trans.* p. 797 (1972).
15. Ahrland, S., *Struct. Bonding (Berlin)* **5**, 118 (1968).
16. Ahrland, S., and Kullberg, L., *Acta Chem. Scand.* **25**, 3677 (1971).
17. Ahrland, S., and Kullberg, L., *Acta Chem. Scand.* **25**, 3692 (1971).
18. Ahuja, I. S., and Garg, A., *J. Inorg. Nucl. Chem.* **34**, 1929 (1972).
19. Ahuja, I. S., and Rastogi, P., *Indian J. Chem.* **8**, 88 (1970).
20. Ahuja, I. S., and Rastogi, P., *J. Inorg. Nucl. Chem.* **32**, 1381 (1970).
21. Akers, C., Peterson, S. W., and Willet, R. D., *Acta Crystallogr., Sect. B* **24**, 1125 (1968).
22. Alasaniya, R. M., Skopenko, V. V., and Tsintsadze, G. V., *Tr. Gruz. Politekh. Inst.* **7**, 21 (1967).
23. Alasaniya, R. M., Skopenko, V. V., and Tsintsadze, G. V., *Ukr. Khim. Zh.* **35**, 568 (1969).
24. Alasaniya, R. M., Tsintsadze, G. V., and Skopenko, V. V., *Tr. Gruz. Politekh. Inst.* **3**, 28 (1968).

25. Al-Kazzaz, Z. M. S., Bagnall, K. W., and Brown, D., *J. Inorg. Nucl. Chem.* **35**, 1501 (1973).
26. Al-Kazzaz, Z. M. S., Bagnall, K. W., Brown, D., and Whittaker, B., *J. Chem. Soc., Dalton Trans.* p. 2273 (1972).
27. Allen, D. W., Millar, I. T., and Mann, F. G., *J. Chem. Soc., A* p. 1101 (1969).
28. Allen, D. W., Millar, I. T., Mann, F. G., Canadine, R. M., and Walker, J., *J. Chem. Soc., A* p. 1097 (1969).
29. Altshuller, A. P., *J. Chem. Phys.* **28**, 1254 (1958).
30. Anand, M. L., *J. Indian Chem. Soc.* **44**, 545 (1967).
31. Andreetti, G. D., Coghi, L., Nardelli, M., and Sgarabotto, P., *J. Cryst. Mol. Struct.* **1**, 147 (1971).
32. Andreetti, G. D., Jain, P. C., and Lingafelter, E. C., *J. Amer. Chem. Soc.* **91**, 4112 (1969).
33. Angelici, R. J., and Charley, L. M., *Inorg. Chem.* **10**, 868 (1971).
34. Angelici, R. J., and Faber, G. C., *Inorg. Chem.* **10**, 514 (1971).
35. Anderson, S. J., and Norbury, A H., unpublished results.
36. Anderson, S. J., and Norbury, A. H., to be published.
37. Anderson, S. J., Norbury, A. H., and Sonstad, J., *J. Chem. Soc., Chem. Comm., p. 37* (1974).
38. Antsyshkina, A. S., and Porai-Koshits, M. A., *Kristallografia* **3**, 686 (1958).
39. Armor, J. N., and Haim, A., *J. Amer. Chem. Soc.* **93**, 867 (1971).
40. Ashley, K. R., and Kulprathipanja, S., *Inorg. Chem.* **11**, 444 (1972),
41. Ashworth, M. R. F., "The Determination of Sulphur-Containing Groups," Vol. 1. Academic Press, New York, 1972.
42. Aslam, M., and Massie, W. H. S., *Inorg. Nucl. Chem. Lett.* **7**, 961 (1971).
43. Austad, T., Engemyr, L. B., and Songstad, J., *Acta Chem. Scand.* **25**, 3535 (1971).
44. Austad, T., Songstad, J., and Åse, K., *Acta Chem. Scand.* **25**, 331 (1971).
45. Austad, T., Songstad, J., and Åse, K., *Acta Chem. Scand.* **25**, 1136 (1971).
46. Babaeva, A. V., Evstaf'eva, O. N., and Kharitonov, Yu, Ya., *Fiz. Probl. Spektrosk., Mater. Soveshch., 13th, 1960* Vol. 1, p. 415 (1962–1963).
47. Baddley, W. H., *J. Amer. Chem. Soc.* **90**, 3705 (1968).
48. Bagbanly, I. L., Shirai, M. V., and Makov, N. N., *Dokl. Akad. Nauk. Azerb. SSR* **26**, 41 (1970).
49. Baggio, S., and Becka, L. N., *Acta Crystallogr., Sect.B* **25**, 946 (1969).
50. Bagnall, K. W., Brown, D., and Colton, R., *J. Chem. Soc., London* p. 2527 (1964).
51. Bagnall, K. W., Brown, D., Jones, P. J., and Robinson, P. S., *J. Chem. Soc., London* p. 2531 (1964).
52. Bailey, N. A., and McKenzie, E. D., *J. Chem. Soc., Dalton Trans.* p. 1566 (1972).
53. Bailey, N. A., McKenzie, E. D., and Mullins, J. R., *J. Chem. Soc., D* p. 1103 (1970).
54. Bailey, R. A., and Kozak, S. L., *Inorg. Chem.* **6**, 419 (1967).
55. Bailey, R. A., and Kozak, S. L., *Inorg. Chem.* **6**, 2155 (1967).
56. Bailey, R. A., and Kozak, S. L., *J. Inorg. Nucl. Chem.* **31**, 689 (1969).
57. Bailey, R. A., Kozak, S. L., Michelsen, T. W., and Mills, W. N., *Coord. Chem. Rev.* **6**, 407 (1971).
58. Bailey, R. A., and Michelsen, T. W., *J. Inorg. Nucl. Chem.* **34**, 2671 (1972).
59. Bailey, R. A., and Michelsen, T. W., *J. Inorg. Nucl. Chem.* **34**, 2935 (1972).

60. Bailey, R. A., Michelsen, T. W., and Mills, W. N., *J. Inorg. Nucl. Chem.* **33**, 3206 (1971).
61. Bailey, R. A., Michelsen, T. W., and Nobile, A. A., *J. Inorg. Nucl. Chem.* **32**, 2427 (1970).
62. Baker, B. R., Orhanovic, M., and Sutin, N., *J. Amer. Chem. Soc.* **89**, 722 (1967).
63. Baker, B. R., Sutin, N., and Welch, T. J., *Inorg. Chem.* **6**, 1948 (1967).
64. Baker, W. A. and Bobonich H. M., *Inorg. Chem.* **3**, 1184 (1964).
65. Balahura, R. J., and Jordan, R. B., *Inorg. Chem.* **9**, 1567 (1970).
66. Balahura, R. J., and Jordan, R. B., *Inorg. Chem.* **10**, 198 (1971).
67. Baltisberger, R. J., and Hanson, J. V., *Inorg. Chem.* **9**, 1573 (1970).
68. Barakat, T. M., Nelson, J., Nelson, S. M., and Pullin, A. D. E., *Trans. Faraday Soc.* **65**, 41 (1969).
69. Barbucci, R., Cialdi, G., Ponticelli, G., and Paoletti, P., *J. Chem. Soc.*, *A* p. 1775 (1969).
70. Barbucci, R., Fabbrizzi, L., and Paoletti, P., *Coord. Chem. Rev.* **8**, 31 (1972).
71. Barbucci, R., Fabbrizzi, L., and Paoletti, P., *J. Chem. Soc.*, *Dalton Trans.* p. 1099 (1972).
72. Barbucci, R., Paoletti, P., and Ponticelli, G., *J. Chem. Soc.*, *A* p. 1637 (1971).
73. Barclay, G. A., Collard, M. A., Harris, C. M., and Kingston, J. V., *J. Chem. Soc.*, *A* p. 830 (1969).
74. Barcza, L., and Timar, I., *Acta Chim.* (*Budapest*) **67**, 127 (1971).
75. Barnes, J. C., and Day, P., *J. Chem. Soc.*, *London* p. 3886 (1964).
76. Basch, H., *Chem. Phys. Lett.* **5**, 337 (1970).
77. Basolo, F., Baddley, W. H., and Burmeister, J. L., *Inorg. Chem.* **3**, 1202 (1964).
78. Basolo, F., Baddley, W. H., and Weidenbaum, K. J., *J. Amer. Chem. Soc.* **88**, 1576 (1966).
79. Basolo, F., Burmeister, J. L., and Pöe, A. J., *J. Amer. Chem. Soc.* **85**, 1700 (1963).
80. Bauer, R. A., and Basolo, F., *Inorg. Chem.* **8**, 2231 (1969).
81. Beauchamp, A. L., and Goutier, D., *Can. J. Chem.* **50**, 977 (1972).
82. Beauchamp, A. L., Saperas, B., and Rivest, R., *Can. J. Chem.* **49**, 3579 (1971).
83. Beck, M., and Bjerrum, J., *Magy. Kem. Foly.* **69**, 455 1963).
84. Beck, W., *Organometal. Chem. Rev.*, *Sect. A* **7**, 183 (1971).
85. Beck, W., and Fehlhammer, W. P., *Angew. Chem.*, *Int. Ed. Engl.* **6**, 169 (1967).
86. Beck, W., and Fehlhammer, W. P., *MTP Int. Rev. Sci.*, *Inorg. Chem. Ser. 1* **2**, 253 (1972).
87. Beck, W., Fehlhammer, W. P., Pöllman, P., and Schächl, H., *Chem. Ber.* **102**, 1976 (1969).
88. Beck, W., and Lindenberg, B., *Angew. Chem.*, *Int. Ed. Engl.* **9**, 735 (1970).
89. Beck, W., and Lottes, K., *Z. Anorg. Allg. Chem.* **335**, 258 (1965).
90. Beck, W., and Werner, K., *Chem. Ber.* **104**, 2901 (1971).
91. Beck, W., and Werner, K., *Chem. Ber.* **106**, 868 (1973).
92. Beck, W., Werner, H., Engelmann, H., and Smedal, H. S., *Chem. Ber.* **101**, 2143 (1968).
93. Becker, W., and Beck, W., unpublished results quoted in ref. *86* (p. 279, ref. 311).

94. Beech, G., and Kauffman, G. B., *Thermochim. Acta* **1**, 93 (1970).
95. Behrens, H., and Herrmann, D., *Z. Naturforsch. B* **21**, 1236 (1966).
96. Behrens, H., Schwab, R., and Herrmann, D., *Z. Naturforsch. B* **21**, 590 (1966).
97. Below, J. F., Connick, R. E., and Coppel, C. P., *J. Amer. Chem. Soc.* **80**, 2961 (1958).
98. Bennett, M. A., Clark, R. J. H., and Goodwin, A. D. J., *Inorg. Chem.* **6**, 1625 (1967).
99. Bennett, M. A., Kneen, W. R., and Nyholm, R. S., *Inorg. Chem.* **7**, 556 (1968).
100. Bennett, M. A., and Wild, J. D., *J. Chem. Soc., A* p. 545 (1971).
101. Beran, G., Carty, A. J., Chieh, P. C., and Patel, H. A., *J. Chem. Soc., Dalton Trans.* p. 488 (1973).
102. Beran, G., and Palenik, G. J., *Chem. Commun.* p. 1354 (1970).
103. Bertini, I., and Sabatini, A., *Inorg. Chem.* **5**, 1025 (1966).
104. Bianchi, A., and Ghilardi, C. A., *J. Chem. Soc., A* p. 1096 (1971).
105. Bigoli, F., Braibanti, A., Pellinghelli, M. A., and Tiripicchio, A., *Acta Crystallogr., Sect. B* **28**, 962 (1972).
106. Birckenbach, L., and Kellermann, K., *Ber. Deut. Chem. Sec. B* **58**, 786 and 2377 (1925).
107. Birnbaum, E. R., *J. Inorg. Nucl. Chem.* **34**, 3499 (1972).
108. Blasius, E., Augustin, H., and Wenzel, U., *J. Chromatogr.* **50**, 319 (1970).
109. Boehland, H., *Z. Chem.* **12**, 148 (1972).
110. Boehland, H., and Malitzke, P., *Z. Anorg. Allg. Chem.* **350**, 70 (1967).
111. Boehland, H., and Mühle, E., *Z. Anorg. Allg. Chem.* **379**, 273 (1970).
112. Boehland, H., and Niemann, E., *Z. Anorg. Allg. Chem.* **373**, 217 (1970).
113. Boehland, H., and Tiede, E., *J. Less-Common Metals* **13**, 224 (1967).
114. Boehland, H., Tiede, E., and Zenker, E., *J. Less-Common Metals* **15**, 89 (1968).
115. Boehland, H., and Zenker, E., *J. Less-Common Metals* **14**, 397 (1968).
116. Bohland, V., and Schneider, F. M., *Z. Anorg. Allg. Chem.* **390**, 53 (1972).
117. Bonaccorsi, R., Petrongolo, C., Scrocco, E., and Tomasi, J., *J. Chem. Phys.* **48**, 1500 (1968).
118. Bonaccorsi, R., Scrocco, E., and Tomasi, J., *J. Chem. Phys.* **50**, 2940 (1969).
119. Bose, K. S., and Patel, C. C., *J. Inorg. Nucl. Chem.* **33**, 755 (1971).
120. Bosnitch, B., *Nature (London)* **196**, 1196 (1962).
121. Bosnitch, B., and Dwyer, F. P., *Aust. J. Chem.* **19**, 2229 (1966).
122. Bosnitch, B., and Dwyer, F. P., *Aust. J. Chem.* **19**, 2235 (1966).
123. Boughton, J. H., and Keller, R. N., *J. Inorg. Nucl. Chem.* **28**, 2851 (1966).
124. Bounsall, E. J., and Koprich, S. R., *Can. J. Chem.* **48**, 1481 (1970).
125. Britton, D., and Dunitz, J. D., *Acta Crystallogr.* **18**, 424 (1965).
126. Brookes, P. R., and Shaw, B. L., *J. Chem. Soc., A* p. 1079 (1967).
127. Brown, B. W., and Lingafelter, E. C., *Acta Crystallogr.* **16**, 753 (1963).
128. Brown, B. W., and Lingafelter, E. C., *Acta Crystallogr.* **17**, 254 (1964).
129. Brown, L. D., and Pennington, D. E., *Inorg. Chem.* **10**, 2117 (1971).
130. Brown, T. M., and Horn, C. J., *Inorg. Nucl. Chem. Lett.* **8**, 377 (1972).
131. Brown, T. M., and Knox, G. F., *J. Amer. Chem. Soc.* **89**, 5296 (1967).
132. Brusilovets, A. I., Skopenko, V. V., and Tsintsadze, G. V., *Russ. J. Inorg. Chem.* **14**, 239 (1969).
133. Buckingham, D. A., Creaser, I. I., Marty, W., and Sargeson, A. M., *Inorg. Chem.* **11**, 2738 (1972).

134. Buckingham, D. A., Creaser, I. I., and Sargeson, A. M., *Inorg. Chem.* **9**, 655 (1970).
135. Buckley, R. C., and Wardeska, J. G., *Inorg. Chem.* **11**, 1723 (1972).
136. Buijs, K., and Choppin, G. R., *J. Chem. Phys.* **39**, 2042 (1963).
137. Burbridge, C. D., and Goodgame, D. M. L., *Inorg. Chim. Acta* **4**, 231 (1970).
138. Burger, H., and Schmid, W., *Z. Anorg. Allg. Chem.* **388**, 67 (1972).
139. Burmeister, J. L., *Inorg. Chem.* **3**, 919 (1964).
140. Burmeister, J. L., *Coord. Chem. Rev.* **7**, 205 (1966).
141. Burmeister, J. L., *Coord. Chem. Rev.* **3**, 225 (1968).
142. Burmeister, J. L., and Al-Janabi, M. Y., *Inorg. Chem.* **4**, 962 (1965).
143. Burmeister, J. L., and Basolo, F., *Inorg. Chem.* **3**, 1587 (1964).
144. Burmeister, J. L., and Deardoff, E. A., *Inorg. Chim. Acta* **4**, 97 (1970).
145. Burmeister, J. L., Deardoff, E. A., Jensen, A., and Christiansen, V. H., *Inorg. Chem.* **9**, 58 (1970).
146. Burmeister, J. L., Deardoff, E. A., and Van Dyke, C. E., *Inorg. Chem.* **8**, 170 (1969).
147. Burmeister, J. L., and DeStefano, N. J., *Inorg. Chem.* **8**, 1546 (1969).
148. Burmeister, J. L., and DeStefano, N. J., *Chem. Commun.* p. 1698 (1970).
149. Burmeister, J. L., and DeStefano, N. J., *Inorg. Chem.* **9**, 972 (1970).
150. Burmeister, J. L., and DeStefano, N. J., *Inorg. Chem.* **10**, 998 (1971).
151. Burmeister, J. L., and Gysling, H. J., *Chem. Commun.* p. 543 (1967).
152. Burmeister, J. L., and Gysling, H. J., *Inorg. Chim. Acta* **1**, 100 (1967).
153. Burmeister, J. L., Gysling, H. J., and Lim, J. C., *J. Amer. Chem. Soc.* **91**, 44 (1969).
154. Burmeister, J. L., Hassel, R. L., and Phelan, R. J., *Chem. Commun.* p. 679 (1970).
155. Burmeister, J. L., Hassel, R. L., and Phelan, R. J., *Inorg. Chem.* **10**, 2032 (1971).
156. Burmeister, J. L., and Lim, J. C., *Chem. Commun.* p. 1346 (1968).
157. Burmeister, J. L., and Lim, J. C., *J. Chem. Soc., D* p. 1154 (1969).
158. Burmeister, J. L., and O'Sullivan, T. P., *Inorg. Chim. Acta* **3**, 479 (1969).
159. Burmeister, J. L., O'Sullivan, T. P., and Johnson, K. A., *Inorg. Chem.* **10**, 1803 (1971).
160. Burmeister, J. L., Patterson, S. D., and Deardorff, E. A., *Inorg. Chim. Acta* **3**, 105 (1969).
161. Burmeister, J. L., and Weleski, E. T., *Syn. Inorg. Metal.-Org. Chem.* **2**, 295 (1972).
162. Burmeister, J. L., and Williams, L. E., *Inorg. Chem.* **5**, 1113 (1966).
163. Burmeister, J. L., and Williams, L. E., *J. Inorg. Nucl. Chem.* **29**, 839 (1967).
164. Busetto, L., Palazzi, A., and Ros, R., *Inorg. Chem.* **9**, 2792 (1970).
165. Bush, M. A., and Sim, G. A., *J. Chem. Soc., A* p. 605 (1970).
166. Butcher, A. V., and Chatt, J., *J. Chem. Soc., A* p. 2652 (1970).
167. Butcher, A. V., Phillips, D. J., and Redfern, J. P., *J. Chem. Soc., A* p. 1640 (1971).
168. Caldwell, S. M., and Norris, A. R., *Inorg. Chem.* **7**, 1667 (1968).
169. Cannas, M., Carta, G., and Marongiu, G., *J. Chem. Soc., Dalton Trans.* p. 251 (1973).
170. Cannon, R. D., Chiswell, B., and Venanzi, L. M., *J. Chem. Soc., A* p. 1277 (1967).
171. Capacchi, L. C., Gasparri, G. F., Ferrari, M., and Nardelli, M., *Chem. Commun.* p. 910 (1968).

172. Capacchi, L. C., Gasparri, G. F., Nardelli, M., and Pelizzi, G., *Acta Crystallogr. Sect. B* **24**, 1199 (1968).

173. Carbacho, H., Ungere, B., and Contreras, G., *J. Inorg. Nucl. Chem.* **32**, 579 (1970).

174. Cariati, F., Galizzioli, D., and Naldini, L., *Chim. Ind. (Milan)* **52**, 995 (1970).

175. Cariati, F., and Naldini, L., *Inorg. Chim. Acta* **5**, 172 (1971).

176. Cariati, F., Ugo, R., and Bonati, F., *Inorg. Chem.* **5**, 1129 (1966).

177. Carlin, R. L., and Edwards, J. O., *J. Inorg. Nucl. Chem.* **6**, 217 (1958).

178. Carlyle, D. W., and King, E. L., *Inorg. Chem.* **9**, 2333 (1970).

179. Carty, A. J., and Efraty, A., *Can. J. Chem.* **47**, 2573 (1969).

180. Carty, A. J., and Efraty, A., *Inorg. Chem.* **8**, 543 (1969).

181. Cash, D. N., and Harris, R. O., *Can. J. Chem.* **49**, 867 (1971).

182. Cavalca, L., Nardelli, M., and Fava, G., *Proc. Chem. Soc., London* p. 159 (1959).

183. Cavalca, L., Nardelli, M., and Fava, G., *Acta Crystallogr.* **13**, 125 (1960).

184. Cerrini, S., Colapietro, M., Spagna, R., and Zambonelli, L., *J. Chem. Soc., A* p. 1375 (1971).

185. Chamberlain, M. M., and Bailar, J. C., *J. Amer. Chem. Soc.* **81**, 6412 (1959).

186. Chatt, J., and Duncanson, L. A., *Nature (London)* **178**, 997 (1956).

187. Chatt, J., Duncanson, L. A., Hart, F. A., and Owston, P. G., *Nature (London)* **181**, 43 (1958).

188. Chatt, J., and Hart, F. A., *Nature (London)* **169**, 673 (1952).

189. Chatt, J., and Heaton, B. T., *J. Chem. Soc., A* p. 2745 (1968).

190. Chatt, J., and Shaw, B. L., *J. Chem. Soc., London* p. 5075 (1962).

191. Chew, K. F., Derbyshire, W., and Logan, N., *J. Chem. Soc., Faraday Trans.* **2**, 594 (1972).

192. Chew, K. F., Derbyshire, W., Logan, N., Norbury, A. H., and Sinha, A. I. P., *J. Chem. Soc., D* p. 1708 (1970).

193. Chiang, H.-C., and Wilmarth, W. K., *Inorg. Chem.* **7**, 2535 (1968).

194. Chow, K. K., and McAuliffe, C. A., *Inorg. Nucl. Chem. Lett.* **8**, 1031 (1972).

195. Chughtai, A. R., and Keller, R. N., *J. Inorg. Nucl. Chem.* **31**, 633 (1969).

196. Ciampolini, M., and Paoletti, P., *Inorg. Chem.* **6**, 1261 (1967).

197. Ciavatta, L., and Grimaldi, M., *Inorg. Chim. Acta* **4**, 312 (1970).

198. Clark, G. R., Palenik, G. J., and Meek, D. W., *J. Amer. Chem. Soc.* **92**, 1077 (1970).

199. Clark, G. R., Palenik, G. J., and Meek, D. W., *Inorg. Chem.* **9**, 2754 (1970).

200. Clark, R. J. H., Bradford, C. W., Van Bronswijk, W., and Nyholm, R. S., *J. Chem. Soc., A* p. 2889 (1970).

201. Clark, R. J. H., and Williams, C. S., *Spectrochim. Acta* **22**, 1081 (1966).

202. Cleare, M. J., and Griffith, W. P., *J. Chem. Soc., A* p. 1117 (1970).

203. Clifford, A. F., *J. Phys. Chem.* **63**, 1227 (1959).

204. Cohen, M. A., Melpolder, J. B., and Burmeister, J. L., *Inorg. Chim. Acta* **6**, 188 (1972).

205. Colquhoun, I., Greenwood, N. N., McColm, I. J., and Turner, G. E., *J. Chem. Soc., Dalton Trans.* p. 1337 (1972).

206. Colton, R., and Scollary, G. R., *Aust. J. Chem.* **21**, 1435 (1968).

207. Conocchioli, T. J., and Sutin, N., *J. Amer. Chem. Soc.* **89**, 282 (1967).

208. Contreras, G., and Schmidt, R., *J. Inorg. Nucl. Chem.* **32**, 127 (1970).

209. Contreras, G., and Schmidt, R., *J. Inorg. Nucl. Chem.* **32**, 1295 (1970).

210. Cooney, R. P. J., and Hall, J. R., *Aust. J. Chem.* **22**, 2117 (1969).

211. Cotton, F. A., and Goodgame, M., *J. Amer. Chem. Soc.* **83**, 1777 (1961).

212. Cotton, F. A., Goodgame, D. M. L., Goodgame, M., and Haas, T. E., *Inorg. Chem.* **1**, 565 (1962).
213. Cotton, F. A., Goodgame, D. M. L., Goodgame, M., and Sacco, A., *J. Amer. Chem. Soc.* **83**, 4157 (1961).
214. Cotton, F. A., Robinson, W. R., Walton, R. A., and Whyman, R., *Inorg. Chem.* **6**, 929 (1967).
215. Cotton, S. A., and Gibson, J. F., *J. Chem. Soc., A* p. 859 (1971).
216. Countryman, R., and McDonald, W. S., *J. Inorg. Nucl. Chem.* **33**, 2213 (1971).
217. Courtot-Coupez, J., and L'Her, M., *Bull. Soc. Chim. Fr.* [6] **2**, 675 (1969).
218. Cousins, D. R., and Hart, F. A., *J. Inorg. Nucl. Chem.* **30**, 3009 (1968).
219. Coutts, R. S. P., and Wailes, P. C., *Aust. J. Chem.* **19**, 2069 (1966).
220. Coutts, R. S. P., and Wailes, P. C., *Inorg. Nucl. Chem. Lett.* **3**, 1 (1967).
221. Cox, J. L., and Howatson, J., *Inorg. Chem.* **12**, 1205 (1973).
222. Craciunescu, D., and Ben-Bassatt, A. H. I., *J. Less-Common Metals* **25**, 11 (1971).
223. Crawford, N. P., and Melson, G. A., *Inorg. Nucl. Chem. Lett.* **4**, 399 (1968).
224. Crawford, N. P., and Melson, G. A., *J. Chem. Soc., A* p. 427 (1969).
225. Crawford, N. P., and Melson, G. A., *J. Chem. Soc., A* p. 1049 (1969).
226. Cristini, A., and Ponticelli, G., *J. Inorg. Chem. Nucl.* **35**, 2691 (1973).
227. Cummings, S. C., and Busch, D. H., *Inorg. Chem.* **10**, 1220 (1971).
228. Curtis, N. F., and Curtis, Y. M., *Aust. J. Chem.* **19**, 1423 (1966).
229. Curtis, N. F., and Reader, G. W., *J. Chem. Soc., A* p. 1771 (1971).
230. Czakis, M., *Rocz. Chem.* **33**, 957 (1959).
231. Dahlhoff, W. V., and Nelson, S. M., *J. Chem. Soc., A* p. 2184 (1971).
232. Dainton, F. S., Laurence, G. S., Schneider, W., Stranks, D. R., and Vaidya, M. S., *Radioisotop. Sci. Res., Proc. Int. Conf., 1957* vol. 2, p. 305 (1958).
233. Dalley, N. K., Smith, D. E., Izatt, R. M., and Christensen, J. J., *Chem. Commun.* 90 (1972).
234. Da Mota, M. M., Rodgers, J., and Nelson, S. M., *J. Chem. Soc., A* p. 2036 (1969).
235. Dapporto, P., and Di Vaira, M., *J. Chem. Soc., A* p. 1891 (1971).
236. David, P. G., *J. Inorg. Nucl. Chem.* **35**, 1463 (1973).
237. Davies, N. R., and Mullins, T. L., *Aust. J. Chem.* **21**, 915 (1968).
238. Davies, R., and Fergusson, J. E., *Inorg. Chim. Acta* **4**, 23 (1970).
239. Davis, A. R., Murphy, C. J., and Plane, R. A., *Inorg. Chem.* **9**, 423 (1970).
240. Davis, T. L., and Logan, A. V., *J. Amer. Chem. Soc.* **50**, 2493 (1928).
241. Davis, T. L., and Ou, C. W., *J. Amer. Chem. Soc.* **56**, 1061 (1934).
242. Day, P., *Inorg. Chem.* **5**, 1619 (1966).
243. De Bolster, M. W. G., and Groeneveld, W. L., *Rec. Trav. Chim. Pays-Bas* **90**, 687 (1971).
244. Decius, J. C., and Gordon, D. J., *J. Chem. Phys.* **47**, 1286 (1967).
245. De-Cugnac-Paill, A., and Pouradier, J., *C.R. Acad. Sci., Ser. C* **273**, 1565 (1971).
246. Diebler, H., *Z. Phys. Chem. (Frankfurt am Main)* [N.S.] **68**, 64 (1969).
247. Dieck, H. T., and Friedel, H., *J. Organometal. Chem.* **14**, 375 (1968).
248. Dillard, J. G., and Franklin, J. L., *J. Chem. Phys.* **48**, 2353 (1968).
249. Di Sipio, L., Oleary, L., and de Michelis, G., *Coord. Chem. Rev.* **1**, 7 (1966).
250. Di Vaira, M., and Orlandini, A. B., *J. Chem. Soc., Dalton Trans.* p. 1704 (1972).
251. Di Vaira, M., and Sacconi, L., *J. Chem. Soc., D* p. 10 (1969).

252. Di Vaira, M., and Sacconi, L., *J. Chem. Soc.*, *A* p. 148 (1971).
253. Dodd, D., and Johnson, M. D., *J. Chem. Soc.*, *Dalton Trans.* p. 1218 (1973).
254. Dodgen, H. W., Murray, R., and Hunt, J. P., *Inorg. Chem.* **4**, 1820 (1965).
255. Dolcetti, G., Peloso, A., and Sindellari, L., *Gazz. Chim. Ital.* **96**, 1648 (1966).
256. Domiano, P., Manfredotti, A. G., Grossóni, G., Nardelli, M., and Tani, M. E. V., *Acta Crystallogr., Sect. B* **25**, 591 (1969).
257. Dori, Z., and Gray, H. B., *Inorg. Chem.* **7**, 889 (1968).
258. Dorm, E., and Lindh, B., *Acta Chem. Scand.* **21**, 1661 (1967).
259. Dost, J., Steinike, U., and Paudert, R., *Krist. Tech.* **3**, 247 (1968).
260. Downs, A. W., *Chem. Commun.* p. 1290 (1968).
261. Doyle, G., and Tobias, R. S., *Inorg. Chem.* **7**, 2479 (1968).
262. Drickamer, H. G., Lewis, G. K., and Fung, S. C., *Science* **163**, 885 (1969).
263. DuBois, T. D., and Meek, D. W., *Inorg. Chem.* **6**, 1395 (1967).
264. Duffy, N. V., and Earley, J. E., *J. Amer. Chem. Soc.* **89**, 272 (1967).
265. Duggan, D. M., and Hendrickson, D. N., *J. Chem. Soc., Chem. Commun.* p. 411 (1973).
266. Dutta, R. L., and De, D., *J. Indian Chem. Soc.* **46**, 1 (1969).
267. Ellestad, O. H., Klaeboe, P., and Songstad, J., *Acta Chem. Scand.* **26**, 1724 (1972).
268. Emori, S., Inoue, M., and Kubo, M., *Bull. Chem. Soc. Jap.* **44**, 3299 (1971).
269. Epps, L. A., and Marzilli, G., *Chem. Commun.* p. 109 (1972).
270. Eremin, Yu. G., Katochkina, V. S., and Komissarova, L. N., *Zh. Neorg. Khim.* **15**, 1248 (1970).
271. Eriks, K., and Knox, J. R., *Acta Crystallogr.* **21**, A140 (1966).
272. Eriks, K., and Knox, J. R., *Inorg. Chem.* **7**, 84 (1968).
273. Espenson, J. H., *Inorg. Chem.* **4**, 121 (1965).
274. Eyster, E. H., Gillette, R. H., and Brockway, L. O., *J. Amer. Chem. Soc.* **62**, 3236 (1940).
275. Fackler, J. P., Dolbear, G. E., and Coucouvanis, D., *J. Inorg. Nucl. Chem.* **26**, 2035 (1964).
276. Falk, L. C., and Linck, R. G., *Inorg. Chem.* **10**, 215 (1971).
277. Farago, M. E., and James, J. M., *Inorg. Chem.* **4**, 1706 (1965).
278. Farona, M. F., Frazee, L. M., and Bremer, N. J., *J. Organometal. Chem.* **19**, 225 (1969).
279. Farona, M. F., and Wojcicki, A., *Inorg. Chem.* **4**, 857 (1965).
280. Farona, M. F., and Wojcicki, A., *Inorg. Chem.* **4**, 1402 (1965).
281. Fay, D. P., and Sutin, N., *Inorg. Chem.* **9**, 1291 (1970).
282. Fee, W. W., Harrowfield, J. N. MacB., and Jackson, W. G., *J. Chem. Soc.*, *A* p. 2612 (1970).
283. Ferrari, A., Braibanti, A., Bigliardi, G., and Lanfredi, A. M., *Acta Crystallogr.* **18**, 367 (1965).
284. Finta, Z., Varhelyi, C., and Zsako, I., *J. Inorg. Nucl. Chem.* **32**, 3013 (1970).
285. Fleischer, E. B., and Hawkinson, S., *J. Amer. Chem. Soc.* **89**, 720 (1967).
286. Flint, C. D., and Goodgame, M., *J. Chem. Soc.*, *A* p. 2178 (1968).
287. Flint, C. D., and Goodgame, M., *Inorg. Chem.* **8**, 1833 (1969).
288. Flint, C. D., and Goodgame, M., *J. Chem. Soc.*, *A* p. 442 (1970).
289. Foong, S. W., Stevenson, M. B., and Sykes, A. G., *J. Chem. Soc.*, *A* p. 1064 (1970).
290. Ford, P. C., *Inorg. Chem.* **10**, 2153 (1971).

291. Forster, D., *Inorg. Chim. Acta* **2**, 116 (1968).
292. Forster, D., and Goodgame, D. M. L., *J. Chem. Soc., London* p. 2790 (1964).
293. Forster, D., and Goodgame, D. M. L., *Inorg. Chem.* **4**, 715 (1965).
294. Forster, D., and Goodgame, D. M. L., *Inorg. Chem.* **4**, 823 (1965).
295. Forster, D., and Goodgame, D. M. L., *Inorg. Chem.* **4**, 1712 (1965).
296. Forster, D., and Goodgame, D. M. L., *J. Chem. Soc., London* p. 262 (1965).
297. Forster, D., and Goodgame, D. M. L., *J. Chem. Soc., London* p. 268 (1965).
298. Forster, D., and Goodgame, D. M. L., *J. Chem. Soc., London* p. 454 (1965).
299. Forster, D., and Goodgame, D. M. L., *J. Chem. Soc., London* p. 1286 (1965).
300. Forster, D., and Goodgame, D. M. L., *J. Chem. Soc., A* p. 170 (1966).
301. Forster, D., and Horrocks, W. DeW., *Inorg. Chem.* **6**, 339 (1967).
302. Fouche, K. F., and Lessing, J. G., *J. Inorg. Nucl. Chem.* **32**, 2357 (1970).
303. Fraser, R. T. M., *Advan. Chem. Ser.* **62**, 295 (1965).
304. Freni, M., Giusto, D., and Romiti, P., *Gazz. Chim. Ital.* **99**, 641 (1969).
305. Fronaeus, S., and Larsson, R., *Acta Chem. Scand.* **16**, 1447 (1962).
306. Funk, H., and Böhland, H., *Z. Anorg. Allg. Chem.* **324**, 168 (1963).
307. Galliart, A., and Brown, T. M., *J. Inorg. Nucl. Chem.* **34**, 3568 (1972).
308. Ganescu, I., and Varhelyi, C., *Rev. Roum. Chim.* **12**, 395 (1967).
309. Ganescu, I., Varhelyi, C., and Oprescu, D., *Rev. Chim. Miner.* **6**, 765 (1969).
310. Ganescu, I., Varhelyi, C., and Oprescu, D., *Studia* **14**, 113 (1969).
311. Gans, P., and Marriage, J., *J. Chem. Soc., Dalton Trans.* p. 1738 (1972).
312. Garaj, J., *Inorg. Chem.* **8**, 304 (1969).
313. Garaj, J., and Dunaj-Jurco, M., *Chem. Commun.* p. 518 (1968).
314. Garaj, J., Dunaj-Jurco M., and Lindgren, O., *Collect. Czech. Chem. Commun.* **36**, 3863 (1971).
315. Gaughan, A. P., Ziolo, R. F., and Dori, Z., *Inorg. Chim. Acta* **4**, 640 (1970).
316. Ghilardi, C. A., and Orlandini, A. B., *J. Chem. Soc., Dalton Trans.* p. 1698 (1972).
317. Ghosh, S. P., and Mishra, A., *J. Inorg. Nucl. Chem.* **33**, 4199 (1971).
318. Giacometti, G., and Turco, A., *J. Inorg. Nucl. Chem.* **15**, 242 (1960).
319. Gibson, J. G., and McKenzie, E. D., *J. Chem. Soc., A* p. 1029 (1971).
320. Giddings, S. A., *Inorg. Chem.* **6**, 849 (1967).
321. Gillard, R. D., and Laurie, S. H., *J. Inorg. Nucl. Chem.* **33**, 947 (1971).
322. Gillard, R. D., and Twigg, M. W., *Inorg. Chim. Acta* **6**, 150 (1972).
323. "Gmelin's Handbuch der Anorganischen Chemie" (Kohlenstoff, D1), p. 316. Verlag Chemie, Weinheim, 1971.
324. Golub, A. M., and Borshch, A. N., *Ukr. Khim. Zh.* **32**, 923 (1966).
325. Golub, A. M., and Borshch, A. N., *Ukr. Khim. Zh.* **34**, 1195 (1968).
326. Golub, A. M., and Kalibabchuk, V. A., *Russ. J. Inorg. Chem.* **12**, 1249 (1967).
327. Golub, A. M., Kopa, M. V., Skopenko, V. V., arid Tsintsadze, G. V., *Ukr. Khim. Zh.* **36**, 871 (1970).
328. Golub, A. M., Kopa, M. V., Skopenko, V. W., and Zinzadse, G. W., *Z. Anorg. Allg. Chem.* **375**, 302 (1970).
329. Golub, A. M., and Lishho, T. P., *Zh. Neorg. Khim.* **15**, 1527 (1970).
330. Golub, A. M., and Luong, A. V., *Zh. Neorg. Khim.* **13**, 3372 (1968).
331. Golub, A. M., and Luong, A. V., *Zh. Neorg. Khim.* **14**, 90 (1969).
332. Golub, A. M., Olevinskii, M. I., and Zhigulina, N. S., *Russ. J. Inorg. Chem.* **11**, 841 (1966).
333. Golub, A. M., and Sergunkin, V. N., *Russ. J. Inorg. Chem.* **11**, 419 (1966).

334. Golub, A. M., and Skopenko, V. V., *Russ. Chem. Rev.* **34**, 901 (1965).
335. Golub, A. M., Tsintsadze, G. V., and Mamulashvili, A. M., *Russ. J. Inorg. Chem.* **14**, 1589 (1969).
336. Goodgame, D. M. L., and Malerbi, B. W., *Spectrochim. Acta, Part A* **24**, 1254 (1968).
337. Goodwin, H. A., and Sylva, R. N., *Inorg. Chim. Acta* **4**, 197 (1970).
338. Gottardi, W., *Monatsh. Chem.* **102**, 264 (1971).
339. Graddon, D. P., and Watton, E. C., *Aust. J. Chem.* **18**, 507 (1965).
340. Graham, A. J., and Fenn, R. H., *J. Organometal. Chem.* **17**, 405 (1969).
341. Graham, A. J., and Fenn, R. H., *J. Organometal. Chem.* **25**, 173 (1970).
342. Graham, B. W., Laing, K. R., O'Connor, C. J., and Roper, W. R., *Chem. Commun.* p. 1272 (1970).
343. Grall, J. M., Kergoat, R., and Guerchais, J. E., *C.R. Acad. Sci., Ser. C* **267**, 1410 (1968).
344. Green, M. L. H., and Lindsell, W. E., *J. Chem. Soc., A* p. 2150 (1969).
345. Green, M. L. H., Munakata, H., and Saito, H., *J. Chem. Soc., A* p. 469 (1971).
346. Greenwood, N. N., and Hunter, G., *J. Chem. Soc., A* p. 929 (1969).
347. Greenwood, N. N., Little, R., and Sprague, M. J., *J. Chem. Soc., London* p. 1292 (1964).
348. Gregory, U. A., Jarvis, J. A. J., Kilbourn, B. T., and Owston, P. J., *J. Chem. Soc., A* p. 2770 (1970).
349. Grey, I. E., and Smith, P. W., *Aust. J. Chem.* **22**, 311 (1969).
350. Grinberg, A. A., and Borzakova, S. S., *Russ. J. Inorg. Chem.* **14**, 1044 (1969).
351. Grønbaek, R., and Dunitz, J. D., *Helv. Chim. Acta* **47**, 1889 (1964).
352. Gulia, V. G., Komissarova, L. N., Krasnoyarskaya, A. A., and Sas, T. M., *Zh. Neorg. Khim.* **15**, 966 (1970).
353. Gusarsky, E., and Treinin, A., *J. Phys. Chem.* **69**, 3176 (1965).
354. Gutmann, V., and Buhonovsky, O., *Monatsh. Chem.* **99**, 751 (1968).
355. Gutmann, V., and Laussegger, H., *Monatsh. Chem.* **99**, 947 (1968).
356. Gutmann, V., and Laussegger, H., *Monatsh. Chem.* **99**, 963 (1968).
357. Gutmann, V., and Mayer, U., *Monatsh. Chem.* **99**, 1383 (1968).
358. Gutterman, D. F., and Gray, H. B., *J. Amer. Chem. Soc.* **91**, 3105 (1969).
359. Gutterman, D. F., and Gray, H. B., *J. Amer. Chem. Soc.* **93**, 3364 (1971).
360. Gutterman, D. F., and Gray, H. B., *Inorg. Chem.* **11**, 1727 (1972).
361. Hackel-Wenzel, B., and Thomas, G., *J. Less-Common Metals* **23**, 185 (1971).
362. Hadzi, D., Detoni, S., Smerkolj, R., Hawranek, J., and Sobczyk, L., *J. Chem. Soc., A* p. 2851 (1970).
363. Haim, A., and Sutin, N., *J. Amer. Chem. Soc.* **87**, 4210 (1965).
364. Haim, A., and Sutin, N., *J. Amer. Chem. Soc.* **88**, 434 (1966).
365. Hamada, S., *Nippon Kagaku Zasshi* **82**, 1327 (1961).
366. Harmon, H. D., Peterson, J. R., Bell, J. T., and McDowell, W. J., *J. Inorg. Nucl. Chem.* **34**, 1711 (1972).
367. Harmon, H. D., Peterson, J. R., McDowell, W. J., and Coleman, C. F., *J. Inorg. Nucl. Chem.* **34**, 1381 (1972).
368. Harris, C. M., and Lockyer, T. N., *Aust. J. Chem.* **23**, 1703 (1970).
369. Hart, F. A., and Laming, F. P., *J. Inorg. Nucl. Chem.* **26**, 579 (1964).
370. Hassel, R. L., and Burmeister, J. L., *Chem. Commun.* p. 568 (1971).
371. Hathaway, B. J., Billing, D. E., Dudley, R. J., Fereday, R. J., and Tomlinson, A. A. G., *J. Chem. Soc., A* p. 806 (1970).
372. Hauck, J., and Schwochau, K., *Inorg. Nucl. Chem. Lett.* **9**, 303 (1973).
373. Hawkes, M. J., and Ginsberg, A. P., *Inorg. Chem.* **8**, 2189 (1969).

374. Hazell, A. C., *J. Chem. Soc., London* p. 5745 (1963).
375. Headley, O. St. C., Nyholm, R. S., McAuliffe, C. A., Sindellari, L., Tobe, M. L., and Venanzi, L. M., *Inorg. Chim. Acta* 4, 93 (1970).
376. Hendrickson, D. N., Hollander, J. M., and Jolly, W. L., *Inorg. Chem.* 8, 2642 (1969).
377. Hendriksma, R. R., *Inorg. Nucl. Chem. Lett.* 8, 1035 (1972).
378. Herzberg, G., and Reid, C., *Discuss. Faraday Soc.* 9, 92 (1950).
379. Hewkin, D. J., and Poë, A. J., *J. Chem. Soc., A* p. 1884 (1967).
380. Hodgkin, D. C., *Fortschr. Chem. Org. Naturst.* 15, 167 (1958).
381. Holba, V., *Collect. Czech. Chem. Commun.* 32, 2469 (1967).
382. Holba, V., *Collect. Czech. Chem. Commun.* 35, 1506 (1970).
383. Holsboer, F. J., and Beck, W., *J. Chem. Soc., D* p. 262 (1970).
384. Holsboer, F. J., and Beck, W., *Z. Naturforsch. B* 27, 884 (1972).
385. Homan, J. M., Kawamato, J. M., and Morgan, G. L., *Inorg. Chem.* 9, 2533 (1970).
386. Horn, C. J., and Brown, T. M., *Inorg. Chem.* 11, 1970 (1972).
387. Horrocks, W. DeW., and Hutchison, J. R., *J. Chem. Phys.* 46, 1703 (1967).
388. Hougen, J. T., Schug, K., and King, E. L., *J. Amer. Chem. Soc.* 79, 519 (1957).
389. House, J. E., and Bailar, J. C., *J. Amer. Chem. Soc.* 91, 67 (1969).
390. Howarth, O. W., Richards, R. E., and Venanzi, L. M., *J. Chem. Soc., London* p. 3335 (1964).
391. Howatson, J., and Morosin, B., *Cryst. Struct. Commun.* 2, 51 (1973).
392. Hudson, M. J., Nyholm, R. S., and Stiddard, M. H. B., *J. Chem. Soc., A* p. 40 (1968).
393. Hughes, M. N., and Rutt, K. J., *J. Chem. Soc., A* p. 3015 (1970).
394. Huheey, J. E., *J. Phys. Chem.* 70, 2086 (1966).
395. Hunter, J. A., Massie, W. H. S., Meiklejohn, J., and Reid, J., *Inorg. Nucl. Chem. Lett.* 5, 1 (1969).
396. Iizuka, M., and Sudo, T., *Z. Kristallogr.* 126, 376 (1968).
397. Ikeda, R., Nakamura, D., and Kubo, M., *Bull. Chem. Soc. Jap.* 40, 701 (1967).
398. Iqbal, Z., *Struct. Bonding (Berlin)* 10, 25 (1972).
399. Isslieb, K., and Tzschach, A., *Z. Anorg. Allg. Chem.* 297, 121 (1958).
400. Jain, P. C., and Lingafelter, E. C., *J. Amer. Chem. Soc.* 89, 724 (1967).
401. Jain, P. C., and Lingafelter, E. C., *J. Amer. Chem. Soc.* 89, 6131 (1967).
402. Jain, P. C., Lingafelter, E. C., and Paoletti, P., *J. Amer. Chem. Soc.* 90, 519 (1968).
403. Jain, S. C., and Rivest, R., *Inorg. Chim. Acta* 3, 552 (1969).
404. Jain, S. C., and Rivest, R., *Inorg. Chim. Acta* 4, 291 (1970).
405. Jaselskis, B., *J. Amer. Chem. Soc.* 83, 1082 (1961).
406. Jeffery, J. W., *Acta Crystallogr., Suppl.* 16, A66 (1963).
407. Jeffery, J. W., and Rose, K. M., *Acta Crystallogr., Sect. B* 24, 653 (1968).
408. Jennings, M. A., and Wojcicki, A., *Inorg. Chem.* 6, 1854 (1967).
409. Jennings, M. A., and Wojcicki, A., *J. Organometal. Chem.* 14, 231 (1968).
410. Jennings, M. A., and Wojcicki, A., *Inorg. Chim. Acta* 3, 335 (1969).
411. Jensen, A., Christiansen, V. H., Hansen, J. F., Likowski, T., and Burmeister, J. L., *Acta Chem. Scand.* 26, 2898 (1972).
412. Jensen, K. A., and Dahl, O., *Acta Chem. Scand.* 22, 1044 (1968).
413. Johnson, E. C., Meyer, T. J., and Winterton, N., *Inorg. Chem.* 10, 1673 (1971).

414. Johnson, K. A., Lim, J. C., and Burmeister, J. L., *Inorg. Chem.* **12**, 124 (1973).
415. Johnson, S. A., and Basolo, F., *Inorg. Chem.* **1**, 925 (1962).
416. Jona, E., Sramko, T., Ambrovic, P., and Gazo, J., *J. Therm. Anal.* **4**, 153 (1972).
417. Jona, E., Sramko, T., and Gazo, J., *J. Therm. Anal.* **4**, 61 (1972).
418. Jones, L. H., *J. Chem. Phys.* **25**, 1069 (1956).
419. Jones, L. H., Shoolery, J. N., Shulman, R. G., and Yost, D. M., *J. Chem. Phys.* **18**, 990 (1950).
420. Jörgensen, C. K., *Inorg. Chem.* **3**, 1201 (1964).
421. Jörgensen, C. K., *Inorg. Chim. Acta Rev.* **2**, 65 (1968).
422. Kabesova, M., Garaj, J., and Gazlo, J., *Collect. Czech. Chem. Commun.* **37**, 942 (1972).
423. Kalsotra, B. L., Multani, R. K., and Jain, B. D., *J. Inorg. Chem. Nucl.* **34**, 2265 (1972).
424. Kamenar, B., and Proutt, C. K., *J. Chem. Soc.*, *A* p. 2379 (1970).
425. Kauffman, G. B., and Beech, G., *Thermochim. Acta* **1**, 99 (1970).
426. Kawaguchi, H., and Kawaguchi, S., *Bull. Chem. Soc. Jap.* **43**, 2103 (1970).
427. Kay, J., Moore, J. W., and Glick, M. D., *Inorg. Chem.* **11**, 2818 (1972).
428. Kazakov, V. P., and Konovalova, M. V., *Russ. J. Inorg. Chem.* **13**, 231 (1968).
429. Keeton, M., Lever, A. B. P., and Ramaswarmy, B. S., *Spectrochim. Acta, Part A* **26**, 2173 (1970).
430. Keller, P., and Harrington, G. W., *Anal. Chem.* **41**, 523 (1969).
431. Kent, J. E., and Wagner, E. L., *J. Chem. Phys.* **44**, 3590 (1966).
432. Kergoat, R., Guerchais, J. E., and Genet, F., *Bull. Soc. Fr. Mineral Crystallogr.* **93**, 166 (1970).
433. Kerridge, D. H., and Mosley, M., *J. Chem. Soc.*, *A* p. 1875 (1967).
434. Kewley, R., Sastry, K. V. L. N., and Winnewisser, M., *J. Mol. Spectrosc.* **10**, 418 (1963).
435. Kharitonov, Yu. Ya., Shul'gina, I. M., Traggeim, E. N., and Babaeva, A. V., *Russ. J. Inorg. Chem.* **8**, 390 (1963).
436. Kharitonov, Yu. Ya., and Skopenko, V. V., *Russ. J. Inorg. Chem.* **10**, 984 (1965).
437. Kharitonov, Yu. Ya., and Skopenko, V. V., *Zh. Neorg. Khim.* **10**, 1803 (1965).
438. Kharitonov, Yu. Ya., and Tsintsadze, G. V., *Russ. J. Inorg. Chem.* **10**, 18 (1965).
439. Kharitonov, Yu. Ya., and Tsintsadze, G. V., *Russ. J. Inorg. Chem.* **10**, 645 (1965).
440. Kharitonov, Yu. Ya., Tsintsadze, G. V., and Porai-Koshits, M. A., *Dokl. Akad. Nauk SSSR* **160**, 1351 (1964).
441. Kharitonov, Yu. Ya., Tsintsadze, G. V., and Tsivadze, A Yu., *Inorg. Nucl. Chem. Lett.* **2**, 201 (1970).
442. Kharitonov, Yu. Ya., Tsintsadze, G. V., and Tsivadze, A. Yu., *Russ. J. Inorg. Chem.* **15**, 204 (1970).
443. Kharitonov, Yu. Ya., Tsintsadze, G. V., and Tsivadze, A. Yu., *Russ. J. Inorg. Chem.* **15**, 364 (1970).
444. Kharitonov, Yu. Ya., Tsintsadze, G. V., and Tsivadze, A. Yu., *Russ. J. Inorg. Chem.* **15**, 484 (1970).

445. Kharitonov, Yu. Ya., Tsintsadze, G. V., and Tsivadze, A. Yu., *Russ. J. Inorg. Chem.* **15**, 614 (1970).
446. Kharitonov, Yu. Ya., Tsintsadze, G. V., and Tsivadze, A. Yu., *Russ. J. Inorg. Chem.* **15**, 776 (1970).
447. Kharitonov, Yu. Ya., Tsintsadze, G. V., and Tsivadze, A. Yu., *Zh. Neorg. Khim.* **15**, 1563 (1970).
448. Kharitonov, Yu. Ya., Tsintsadze, G. V., and Tsivadze, A. Yu., *Russ. J. Inorg. Chem.* **15**, 931 (1970).
449. Kigoshi, K., and Murata, H., *Nippon Kagaku Zasshi* **78**, 526 (1957).
450. King, H. C. A., Körös, E., and Nelson, S. M., *Nature (London)* **196**, 572 (1962).
451. King, H. C. A., Körös, E., and Nelson, S. M., *J. Chem. Soc. (London* p. 5449 (1963).
452. Kinell, P. O., and Strandberg, B., *Acta Chem. Scand.* **13**, 1607 (1959).
453. Kiss, A. I., Csáazár, J., and Lehotsi, L., *Acta Chim. Acad. (Budapest)* **14**, 225 (1958).
454. Klopman, G., *J. Amer. Chem. Soc.* **90**, 223 (1968).
455. Knox, G. F., and Brown, T. M., *Inorg. Chem.* **8**, 1401 (1969).
456. Koenig, E., and Madeja, K., *Inorg. Chem.* **6**, 48 (1967).
457. Koenig, E., Madeja, K., and Boehmer, W. H., *J. Amer. Chem. Soc.* **91**, 4582 (1969).
458. Koepf, H., and Block, B., *Z. Naturforsch. B* **23**, 1534 (1968).
459. Koepf, H., Block, B., and Schmidt, M., *Z. Naturforsch. B* **22**, 1077 (1967).
460. Kohout, J., Kohutova, M., and Jona, E., *Z. Naturforsch. B* **25**, 1054 (1970).
461. Kohout, J., Quastelerova-Hvastijova, M., and Gazo, J., *Collect. Czech. Chem. Commun.* **36**, 4026 (1971).
462. Kohout, J., Quastelerova-Hvastijova, M., Kohutova, M., and Gazo, J., *Monatsh. Chem.* **102**, 350 (1971).
463. Kohout, J., Quastelerova-Hvastijova, M., and Kohutova, M., *Z. Naturforsch. B* **26**, 1366 (1971).
464. Komissarova, L. N., Eremin, U. G., Katochkina, V. S., and Sas, T. M., *Russ. J. Inorg. Chem.* **16**, 1570 (1971).
465. Komissarova, L. N., Sas, T. M., Krasnoyarskaya, A. A., and Gagarina, V. A., *Russ. J. Inorg. Chem.* **15**, 1537 (1970).
466. König, E., and Watson, K. J., *Chem. Phys. Lett.* **6**, 457 (1970).
467. Korvenranta, J., and Pajunen, A., *Suom. Kemistilehti B* **43**, 119 (1970).
468. Kowalski, M. A., and Harrington, G. W., *Anal. Chem.* **44**, 479 (1972).
469. Kratochvil, B., and Long, R., *Can. J. Chem.* **48**, 1414 (1970).
470. Krestov, G. A., *Russ. J. Phys. Chem.* **42**, 452 (1968).
471. Kruse, W., and Thusius, D., *Inorg. Chem.* **7**, 464 (1968).
472. Kulasingam, G. C., and McWhinnie, W. R., *Chem. Ind. (London)* p. 2200 (1966).
473. Kulasingam, G. C., and McWhinnie, W. R., *J. Chem. Soc., A* p. 254 (1968).
474. Kuroda, K., *Kagaku No Ryoiki* **25**, 80 (1971).
475. Kustin, K., and Hurwitz, P., *J. Phys. Chem.* **71**, 324 (1967).
476. Laing, K. R., and Roper, W. R., *J. Chem. Soc., A* p. 2149 (1970).
477. Langford, C. H., and Chung, F. M., *J. Amer. Chem. Soc.* **90**, 4485 (1968).
478. Langford, C. H., and Chung, F. M., *Can. J. Chem.* **48**, 2969 (1970).
479. Lappert, M. F., and Pyszora, H., *Advan. Inorg. Chem. Radiochem.* **9**, 133 (1966).

480. Larsson, R., *Acta Chem. Scand.* **14**, 697 (1960).

481. Larsson, R., and Miezis, A., *Acta Chem. Scand.* **23**, 37 (1969).

482. Lauer, J. L., Peterkin, M. E., Burmeister, J. L., Johnson, K. A., and Lim, J. C., *Inorg. Chem.* **11**, 907 (1972).

483. Lawrence, G. S., *Trans. Faraday Soc.* **53**, 1326 (1957).

484. Le Coz, E., and Guerchais, J. E., *Bull. Soc. Chim. Fr.* [6] p. 80 (1971).

485. Le Coz, E., Guerchais, J. E., and Goodgame, D. M. L., *Bull. Soc. Chim. Fr.* [6] p. 3855 (1969).

486. Lenz, W., Schlaefer, H. L., and Ludi, A., *Z. Anorg. Allg. Chem.* **365**, 55 (1969).

487. Leopold, J., Shapira, D., and Treinin, A., *J. Phys. Chem.* **74**, 4585 (1970).

488. Lever, A. B. P., and Nelson, S. M., *J. Chem. Soc., A* p. 859 (1966).

489. Lewis, J., Nyholm, R. S., and Smith, P. W., *J. Chem. Soc., London* p. 4590 (1961).

490. Lin, S. W., and Schreiner, A. F., *Inorg. Chim. Acta* **5**, 290 (1971).

491. Lindner, E., and Behrens, H., *Spectrochim. Acta, Part A* **23**, 3025 (1967).

492. Lindqvist, I., *Acta Crystallogr.* **10**, 29 (1957).

493. Lindqvist, I., and Strandberg, B., *Acta Crystallogr.* **10**, 173 (1957).

494. Lingafelter, E. C., private communication, cited by Farago and James (*277*).

495. Lingafelter, E. C., *Nature (London)* **182**, 1730 (1958).

496. Liptay, G., Burger, K., Mocsari-Fulop, E., and Porubszky, I., *J. Therm. Anal.* **2**, 25 (1970).

497. Livingstone, S. E., *Quart. Rev., Chem. Soc.* **19**, 386 (1965).

498. Lobanov, N. I., *Zh. Neorg. Khim.* **4**, 337 (1959).

499. Lodzinska, A., *Stud. Soc. Sci. Torun., Sect. B* **3**, 53 (1961).

500. Logan, A. V., Bush, D. C., and Rogers, C. J., *J. Amer. Chem. Soc.* **74**, 4194 (1952).

501. Logan, N., *in* "Nitrogen NMR" (M. Witanowski and G. A. Webb, eds.), p. 319. Plenum, New York, 1973.

502. Macarovici, C. G., and Micu-Semeniuc, R., *Rev. Roum. Chim.* **14**, 357 (1969).

503. McDonald, J. R., Scherr, V. M., and McGlynn, S. P., *J. Chem. Phys.* **51**, 1723 (1969).

504. Maciel, G. E., and Beatty, D. A., *J. Phys. Chem.* **69**, 3920 (1965).

505. McKenzie, E. D., *Coord. Chem. Rev.* **6**, 187 (1971).

506. McLean, A. D., and Yoshimine, M., *I.B.M. J. Res. Develop.* **12** Suppl., 206 (1968).

507. McPartlin, M., Mason, R., and Malatesta, L., *Chem. Commun.* p. 334 (1969).

508. McPhail, A. T., Knox, G. R., Robertson, C. G., and Sim, G. A., *J. Chem. Soc., A* p. 205 (1971).

509. McWhinnie, W. R., *J. Inorg. Nucl. Chem.* **27**, 1619 (1965).

510. Makhija, R. C., Beauchamp, A. L., and Rivest, R., *J. Chem. Soc., Chem. Commun.* p. 1043 (1972).

511. Makhija, R. C., Pazdernik, L., and Rivest, R., *Can. J. Chem.* **51**, 438 (1973).

512. Maki, A., and Decius, J. C., *J. Chem. Phys.* **31**, 772 (1959).

513. Maki, N., and Sakuraba, S., *Bull. Chem. Soc. Jap.* **42**, 579 (1969).

514. Manolov, K., and Stamatova, V., *Natura (Plovdiv)* **2**, 69 (1968).

515. Mardirossian, J., and Skinner, J. F., *Inorg. Chem.* **10**, 411 (1971).

516. Marks, D. R., Phillips, D. J., and Redfern, J. P., *J. Chem. Soc., A* p. 2013 (1968).

517. Marongiu, G., Lingafelter, E. C., and Paoletti, P., *Inorg. Chem.* **8**, 2763 (1969).
518. Martin, J. L., Thompson, L. C., Radonovich, L. J., and Glick, M. D., *J. Amer. Chem. Soc.* **90**, 4493 (1968).
519. Marzilli, L. G., *Inorg. Chem.* **11**, 2504 (1972).
520. Mason, R., Rusholme, G. A., Beck, W., Engelmann, H., Joos, K., Lindenberg, B., and Smedal, H. S., *Chem. Commun.* p. 496 (1971).
521. Mason, W. R., Berger, E. R., and Johnson, R. C., *Inorg. Chem.* **6**, 248 (1967).
522. Mathias, A., *Tetrahedron* **21**, 1073 (1965).
523. Maurey, J. R., and Wolff, J., *J. Inorg. Nucl. Chem.* **25**, 312 (1963).
524. Mawby, A., and Pringle, G. E., *Chem. Commun.* p. 385 (1970).
525. Mawby, A., and Pringle, G. E., *J. Inorg. Nucl. Chem.* **34**, 2213 (1972).
526. Meek, D. W., Nicpon, P. E., and Meek, V. I., *J. Amer. Chem. Soc.* **92**, 5351 (1970).
527. Melpolder, J. B., and Burmeister, J. L., *Inorg. Chem.* **11**, 911 (1972).
528. Merkusheva, S. A., Skorik, N. A., and Serebrennikov, V. V., *Radiokhimiya* **10**, 591 (1968).
529. Merle, A., Dartiguanave, M., and Dartiguanave, M. X., *Bull. Soc. Chim. Fr.* [6] **1**, 87 (1972).
530. Michelsen, K., *Acta Chem. Scand.* **17**, 1811 (1963).
531. Mikulin, G. I., Reznik, F. Ya., and Viteeva, L. N., *Ukr. Khim. Zh.* **33**, 555 (1967).
532. Misra, M. K., and Rao, D. V. R., *J. Indian Chem. Soc.* **46**, 672 (1969).
533. Mitchell, P. C. H., and Williams, R. J. P., *J. Chem. Soc., London* p. 1912 (1960).
534. Mitchell, P. C. H., and Williams, R. J. P., *J. Chem. Soc., London* p. 4570 (1962).
535. Moelwyn-Hughes, J. T., Garner, A. W. B., and Howard, A. S., *J. Chem. Soc., A* p. 2361 (1971).
536. Moelwyn-Hughes, J. T., Garner, A. W. B., and Howard, A. S., *J. Chem. Soc., A* p. 2370 (1971).
537. Mohapatra, B. K., and Rao, D. V. R., *Indian J. Chem.* **8**, 564 (1970).
538. Molodkin, A. K., Ivanova, O. M., Kuchumova, N., and Kozina, L. E., *Russ. J. Inorg. Chem.* **12**, 963 (1967).
539. Molodkin, A. K., and Skotnikova, G. A., *Russ. J. Inorg. Chem.* **7**, 800 (1962).
540. Morassi, R., and Sacconi, L., *J. Amer. Chem. Soc.* **92**, 5241 (1970).
541. Morassi, R., and Sacconi, L., *J. Chem. Soc., A* p. 492 (1971).
542. Morassi, R., and Sacconi, L., *J. Chem. Soc., A* p. 1487 (1971).
543. Morgan, H. W., *J. Inorg. Nucl. Chem.* **16**, 367 (1961).
544. Morris, D. F. C., *J. Inorg. Nucl. Chem.* **6**, 295 (1958).
545. Muirhead, K. A., and Haight, G. P., *Inorg. Chem.* **12**, 1116 (1973).
546. Mulliken, R. S., *J. Chem. Phys.* **3**, 720 (1935).
547. Mulliken, R. S., *J. Chem. Phys.* **3**, 735 (1935).
548. Mureinik, R. J., and Robb, W., *Spectrochim. Acta, Part A* **24**, 837 (1968).
549. Mureinik, R. J., and Robb, W., *Spectrochim. Acta, Part A* **26**, 811 (1970).
550. Nagarajan, G., and Hariharan, T. A., *Acta Phys. Austr.* **19**, 349 (1965).
551. Napper, R., and Page, F. M., *Trans. Faraday Soc.* **59**, 1086 (1963).
552. Nardelli, M., Cavalca, L., and Braibanti, A., *Gazz. Chim. Ital.* **87**, 917 (1957).
553. Nardelli, M., and Chierici, I., *Gazz. Chim. Ital.* **88**, 359 (1958).
554. Nardelli, M., Fava, G. G., Giraldi, B. G., and Domiano, P., *Acta Crystallogr.* **20**, 349 (1966).

555. Nardelli, M., Fava, G. G., Musatti, A., and Manfredotti, A., *Acta Crystallogr.* **21**, 910 (1966).
556. Negoiu, D., and Baloiu, L. M., *Z. Anorg. Allg. Chem.* **382**, 92 (1971).
557. Nelson, J., and Nelson, S. M., *J. Chem. Soc., A* p. 1597 (1969).
558. Nelson, S. M., *Proc. Chem. Soc., London* p. 372 (1961).
559. Nelson, S. M., Kelly, W. S. J., and Ford, G. H., *J. Chem. Soc., A* p. 388 (1971).
560. Nelson, S. M., and Shepherd, T. M., *Inorg. Chem.* **4**, 813 (1965).
561. Nelson, S. M., and Shepherd, T. M., *J. Chem. Soc., London* p. 3276 (1965).
562. Niac, G., Baldea, I., and Lungu, M., *Stud. Univ. Babes-Bolyai, Ser. Chem.* **14**, 83 (1969).
563. Nicolini, M., Pecile, C., and Turco, A., *J. Amer. Chem. Soc.* **87**, 2379 (1965).
564. Nicpon, P., and Meek, D. W., *Inorg. Chem.* **5**, 1297 (1966).
565. Nicpon, P., and Meek, D. W., *Inorg. Chem.* **6**, 145 (1967).
566. Norbury, A. H., *Chem. Ind. (London)* p. 744 (1970).
567. Norbury, A. H., *J. Chem. Soc., A* p. 1089 (1971).
568. Norbury, A. H., unpublished results.
569. Norbury, A. H., Ryder, E. A., and Williams, R. F., *J. Chem. Soc., A* p. 1439 (1967).
570. Norbury, A. H., and Shaw, P. E., unpublished results.
571. Norbury, A. H., Shaw, P. E., and Sinha, A. I. P., *J. Chem. Soc., D* p. 1080 (1970).
572. Norbury, A. H., Shaw, P. E., and Sinha, A. I. P., *J. Chem. Soc. Dalton Trans.* (in press).
573. Norbury, A. H., and Sinha, A. I. P., *Inorg. Nucl. Chem. Lett.* **3**, 355 (1967).
574. Norbury, A. H., and Sinha, A. I. P., *Inorg. Nucl. Chem. Lett.* **4**, 617 (1968).
575. Norbury, A. H., and Sinha, A. I. P., *J. Chem. Soc., A* p. 1598 (1968).
576. Norbury, A. H., and Sinha, A. I. P., *Quart. Rev., Chem. Rev.* **24**, 69 (1970).
577. Norbury, A. H., and Sinha, A. I. P., *J. Inorg. Nucl. Chem.* **33**, 2683 (1971).
578. Norbury, A. H., and Sinha, A. I. P., *J. Inorg. Nucl. Chem.* **35**, 1211 (1973).
579. Norbury, A. H., Thompson, M., and Songstad, J., *Inorg. Nucl. Chem. Lett.* **9**, 347 (1973).
580. Norris, A. R., and Patterson, D., *J. Inorg. Nucl. Chem.* **31**, 3680 (1969).
581. Notley, J. M., and Spiro, M., *J. Phys. Chem.* **70**, 1502 (1966).
582. Nyholm, R. S., Gill, N. S., Barclay, G. A., Christie, T. I., and Pauling, P. J., *J. Inorg. Nucl. Chem.* **18**, 88 (1961).
583. Oki, H., Kyuno, E., and Tsuchiya, R., *Bull. Chem. Soc. Jap.* **41**, 2357 (1968).
584. Oki, H., Kyuno, E., and Tsuchiya, R., *Bull. Chem. Soc. Jap.* **41**, 2362 (1968).
585. Oleari, L., *Stereochim. Inorg. Accad. Naz. Lincei, Corso Estivo Chim., 9th* pp. 421–432 (1965).
586. Orhanovic, M., and Sutin, N., *J. Amer. Chem. Soc.* **90**, 538 (1968).
587. Orhanovic, M., and Sutin, N., *J. Amer. Chem. Soc.* **90**, 4286 (1968).
588. Padova, J., *Bull. Res. Counc. Isr., Sect. A* **10**, 63 (1961).
589. Pajumen, A., and Haemelaeinen, R., *Suom. Kemistilehti B* **45**, 122 (1972).
590. Palade, D. M., Ablov, A. V., and Zubarev, V. N., *Zh. Neorg. Khim.* **14**, 445 (1969).
590a. Palenik, G. J., Steffen, W. L. Mathew, M., Li, M., and Meek, D. W., *Inorg. Nucl. Chem. Lett.* **10**, 125 (1974).
591. Panattoni, C., and Frasson, E., *Gazz. Chim. Ital.* **93**, 601 (1963).
592. Pascal, P., *Bull. Soc. Chim. Fr.* [4] **15**, 11 (1914).

593. Patel, K. C., and Goldberg, D. E., *J. Inorg. Nucl. Chem.* **34**, 637 (1972).

594. Patel, R. N., and Rao, D. V. R., *Indian J. Chem.* **5**, 390 (1967).

595. Paul, R. C., and Sreenathan, B. R., *Indian J. Chem.* **4**, 382 (1966).

596. Pearson, R. G., *J. Amer. Chem. Soc.* **85**, 3533 (1963).

597. Pearson, R. G., "Hard and Soft Acids and Bases." Dowden, Hutchinson & Ross Inc., Stroudsberg, Pennsylvania, 1973.

598. Pearson, R. G., *Inorg. Chem.* **12**, 712 (1973).

599. Pecile, C., *Inorg. Chem.* **5**, 210 (1966).

600. Pecile, C., Turco, A., and Pizzolotto, G., *Ric. Sci., Parte 2: Sez. A* [2] **1**, 247 (1961).

601. Perrier, M., and Vicentini, G., *J. Inorg. Nucl. Chem.* **35**, 555 (1973).

602. Pfieffer, P., and Tilgner, M., *Ber. Deut. Chem. Ges.* **36**, 1436 (1903).

603. Phipps, A. L., and Plane, R. A., *J. Amer. Chem. Soc.* **79**, 2458 (1957).

604. Pidcock, A., *J. Chem. Soc., Chem. Commun.* p. 249 (1973).

605. Pidcock, A., Richards, R. E., and Venanzi, L., *J. Chem. Soc., A* p. 1707 (1966).

606. Plowman, R. A., and Power, L. F., *Aust. J. Chem.* **24**, 309 (1971).

607. Pollak, P., and Cave, G. C. B., *Can. J. Chem.* **45**, 1051 (1967).

608. Poon, C. K., and Tobe, M. L., *J. Chem. Soc., A* p. 1549 (1968).

609. Porai-Koshits, M. A., *Acta Crystallogr.* **16**, A42 (1963).

610. Porai-Koshits, M. A., and Tishchenko, G. N., *Sov. Phys.—Crystallogr.* **4**, 216 (1959).

611. Porai-Koshits, M. A., and Tischenko, G. N., *Kristallografia* **4**, 239 (1959).

612. Porai-Koshits, M. A., and Tsintsadze, G. V., *Itogi Nauki, Kristallokhim.* p. 168 (1967).

613. Powell, J., and Shaw, B. L., *J. Chem. Soc., London* p. 3879 (1965).

614. Preer, J. R., and Gray, H. B., *J. Amer. Chem. Soc.* **92**, 7306 (1970).

615. Pruchnik, F., and Wajda, S., *Rocz. Chem.* **44**, 933 (1970).

616. Pruchnik, F., Wajda, S., and Kwaskowska-Clec, E., *Rocz. Chem.* **45**, 547 (1971).

617. Purohit, D. N., and Bjerrum J., *J. Inorg. Nucl. Chem.* **33**, 2067 (1971).

618. Quastlerova-Hvas, M., Kohout, J., and Gajo, J., *Collect. Czech. Chem. Commun.* **37**, 2891 (1972).

619. Quastlerova-Hvas, M., Kohout, J., and Gajo, J., *Z. Anorg. Allg. Chem.* **396**, 341 (1973).

620. Quastlerova-Hvas, M., and Valtr, Z., *Chem. Zvesti* **20**, 795 (1966).

621. Rabalais, J. W., McDonald, J. R., and McGlynn, S. P., *J. Chem. Phys.* **51**, 5095 (1969).

622. Rabalias, J. W., McDonald, J. R., and McGlynn,, S. P., *J. Chem. Phys.* **51**, 5103 (1969).

623. Rabalais, J. W., McDonald, J. R., Scherr, V., and McGlynn, S. P., *Chem. Rev.* **71**, 73 (1971).

624. Ramaswarmy, H. N., and Jonassen, H. B., *Inorg. Chem.* **4**, 1595 (1965).

625. Ramsay, D. A., *J. Amer. Chem. Soc.* **74**, 72 (1952).

626. Randall, W. C., and Alberty, R. A., *Biochemistry* **6**, 1520 (1967).

627. Rao, V. P. R., Sarma, P. V. R. B., and Rao, D. V. R., *Curr. Sci.* **39**, 371 (1970).

628. Rasmussen, S. E., *Acta Chem. Scand.* **13**, 2009 (1959).

629. Rasmussen, S. E., *Acta Crystallogr.* **13**, A51 (1960).

630. Raymond, K. N., and Basolo, F., *Inorg. Chem.* **5**, 1632 (1966).

631. Reedijk, J., *Rec. Trav. Chim. Pays-Bas* **90**, 1249 (1971).

632. Relf, J., Cooney, R. P., and Henneike, H. F., *J. Organometal. Chem.* **39**, 75 (1972).

633. Ripan, R., Ganescu, I., and Varhelyi, C., *Z. Anorg. Allg. Chem.* **357**, 140 (1968).

634. Robb, W., Steyn, M. M. de V., and Krueger, H., *Inorg. Chim. Acta* **3**, 383 (1969).

635. Roeder, N., and Dehnicke, K., *J. Organometal. Chem.* **33**, 281 (1971).

636. Ros, R., and Tondello, E., *J. Inorg. Nucl. Chem.* **33**, 245 (1971).

637. Rozycki, C., *Chem. Anal.* (*Warsaw*) **11**, 447 (1966).

638. Ruff, J. K., *Inorg. Chem.* **8**, 86 (1969).

639. Sabatini, A., and Bertini, I., *Inorg. Chem.* **4**, 959 (1965).

640. Sabatini, A., and Bertini, I., *Inorg. Chem.* **4**, 1665 (1965).

641. Sacconi, L., and Gelsomini, J., *Inorg. Chem.* **7**, 291 (1968).

642. Sacconi, L., and Morassi, R., *J. Chem. Soc., A* p. 492 (1971).

643. Sadasivan, N., Kernohan, J. A., and Endicott, J. F., *Inorg. Chem.* **6**, 770 (1967).

644. Saha, H. K., Banerjee, A. K., and Mookherjea, S., *J. Indian Chem. Soc.* **47**, 1196 (1970).

645. Sahay, S., and Sinha, P. C., *J. Indian Chem. Soc.* **44**, 724 (1967).

646. Saillant, R. B., *J. Organometal. Chem.* **39**, 71C (1972).

647. Salzmann, J. J., and Schmidtke, H.-H., *Inorg. Chim. Acta* **3**, 207 (1969).

648. Samuel, E., *Bull. Soc. Chim. Fr.* [5] p. 3548 (1966).

649. Samus, I. D., *Acta Crystallogr.* **21**, A150 (1966).

650. Sargeson, A. M., and Taube, H., *Inorg. Chem.* **5**, 1094 (1966).

651. Sarkar, S., *J. Indian Chem. Soc.* **46**, 871 (1969).

652. Sas, T. M., Komissarova, L. N., and Anatskaya, N. I., *Russ. J. Inorg. Chem.* **16**, 45 (1971).

653. Sas, T. M., Komissarova, L. N., Gulia, V. G., and Grigor'ev, A. I., *Russ. J. Inorg. Chem.* **12**, 1090 (1967).

654. Scaife, D. E., *Inorg. Chem.* **6**, 625 (1967).

655. Schafer, M., and Curran, C., *Inorg. Chem.* **4**, 623 (1965).

656. Schäffer, C. E., *J. Inorg. Nucl. Chem.* **8**, 149 (1958).

657. Schettino, V., and Hisatsune, I. C., *J. Chem. Phys.* **52**, 9 (1970).

658. Schilt, A. A., and Fritsch, K., *J. Inorg. Nucl. Chem.* **28**, 2677 (1966).

659. Schmidtke, H.-H., *J. Amer. Chem. Soc.* **87**, 2522 (1965).

660. Schmidtke, H.-H., *Z. Phys. Chem.* (*Frankfurt am Main*) [N.S.] **45**, 305 (1965).

661. Schmidtke, H.-H., *Inorg. Chem.* **5**, 1682 (1966).

662. Schmidtke, H.-H., *J. Inorg. Nucl. Chem.* **28**, 1735 (1966).

663. Schmidtke, H.-H., *Ber. Bunsenges. Phys. Chem.* **71**, 1138 (1967).

664. Schmidtke, H.-H., *in* "Physical Methods in Advanced Inorganic Chemistry" (H. A. O. Hill and P. Day, eds.), p. 107. Wiley, New York, 1968.

665. Schmidtke, H.-H., and Garthoff, D., *Helv. Chim. Acta* **50**, 1631 (1967).

666. Schmidtke, H.-H., and Garthoff, D., *Z. Naturforsch. A* **24**, 126 (1969).

667. Schmitz-Dumont, O., and Füchtenbusch, F., *Anorg. Allg. Chem.* **284**, 278 (1956).

668. Schmitz-Dumont, O., and Ross, B., *Angew. Chem., Int. Ed. Engl.* **3**, 315 (1964).

669. Schmitz-Dumont, O., and Ross, B., *Angew. Chem., Int. Ed. Engl.* **3**, 586 (1964).

670. Schug, K., Gilmore, M. D., and Olson, L. A., *Inorg. Chem.* **6**, 2180 (1967).
671. Schutte, C. J. H., *Z. Naturforsch. A* **18**, 525 (1963).
672. Schwochau, K., and Pieper, H. H., *Inorg. Nucl. Chem. Lett.* **4**, 711 (1968).
673. Scouloudi, H., *Acta Crystallogr.* **6**, 651 (1953).
674. Scovell, W. M., Stocco, G. C., and Tobias, R. S., *Inorg. Chem.* **9**, 2682 (1970).
675. Shchelokov, R. N., Shul'gina, I. M., and Chernyaev, I. I., *Dokl. Akad. Nauk SSSR* **168**, 1338 (1966).
676. Shchelokov, R. N., Shul'gina, I. M., and Chernyaev, I. I., *Russ. J. Inorg. Chem.* **11**, 1424 (1966).
677. Shea, C., and Haim, A., *J. Amer. Chem. Soc.* **93**, 3055 (1971).
678. Shepherd, T. M., and Woodward, I., *Acta Crystallogr.* **19**, 479 (1965).
679. Shevelev, S. A., *Russ. Chem. Rev.* **39**, 844 (1970).
680. Shibahara, T., and Mori, M., *Bull. Chem. Soc. Jap.* **45**, 1433 (1972).
681. Sieckhaus, J. F., and Layloff, T., *Inorg. Chem.* **6**, 2185 (1967).
682. Skopenko, V. V., *Ukr. Khim. Zh.* **38**, 1196 (1972).
683. Skopenko, V. V., Brusilovets, A. I., and Tsintsadze, G. V., *Ukr. Khim. Zh.* **35**, 489 (1969).
684. Skopenko, V. V., and Ivanova, E. I., *Visn. Kiiv. Univ., Ser. Khim.* **10**, 24 (1969).
685. Skopenko, V. V., and Ivanova, E. I., *Russ. J. Inorg. Chem.* **14**, 388 (1969).
686. Skopenko, V. V., and Ivanova, E. I., *Ukr. Khim. Zh.* **36**, 16 (1970).
687. Skopenko, V. V., Ivanova, E. I., and Tsintsadze, G. V., *Ukr. Khim. Zh.* **34**, 1000 (1968).
688. Skopenko, V. V., and Tsintsadze, G. V., *Russ. J. Inorg. Chem.* **9**, 1442 (1964).
689. Skopenko, V. V., Tsintsadze, G. V., and Brusilovets, A. I., *Ukr. Khim. Zh.* **36**, 474 (1970).
690. Sloan, T. E., and Wojcicki, A., *Inorg. Chem.* **7**, 1268 (1968).
691. Smith, D. F., Overend, J., Decius, J. C., and Gordon, D. J., *J. Chem. Phys.* **58**, 1636 (1973).
692. Smith, S. R., and Jonassen, H. B., *J. Inorg. Nucl. Chem.* **29**, 860 (1967).
693. Snow, M. R., and Boomsa, R. F., *Acta Crystallogr., Sect. B* **28**, 1908 (1972).
694. Söderbäck, E., *Acta Chem. Scand.* **11**, 1622 (1957).
695. Songstad, J., and Stangeland, L. J., *Acta Chem. Scand.* **24**, 804 (1970).
696. Sotofte, I., and Rasmussen, S. E., *Acta Chem. Scand.* **21**, 2028 (1967).
697. Spacu, P., Teodorescu, M., and Ciornartan, D., *Monatsch. Chem.* **103**, 1 (1972).
698. Spacu, P., Teodorescu, M., Filotti, G., and Telnic, P., *Z. Anorg. Allg. Chem.* **392**, 88 (1972).
699. Sramko, T., and Jona, E., *Collect. Czech. Chem. Commun.* **37**, 1645 (1972).
700. Stancheva, P., Skopenko, V. V., and Tsintsadze, G. V., *Ukr. Khim. Zh.* **35**, 166 (1969).
701. Staples, P. J., *J. Chem. Soc., A* p. 2213 (1971).
702. Stasiw, R., and Wilkins, R. G., *Inorg. Chem.* **8**, 156 (1969).
703. Stefanović, G., and Janjić, T., *Anal. Chim. Acta* **19**, 488 (1958).
704. Stephan, G., and Specker, H., *Naturwissenschaften* **55**, 443 (1968).
705. Stephens, F. S., *J. Chem. Soc., A* p. 2377 (1970).
706. Stocco, F., Stocco, G. C., Scovell, W. M., and Tobias, R. S., *Inorg. Chem.* **10**, 2639 (1971).
707. Stocco, G. C., and Tobias, R. S., *J. Amer. Chem. Soc.* **93**, 5057 (1971).
708. Stotz, I., Wilmarth, W. K., and Haim, A., *Inorg. Chem.* **7**, 1250 (1968).
709. Sutton, G. E., *Aust. J. Chem.* **12**, 122 (1959).

710. Swank, D. D., and Willett, R. D., *Inorg. Chem.* **4**, 499 (1965).
711. Swartz, W. E., Ruff, J. K., and Hercules, D. M., *J. Amer. Chem. Soc.* **94**, 5227 (1972).
712. Swartz, W. E., Wynne, K. J., and Hercules, D. M., *Anal. Chem.* **43**, 1884 (1971).
713. Swihart, D. L., and Mason, W. R., *Inorg. Chem.* **9**, 1749 (1970).
714. Swinarski, A., and Lodzinska, A., *Rocz. Chem.* **32**, 1053 (1958).
715. Syrtsova, G. P., and Nguyen, S. L., *Zh. Neorg. Khim.* **15**, 1027 (1970).
716. Takemoto, J. H., and Hutchinson, B., *Inorg. Nucl. Chem. Lett.* **8**, 769 (1972).
717. Takeuchi, Y., and Saito, Y., *Bull. Chem. Soc. Jap.* **29**, 319 (1957).
718. Tananaiko, M. M., and Lozovik, A. S., *Zh. Neorg. Khim.* **15**, 1070 (1970).
719. Tarayan, V. M., and Ovsepyan, E. N., *Dokl. Akad. Nauk. Arm. SSR* **25**, 7 (1957).
720. Taylor, K. A., Long, T. V., and Plane, R. A., *J. Chem. Phys.* **47**, 138 (1967).
721. Taylor, L. T., Rose, N. J., and Busch, D. H., *Inorg. Chem.* **7**, 785 (1968).
722. Thayer, J. S., and West, R., *Advan. Organometal. Chem.* **5**, 169 (1967).
723. Thomas, G., *Z. Anorg. Allg. Chem.* **360**, 15 (1968).
724. Thomas, T. R., and Crosby, G. A., *J. Mol. Spectrosc.* **38**, 118 (1971).
725. Tibbetts, D. L., and Brown, T. L., *J. Amer. Chem. Soc.* **91**, 1108 (1969).
726. Toeniskoetter, R. H., and Solomon, S., *Inorg. Chem.* **7**, 617 (1968).
727. Tramer, A., *C. R. Acad. Sci.* **250**, 3150 (1960).
728. Tramer, A., *J. Chim. Phys.* **59**, 232 (1962).
729. Treichel, P. M., and Douglas, W. M., *J. Organometal. Chem.* **19**, 221 (1969).
730. Treindl, L., *Collect. Czech. Chem. Commun.* **33**, 2814 (1968).
731. Tronov, B. V., Gurnitskaya, T. S., and Tronov, A. B., *Zh. Obshch. Khim.* **38**, 2716 (1968).
732. Tsintsadze, G. V., *Soobshch. Akad. Nauk. Gruz. SSR* **57**, 57 (1970).
733. Tsintsadze, G. V., Golub, A. M., and Kopa, M. V., *Russ. J. Inorg. Chem.* **14**, 1444 (1969).
734. Tsintsadze, G. V., Kharitonov, Yu. Ya., Ysivadze, A. Yu., Golub, A. M., and Managadze, A. S., *Russ. J. Inorg. Chem.* **15**, 2110 (1970).
735. Tsintsadze, G. V., Mamulashvili, A. M., and Demchenko, L. P., *Russ. J. Inorg. Chem.* **15**, 145 (1970).
736. Tsintsadze, G. V., Porai-Koshits, M. A., and Antsyshkina, A. S., *Zh. Strukt. Khim.* **8**, 296 (1967).
737. Tsintsadze, G. V., Shvelashvili, A. E., and Skopenko, V. V., *Tr. Gruz. Politekh. Inst.* **4**, 53 (1967).
738. Tsintsadze, G. V., Skopenko, V. V., and Brusilovets, A. I., *Soobshch. Akad. Nauk Gruz. SSR* **50**, 109 (1968).
739. Tsintsadze, G. V., Skopenko, V. V., and Ivanova, E. I., *Soobshch. Akad. Nauk Gruz. SSR* **51**, 107 (1968).
740. Tsintsadze, G. V., Tsivadze, A. Yu., and Borshch, A. N., *Soobshch. Akad. Nauk Gruz. SSR* **56**, 565 (1969).
741. Tsivadze, A. Yu., Kharitonov, Yu. Ya., and Tsintsadze, G. V., *Russ. J. Inorg. Chem.* **15**, 1094 (1970).
742. Tsivadze, A. Yu., Kharitonov, Yu. Ya., and Tsintsadze, G. V., *Russ. J. Inorg. Chem.* **17**, 1417 (1972).
743. Tsivadze, A. Yu., Tsintsadze, G. V., Kharitonov, Yu. Ya., Golub, A. M., and Mamulashvili, A. M., *Russ. J. Inorg. Chem.* **15**, 934 (1970).
744. Turco, A., Panottoni, C., and Frasson, E., *Nature (London)* **187**, 772 (1960).

745. Turco, A., and Pecile, C., *Nature (London)* **191**, 66 (1961).
746. Turco, A., Pecile, C., and Niccolini, M., *Proc. Chem. Soc., London* p. 213 (1961).
747. Turco, A., Pecile, C., and Niccolini, M., *J. Chem. Soc., London* p. 3008 (1962).
748. Udupa, M. R., *Z. Kristallogr. Kristallogeometrie, Kristallphys., Kristallchem.* **134**, 311 (1971).
749. Ugo, R., Beck, W., Bauder, M., La Monica, G., and Cenini, S., *J. Chem. Soc., A* p. 113 (1971).
750. Ul'ko, N. V., and Parubocha, M. L., *Visn. Kiiv. Univ., Ser. Khim.* **10**, 18 (1969).
751. Ul'ko, N. V., and Savchencko, R. A., *Russ. J. Inorg. Chem.* **12**, 169 (1967).
752. Valtr, Z., and Quastlerova-Hvas, M., *Z. Chem.* **6**, 348 (1966).
753. Vanderzee, C. E., and Myers, R. A., *J. Phys. Chem.* **65**, 153 (1961).
754. Vanderzee, C. E., and Westrum, E. F., *J. Chem. Thermodyn.* **2**, 417 (1970).
755. Varhelyi, C., Finta, Z., and Zsabo, J., *Z. Anorg. Allg. Chem.* **374**, 326 (1970).
756. Varhelyi, C., Ganescu, I., and Oprescu, D., *Stud. Univ. Babes-Bolyai, Ser. Chem.* **13**, 41 (1968).
757. Varhelyi, C., Ganescu, I., and Oprescu, D., *Monatsh. Chem.* **100**, 106 (1969).
758. Varhelyi, C., Zsako, I., and Finta, Z., *Stud. Univ. Babes-Bolyai, Ser. Chem.* **15**, 81 (1970).
759. Varhelyi, C., Zsako, I., and Finta, Z., *Monatsh. Chem.* **101**, 1013 (1970).
760. Varshavskii, Yu. S., Cherkasova, T. G., Singh, M. M., and Buzina, N. A., *Zh. Neorg. Khim.* **15**, 2746 (1970).
761. Vasil'ev, V. P., Zolotarev, E. K., Kapustinskii, A. F., Mishchenko, K. P., Podgornaya, E. A., and Yatsimirskii, K. B., *Zh. Fiz. Khim.* **34**, 1763 (1960).
762. Vaska, L., and Peone, J., *Chem. Commun.* p. 419 (1971).
763. Vdovenko, V. M., Skoblo, A. I., and Suglobov, D. N., *Radiokhimiya* **9**, 119 (1967).
764. Vdovenko, V. M., Skoblo, A. I., and Suglobov, D. N., *Russ. J. Inorg. Chem.* **13**, 1577 (1968).
765. Venanzi, L. M., *J. Chem. Soc., London* p. 719 (1958).
766. Venanzi, L. M., *Chem. Brit.* **4**, 162 (1968).
767. Venkateswarlu, K., and Devi, V. M., *Curr. Sci.* **36**, 118 (1967).
768. Venkateswarlu, K., Mariam, S., and Rajalakshmi, K., *Bull. Cl. Sci., Acad. Roy. Belg.* **51**, 359 (1965).
769. Vincentini, G., Forneris, R., and Perrier, M., *An. Acad. Brasil. Cienc.* **38**, 261 (1966).
770. Vincentini, G., Najjar, R., and Airoldi, C., *An. Acad. Brasil. Cienc.* **41**, 375 (1969).
771. Vuletic, N., and Djordjevic, C., *J. Chem. Soc., Dalton Trans.* p. 2322 (1972).
772. Wagner, E. L., *J. Chem. Phys.* **43**, 2728 (1965).
773. Wagner, W. C., Mattern, J. A., and Cartledge, G. H., *J. Amer. Chem. Soc.* **81**, 2958 (1959).
774. Wajda, S., and Pruchnik, F., *Nukleonika* **11**, 673 (1966).
775. Wajda, S., and Pruchnik, F., *Nukleonika* **12**, 821 (1967).
776. Wajda, S., and Pruchnik, F., *Rocz. Chem.* **41**, 1473 (1967).
777. Wajda, S., Pruchnik, F., and Zarzeczny, A., *Nukleonika* **13**, 789 (1968).
778. Walsh, A. D., *J. Chem. Soc., London* p. 2260 (1953).
779. Wasson, J. R., and Trapp, C., *J. Inorg. Nucl. Chem.* **30**, 2437 (1968).
780. Weick, F., and Basolo, F., *Inorg. Chem.* **5**, 576 (1966).

781. Wenschuh, E., and Rudolph, K. P., *Z. Anorg. Chem.* **380**, 7 (1971).
782. Werner, H., Beck, W., and Engelmann, H., *Inorg. Chim. Acta* **3**, 331 (1969).
783. Werner, K. V., and Beck, W., *Chem. Ber.* **105**, 3947 (1972).
784. Whimp, P. O., and Curtis, N. F., *J. Chem. Soc., A* p. 867 (1966).
785. Williams, R. J. P., and Hale, J. D., *Struct. Bonding (Berlin)* **1**, 207 (1966).
786. Wilmshurst, J. K., *J. Chem. Phys.* **28**, 733 (1958).
787. Wojcicki, A., and Farona, M. F., *J. Inorg. Nucl. Chem.* **26**, 2289 (1964).
788. Worley, S. D., *Chem. Rev.* **71**, 295 (1971).
789. Wulfman, C. E., *J. Chem. Phys.* **33**, 1567 (1960).
790. Yoneda, H., *Bull. Chem. Soc. Jap.* **28**, 125 (1955).
791. Yukhno, E. K., and Porai-Koshits, M. A., *Kristallografiya* **2**, 239 (1957).
792. Zavodnik, V. E., Zvonkova, Z. V., Zhdanov, G. S., and Mirevich, E. G., *Kristallografiya* **17**, 107 (1972).
793. Zhdanov, G. S., Zvonkova, Z. V., and Glushkova, V. P., *Zh. Fiz. Khim.* **27**, 106 (1953).
794. Ziolo, R. F., and Dori, Z., *J. Amer. Chem. Soc.* **90**, 6560 (1968).
795. Zsaki, I., *Stud. Univ. Babes-Bolyai, Ser. Chem.* **15**, 93 (1970).
796. Zvonkova, Z. V., *Russ. J. Phys. Chem.* **26**, 1798 (1952).
797. Zvonkova, Z. V., and Zhdanov, G. S., *Zh. Fiz. Khim.* **26**, 586 (1952).

Appendix

The following represents a brief outline of the more important of the many papers published since this review was completed. The references cited are only those which have a direct bearing on the more general themes discussed previously.

A. Cyanates

The crystal standard of $(\pi\text{-cp})_2\text{Ti(NCO)}_2$ has been determined and confirms that, in the solid state, the cyanate group is here N-bonded; it was only in the last stages of the analysis that the O-bonded alternative could be finally eliminated (1). Hexa-N-cyanato complexes of ytterbium, erbium, and neodymium have been reported as quaternary onium salts (16) and reference made to the series of tetraethylammonium salts of $[\text{Ln(NCO)}_6]^{3-}$ (Ln = Eu–Yb) (22); these results extend Table XLII. The ESR spectra of series of complexes $\text{CuL}_2(\text{NCO})_2$ (L = an, or substituted an) have been interpreted to show Cu—N(CO)—Cu bridges with no indication of any Cu—O interactions (19). The ESCA spectra of $[\text{M(NCO)}_4]^{2-}$ (M = Mn, Co, Zn) have been recorded (12).

B. Thiocyanates

Crystal structures have been determined by X-ray methods for a number of compounds, among the more important being the following:

the dimer [CoPPQ(NCS)$_2$]$_2$ (PPQ = 2-(2'-pyridyl)-3-(N-2-picolylimino)-4-oxo-1,2,3,4-tetrahydroquinazoline) in which each cobalt exists in a distorted octahedron of five nitrogen atoms, three from the organic ligand, and two from thiocyanate groups, and a sulfur atom from a bridging thiocyanate group (20); the different salts of the linkage isomers [Ir(NH$_3$)$_5$(SCN)](ClO$_4$)$_2$ (14) and [Ir(NH$_3$)$_5$(NCS)]Cl$_2$ (13); the complex NiL$_2$(NCS)$_2$ (L = thiosemicarbazide) where the ligand is bidentate and the thiocyanate groups complete octahedral coordination around the nickel atom by forming two Ni–NCS linkages (10); the phosphite complex Pd(P(OPh)$_3$)$_2$(SCN)$_2$ unlike the corresponding phosphine, contains only Pd–SCN linkages (18); analysis of repulsive contacts in cis-Pt(Ph$_2$PC≡C-t-Bu)$_2$(NCS)(SCN) leads to the conclusion that the mixed bonding is mainly due to steric inhibition of dithiocyanato coordination (28); the series of polyamine copper(II) complexes has been extended with [Cu(H$_2$NCH$_2$CH$_2$CH$_2$NH$_2$)$_2$(NCS)]ClO$_4$ in which the metal has trigonal bipyramidal coordination (5), [CuHN-(CH$_2$CH$_2$NH$_2$)$_2$(NCS)$_2$] in which the metal has tetragonal pyramidal coordination (6), and [CuHN(CH$_2$CH$_2$NH$_2$)$_2$(NCS)]$_2$(ClO$_4$)$_2$ in which Cu—NCS—Cu bridges lead to a tetragonally distorted octahedron with a long copper–perchlorate interaction (7).

Various geometric isomers can be obtained in the chromium(III)–amine–thiocyanate system. The isothermal decompositions of [Cr(NH$_3$)$_6$](NCS)$_3$ and [Cr(NH$_3$)$_5$(NCS)](NCS)$_2$ both proceed via trans-[Cr(NH$_3$)$_4$(NCS)$_2$]NCS to give mer-[Cr(NH$_3$)$_3$(NCS)$_3$] (24), while the aqueous photolysis of trans-[Cren$_2$(NCS)$_2$]$^+$ gives a mixture of cis-[Cren$_2$(H$_2$O)(NCS)]$^{2+}$ and [Cren(enH)(H$_2$O)(NCS)$_2$]$^{2+}$, where the thiocyanate groups are still trans to each other (2). Similarly, aqueous photolysis of trans-[Cren$_2$Cl(NCS)]$^+$ leads to the aqueous replacement of either anion, while use of the corresponding cis-isomer gives initially [Cren(enH)(H$_2$O)Cl(NCS)]$^{2+}$ (15).

The ESCA spectra of [M(NCS)$_4$]$^{2-}$ (M = Co, Ni, Zn), [Ni(NCS)$_6$]$^{4-}$ (12), [Re(NCS)$_6$]$^{2-}$, and [Re$_2$(NCS)$_8$]$^{2-}$ (27) have been recorded. Further details have been published on the compound (Me$_4$N)$_2$[Tc(NCS)$_6$] together with the new compound (Me$_4$N)[Tc(NCS)$_6$] which extends Table XXXVIII. The two anions form a redox system with $E° = 0.53$ V in 1M H$_2$SO$_4$, and their infrared spectra support their formulation as N-thiocyanates (26).

The complex (Ph$_4$As)[Co(DH)$_2$(SCN)$_2$] has been obtained in three forms, and the S,S-, N,S-, and N,N-isomers all characterized (11). Spectrophotometric measurements on the equilibrium

$$Co(DH)_2py(SCN) \rightleftharpoons Co(DH)_2py(NCS)$$

indicate that in chloroform at 298°K, $\Delta G = 3.1$ kJ mol^{-1}, $\Delta H = 26.4$ kJ mol^{-1}, and $\Delta S = -78$ J K^{-1} mol^{-1} (23); energy differences which are so small that they more than adequately illustrate the force of the closing sentence of the main body of this review. The consequences of modifying the activation energy for the isomerization by adding cobalt(II) species to similar systems and the relative importance of adjacent and remote attack during the electron transfer reaction have been discussed (21).

The recognition of the importance of steric effects in determining thiocyanate coordination to palladium(II) and platinum(II) continues to gain ground. Thus, the mixed complexes cis-Pt(Ph$_2$PC≡CR)$_2$(NCS)-(SCN), (R = Ph, Et, i-Pr, t-Bu) have been characterized by infrared spectroscopy and, when R = t-Bu, by X-ray crystallography, and the authors suggest, on the basis of a careful analysis of intermolecular repulsions, that these results are consistent with the reduced steric demands of the phosphinoacetylene increasing the possibility of S-coordination (28). Similarly, X-ray crystallography showed that trans-Pd[P(OPh)$_3$]$_2$(SCN)$_2$ is S-bonded as a solid while infrared measurements showed no evidence for complete isomerization in solution. These results are in important contrast to the behavior of Pd(PPh$_2$)$_2$(NCS)$_2$ since P(OPh$_3$)$_3$ is the better π-acceptor and would be expected to promote N-thiocyanate coordination if the π-bonding hypothesis were of major importance (18). The solid state isomerization of the unstable linkage isomers cis-PtL$_2$(SCN)$_2$ (L = Ph$_3$P, Ph$_3$, As) is not accompanied by any geometric isomerization (17). The steric effects which encourage the isomerization of [PdET$_4$dien(SCN)]$^+$ in solution to give the N-bonded isomer, which in turn reisomerizes as the solid S-bonded tetraphenylborate have been described previously, and these have been reiterated with further examples. Interestingly though the corresponding platinum(II) complex exists only as the S-thiocyanate in solid and solution (3).

Electronic effects of one sort or another do, however, appear to have some part to play in certain complexes as indicated previously. Recent examples are [Pdtripy(NCS)]BPh$_4$ which exists only in the N-bonded form (3), whereas steric effects of the compact tripy ligand might be expected to be less than of ET$_4$dien. Similarly, the steric effects of Ph$_2$PCH:CHPPh$_2$ and Ph$_2$PCF:CFPPh$_2$ would appear to be comparable yet with the former ligand PdL(SCN)$_2$ exists while the latter gives PdL(NCS)(SCN) (8). In contrast, the formation of a series of isonitrile complexes Pd(CNR)$_2$(SCN)$_2$ (R = Ph, C$_6$H$_{11}$), Pd(CNR)(PPh$_3$)(SCN)$_2$ (R = Ph, C$_6$H$_{11}$, p-NO$_2$–C$_6$H$_4$), and Pd(CNPh)(AsPh$_3$)(SCN)$_2$ lead to the suggestion that the isonitriles function only as σ-donors in this case, and that their trans influence is insufficient to modify the mode of coordination of the thiocyanate group (9), whereas the reduced steric effects of the linear isonitriles appear to offer an explanation which is more consistent with the larger body of data.

The Au–NCS/Au–SCN ratios of a series of LAu (thiocyanate) complexes in solution have been measured and the authors (4) claim that their results support Pearson's concept of antisymbiosis (25a), but it has been pointed out that the ligands used give ratios in an order which does not follow that accepted for the trans effect (18).

C. SELENOCYANATE

The crystal standard of bis(N,N-dimethylethylenediamine)-copper(II) selenocyanate has been determined (25). The ESCA spectra of $[M(NCSe)_4]^{2-}$ (M = Co, Zn) and $[M(NCSe)_6]^{4-}$ (M = Co, Ni) have been measured (12).

REFERENCES

1. Anderson, S. J., Brown, D. S., and Norbury, A. H., unpublished results.
2. Bifano, C., and Linck, R. G., *Inorg. Chem.* **13**, 609 (1974).
3. Burmeister, J. L., Hassel, R. L., Johnson, K. A., and Lim, J. C., *Inorg. Chim. Acta* **9**, 23 (1974).
4. Burmeister, J. L., Melpolder, J. B., *J. Chem. Soc., Chem. Commun.* p. 613 (1973).
5. Cannas, M., Carta, G., and Marongiu, G., *J. Chem. Soc., Dalton Trans.* p. 550 (1974).
6. Cannas, M., Carta, G., and Marongiu, G., *J. Chem. Soc., Dalton Trans.* p. 553 (1974).
7. Cannas, M., Carta, G., and Marongiu, G., *J. Chem. Soc., Dalton Trans.* p. 556 (1974).
8. Chow, K. K., and McAuliffe, C. A., *Inorg. Nucl. Chem. Lett.* **9**, 1189 (1973).
9. Cooke, R. R., and Burmeister, J. L., *J. Organometal. Chem.* **63**, 471 (1973).
10. Dunaj-Jurco, M., Garaj, J., and Sirota, A., *Collect. Czech. Chem. Commun.* **39**, 236 (1974).
11. Epps, L. A., and Marzilli, L. G., *Inorg. Chem.* **12**, 1514 (1973).
12. Escard, J., Mavel, G., Guerchais, J. E., and Kergoat, R., *Inorg. Chem.* **13**, 695 (1974).
13. Flack, H., *Acta Crystallogr. Sect. B.* **29**, 2610 (1973).
14. Flack, H. D., and Parthe, E., *Acta Crystallogr. Sect. B.* **29**, 1099 (1973).
15. Gandolfi, M. T., Manfrin, M. F., Juris, A., Moggi, L., and Balzani, V., *Inorg. Chem.* **13**, 1342 (1974).
16. Harris, M. B., and Thompson, L. C., *J. Inorg. Nucl. Chem.* **36**, 212 (1974).
17. Hassel, R. L., and Burmeister, J. L., *Inorg. Chem. Acta* **8**, 155 (1974).
18. Jacobson, S., Wong, Y. S., Chieh, P. C., and Carty, A. J., *J. Chem. Soc., Chem. Commun.* p. 520 (1974).
19. Kohout, J., Quastlerova-Hvastijova, M., and Gazo, J., *Inorg. Chim. Acta* **8**, 241 (1974).
20. Mangia, A., Nardelli, M., and Pelizzi, G., *Acta Crystallogr. Sect. B.* **30**, 487 (1974).
21. Marzilli, L. G., Stewart, R. C., Epps, L. A., and Allen, J. B., *J. Amer. Chem. Soc.* **95**, 5796 (1973).

22. Moeller, T., and Dieck, R. L., private communication cited in *16*.

23. Norbury, A. H. and Raghunathan, S., unpublished results.

24. Oki, H., Yasuoka, M., and Tsuchiya, R., *Bull. Chem. Soc. Jap.* **47**, 652 (1974).

25. Pajumen, A., and Korvenranta, J., *Suom. Kemistilehti B* **46**, 139 (1973).

25a. Pearson, R. G., *Inorg. Chem.* **12**, 712 (1973).

26. Schwochau, K., Astheimer, L., and Shemk, H. J., *J. Inorg. Nucl. Chem.* **35**, 2249 (1973).

27. Tisley, D. G., and Walton, R. A., *J. Chem. Soc., Dalton Trans.* p. 1039 (1973).

28. Wong, Y. S., Jacobson, S., Chieh, P. C., and Carty, A. J., *Inorg. Chem.* **13**, 284 (1974).

SUBJECT INDEX

A

Absorption spectra, *see* specific compounds
Actinides, *see also* specific elements
  cyanates, 274
  selenocyanates, 275
  thiocyanates, 274, 275
Alkyl copper compounds, 140, 142, 143
Alumina spheroids, 107, 108
Aluminum–chromium oxides, 95
Aluminum oxide
  decomposition of, in radio-frequency
    plasma, 108, 109
  reaction in radio-frequency plasma, 93
  spheroids, 106
Aluminum trifluoroacetates, 5, 9, 10
Ammonium trifluoroacetates, 5, 8, 27–29,
  32
Antimony
  "heavy" donor atom, 165
  homopolyatomic cations, 78, 79
Antimony oxide, reaction in radio-
  frequency plasma, 97, 98
Antimony trifluoroacetates, 12, 13
Aquocopper(I) complexes, 117, 118
Arsenic, "heavy" donor atom, 165
Arsenic trifluoroacetates, 12, 13, 27, 29, 32,
  33
Aryl copper compounds, 140, 142, 143
Autoprotolysis, *see* Solvents, self-ionization

B

Barium, homopolyatomic cations, 82
Barium oxide, spheroids, 107
Barium trifluoroacetates, 5, 33
Beryllium trifluoroacetate, 9
Bismuth, homopolyatomic cations, 77,
  78
  Bi_5^{3+}, 77, 78
  Bi_8^{2+}, 77, 78
  Bi_9^{5+}, 77
Bismuth trifluoroacetates, 12, 13
Bis(triphenylphosphine)copper(I)
  complexes, 126–129

Bis(triphenylphosphine)nitratocopper(I),
  126
Bond energy, 213
Boron, spheroids, 107
Boron carbide, preparation from radio-
  frequency plasma, 99
Boron trichloride, reaction in radio-
  frequency plasma, 94, 95
Boron trifluoroacetate, 9, 10
Bromine, homopolyatomic cations, 50, 54,
  55
  Br_2^+, 54–58
  Br_3^+, 54, 58
Butene, pyrolysis of, using radio-frequency
  plasma, 92

C

Cadmium, homopolyatomic cations, 50,
  79, 82
Cadmium cyanates, 333
Cadmium selenocyanates, 337
Cadmium thiocyanates, 333–336
Cadmium trifluoroacetates, 25, 26
Calcium, homopolyatomic cations, 82
Cerium trifluoroacetates, 5, 26
Cesium trifluoroacetates, 8, 31, 32
Chalcogenocyanates, *see also* specific compounds
  absorption spectra, 240, 241
  binding energies for, 242
  bond angles, 236
  chemical methods, 269, 270
  comparative coordination chemistry,
    359–361
  dipole moments, 271
  electron spectroscopy for chemical
    analysis, 242, 243
  electronic structure, 238, 240
  factors affecting mode of coordination,
    337–359
  force constants, 237–239
  frequency ranges, 254, 255
  infrared spectroscopy, 246–261

CONTENTS OF PREVIOUS VOLUMES

A
B   5
C   6
D   7
E   8
F   9
G   0
H   1
I   2
J   3